Developments in Environmental Modelling

1. ENERGY AND ECOLOGICAL MODELLING edited by W.J. Mitsch, R.W. Bossermann and J.M. Klopatek, 1981
2. WATER MANAGEMENT MODELS IN PRACTICE: A CASE STUDY OF THE ASWAN HIGH DAM by D. Whittington and G. Guariso, 1983
3. NUMERICAL ECOLOGY by L. Legendre and P. Legendre, 1983

4A. APPLICATION OF ECOLOGICAL MODELLING IN ENVIRONMENTAL MANAGEMENT PART A edited by S.E. Jørgensen, 1983

4B. APPLICATION OF ECOLOGICAL MODELLING IN ENVIRONMENTAL MANAGEMENT PART B edited by S.E. Jørgensen and W.J. Mitsch, 1983

5. ANALYSIS OF ECOLOGICAL SYSTEMS: STATE-OF-THE-ART IN ECOLOGICAL MODELLING edited by W.K. Lauenroth, G.V. Skogerboe and M. Flug, 1983
6. MODELLING THE FATE AND EFFECT OF TOXIC SUBSTANCES IN THE ENVIRONMENT edited by S.E. Jørgensen, 1984
7. MATHEMATICAL MODELS IN BIOLOGICAL WASTE WATER TREATMENT edited by S.E. Jørgensen and M.J. Gromiec, 1985
8. FRESHWATER ECOSYSTEMS: MODELLING AND SIMULATION by M. Straškraba and A.H. Gnauck, 1985
9. FUNDAMENTALS OF ECOLOGICAL MODELLING by S.E. Jørgensen, 1986
10. AGRICULTURAL NONPOINT SOURCE POLLUTION: MODEL SELECTION AND APPLICATION edited by A. Giorgini and F. Zingales, 1986
11. MATHEMATICAL MODELLING OF ENVIRONMENTAL AND ECOLOGICAL SYSTEMS edited by J.B. Shukla, T.G. Hallam and V. Capasso, 1987
12. WETLAND MODELLING edited by W.J. Mitsch, M. Straškraba and S.E. Jørgensen, 1988
13. ADVANCES IN ENVIRONMENTAL MODELLING edited by A. Marani, 1988
14. MATHEMATICAL SUBMODELS IN WATER QUALITY SYSTEMS edited by S.E. Jørgensen and M.J. Gromiec, 1989
15. ENVIRONMENTAL MODELS: EMISSIONS AND CONSEQUENCES edited by J. Fenhann, H. Larsen, G.A. Mackenzie and B. Rasmussen, 1990

16 MODELLING IN ECOTOXICOLOGY edited by S.E. Jørgensen, 1990

17. MODELLING IN ENVIRONMENTAL CHEMISTRY edited by S.E. Jørgensen, 1991
18. INTRODUCTION TO ENVIRONMENTAL MANAGEMENT edited by P.E. Hansen and S.E. Jørgensen
19. FUNDAMENTALS OF ECOLOGICAL MODELLING, 2nd Edition by S.E. Jørgensen, 1994
20. NUMERICAL ECOLOGY, 2nd Edition by P. Legendre and L. Legendre, 1998
21. FUNDAMENTALS OF ECOLOGICAL MODELLING, 3rd Edition by S.E. Jørgensen and G. Bendoricchio, 2001

Developments in Environmental Modelling, 22

ENVIRONMENTAL FORESIGHT AND MODELS: A MANIFESTO

Developments in Environmental Modelling, 22

ENVIRONMENTAL FORESIGHT AND MODELS: A MANIFESTO

Edited by

M.B. Beck

University of Georgia,
Athens, Georgia, USA

2002

ELSEVIER

Amsterdam — London — New York — Oxford — Paris — Shannon — Tokyo

ELSEVIER SCIENCE Ltd
The Boulevard, Langford Lane
Kidlington, Oxford OX5 1GB, UK

First edition 2002

Library of Congress Cataloging in Publication Data
A catalog record from the Library of Congress has been applied for.

ISBN: 0-080-44086-X

The paper used in this publication meets the requirements of ANSI/NISO Z39.48-1992 (Permanence of Paper).

Printed in The Netherlands.

Contents

Preface

Everyone – or so it seems at times – has an opinion on modelling. This belittles the subject, I fear. For it suggests the principles of constructing a mathematical model, evaluating it against whatever experience one has of the behaviour of the prototype, and subsequently applying it for the purposes of exploring the future, are trivially straightforward – easily grasped by the non-expert. There is, of course, no need for us to make the subject full of spurious jargon and complication, just to give ourselves a sense of what Simon Schaffer has called our "own private world where ignorant outsiders cannot penetrate".[1] My point is rather this. I acknowledge I know little in detail of the subjects of, say, meteorology, forest ecology, or aluminium speciation in soils; I do not presume therefore to tell those who do, how best they should go about their business; I am not the one to judge whether good science has been done in these subject areas; I have respect for their subtleties and complexities, which I know I am unable to fathom. So it should be with modelling. This is a free-standing academic discipline and one in which profound questions of the appropriateness of our premises and principles should be being asked, not least because of the times in which we live and the momentous issues with which we must struggle in order to cope with environmental change. Some of these questions we shall try to articulate in this monograph.

But first I must confess the book has a somewhat unusual origin. In 1992 a small group of twenty or so scientists and engineers was assembled to form the *International Task Force on Forecasting Environmental Change*. With financial support from the National Water Research Institute of Environment Canada, the National Institute of Public Health and Environmental Protection (RIVM) of the Netherlands, and the International Institute for Applied Systems Analysis (IIASA), Laxenburg, Austria, the purpose of the Task Force was to deliberate on the methodological, and to some extent theoretical and philosophical, problems of forecasting the behaviour of the environment. This concern, to mount a sustained attack on the question of whether models may be trusted in predicting a change in the climate or some part of the environment more generally, was born of a less formally gathered collection of papers appearing under the same title in a special issue of the *Journal of Forecasting*.[2]

---

[1] Schaffer, S.. 1993. Comets and the world's end, in: *Predicting the Future* (L. Howe and A. Wain, eds.). Cambridge University Press, Cambridge, pp. 52–76.

[2] Beck, M.B. (ed.), 1991. Forecasting environmental change. *J. Forecasting*, **10** (1&2).

The Task Force was designed from the outset to generate a single product: the monograph you are now reading. This was to be achieved essentially as a networked activity, focused and catalysed, however, through three Workshops to be held at IIASA (in February, 1993, July, 1994, and July, 1996). The deliberations of the Task Force have thus far received no public exposure, as intended; and sincerely, I hope I have not buried the freshness of everyone's ideas in the time it has taken me to wrestle with making a coherent whole of the parts. I must further confess to an abiding anxiety to escape from appearing naïve in my expression of the problems and solutions before us. Substantially new thinking is presented for the first time in the monograph. It is, in short, the integral of almost a decade of research across the network of the Task Force.

A manifesto is a public declaration of intentions, and that is how I feel about this book, especially Chapter 5. When I had finished drafting that chapter I could not get rid of the idea that we were "all dressed up with nowhere to go". We had thought hard and long about how we might use computational analyses for detecting, exploring, and coping with a future in which there might well be "structural change"; yet, within the span of the Task Force, we had no all-embracing case study on which to demonstrate the worth of our entire argument. In some ways, in the time it has taken me to go from Chapter 5 to the rest of the book, I am pleased to say I have been overtaken by events. Many of the intentions expressed herein have begun to be put into practice in a current research project, on the watershed of Lake Lanier, Georgia, in the south-eastern United States. But that might be the subject of another book – and I can already hear the patter of feet, of any potential co-authors running away from the prospect of a decade-long moratorium on their publishing plans. Chapter 11 will have to suffice for the time being: as an inkling of where I (rather than my much loved "royal we") might next proceed from this book, dressed up and all.

I have in mind an imagined reader of this book, who would begin at the beginning and end at the end. If such a reader exists, s/he would find the text cycles through its subject more than once. The first six chapters (Part I) are a miniature of the remaining thirteen. Thus, Chapter 2 has its larger counterpart in the Case Histories of Part II; Chapter 5 has its counterpart in The Approach of Chapters 10 through 17 (Part III); and the epilogue of Chapter 6 is echoed in Chapters 18 and 19, which form Part IV (the Epilogue). More specifically, there is one turn of the cycle in Chapter 5, another in Chapter 6, and yet another in Part III. Chapter 2 reflects my personal view on some of the major studies of the past in modelling the behaviour of the environment. Naturally, my views may be read as diverging from those expressed by other authors in the case histories of Chapters 7, 8, and 9. There is a less obvious cycle rotating through the sequence of chapters in Part III (The Approach). It has to do with the enduring tension between low-order and high-order models, or between the "small" and the "large" in the vernacular. We begin with the relatively large in Chapter 11 and continue likewise through Chapter 12 until Chapter 13, which is pivotal in dealing with both the large and the small, thereby launching the reader into the smallness of the models in Chapters 14 and 15, and even 16, which nevertheless points us back to the very large indeed, once more, in Chapter 17. And then there is my idiosyncratic use of the metaphor of branch-node network diagrams, to explain to myself, and to others, how we might conceive of structural change, its component problems, and the potential avenues of approach to their possible solution. Tracking through Chapters 4, 5, 11, and 15, will tell something of this pictorial story.

I owe many people, places, and institutions my gratitude for now having the pleasure of writing this Preface. If the length and style of what follows seems to suggest this monograph has been exhausting, you are correct. It has. And for that reason alone, I wish to grab this opportunity while I have it. It may not come along again.

Let me start by thanking some of the people. First, there are my fellow authors, who truly have had to have a good measure of patience to wait so long for their labours to see the light of day. Some have had to put up with my interfering with their writing. I wanted so very much to give the reader the smoothest possible ride from start to finish. My co-authors still associate with me, this notwithstanding. Then there are those who took part in the Task Force and influenced its direction, though they do not appear as authors. I am indebted to Jerry Ravetz, in particular; and to Tom Barnwell, Lin Brown, Peter Janssen, Olivier Klepper, Todd Rasmussen, Wout Slob, John Taylor, and Howard Wheater. Jenny Yearwood had everything to do with producing the figures and diagrams for the book; few may come to know those idiosyncratic diagrams – as I have called them – as intimately as she has.

Then there are the places and institutions to which this book owes its existence. I was with the Department of Civil Engineering at Imperial College when the Task Force began. Since 1993, however, I have been with the University of Georgia. The workshops of the Task Force were held at IIASA, while Chapter 15, which in many ways betrays the origins of this monograph (going back to my first time in Cambridge), was completed – appropriately enough – during a visit to the Isaac Newton Institute for Mathematical Sciences (in 1998). I recall I hatched the plan for the Task Force while a Visiting Scientist with the US Environmental Protection Agency (EPA) in Athens and, if one is looking for a certain symmetry, it is fitting that the EPA is currently supporting the case study material of Chapter 11 (with a hint of what is to come) through a grant from its Water and Watersheds Program. To all of these institutions I am deeply indebted for the freedom they have given me to work on the subjects of this monograph.

Most of all, however, this book is about IIASA and the privilege it was for me to spend my defining years there.

A friend of mine, who shall remain nameless, put this in the Preface to his book:

> "My wife has asked me not to write one of those embarrassing acknowledgements, saying how impossible this research would have been without her constant encouragement and support; consequently I shall leave this to the reader's imagination."

No such request has been put to me. And in any case, why should I presume to dedicate an edited book in a personal manner, when so many others have invested so much of their effort in its production? But since I doubt I shall edit, let alone write, many *real* books (the ones without multiple contributing authors), I feel compelled nevertheless to say something now – rather than never – of a strictly personal nature.

> Thank you – so very much. It was fine for me; how about you? I think I know the answer.

On a lighter note, for those who know only of my work as though it dealt narrowly with the subject of "sludge" alone – of the sewage-derived sort, that is – welcome to my spare-time hobby!

M.B. Beck
Athens, Georgia
July, 2001

PART I
THE MANIFESTO

CHAPTER 1

Introduction

M.B. Beck

1.1 AN OLD FAMILIAR PROBLEM

We have always striven to develop models for extrapolating beyond conditions hitherto observed, to escape the confines of the record of the past. Had we come to the subject of this monograph from the context of Geophysics, we might have thought substantial extrapolation possible. The jibe once was that any model requiring "calibration" against past observations must thereby be rendered inferior, certainly incapable of extrapolation to conditions not previously observed. But a model based *truly* on the Laws of Physics, we supposed, would not suffer from such difficulties. For these models would employ constants whose values are known *a priori* and which are universally applicable. In the ideal, they would contain no parameters without immediate physical meaning. All their parameters would be capable of independent measurement in the field; none would require calibration. Such a happy state of affairs does not obtain in practice. To believe it would be so has been an illusion, albeit an extraordinarily useful one, even now.

Today's problems of environmental protection differ significantly from those of the past in several respects. Most obviously, the scale of the current problems is often global (not local) and their dynamics are evolving with relatively long (as opposed to short) time constants. Analysis of such problems will require extrapolation of perhaps staggering proportions: of making statements about the entire mosaic having inspected just the nature of a single tile. Perhaps less obviously, but more directly indicative of the distinctive character of this monograph, we must find solutions that are based on *inconclusive model evidence*, not conclusive field evidence. Our research must be conducted in a setting of *policy proximity and data poverty*, as opposed to policy remoteness and data richness (Funtowicz and Ravetz, 1990). And we shall be less concerned with optimising recovery under low costs of failure, rather with *avoiding disasters with high costs of failure*.

It has been said that policy-makers and the public tend to be more fearful of the possibilities of nonlinear dislocations and surprises in the future behaviour of environmental systems than of the likelihood of smooth extrapolations of current trends (Brooks, 1986). Most of us *are* disturbed by the thought that our environment may yet come to differ significantly from what we have known in our life-times. Indeed, we are really rather creative in imagining what may go wrong (Leggett, 1996). But what may appear as radical changes of behaviour in the state space – of quantities that we can observe – may well be a function of relatively modest changes in a model's parameter space (Doveri et al., 1993). What may have been a minor mode of behaviour in the past, submerged almost under the residual noise of the empirical record, may come to dominate the system's behaviour in the future. How then shall we deal with our fears of substantial change in the future? What models might we construct and manipulate for this purpose? Have we, in short, an approach appropriate to the development and application of models for dealing with *structural change* in the behaviour of the environment? This, then, is the central concern of the monograph.

But this is a grand concern, and somewhat abstract. In terms that can be grasped more readily, these too are our concerns:

(i) Given the lay person's apprehension at the prospect of surprises and qualitative dislocations in future scenarios of environmental change, what computational methods are available for generating these potentially radically different patterns of behaviour in a more organised manner?

(ii) In the presence of gross uncertainty, what are the key constituent hypotheses about the system's behaviour, i.e., what are the key model mechanisms, on which the reachability of these surprises might turn?

(iii) Is the system in imminent danger of a major dislocation which would lead to a feared pattern of future behaviour?

(iv) Is there any evidence in the narrow window of the empirical record to suggest that the system may already have embarked upon a path of collapse into a feared future; can we identify and diagnose the seeds of any imminent change in this record of the past?

(v) Can we design our models for the earliest discovery of our ignorance?

(vi) Can we make useful forecasts of possible behaviour patterns in the face of these extrapolated dislocations?

(vii) And how should we design regulatory policies for minimising the reachability of feared future states while maximising the reachability of desired states?

Yet even these are still not trivial questions; and demonstrable progress in their resolution over the course of this monograph will be but modest.

For our task, it must be said, is immodestly large indeed. We judge that two programmes for modelling environmental systems have been predominant in the past. To caricature them, they were (in chronological order):

(i) **Programme (1)**: Include in the model everything of conceivable relevance; in particular, maximise use of the Laws of Physics in this composition; maximise thus the number of constants (parameters, coefficients) that are either universally known or measurable, independently, in the field or laboratory; minimise thereby the freedom arbitrarily to adjust the parameterisation of the model in the light of the observed behaviour of the prototype; juxtapose the output of the model with this empirical evidence; and make a singular (deterministic) prediction.

(ii) **Programme (2)**: Acknowledge the uncertainty attaching to the set of prior model concepts and the empirical record of observed behaviour; presume therein that the parameters of the model are constants, but not known precisely; in the process of reconciling observed and estimated system responses, employ the past observations to adjust and quantify the uncertainty attaching to the model, i.e., to its constituent parameters; and make an ensemble of predictions that are therefore intrinsically uncertain.

While there are many aspects of these programmes we wish to employ, in particular, those of the latter, our strategic goal has been to articulate a framework for a third programme of modelling, which we state as follows:

(iii) **Programme (3)**: Derive qualitative statements about possible future patterns of behaviour from systematic organisation and manipulation of current (non-quantitative) beliefs; acknowledge that the model's parameters are the focal points in a map of the scientific partial knowns and unknowns about the behaviour of the system; presume that these parameters will, in general, change with time; assess the candidate parameterisations and parametric changes enabling the given futures to be reached; and direct interpretation of the observations of past behaviour in the light of performing this predictive task.

Put another way, our goal is the development of an approach and a set of methods that will enable us to examine the record of the past in a manner guided expressly by prior contemplation of the reachability of certain feared patterns of future behaviour. We wish to contribute to the process of generating environmental foresight (Science Advisory Board, 1995), albeit in literally a round-about manner[1].

---

[1] And using a rear-view mirror to complement Schellnhuber's (1999) forward-view mirror, which he defines as "contemplation of the future by reflection on the past".

From the outset, however, let it be said that we do not believe the mere development of more "comprehensive" models will of itself reduce or resolve all the uncertainties in understanding and coping with environmental change. All models are bound to be approximative in some way; the parameters of a model reflect the seat of this approximation; changes in the structure of a system's behaviour take place in the parameter space of the model – *not* its state space; and, up to a point, our expectation is that parametric change is relatively slow, smooth, and, perhaps, even predictable. Our models are *not* inhabited by *constants*. In this monograph the humble coefficient, hand-maiden for so long of the relationship between the more important variables, will be elevated to a superior position in the scheme of things. This perspective, then, is both a distinctive feature and a cardinal point of our approach.

A second cardinal point in our approach is that under Programme (3), the tasks of modelling will differ from those of the other programmes. Our models will have to be designed around criteria different from those of Programme (1), in particular. They will be designed to answer questions *not* of simple prediction: not of generating statements about the precise location of the system at some point in the future over the entire space of all possible behaviours; nor of making predictions that aspire to being a fact (which subsequent observation may reveal, in the event, as true or false). Their purpose, and the purpose in using models to make statements about future behaviour, is to reveal something about the features of the model – and therefore about our *current understanding* of the system's behaviour – in relation to the reachability of certain domains of the state space.

Third, we shall free ourselves from a restrictive mind-set: that what constitutes the description of a system's behaviour can only be but a set of (conventional) observations. In just the same manner as formal statements can be made about what is believed to have been the observed behaviour of the system in the past, so formal statements may be made about what we fear may happen in the future. The question of whether that future is reachable can thus be made formally similar to the familiar notion of reconciling theory with observation. Figuratively speaking, the significance of the present as the fundamental dividing line between past and future is thus diminished: we shall have merely a string of statements about where the state of the system should be at any time from the past to the future. This will be the third cardinal point of our approach.

But at the heart of what we wish to say lies a need to exploit to the full the panoply of computational techniques already available to us, and to do so in an organised manner. Contrary to the expectation we may already have engendered, this monograph seeks not to demonstrate the poverty of some supposedly reductionist geophysical approach to forecasting environmental change. Poverty, in our view, lies in the single-minded adoption of but one perspective on this problem. Resolving the problem, if this is possible, needs all the richness of perspective it can get.

1.2 EMBARKING ON A CHANGE OF PERSPECTIVE

Who are we in this group, then, and on what basis do we presume to contribute to the debate on forecasting environmental change?

"We" are simply the authors of the chapters of this monograph; our interests and experience can be found at the back of the book. There is no orthodoxy to which we would all subscribe, except – as declared above – that a variety of perspective and approach is essential. We indeed debated the issue of what we were about for over six years and through three Workshops. Our greatest difficulty was the fear we were about to re-invent the problem to which the wheel, so to speak, would be the proverbial solution – in the centuries still come. In the event, this has not proved to be the case.

Perhaps somewhat unconventionally for our subject, we shall be drawing in part upon reality-structuring rules having to do with our more subjective beliefs and fears. These, in fact, may not be so much "our" beliefs and fears as those of the public and the policy-maker. There is an obligation for us therefore to reflect on the social and cultural context of forecasting. In his book (*Reality Rules: Picturing the World in Mathematics*) Casti (1992) offers the following rather pithy reminder of our predicament:

> ...Western science is a johnny-come-lately in the reality-generation game, having arisen historically in the Middle Ages as a response to the inability of the competition to offer a satisfactory explanation of the Black Death. (Casti, 1992; p 379)
>
> In the Reality Game, religion has always been science's toughest opponent... (Casti, 1992; p 398)

We cannot entirely divorce the way in which we picture "reality" from the social and cultural milieu of our time and place. Of particular importance is thus the following. Taken from Simon Schaffer's contribution to the book of Howe and Wain (1993) on *Predicting the Future*, it has to do with the role of comets in marking a pair of historical changes in the way we predict (Schaffer, 1993):

> First, we see a change from the use of cometary appearances as a chance for soothsayers to predict the future course of politics, the weather and other features of immediate interest, to cometary transits as themselves foreseeable threats to this planet. Comets had been occasions for prediction; then they became predictable. Second, there were complicated changes in the kind of person who could be trusted to make predictions. Court astrologers and popular almanac makers gave way to expert advisers brandishing tools of statistics and physical science. So comets provide us with a good way of seeing how prediction depends on politics. Comets were publicly available signs, representations of heavenly wrath. Somehow or other Europeans learnt to stop worrying and trust the astronomers. In this process, the great divide between traditional and modern culture was created.

Science has been an essential part of this "modern culture". Yet since the close of the Second World War the basis for the special place of science in our affairs – being supposedly immune to fashion and subjectivity in its conduct – has been eroded (Funtowicz and Ravetz, 1990, 1993). Indeed, the rate of such erosion seems almost to have been accelerated through the very desire to put science to work in the formulation of environmental policy, witness Jasanoff (1992):

> The content of science, and even what constitutes "science" in a given situation, is now seen as lending itself to a multiplicity of interpretations.

This so-called pluralism is blossoming: in the democracy of engaging and working with the perspectives of all those holding a stake in the outcome of policies and decisions affecting the environment (Darier et al., 1999; Korfmacher, 2001); and just as much, in contemporary studies of the use of Cultural Theory (for example, Thompson, 1997) to patch together alternative combinations of both science and policy preferences to produce a multiplicity of model structures and policy forecasts (van Asselt and Rotmans, 1996). Such pluralism in our perspectives on the man–environment relationship can be computationally replicated through populating our classical environmental simulation models, as it were, with human agents possessing the requisite variety of "mental models" (Janssen and Carpenter, 1999). Each of these agents – in the computer world – perceives the state of the simulated environment, interprets this in terms of his/her individual (simulated) mental model of the man–environment relationship, and acts accordingly, implementing thus a course of action affecting the future state of the (simulated) environment.

But this latter is not quite the path we shall take in the monograph. We wish rather to acknowledge that "we know our place" – at the end of the thread of Schaffer's argument, which continues as follows (Schaffer, 1993):

> Each culture privileges a select group of soothsayers...
>
> Experts need their own private world where ignorant outsiders cannot penetrate. The very obscurity of the sums in which the cost-benefit analysts or astrologers engage helps give them impressive authority. But the expert predictors also need outsiders' trust: they need to show that the terms they use are, in some way, connected with what matters to their customers.
>
> So predictors have to move between specialist technical work and public, widely accessible, concerns. One mistake is to suppose that the culture of the wider public has no effect on the specialist predictors: it does. A lesson of the comet stories is that the most apparently technical estimates of cometary science are very sensitive indeed to public needs and attitudes.

Are we, the authors of this monograph, a "select group of soothsayers"? We are not. Our goal, as already stated, is not that of foretelling the truth; it is rather the earliest discovery of whether we stand on the threshold of change. And in this respect, our legitimacy rests on careers built on trying to reconcile our models with the observations of past behaviour. We have employed the reasonably rational methods of scientific enquiry for these purposes. But in this monograph we acknowledge a growing awareness of the manner in which the quantitative techniques of such enquiry must serve public needs and attitudes.

REFERENCES

Brooks, H., 1986. The typology of surprises in technologies, institutions, and development. In: *Sustainable Development of the Biosphere* (W.C. Clark and R.E. Munn, eds). Cambridge University Press, Cambridge, pp 325–348.

Casti, J.L., 1992. *Reality Rules: II. Picturing the World in Mathematics – The Frontier*. Wiley, Chichester.

Darier, E., Gough, C., De Marchi, B., Funtowicz, S., Grove-White, R., Kitchener, D., Guimarães-Pereira, Â., Shackley, S. and Wynne, B., 1999. Between democracy and expertise? Citizens' participation and environmental integrated assessment in Venice (Italy) and St. Helens (UK). *J. Environ. Policy & Planning*, **1**, 103–120.

Doveri, F., Scheffer, M., Rinaldi, S., Muratori, S. and Kuznetsov, Y., 1993. Seasonality and chaos in plankton-fish model. *Theoretical Popul. Biol.*, **43**(2), 159–183.

Funtowicz, S.O. and Ravetz, J.R., 1990. *Uncertainty and Quality in Science for Policy*. Kluwer, Dordrecht.

Funtowicz, S.O. and Ravetz, J.R., 1993. Science for the post normal age. *Futures*, **25**(7) 739–755.

Howe, L. and Wain, A. (eds.), 1993. *Predicting the Future*. Cambridge University Press, Cambridge.

Janssen, M.A. and Carpenter, S.R., 1999. Managing the resilience of lakes: a multi-agent modeling approach. *Conservation Ecol.*, **3**(2), 15 [online].

Jasanoff, S., 1992. Pluralism and convergence in international science. In: *Science and Sustainability*, International Institute for Applied Systems Analysis, Laxenburg, Austria, pp. 157–180.

Korfmacher, K.F., 2001. The politics of participation in watershed modeling. *Environ. Manage.*, **27**(2) 161–176.

Leggett, J.K., 1996. The threats of climate change: a dozen reasons for concern. In: *Climate Change and the Financial Sector: The Emerging Threat, the Solar Solution* (J.K. Leggett, ed.). Gerling Akademie Verlag, Munich, pp. 27–57.

Schaffer, S., 1993. Comets and the world's end. In: *Predicting the Future* (L. Howe and A. Wain, eds). Cambridge University Press, Cambridge, pp. 52–76.

Schellnhuber, H.J., 1999. 'Earth system' analysis and the second Copernican revolution. *Nature*, **402** (Supplement, 2 December), C19–C23.

Science Advisory Board, 1995. *Beyond the Horizon: Using Foresight to Protect the Environmental Future*, Report EPA-SAB-EC-95-007. Science Advisory Board, US Environmental Protection Agency, Washington, DC.

Thompson, M., 1997. Cultural theory and integrated assessment. *Environ. Modeling and Assess.*, **2**, 139–150.

van Asselt, M.B.A. and Rotmans, J., 1996. Uncertainty in perspective. *Global Environ. Change*, **6**(2) 121–157.

Environmental Foresight and Models: A Manifesto
M.B. Beck (editor)

CHAPTER 2

We Have A Problem

M.B. Beck

2.1 LIMITS TO THE SCIENCE BASE

In one sense "scientific uncertainty" might be understood as follows. We are engaged in the enterprise of developing computational models capable of replicating the behaviour of environmental systems. At any stage we are unable to include in these models all that we "know" – or believe we know – of the things that must be affecting the behaviour of the system "in reality". According to conventional wisdom, uncertainty is reduced by including greater refinement of description in the model, that is, the inclusion of more state variables to represent both a greater diversity of chemical and biological species and the spatial distribution of these species in greater detail.

Few things capture better the spirit of this conventional wisdom than the "bold and exciting scientific venture" of the International Geosphere–Biosphere Programme (IGBP). In the words of the then President of the International Council of Scientific Unions (Menon, 1992):

> [w]hilst complete understanding of every aspect of the Earth system is unrealistic, the concerted scientific effort of IGBP will significantly reduce uncertainties in predicting the future consequences of current trends.

Uncertainty in this sense should, in principle, be reduced by unfolding the coarser-scale parts of the model, specifically some of its *parameters*, or coefficients ($\boldsymbol{\alpha}$), into sets of more refined state variables ($\boldsymbol{x}$), and inter-relationships among them containing parameters that are *truly invariant* at this finer scale of resolution. Though "known about" before, albeit perhaps in outline alone, the nonstationary behaviour of these finer-scale parts would previously have been subsumed under a parameter, assumed to be constant. The finer-scale variability would have had to have been "parameterised" in the coarser-scale model, because in principle the resulting size of that model would otherwise not have permitted reasonable computational times and costs. For as long as unit computational times and costs are decreasing, such uncertainty might continue to be happily resolved in this manner. It would be prudent, nevertheless, to acknowledge that reducing uncertainty through including more science in the model is not the same as eliminating uncertainty from what has already been included.

For most of us, working in the relatively ill-defined world of Environmental Science, the world of Physics seems so secure in comparison, even to the extent of being intimidating. One simply does not challenge the Laws of Physics, unless one numbers among those who are writing and re-writing them. We, the authors of this monograph, do not count among those people. To us, theoretical Physics has the kind of obscurity that (in Schaffer's words) bestows impressive authority on its practitioners. If the prevailing mood in one's own field derives from the appeal of the seeming security and perfection of Physics (as seen from a distance), the audacity to challenge the accepted norm will continue to fail us. For if we could only base our models on the Laws of Physics, true extrapolation beyond what has been observed in the past could be achieved. In the ideal, this is how it was thought we should go about the business of modelling (Mason, 1986):

> It had already become apparent, a quarter of a century ago [the early 1960s], that the traditional, empirical methods of weather forecasting, based mainly on the extrapolation of very recent developments and the experience of individual human forecasters, were unlikely to improve significantly or produce reliable forecasts for more than about 24 hours ahead. Fortunately, with the arrival of powerful digital computers, it became possible to replace these highly subjective methods by objective mathematical predictions based on a firm structure of physical theory that treats the atmosphere as a rotating turbulent fluid with energy sources and sinks.
>
> This involves the building of very large and complex mathematical models of the atmosphere based on the physical and dynamical laws that govern the birth, growth, decay and movement of the main weather systems that can be resolved by the model. In other words, the models must properly represent the relevant or significant scales of motion and their nonlinear interactions, but smooth out all the smaller scale motions that cannot be adequately observed or represented individually while allowing for their overall contribution to transport and energy conversion processes by representing their statistically averaged properties in terms of larger-scale parameters that can be measured.

This has been the prevailing view throughout our working lives. The problem is not so much that this view is flawed. Nor does the problem lie so much in the futility of a search for "parameters that can be measured"; *measured*, we must presume, independently and in a manner that is free of any assumption, i.e., free of any assumed structure of a model, for the way in which the parameter relates the observable quantities to each other.[1] It is rather that this view cannot provide the only route to insight and policy formulation (Shackley et al., 1998). The certitudes of this programme are perhaps more apparent than real.

For writ very large indeed, contemporary studies in Physics suggest that the behaviour of the universe is not chaotic, but quite predictable (Hawking, 1993). Yet the question of whether the universe will continue to expand for ever or eventually recollapse hinges rather precisely on whether its current density is above or below a critical value – arguably thus a candidate for the most vital parameter in the universe. Whatever the means of deriving this value (the theory from which it emerges appears to be unassailably sound), attempts at then establishing whether the magnitude of the present density of the universe is close to this value demands some substantial theorising: a theory of inflation of the universe (perhaps now in the throes of being overturned; Hogan et al., 1999), coupled with the possibility of neutrinos having a non-zero mass (if they exist) and of there being black holes uniformly distributed throughout the universe. The existence of these latter, as it happens, can only be inferred from not entirely precise observations of the current expansion of the universe. In particular, a single observation at one end of the range of observed values (relating to the most distant galaxies) motivates the turning of a straight-line empirical fit into a higher-order curve that is straight but for an upward turning cusp

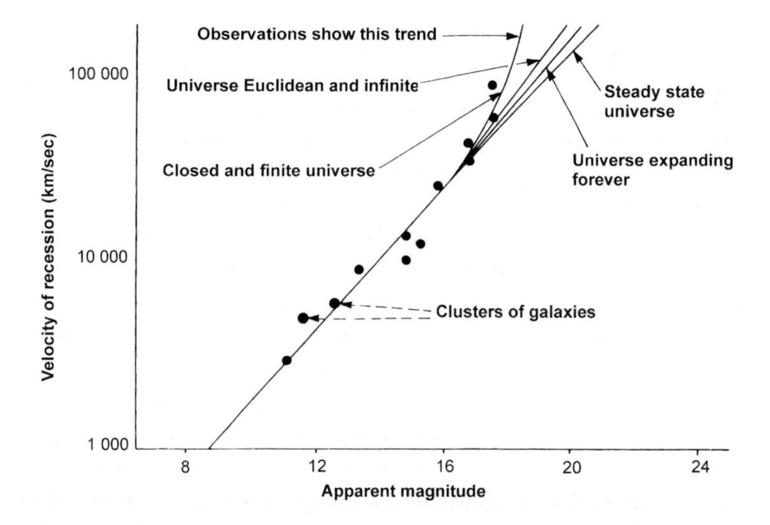

Fig. 2.1. Future of the universe: velocity *versus* distance of galaxies (reproduced with permission from Hawking (1993)).

---

[1] We do not measure a parameter of hydraulic conductivity in an aquifer. We measure pressure heads, assume Darcy's Law relates the heads to each other, and therefrom back-calculate, i.e., estimate, the conductivity parameter, assuming further that Darcy's Law is inviolable.

at one extremity (Figure 2.1) (Hawking, 1993). The sought-for extrapolation will be dramatically different depending on which of the two fitted curves is chosen for prediction: the universe will either continue to expand or collapse back to the "Big Crunch". The figure shows just what a fine line there may be between predictions of such cosmically different proportions.

Writ infinitesimally small, Gell-Mann (1994; p 197) would assign to us lay observers the presumed view of particle Physics that it must "... seem crazy to suppose that the basic law of all matter in the universe could rest on such a large and heterogeneous collection of fundamental particles", which he numbers as sixty-one. He goes on to dismiss the supposition of an ultimate description of matter involving a "smaller number of new, truly elementary constituents" as being without any theoretical or experimental evidence. And the notion that there might be no end to the unfolding of yet more layers of ever more truly elementary particles is dismissed on the circuitous basis that this view had only prospered among scientists in the People's Republic of China under the dictate of Chairman Mao (Gell-Mann, 1994; p 197/8). Instead, there should exist a simple, more fundamental theory (superstring theory) in which the number of such elementary particles can be regarded as infinite, with only a finite number accessible to experimental detection at available energies. In much the same manner as sub-grid scale variation is "parameterised" in conventional general circulation models – but could be correctly described if only we could resolve the system at the finer-scale at which the Laws of Physics truly apply – this theory would allow deduction of the "dozen arbitrary numbers [parameters] that cannot be calculated and must be taken instead from experiment" (Gell-Mann, 1994; p 200). In other words, the new (and simpler) theory would provide a logical and reasoned means of deriving values for the parameters of the less refined (and older) model, which hitherto would have suffered the ignominy of having to be "calibrated" in some way. Nonetheless, these parameters do appear to be well identifiable (in the terms of the present discussion).

From a distance, what are we environmental scientists to make of these things? One might conclude that conjecture about the extrapolated fate of the universe is still captive of the need to fit a curve to the past observations; that a constant (the deduced threshold value for the density of the universe) may be the spur to great theorising; that at any given scale of representation, the model's parameters are commonly used as interim, empirical expedients – temporary parking places, as it were, for knowledge of a too detailed or too speculative nature; that these parameters are the ports of entry into the search for invariance – at a finer scale of resolution; and that this search, the very essence of progress in understanding, may one day culminate in use alone of the irreducible set of the three universal constants of Nature (the velocity of light in empty space; the quantum constant of Max Planck; and the gravitational constant of Isaac Newton).

Such is our *perception* of some of the parts of Physics, inevitably skewed by our lack of insider expertise, to which we freely admit. What follows in the monograph is an account of how we have wrestled with recording those things of professional interest to us, where we do indeed consider our expertise lies: in understanding how the environment functions. In our world, accessible by just the more common senses (at the meso-scale, neither infinitesimally small, nor infinitely large), most environmental scientists still work with an eye on the apparent security and solidity of the dominant programme of Geophysics. Yet our everyday experience has *not* been one of uninterrupted success in this programme of model-building. Access to a scale of refinement in the resolving power of our models permitting application of the Laws of Physics – using only parameters with physical meaning that are somehow independently measurable in the field as isolates cut away from the whole – has not been achieved (Beven, 1989). All our models in principle contain parameters whose values will have to be assigned on the basis of attempts at reconciling the performance of the model with observations of the *in situ* system, behaving, as it does, as a complex whole. All our models, including those dealing with the *physics* of water flow over and under the surface of the earth, will be the subjects of calibration.

Even at the core of the programme, in meteorology (Lorenc, 1986), oceanography (Evensen, 1994), and the study of global biogeochemical cycles (Kasibhatla et al., 2000), something curious is happening, arguably latent in the rise to popularity there of data assimilation (or state reconstruction). Why should we say this; what precisely is "data assimilation"? The most intensive of today's data-acquisition platforms cannot provide sufficient access to the spatial heterogeneity of geophysical systems or the biological states of the environment. Our models are laden, therefore, with a profusion of unobserved state variables. Reconciling a model with observed behaviour in order to improve understanding is quintessentially an issue of demonstrating, beyond reasonable doubt, that matching of the two approximations of the truth has not been achieved at the expense of imposing absurd values on the model's parameters (Beck, 1994). For as long as estimates of the unobserved state variables may dance around to the tune of any significant mismatch between the model and the data, understanding will not move forward. To put it bluntly, if one wants to fit data, an unobserved state variable in the model is worth any number of parameters. Data assimilation is therefore what its name suggests: an act of assimilating relatively sparse data into a theoretical structure presumed *a priori* to be secure. It is not an activity capable of ruthlessly rooting out the inadequacies of that structure. It may not even be all that successful a means of tuning the subgrid-scale variability and mechanisms lodged under parts of the parameterisation of the model's structure.

Could it be, then, that all of our models must inevitably be subject to calibration, even those designed deliberately to circumvent this perceived limitation? And could it be that achieving such calibration is becoming an ever more distant prospect, especially for these latter kinds of models? How shall we escape from this impasse, without abandoning a programme of model-building that has served us well, yet benefiting from the innovation released by confessing to its inherent flaws?[2] We

[2] When Oreskes (1998) argues that "predictive power is itself a fallible judge of scientific knowledge" she is granting us an entirely reasonable basis for just such innovation.

believe the answer may lie in Chapter 17: in thinking of models *not* merely as succinct archives of knowledge, nor as devices for the assimilation of data, nor even as instruments of scientific prediction, but as vehicles to be designed for discovering our ignorance – and at the earliest possible moment.

2.2 RIPPLES ACROSS THE COMMUNITY OF ENVIRONMENTAL SCIENTISTS

When the Meadows and their colleagues (1972) reported on the (then) "predicament of mankind" to the Club of Rome, their work was a celebration of exponential growth and the determinism of computer simulation. Predictions of "boom and bust" stride confidently across the pages of their text, with no sign of any bound on the possible errors of these forecasts.

We were experiencing then – at the turn of the decade from the 1960s to 1970s – the first wave of quantitative and systematic analyses of global environmental problems. In retrospect, the ripples that can be seen to have propagated across the community of environmental scientists are revealing. The engineering community – those, in particular, who had been professionally involved in simulating and forecasting the behaviour of aerospace and chemical process industrial systems – were both sceptical and critical of the application of these techniques in the less well defined areas of the environment and the economy (Cuypers and Rademaker, 1974). And those scientists who had already made their careers through studies of the environment, the economy and the social system were just as caustic in their criticisms of the engineers who were entering their subjects (Hoos, 1972).

It was those professing themselves to be "systems analysts" or "systems engineers" who attracted the greatest disapproval from those already in the field. Sadly, the written word does not reflect the actual tenor of the debate of the time. Still, one does not have to read too deeply between the lines that follow to gain a sense of this disapproval (Philip, 1975):

> Let me explain at the outset that, in the vocabulary of this Conference, I come 'from a pure physical orientation' – though my friends in particle physics doubt that I am a *physicist* and my mathematical friends are all quite certain I am not *pure*.
>
> I must admit that, personally, I have trouble in comprehending how the great new human enterprise of *simulation* aims to do anything different from what *natural science* has been trying to do for the last 300 years. Perhaps, after all, Newton and Einstein were simply 'simulation specialists'. Maybe the only thing which sets them apart is that they were especially wise and especially humble.
>
> In my opinion, there is a deep methodological difficulty here. One hears sometimes that 'systems analysis', 'modelling', 'simulation' – call it what you will – is the panacea[†]. [[†] (footnote). The role of systems analysis as a panacea is very understandable. Persons possessing great responsibility, authority, and power remain, nevertheless, human beings; and, like the rest of us, they love to hear good news. And the good news which

> the messengers have been bringing over the last decade or so is that systems analysis is alive and well and is ready and able to solve their problems.]
>
> There seems to me a danger that, when systems analysis is applied in very broad fields, it is prone to the forms of intellectual dishonesty noted by Andreski... : the disguising of simplistic and/or old ideas, and of slip-shod work, in space-age jargon and in spurious mathematisation.
>
> Beyond this, there are further problems connected with the economic, political and sociological elements in very large systems. These involve the general philosophical and methodological difficulties of the social sciences, particularly those concerned with *values*... A value-free social study is all too prone to home in happily on trivia, but to be impotent *vis à vis* the basic human issues.

Here we see, in biting terms, an objection to the "wetting front" of a Physics-based paradigm naïvely over-reaching itself as it sought to move outwards from its core domain.[3] But these were exhilarating times in which there seemed no limit to what could be encoded and manipulated on the computer. People were sincere in the belief that a model, as in CLEAN (Comprehensive Lake Ecosystem ANalyzer; Park et al., 1974), could be pushed – through CLEANER (Comprehensive Lake Ecosystem ANalyzer for Environmental Resources), which *was* produced (Park et al., 1975) – to home in on the truth, presumably the CLEANEST (and for which we would have had to have waited for ever).

2.3 LAKE ERIE AND EUTROPHICATION: YOUTHFUL EXUBERANCE OF SYSTEMS ECOLOGY

Eutrophication of Lake Erie, and therefore the lake's imminent demise, is an icon of the environmental movement of the 1960s and 1970s. Within the global scheme of things, because we had accelerated the production of food on the land surface through the artificial manipulation of the nitrogen and phosphorus cycles, and because we had chosen water as the vehicle for removing wastes from our urban environments, we had inadvertently shifted a balance of nutrients away from the terrestrial and atmospheric environments and into the aquatic environment.

In its simplest form the "problem" of eutrophication may be defined as follows: artificial acceleration of the natural ageing of lakes as a result of an elevated rate of accumulation of nutrients (carbon-, nitrogen-, and phosphorus-bearing substances) within a body of water. What was intended for the beneficial enhancement of primary biological production on the land surface was instead being diverted into enhanced primary production in the aquatic environment. This was manifest in the high-

[3] Now, a quarter of a century on, when the collective behaviour of a stock market populated by economic agents with bounded rationality (people, that is) can be simulated on a computer, with apparent verisimilitude (Arthur et al., 1997; Waldrop, 1992; Kauffman, 1995), we might not wish so to object.

frequency, high-amplitude perturbations of sudden bursts of growth and rapid collapse in the biomass of phytoplankton – microscopic organisms at the base of the food chain in an aquatic ecosystem. Given a ready supply of the principal nutrients, it was well known that under an appropriate combination of environmental factors (solar irradiance, temperature, concentrations of other trace nutrients, and so on) populations of the various species of phytoplankton would grow rapidly, with a succession of species becoming dominant over the annual cycle. The phytoplankton would be preyed upon by zooplankton, which in turn were themselves prey to the fish. Dead, decaying, and faecal matter from all of this activity was equally well known to be re-mineralised by a host of bacterial species – operating either in the water column or in the bottom sediments of the lake. Such re-mineralisation had an influence over the concentration of oxygen dissolved in the water and this, in its turn, exerted an influence over the degradation pathways operative in the overall process of microbially mediated re-mineralisation. Thus the cycling of elemental material fluxes was closed; thus the disturbances of excessive phytoplankton growth would reverberate around these cycles.

In short, worsening water quality in Erie in the 1960s was perceived as a consequence of the primary (anthropogenic) cause of increased discharges of nutrient residuals from the surrounding watershed. Given a sufficient basis of nutrients, solar radiation was able to fuel a significantly larger biological conversion of these inorganic materials into living organic matter (which was then propagated up the food chain). The ramifications and secondary effects of this primary production were a significant increase in the powering of bacterial remineralisation of the resulting detrital organic matter back into inorganic materials. This microbial re-mineralisation was normally sustained by the dissolved oxygen resources of the water, but would switch to alternative patterns of recycle in the event of these resources being exhausted, i.e., in the event of anoxia. Dramatic changes in the species composition of commercial fish catches had also occurred over the years; and, almost as an aside, there was faecal coliform bacterial pollution of the in-shore zones of the lake.

Stepping back from the clutter of the details, it seemed simple and obvious that a lake was little more than a laboratory beaker in which excessive algal growth would occur when the contents of the beaker were over-fertilised. Broadly, therefore, it made sense to restrict the input of these nutrients to the lake; and just such was the action taken.

The practical, good sense of this coarse-scale description and resolution of the problem seems, even now, to puncture any ambition to search for a more complex model of all that we think we know of the fine-grained behaviour of an ecosystem in a lake. The view at the time was much the same:

> The major gaps in our knowledge, be it primary production or hydrodynamic circulation, are still to be addressed by highly specialized disciplinary research. In recent years, computer modeling has developed into a powerful and exciting addition to the more conventional sciences, but the immediate practical benefits with regard to environmental management may have been somewhat oversold. (Simons and Lam, 1980)

> Once again we lack basic biological information. I recommend that N. America pay less heed to the creed of system engineering and move to the (European) tradition of basic biology. (Harris, 1980)

Action was taken: the addition of nutrients to the continuously stirred tank reactor of Lake Erie was curtailed.

But the system did not respond entirely as expected. For the logic of this essentially "biochemical" reasoning of the fertilised phytoplankton reactor cannot be abstracted from the "physical" reasoning about water movements in the lake. Whereas the manipulation of water-borne nutrient input primes (or disables) the lake's propensity to generate algal blooms and crashes, it exerts no control over where this production is concentrated or where recycling of its by-products is concentrated. In retrospect, we can see that water circulation in the lake – driven by solar radiation, wind and other aspects of the lake's atmospheric climate – would cause the connection between the upper (warm) waters and lower (cold) waters to be severed, thus concentrating algal production in the former and bacterial re-mineralisation in the latter. Whether or not the limited resources of oxygen in the isolated bottom waters would be exhausted before the end of this summer stratification, in the absence of scope for replenishment, would be a subtle integral of the relatively high-frequency fluctuations in the weather over the lake (Lam and Schertzer, 1987; Lam et al., 1987). And so whether the primary symptoms of algal blooms and crashes would continue to be observed was "in truth" a still more subtle function of the direct influence of weather on the upper waters and its indirect influence over the recycling (or otherwise) of remineralised nutrients triggered by the secondary symptom of anoxic bottom waters – irrespective of fertilisation from human activities in the watershed.

These things were known from the outset. They were simply too subtle to be geared to the pulling of a crisply defined, albeit blunt, lever of policy. In this case, we could not have known beforehand of the consequences of what seems, with hindsight, to have been an over-simplification. Most of the torch-light of science was being shone in the direction of primary production of phytoplankton from the available nutrients (for example, Thomann and Winfield, 1974). Whereas the consequences of this primary action on dissolved oxygen were of interest, the consequences in turn of the dissolved oxygen on bacterially mediated re-mineralisation of nutrients lay outside the spotlight (for example, Di Toro, 1980). These secondary consequences were subsumed under the model's parameters: the *constant* volumes of water in the upper and lower segments of the lake; the *coefficients* of exchange of water between these segments (reactors); the remineralisation rate *constants* and so forth (Di Toro and Connolly, 1980). The phytoplankton and nutrients achieved the status of state variables, while the biomasses of the bacteria failed to rise above the level of a parameter. Conventional thinking was that these bacteria were not part of the foodweb (phytoplankton, zooplankton, fish) through which the nutrients were propagated; they merely mediated conversion of the residuals of this foodweb activity – the non-living organic detritus – into the inorganic nutrients at the base of the foodweb; they were set apart from the principal web of interactions in the system (Porter, 1996; Porter et al., 1996; Osidele and Beck, 2001; and Chapter 11).

Not surprisingly, we believe we are wiser after the event. But to imagine that the "not entirely expected" could not happen in the future would be to have learned little from the experience. How would the behaviour of Lake Erie change as a function, for example, of a change induced in its thermal regime, and thus its constituent reactor volumes, as a result of climate change (Schertzer and Sawchuk, 1990; Chapter 7)? Less obviously, what changes might result from a substitution of water-borne by air-borne inputs of nutrient residuals (via agricultural dust and car-exhaust emissions)? More subtly, what structural change might have been induced by an inadvertent grafting of the exotic zebra mussel population onto the pre-existing knowledge of the lake's phosphorus cycle? More bizarrely, almost, could we have imagined (thirty years ago) that we would now be contemplating adding phosphorus to the lake in order to compensate for the structural distortions wrought by the ingress of these mussels into the ecosystem? More obscurely, perhaps even outlandishly, what would happen if, in addition, a changed thermal regime were to cause the microbial loop, previously set apart from the foodweb and thereby relegated to insignificance in our thinking, to be drawn into it and indeed come to dominate it? How should we design our models, our monitoring systems, and our policies to maximise the earliest awareness of the possible occurrence of the previously unimagined?

Simplicity of action lies in reversing the (obvious) action that has caused stress on the environmental system. Once this stress has been removed, it becomes much less easy to determine what the problems are and what actions should be taken, and to predict what the consequences of disturbance and action might be (Beck, 1996). We have always worked on stressed systems, allowing (through our actions) an increasingly diverse community of biological species to return to progressively less stressed systems. And as this diversity of the unstressed system increases, it is not hard to imagine that the predictability of behaviour will decline, at least until observation of this newly revealed variety can once again catch up with theory.

2.4 SURFACE WATER ACIDIFICATION: INSUFFICIENCY OF A PHYSICS-BASED PROGRAMME

Once a drop of precipitation had hit the ground, so to speak, the observed phenomenon of surface water acidification was regarded (in the early 1980s) as the province of soil physics, soil chemistry, and sub-surface hydraulics. The applied systems analysis of the 1960s, which had helped set the tone of the biological and ecological studies of eutrophication in the 1970s, had by that time fallen into disrepute. Besides, Physics and Chemistry were much more secure in their knowledge bases than were the biology and ecology upon which one had had to rely for the study of eutrophication. Hydraulics, among all the branches of science, moreover, had had one of the longest and most illustrious traditions of mathematical representation and analysis of its problems. Such tradition and success were not to be set aside without very good cause.

The prototypical model of a catchment's hydrochemical event response was originally due to Christophersen and Wright (1981) and Christophersen et al. (1982). In some ways it withstood the tests of the times, remaining intact and largely in its original form for nearly a decade (although over the same span of time in the 1970s the models of lake ecosystems had exploded in complexity from a handful to thousands of state variables; Thomann, 1982). But this seeming durability of the model was also a measure of the difficulty of the problem – of understanding surface water acidification – and of the inability of field data to provide an effective test of the model's constituent hypotheses (Beck et al., 1990). Indeed, in contradistinction with most modelling programmes, this simple, conceptual model was thought in need of still greater simplification (Hooper et al., 1988; de Grosbois et al., 1988).

It is curious that few were to enter this field of modelling, other than those who had long been its proponents. Most of the scientists working on surface water acidification came from a tradition of laboratory and field science. For them it was the custom to replicate the conditions of the field in the laboratory soil column or to create the conditions of the laboratory in the field, through constructing a roof, guttering, and artificially manipulated precipitation over the natural forest. Few scientists worked on models of the processes of surface water acidification. Fewer still worked on attempting to reconcile these models against *in situ* field observations. Yet everyone – it seemed – had arrogated the authority to volunteer their opinions on how the development and evaluation of a model should be conducted. The text that follows, for example, comes from an invited discussion of the models developed during the *Surface Water Acidification Programme* (SWAP) (Johnson, 1990):

> Conspicuous by its absence is the ILWAS model; mention was made of it once... but it does not seem to have been employed by any of the investigators either for forecasting or for scientific reasons. I must ask why this is so, given its scientific stature and given that it is by far the most rigorously and realistically constructed of all the dynamic models. One could guess as to why it was not used (too complicated, too many knobs to turn, too expensive)...
>
> With a few notable exceptions... there is little evidence [at this symposium] of self-examination and self-criticism in the acid deposition modeling literature, but rather a hard sell for using the models as predictive tools for exactly the purpose of guiding policy. If we are allowed the same standards of accuracy set by economic forecasters, this will present no problems; but expectations seem to be much higher than that.

Criticism, especially constructive criticism, is invaluable. But after more than three decades of inter-disciplinary studies, now is probably no longer the time to indulge a lack of respect for the principles by which other disciplines conduct their scholarly enquiry.

Debate, including that of a dialectical nature, is essential. But the merits of the small – being "beautiful" in the eyes of some who develop models – have long been pitted against the merits of the big – thought "ambiguous" in those same eyes. Given all that has gone before (Young, 1978; Beck, 1981; Beven, 1989; Jakeman and Hornberger, 1993; Grayson et al., 1992; and Smith et al., 1994; as just a small, biased sample), it is perplexing to find still the following (Armstrong et al., 1995):

> [S]uccessful models are... more than simply those models that reproduce a given set of validation data well, but which also reflect the structure of the physical system that they attempt to represent. Where a model is based on a detailed study of the physics of the phenomena under study, then a successful modelling exercise can learn from the lack of fit between model predictions and observed behaviour. The results presented in this paper have shown that accurate modelling of solute behaviour is possible using models that are designed specifically for such soils by the inclusion of the cracks in the conceptual scheme.
>
> In general, where the model is developed entirely from a combination of physical theory and independently measured parameters, there is then no question of "adjusting" or "fitting" the model to reproduce system behaviour; the model can be in some sense verified....
>
> Errors in the observation data are frequently ignored in modelling studies, but can in fact lead to major problems if models attempt to reproduce spurious data. This problem is a major issue for empirically derived models, such as response function models, which derive their parameters from observational data.
>
> Where the model uses, for example, least-squares fitting to "calibrate" the critical parameters, then it is often very difficult to identify the sources of error. Only with independently measured parameters is it possible to verify, or more properly falsify (Popper, 1935), the model.

It is as if the stormy debate has passed by, seen on the horizon, barely ruffling the sails of this ship of thought as it proceeds, apparently steadfast on an unchanged course.

In the end the policy issues of surface water acidification were not quite as easily resolved as were those of eutrophication in Lake Erie. For a start, there were some more strongly competitive candidate hypotheses, at least in the European context, and the system itself was both less accessible to observation and exhibited behaviour that could not be so easily explained in the simple terms of a laboratory experiment. The great studies of the time notwithstanding, nowhere is this history more lucidly recounted than in the thesis of Bishop (1991), on which the following draws heavily. It was not obvious, as Odén had asserted in 1968, that atmospheric pollutants emitted from the distant industrial centres of Europe and carried aloft by the prevailing winds to southern Scandinavia were the principal cause of surface water acidification. After all, as Rosenqvist was to observe a decade later, how could it be that even rainstorms with circumneutral pHs could occasion spates of more acid water in episodically acid streams? He argued then, as later, that changes of land use and vegetation cover were the primary causes of the loss of fish in acidified surface waters (Rosenqvist, 1990):

> Inland fish has always been important in Norwegian rural diets. For this reason, total extinctions of all fish in former good fish brooks and lakes are mentioned in mediaeval folklore. The explanation given was divine punishment for sinful behaviour.

> The former theological hypothesis was replaced *ca*. 1920, by geological explanations, as it was found that pH in the barren lakes was lower than in the rain.
>
> The strong acidity was due to biogeochemical reactions between rain and the soil in the catchments. These reactions are complicated and not readily understood by laymen. Between 1960 and 1970, long distance transport of acid industrial air pollutants was established, and a new dogma became popular among politicians and research workers without backgrounds in geochemistry.
>
> Palaeolimnological tests of former acidity have been done in two lakes with the late mediaeval name Fiskeløstjern, i.e., tarn with no fish. The lakes are situated in areas that lost the human population during the plague of 1349. The catchment became woodland until A.D. 1500 when it was used mostly for grazing up until 1940.
>
> In both lakes, which now have pH 4.4, the pH was found to be below 4.6 around the years 1400–50. Before 1350, and between 1500 and 1940, pH was above 5.2.
>
> There are several reasons for modern surface water acidity: acid rain is one, but biogeochemical changes in the catchment are the overwhelming factors. Nobody has found a lake in Norway where the acidity of modern precipitation contributes more than 10% of the acid–base budget.

In the end it was not the hydrogen ions in the precipitation that mattered so much as the accompanying sulphate ions, which, being arrested but little in their passage through the surface vegetation and soils of the catchment would drag oppositely charged hydrogen ions from these land-surface features into the streams. On the conceptual platform of the weathering of rocks over geological time; the changes of industry over centuries; the changes of vegetation over life-times; the movement of waters along different flowpaths in days and minutes; and the almost instantaneous changes of the chemical signatures of these waters as they come into contact with differently primed soil particles – which reflect all of this complex history – came the insight in understanding (Bishop, 1991).

In the end the puzzle did not yield to a set of partial differential equations for the hydraulics of water flow and the chemistry of soil-water interactions. No such model has been proposed for surface water acidification to this day, as we shall see in Chapter 8. It was as though we had attempted conquest through a parachute drop into the unknown well beyond the frontiers of science, but then failed to establish enduring supply lines into the salient. Those who had objected in the 1970s to the naïve over-reaching of a Physics-based programme appeared to have been right. For here, in the 1980s, was evidence of its inability to cut in-roads into even its core domain.[4] In his own words, this is how Bishop saw what had happened (Bishop, 1990):

---

[4] Perhaps the programme was not lacking in sufficiency but its practitioners lacked a sufficient training in Physics.

> [W]e should not lose sight of the fact that SWAP was launched in 1984 on the assumption that five years of research would in fact lead to models reliable enough to provide an objective basis for environmental policy decisions. It must be considered fortunate that no government waited for SWAP's models and predictions before making difficult policy decisions concerning acidification. If any government had waited, it would have done so in vain. Furthermore, the initial fears of many Scandinavians that SWAP would lead to unwarranted delays in the decision-making process would have been proven telling prescient. Hopefully the scientific community will have learned from the humbling experience of trying to model acidification and plans for future environmental research will be informed by a more realistic expectation of what we can hope to reliably predict about the health of complex ecosystems.

But all this talk of the "end" is deceptive, of course. Where once we sought to unscramble the physics of water flow on the solid, inviolable presumption of a known chemistry – of the dynamics of chloride ions (Hooper et al., 1988) – we have since come to discover the insecurity of this presumption (Chen et al., 2001). We can no more assume the chemistry safely known in order to identify the unknown water flow than we can assume the water flow known in order to disentangle our understanding of the chemistry. The behaviour of the system is an indivisible whole. The problem has not gone away. We shall have to wait a very long time for the *post mortem* examination of the UK Government's decision, in the autumn of 1987, to authorise a larger flue-gas desulphurisation programme for existing and all new power stations in the UK. And on what key current "unknowns", one wonders, might the reachability of some other feared future turn – that in spite of all this the fish do not return to the streams and lakes of southern Norway?

2.5 TROPOSPHERIC OZONE CONTROL: DYNAMICS OF SCIENCE AND POLICY OUT OF PHASE

Figure 2.2 shows a sequence of predicted changes in stratospheric ozone concentration under the assumption that gaseous emissions of chlorofluorocarbons (CFCs) and nitrogen oxides would continue unchanged at mean rates estimated for the period of the late 1970s and early 1980s (adapted from Schneider and Thompson (1985)). In this "laboratory" test, with all but the science of stratospheric chemistry held constant with time, the dial on the instrument of prediction – the model – is subject to considerable fluctuation.[5] Had any policy of control crystallized instantaneously at some point around the contemporary image of the science base during this period, it would have been prone to rather swift obsolescence.

---

[5] Whether such fluctuation is significant, either in terms of the formulation of policy or for discriminating amongst competing candidate hypotheses, is unclear, for no bounds on the uncertainty attaching to these singular predictions are provided.

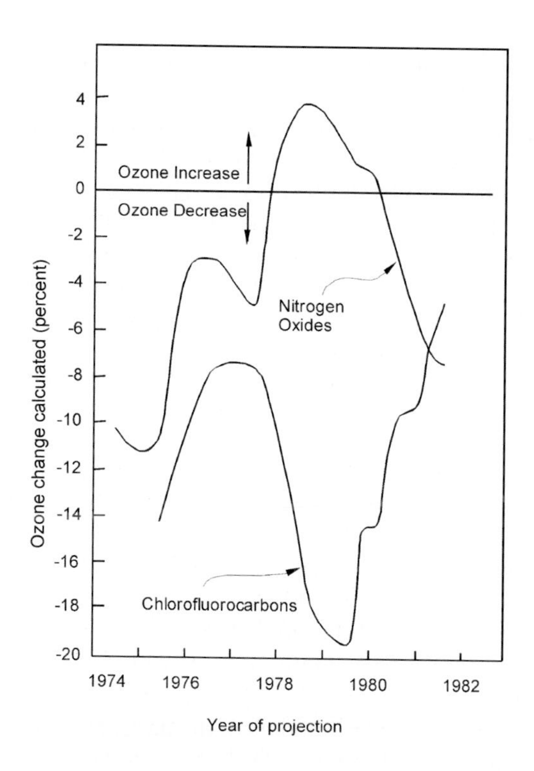

Fig. 2.2. Long-term stratospheric ozone change projections from constant emission rates (reproduced with permission from Schneider and Thompson (1985)).

While regulatory attention to the problems of eutrophication and acidification has ebbed and flowed over the decades, controlling tropospheric ozone has remained consistently within the frame of management concern (Chapter 9). The issue is this: given a set of initial conditions in the lower atmosphere (the troposphere) as the sun rises, how much ozone will be produced during that day and how can the peak daily ozone concentration be reduced, as a function of two levers of policy, i.e., by reducing hydrocarbon and NO_X emissions from ground sources? In other words, how can these high-frequency perturbations be regulated against a changing, lower-frequency dynamic underlying the causal relationships of persistence in the conditions at sunrise from one day to the next? The answer to this question would seem to lie simply in reversing the strain imposed on the system by reversing the (relatively) very low-frequency upward trend in pollutant emissions, just as in the common-sense approach to solving the problems of eutrophication and surface water acidification.

In short, with a clear focus on air masses over urban areas, assuming that rural sources of reactive hydrocarbons would not be significant, and acknowledging the absence of a political will to reduce NO_X emissions, cutting urban hydrocarbon emissions should have returned the system to its original unstressed (equilibrium) condition. But this did not come to pass. What goes up may not always come down in quite the same way, at least not in the nonlinear system of tropospheric chemistry. It was not that the nonlinearity was not appreciated, for it had earlier been found that reductions in NO_X emissions could, down to a point, induce significant increases in the production of ozone (as recounted, in detail, in Chapter 9). Indeed, such nonlinearity had been a strong motivation for having a model of the system's behaviour in the first place. It was rather that at lower initial concentrations of NO_X in the troposphere (at sunrise), the same rate of ozone production would prevail whatever the reduction in hydrocarbon emissions. Yet the apparatus of regulation had been locked on to a programme of reductions in volatile organic carbon (VOC) emissions. Worse still, in the urban centres reactive VOC species were likely to persist at elevated concentrations because of their transport into those zones from natural, i.e., biogenic, emissions in the surrounding rural region. The case of isoprene, in particular, had become something of a *cause célèbre* (Chameides et al., 1988). The science base was evolving faster than the policy framework. For the US Clean Air Amendment Act of 1990 was to confirm the hegemony of VOC control over NO_X control, while the models of the day were forecasting the ineffective and utterly wasted effort of investing in such prescribed action.

Our perception – that the extent of the problems is growing, bringing a greater span of spatial scales, temporal scales, and chemical and biological species into the boundaries of our analyses, as here, in the "ozone problem" – is a commonplace. We are well aware of the fact that solving one problem may compromise our ability to solve other problems; we need no reminders of the dangers of dividing the world into small, clinically isolated, solvable parts. We acknowledge and lament the human condition: of responding to damage and stress only when it is palpably present, after the event. We have come to expect inexorable growth in our knowledge bases, with diminutive or no shifts of paradigm in the sense of Kuhn (1970). We are, after all, dealing with the largely classical Newtonian world of environmental systems (neither the elementary particles of the universe, nor the nature of evolution). In a sense, we might retire merely to watch an unremarkable playing out of precisely what the International Geosphere-Biosphere Programme would refer to as "reducing the uncertainty". How then should we punctuate our account of this monochromatic, perhaps even featureless, development in our thinking? For distinction lies in none of these things.

2.6 A GLOBAL PICTURE: STABILISING GREENHOUSE GAS CONCENTRATIONS (AND STABILISING OUR FORECASTS)

The Inter-governmental Panel on Climate Change (IPCC) concluded in 1992 that (Houghton et al., 1992):

> [T]he evidence from the modelling studies, from observations and sensitivity analyses indicates that the sensitivity of global mean surface temperature to doubling CO_2 is unlikely to lie outside the range 1.5 to 4.5°C.

That is to say, were the (current) global mean atmospheric concentration of CO_2 to be doubled instantaneously, the numerical solution of any one of the several principal General Circulation Models (GCMs) computed over several decades into the future would settle at a new, invariant equilibrium in which atmospheric temperature would lie somewhere between 1.5 and 4.5°C above the equilibrium that would otherwise eventually obtain were the current CO_2 concentration to prevail for ever. Such a range of values derives from the differences among the resolving powers (spatial grids), feedback loops, and parameterisations of finer-scale (sub-grid) variabilities incorporated into the different GCMs. In spite of all the cerebral energy invested in reducing the uncertainties, and quite unlike the wavering science base of stratospheric ozone behaviour, it would *seem* this "headline" forecast has remained impressively stationary over the years. In fact, the constancy of the forecast may have fulfilled instead a sociological role: of maintaining coherence in the fragile process of building a global policy community, while not doing justice to the variegated, evolving understanding of the earth system and indeed permitting different interpretations of the forecast (van der Sluijs et al., 1998).[6] In particular, the predicted change might be read as reflecting merely the extent of the transient response in temperature at the time when the atmospheric CO_2 concentration actually reaches twice its current value, notably well before any corresponding equilibrium state could have been attained.

But this is, above all, a *forecast*. Unlike the problems of eutrophication or acid rain, the damage and stress of an altered climate are not yet manifestly present. We do not know the full extent of the displacement of this global system from its pre-existing equilibrium. We do not have replicate copies of either the fully damaged or pristine states of the local (even regional) sub-systems, to which to make reference in order to determine an appropriate restoring reaction. We are contemplating actions designed to pre-empt departure upon a path towards an imagined and unpalatable future. The key objective of the United Nations (UN) Framework Convention on Climate Change (presented for signatures at the UN Conference on Environment and Development, Rio de Janeiro, 1992) is stated accordingly as:

> Stabilisation of greenhouse gas concentrations in the atmosphere at a level that would prevent dangerous anthropogenic interference with the climate system.

To serve this purpose the IPCC has developed seven trajectories for the transfer of the atmosphere from its present state – as gauged by the concentration of CO_2 – to one of five possible future end states in which this concentration attains an equilibrium by the year 2300 at values ranging from 350 to 750 (ppm). In other words, the "desired" or "feared" evolution of the system's response has been specified. The question is, given a model of the relationships between this output and the exchanges (fluxes) of carbon dioxide between the earth's surface and its atmosphere (Taylor

---

[6] Just as Schaffer (1993) has said: "the most apparently technical estimates of cometary [earth systems] science are very sensitive to public needs and attitudes".

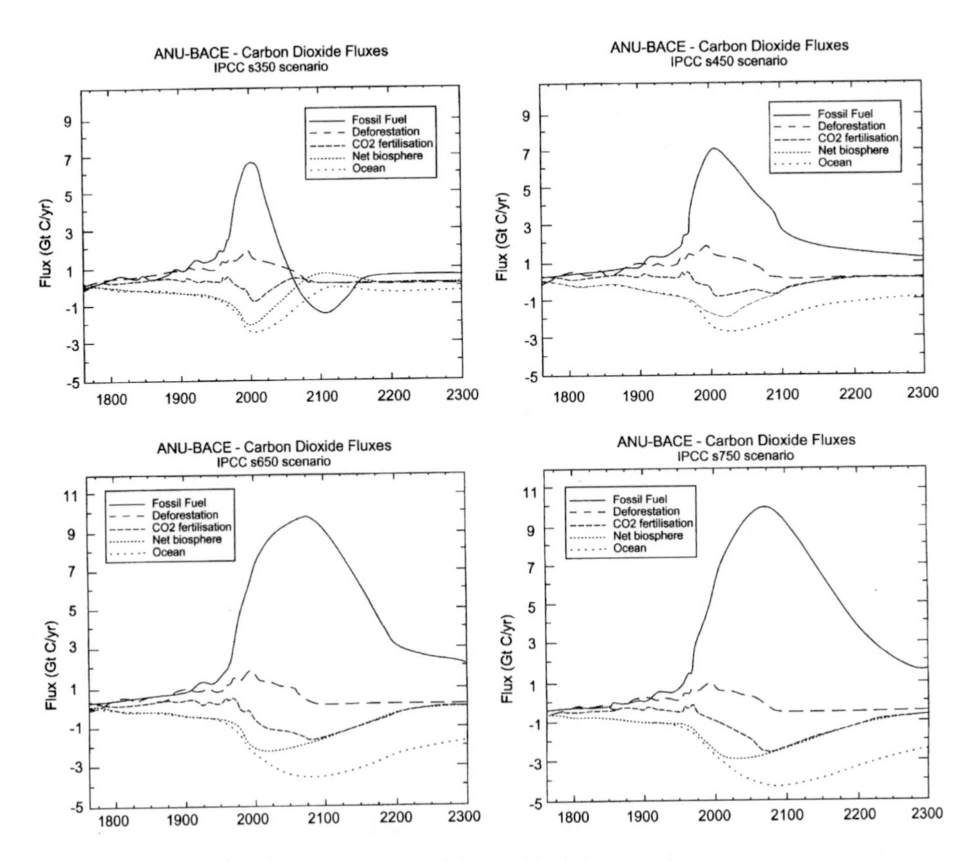

Fig. 2.3. Estimated CO_2 fluxes from various sources required to follow four IPCC scenarios for atmospheric CO_2 concentration; results computed from the ANU-BACE model (reproduced with permission from Taylor (1993)).

and Lloyd, 1992; Siegenthaler and Sarmiento, 1993), how should we manipulate the input flux from fossil fuel combustion in order to follow any paths between the present and end states? Four of the seven control trajectories are shown in Figure 2.3 (Taylor, 1993); one of them implies the large-scale re-conversion of carbon as gaseous CO_2 back into fossil form.

2.7 SCIENCE BASE CHANGING UNDER OUR FEET

In managing the state of the environment we have, by and large, assumed the path by which we arrived in the present state (*P*) to be known, more or less (as in Figure 2.4(a)). Such a "disturbed" state, however, is not something we would consider harmonious or in equilibrium. Looking around therefore, to oligotrophic lakes, neutral catchments, and rural tropospheric air masses elsewhere (outside of the

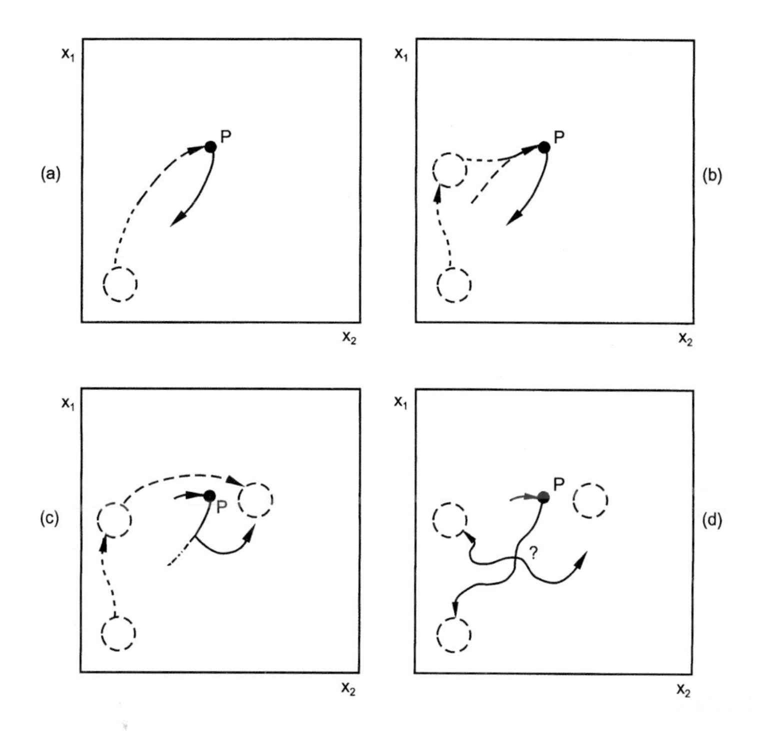

Fig. 2.4. Illustrative state-space (x_1, x_2) representation of changes in the nature of managing the behaviour of environmental systems (point *P* represents present state; open circles represent equilibrium positions): (a) historical path towards present disturbed (or stressed) state with possible future trajectory returning to equilibrium point; (b) apparent movement over time in the presumed position of the equilibrium; (c) still further migration in the presumed position of the equilibrium; (d) uncertainty in possible future trajectories as a function of uncertainty in presumed positions of the equilibrium.

damaged sector), the state of equilibrium to which the environment should be returned was known and clear. And so was the restoring action: simple and unambiguous, it was to reverse the stress that had caused departure from the pre-existing equilibrium and hence arrival at *P* (Figure 2.4(a)). Models were constructed to establish merely the extent to which the lever of policy would have to be pulled back. Such sureness of touch is crumbling away. We might have known, and probably correctly so in the case of eutrophication, whence we had come and whither we should return – it was just that the lake's behaviour was more subtle than first thought (Figure 2.4(a)). But in truth it could have been that we knew well only the most recent part of the path to the present. Because of the limitations in our understanding at this time, moreover, the equilibrium from which the system had originally been disturbed may not have been located where it was supposed to be (Figure 2.4(b)). Perhaps we *had* been correct in this supposition, but the equilibrium position was changing with time, so that the system's response to the restoring action might swerve towards something unexpected (Figure 2.4(c)). Indeed, perhaps the *only* fixed point in this is our knowledge of the disturbed position of the present and immediate past, while the single equilibrium position is slowly describing a trajectory around this point – as the basis of our science evolves (Figures 2.4(a), (b) and (c)).

What therefore is distinctive about the present (at the opening of the 21st century) is not that we know, for sure, that the position of the restored equilibrium is changing with time in the state space, but that we now entertain the possibility of this happening, or that there may be multiple equilibria, or both – though this has long been appreciated in the concept of ecological, as opposed to engineering, resilience (Holling, 1996). In short, we are not entirely sure about where the system is headed (Figure 2.4(d)), not necessarily because of the unexpected consequences of the disturbances and stresses resulting from our actions and reactions, but because of the evolving state of our knowledge – the continual interpretation and re-interpretation of the evidence surrounding our present position and the path to it from the immediate past. The IPCC and the IGBP might wish the framing of their forecasting problems to have the kind of singularity (of present and equilibrium positions) of Figure 2.4(a). They might prefer to suppose the piling on of more scientific enquiry and computing effort will make the image of this singularity more secure. But the terms of the forecasting problem may actually be more fruitfully cast as those of the plurality of Figure 2.4(d); a plurality, at least, of the potential equilibrium positions, which may themselves change as our knowledge base evolves.

REFERENCES

Armstrong, A., Addiscott, T. and Leeds-Harrison, P., 1995. Methods for modelling solute movement in structured soils. In: *Solute Modelling in Catchment Systems* (S.T. Trudgill, ed.). Wiley, Chichester, pp. 133–161.

Arthur, W.B., Holland, J.H., LeBaron, B., Palmer, R. and Taylor, P., 1997. Asset pricing under endogenous expectation in an artificial stock market. *Economic Notes, Banca Monte dei Paschi di Siena SpA*, **26**(2), 297–330.

Beck, M.B., 1981. Hard or soft environmental systems? *Ecol. Modelling*, **11**, 237–251.

Beck, M.B., 1994. Understanding uncertain environmental systems. In: *Predictability and Nonlinear Modelling in Natural Sciences and Economics* (J. Grasman and G. van Straten, eds.). Kluwer, Dordrecht, pp. 294–311.

Beck, M.B., 1996. Transient pollution events: acute risks to the aquatic environment. *Water Sci. Technol.*, **33**(2), 1–15.

Beck, M.B., Kleissen, F.M. and Wheater, H.S., 1990. Identifying flow paths in models of surface water acidification. *Rev. Geophys.*, **28**(2), 207–230.

Beven, K.J., 1989. Changing ideas in hydrology – the case of physically-based models. *J. Hydrol.*, **105**, 157–172.

Bishop, K.H., 1990. General discussion. In: *The Surface Waters Acidification Programme* (B.J. Mason, ed.). Cambridge University Press, Cambridge, p. 507.

Bishop, K.H., 1991. Episodic increases in stream acidity, catchment flow pathways and hydrograph separation, PhD Thesis, Department of Geography, University of Cambridge.

Chameides, W.L., Lindsay, R.W., Richardson, J. and Kiang, C.S., 1988. The role of biogenic hydrocarbons in urban photochemical smog: Atlanta as a case study. *Science*, **241**, 1473–1475.

Chen, J., Wheater, H.S. and Lees, M.J., 2002. Identification of processes affecting stream chloride response in the Hafren Catchment, Mid-Wales. *J. Hydrology*, in press.

Christophersen, N. and Wright, R.F., 1981. Sulphate budget and model for sulphate concentrations in stream water at Birkenes, a small forested catchment in southernmost Norway. *Water Resour. Res.*, **17**, 377–389.

Christophersen, N., Seip, H.M. and Wright, R.F., 1982. A model for streamwater chemistry at Birkenes, Norway. *Water Resour. Res.*, **18**, 977–996.

Cuypers, J.G.M. and Rademaker, O., 1974. An analysis of Forrester's World Dynamics Model. *Automatica*, **10**, 195–201.

de Grosbois, E., Hooper, R.P. and Christophersen, N., 1988. A multisignal automatic calibration methodology for hydrochemical models: a case study of the Birkenes Model. *Water Resour. Res.*, **24**, 1299–1307.

Di Toro, D.M., 1980. The effect of phosphorus loadings on dissolved oxygen in Lake Erie. In: *Phosphorus Management Strategies for Lakes* (R.C. Loehr, C.S. Martin and W.B. Rast, eds.). Ann Arbor Science, Ann Arbor, Michigan, pp. 191–205.

Di Toro, D.M. and Connolly, J.P., 1980. Mathematical models of water quality in large lakes, Part 2: Lake Erie, Report EPA-600/3-80-065, US Environmental Protection Agency, Environmental Research Laboratory, Duluth, Minnesota.

Evensen, G., 1994. Inverse methods and data assimilation in nonlinear ocean models. *Physica D*, **77**, 108–129.

Gell-Mann, M., 1994. *The Quark and the Jaguar*. Little, Brown and Co, London.

Grayson, R.B., Moore, I.D. and McMahon, T.A., 1992. Physically based hydrologic modeling. 2. Is the concept realistic? *Water Resour. Res.*, **26**(10), 2659–2666.

Harris, G.P., 1980. Temporal and spatial scales in phytoplankton ecology. Mechanisms, methods, models, and management. *Can. J Fish. Aquat. Sci.*, **37**, 877–900.

Hawking, S., 1993. The future of the universe. In: *Predicting the Future* (L. Howe and A. Wain, eds.). Cambridge University Press, Cambridge, pp. 8–23.

Hogan, C.J., Kirshner, R.P. and Suntzeff, N.B., 1999. Surveying space-time with supernovae. *Sci. Am.*, **280**(1), 46–51 (January).

Holling, C.S., 1996. Engineering resilience *versus* ecological resilience. In: *Engineering Within Ecological Constraints* (P.C. Schulze, ed.). National Academy of Engineering, Washington DC, pp 31–43.

Hooper, R.P., Stone, A., Christophersen, N., de Grosbois, E. and Seip, H.M., 1988. Assessing the Birkenes Model of stream acidification using a multisignal calibration methodology. *Water Resour. Res.*, **24**, 1308–1316.

Hoos, I., 1972. *Systems Analysis in Public Policy*. University of California Press, Berkeley, California.

Houghton, J.T., Callander, B.A. and Varney, S.K. (eds.), *Climate Change 1992, The Supplementary Report to the IPCC Scientific Assessment.* Cambridge University Press, Cambridge.

Jakeman, A.J. and Hornberger, G.M., 1993. How much complexity is warranted in a rainfall-runoff model? *Water Resour. Res.*, **29**(8), 2637–2649.

Johnson, D.W., 1990. Discussion. In: *The Surface Waters Acidification Programme* (B.J. Mason, ed.). Cambridge University Press, Cambridge, pp. 452–454.

Kasibhatla, P., Heimann, M., Rayner, P., Mahowald, N., Prinn, R.G. and Hartley, D.E. (eds.), 2000. *Inverse Methods in Global Biogeochemical Cycles*, Geophysical Monograph Series, Vol. 114. American Geophysical Union, Washington, DC.

Kauffman, S., 1995. *At Home in the Universe*. Oxford University Press, Oxford.

Kuhn, T.S., 1970. *The Structure of Scientific Revolutions*. University of Chicago Press, Chicago, Illinois.

Lam, D.C.L. and Schertzer, W.M., 1987. Lake Erie thermocline model results: comparison with 1967–1982 data and relation to anoxic occurrences. *J. Great Lakes Res.*, **13**(4), 757–769.

Lam, D.C.L., Schertzer, W.M. and Fraser, A.S., 1987. Oxygen depletion in Lake Erie: modelling the physical, chemical and biological interactions, 1972 and 1979. *J. Great Lakes Res.*, **13**(4), 770–781.

Lorenc, A.C., 1986. Analysis methods for numerical weather prediction. *Q. J. Roy. Meteorol. Soc.*, **112**, 1177–1194.

Mason, B.J., 1986. Numerical weather prediction. In: *Predictability in Science and Society* (B.J. Mason, P. Mathias and J.H. Westcott, eds.), Royal Society and British Academy, London, pp. 51–60.

Meadows, D.H., Meadows, D.L., Randers, J. and Behrens, W.W., 1972. *The Limits to Growth: A Report for the Club of Rome's Project on the Predicament of Mankind*. Universe Books, New York.

Menon, J., 1992. Foreword. In: *Global Change: Reducing Uncertainties*, International Geosphere-Biosphere Programme (IGBP). The Royal Swedish Academy of Sciences, Stockholm.

Oreskes, N., 1998. Evaluation (not validation) of quantitative models. *Environ. Health Perspect.*, **106** (Supplement 6), 1453–1460.

Osidele, O.O. and Beck, M.B., 2001. Identification of model structure for aquatic ecosystems using regionalized sensitivity analysis. *Water Sci. Technol.*, **43**(7), 271–278.

Park, R.A., Scavia, D. and Clesceri, N.L., 1975. CLEANER: The Lake George Model. In: *Ecological Modeling in a Resource Management Framework* (C.S Russell, ed.). Resources for the Future, Washington, DC, pp. 49–81.

Park, R.A., O'Neill, R.V., Bloomfield, J.A., Shugart, H.H., Booth, R.S., Goldstein, R.A., Mankin, J.B., Koonce, J.F., Scavia, D., Adams, M.S., Clesceri, L.S., Colon, E.M., Dettman, E.H., Hoopes, J.A., Huff, D.D., Katz, S., Kitchell, J.F., Kohberger, R.C., LaRow, E.J., McNaught, D.C., Peterson, J.L., Titus, J.E., Weiler, P.R., Wilkinson, J.W. and Zahorcak, C.S., 1974. A generalized model for simulating lake ecosystems. *Simulation*, **23**, 33–50.

Philip, J.R., 1975. Soil-water physics and hydrologic systems. In: *Modeling and Simulation of Water Resources Systems* (G.C. Vansteenkiste, ed.). North-Holland, Amsterdam, pp. 85–101.

Popper, K.R., 1935. *Logik der Forschung* (English translation: *The Logic of Scientific Discovery*). Hutchinson, London (1959).

Porter, K.G., 1996. Integrating the microbial loop and the classic food chain into a realistic planktonic food web. In: *Food Webs: Integration of Patterns and Dynamics* (G.A. Polis and K. Winemiller, eds.). Chapman and Hall, New York, pp. 51–59.

Porter, K.G., Saunders, P.A., Haberyan, K.A., Macubbin, A.E., Jacobsen, T.R. and Hodson, R.E., 1996. Annual cycle of autotrophic and heterotrophic production in a small, monomictic Piedmont Lake (Lake Oglethorpe): analog for the effects of climatic warming on dimictic lakes. *Limnol. Oceanogr.*, **41**(5), 1041–1051.

Rosenqvist, I.T., 1990. Pre-industrial acid water periods in Norway. In: *The Surface Waters Acidification Programme* (B.J. Mason, ed.), Cambridge University Press, Cambridge, pp. 315–319.

Schaffer, S., 1993. Comets and the world's end. In: *Predicting the Future* (L. Howe and A. Wain, eds.). Cambridge University Press, Cambridge, pp. 52–76.

Schertzer, W.M. and Sawchuk, A.M., 1990. Thermal structure of the lower Great Lakes in a warm year: implications for the occurrence of hypolimnion anoxia. *Trans. Am. Fish. Soc.*, **119**(2), 195–209.

Schneider, S.H. and Thompson, S.L., 1985. Future changes in the atmosphere. In: *The Global Possible* (R. Repetto, ed.). Yale University Press, New Haven, Connecticut, pp. 397–430.

Shackley, S,. Young, P.C., Parkinson, S. and Wynne, B., 1998. Uncertainty, complexity and concepts of good science in climate change modelling: are GCMs the best tools? *Climatic Change*, **38**, 159–205.

Siegenthaler, U. and Sarmiento, J.L., 1993. Atmospheric carbon dioxide and the ocean. *Nature*, **365**, 119–125.

Simons, T.J. and Lam, D.C.L., 1980. Some limitations on water quality models for large lakes: a case study of Lake Ontario. *Water Resour. Res.*, **16**(1), 105–116.

Smith, R.E., Goodrich, D.R., Woolhiser, D.A. and Simanton, J.R., 1994. Comment on: Physically based hydrologic modeling. 2. Is the concept realistic? by R.B. Grayson, I.D. Moore and T.A. McMahon. *Water Resour. Res.*, **30**(3), 851–854.

Taylor, J.A. and Lloyd, J., 1992. Sources and sinks of atmospheric CO_2. *Austr. J. Botany*, **40**, 407–418.

Taylor, J.A., 1993. Fossil fuel emissions required to achieve atmospheric CO_2 stabilisation using ANU-BACE: a box diffusion carbon cycle model. In: *Proceedings International Congress on Modelling and Simulation* (M.J. McAleer and A.J. Jakeman, eds.), **2**, pp. 765–770.

Thomann, R.V., 1982. Verification of water quality models. *J. Environ. Eng. Div., Proc. Am. Soc. Civil Eng.*, **108** (EE5), 923–940.

Thomann, R.V. and Winfield, R.P., 1974. Modeling of phytoplankton in Lake Ontario (IFYGL). *Int. Assoc. Great Lakes Res.*, **17**, pp. 135–149.

Waldrop, M.M., 1992. *Complexity*. Touchstone, New York.

van der Sluijs, J.P., van Eijndhoven, J., Shackley, S. and Wynne, B., 1998. Anchoring devices in science for policy: the case of consensus around climate sensitivity. *Social Studies Sci.*, **28**(2), 291–323.

Young, P.C., 1978. General theory of modelling badly defined systems. In: *Modelling, Identification and Control in Environmental Systems* (G.C. Vansteenkiste, ed.). North-Holland, Amsterdam, pp. 103–135.

Environmental Foresight and Models: A Manifesto
M.B. Beck (editor)

CHAPTER 3

Beginnings of a Change of Perspective

M.B. Beck

3.1 STATE VARIABLES MASQUERADING AS PARAMETERS

From the International Geosphere-Biosphere Programme (IGBP), with its desire to "reduce the uncertainties" and to remove ever more detailed knowledge from the expedient of "parameterisation" ($\boldsymbol{\alpha}$), a picture emerges of the model's resolving power unfolding through successive layers of ever greater refinement. We acknowledge this parameterisation as subsuming attributes that may change with time (t), but hope that the net effect of these more detailed variations can be approximated as invariant for the purposes of simulating the behaviour of the more important, more macroscopic state variables ($\boldsymbol{x}$). As the science base evolves, the goal is pursuit of an invariant $\boldsymbol{\alpha}$. In this scheme of things the ethos is one of temporal parametric variation being considered "bad", as too the use of a macroscopic parameterisation, once the more microscopic detail has been shown to be encodable within the ever expanding capacity of our computational platforms. But these things can be "goods" in certain practical engineering contexts.

In particular, in the perhaps small world of system identification, on-line and real-time identification, to which the existence of this monograph owes much, were once famously described as a "fiddler's paradise" (Åström and Eykhoff, 1971). The 1960s had been a decade in which adaptive control of aerospace systems had risen rapidly to prominence. The practical problem was that unmanned projectiles needed very fast computation for the rapid implementation of stabilising and regulating actions. A complex nonlinear model of the system's behaviour could not be solved sufficiently quickly for this purpose and had therefore to be substituted by a simpler approximation, at a less refined scale of description. It was obvious that the parameters in the approximative model would *change with time*, as the trajectory of the projectile evolved. The critical requirement was therefore a procedure for processing the real-time observations of the state of the projectile, in order to adapt the estimates of the parameters in the approximative model and the accompanying feedback control relationships (as in Young, 1984). The projectile's controller, as well as the model of its behaviour, would thus change with time. Rapid re-sampling of the state of the system, i.e., observation, ensured that the errors of poorly estimated model parameters and of imperfect, coarse replication of the system's "true" behaviour, would not propagate far in time. Thus was adaptive management of the system realised. At a later, more mature stage it assumed a form in which the control actions would be designed not only for the purposes of regulation, but also for probing the unknown in a manner intended to reduce some of the uncertainty about the behaviour of the system. And in this form it has since prospered as Adaptive Environmental Assessment and Management (Holling, 1978).

At the time, in the early 1970s, the notion that the parameters of a model might have to vary with time was seen entirely as a pragmatic convenience arising from an inability to work with the true model of the system, which one knew must exist in reality. As in so many fields, these were times of great expectations: it would not be long, we believed, before the admittedly less well defined problems of environmental systems analysis would yield to the vast array of methods available from applied mathematics and control theory (Beck, 1994). Prominent among these methods were those published in the classic text of Box and Jenkins (1970) on *Time Series Analysis, Forecasting and Control*; and central to their development and illustration in that book was a case study of the input–output behaviour of a gas furnace. It was in the analysis of these data that the power of detecting a lack of invariance in the estimates of a model's parameters was first brought to bear in an interpretative or explanatory fashion (Young et al., 1971). This was enabled through the algorithms of recursive estimation, whose potential had earlier been opened up in Kalman's (1960) celebrated contribution to filtering theory (although the basic principle of recursive estimation can be traced back to the work of Gauss in the early part of the 1800s; Young, 1984). In the case of the gas furnace data significant drift and change in the parameter estimates were identified a few sampling instants before the end of the observed record. History has not revealed what was the cause of this anomaly, but one speculation was that something had gone wrong with the experiment, causing it thus to be terminated shortly thereafter, perhaps even prematurely (Young, 1984).

In the many other subsequent illustrations of this principle,[1] some of the interpretations pinned on an anomalous fluctuation in the estimated value for a supposedly time-invariant model parameter have been shot with humour. The mischief in this arises from the way in which we have become so fond of abstracting such analyses from the mundane of everyday life; yet this ordinariness cannot be so excluded from clinical scientific enquiry. On one occasion upstream and downstream time-series from a study of flow in a well managed section of channel in Holland had

[1] Its present climax being evident in Young (1998, 1999).

returned an apparently spurious fluctuation in the model's parameter estimates. The bizarre deflection in the recursive estimate was brought to the attention of the authors of the experiment for comment, whereupon someone recalled that during the experiment the grass alongside the channel had been mown. The wind had blown the grass clippings into the channel and the weir (between the upstream and downstream observing points) had become blocked, a plausible explanation thus of why the channel had appeared to undergo a shift in the structure of its hydraulic behaviour (Young, 1982).

Such an apparent "shift in the structure" of a system's dynamic behaviour is central to this monograph. Put succinctly, we mean simply that the way in which the state variables of the model are interconnected with each other may *appear* to change with time.

When an explanation of phenomena must rest on a presumed constant that can only be identified as varying with time, this is disturbing, no more so today than it was three decades ago. That the fundamental constants of Physics may be exhibiting this trait – of evolving with time – is newsworthy even for the general public, although the fine structure constant to which this is happening (Webb et al., 2001) has always been something of a troublesome "constant" (Funtowicz and Ravetz, 1990). But before the computational realisation of filtering theory and recursive parameter estimation – the efforts of Gauss all those years ago notwithstanding – such features could not have been easily revealed. What these algorithms enabled, as was precisely the intent of their original development and application, was a device for tracking and adapting to observed change in the behaviour of a dynamical system. Whatever twists and turns of the parameter estimates were necessary for the model to track the observations, so was it. But in the absence of the immediate need for real-time control of the system, these twists and turns could in themselves have great diagnostic value. The end, that is, fitting the curve of the model's output to the dots of the observed output, could in this way become the systematic means of revealing inadequacy in the model's structure. For there must be, we presumed, a true structure underlying the observed behaviour of the system in which the values of the model's parameters would indeed be invariant with time. The systematic search for these "invariants" is the very essence of science and its outcome the basis on which extrapolation to conditions previously unencountered is justified.

According to this programme, the structure of the model would be revised (usually amplified) until one failed to demonstrate any further significant change with time of any of its constituent parameters (Beck, 1978; 1979; 1982; 1986; Beck and Young, 1976). As a diagnostic programme of enquiry into the record of the past, the problem became known as model structure identification. Filtering theory provided the framework within which the parts of the problem were unfolded, exposed and defined, but not resolved (Beck, 1994). Much rested on the estimation of parameter values that could change with time. And just as much rested dangerously, like an upturned pyramid, on the narrow basis of a single prototypical case study of the assimilation of easily degradable organic matter in a freshwater river system (Beck, 1973). The weather had been *remarkably* sunny during the period of the observations – this was England (in 1972) before serious public discussion of climate change – and the classical notions of how oxygen-demanding degradation of organic matter would interact with the dissolved oxygen concentration in the river, could not be made to fit the data. Something, arguably the growth and decay of algal (phytoplankton) biomass, had been omitted from the structure of the model, thereby having to reveal itself through a seemingly absurd fluctuation in the recursive estimates of the rate *constant* for the kinetics of biochemical oxygen demand (BOD) degradation (as we shall see in Chapters 15 and 16). This was a novelty.

But the sceptic would say (and did at the time) that there is nothing new under the sun – and certainly not if this is the growth of algae in the summer. What we knew about before should have been included *from the outset* in the model (and in the design of the accompanying experimental observation programme). Likewise, in the example given earlier, perhaps the investigator could have guessed that the grass alongside the Dutch channel, being long, might be due for a cut (because Holland is perceived as a very well ordered country) and that this possibility should therefore have been included in the model – *from the outset* – thus enabling truly time-invariant estimates of the model's parameters to be obtained. But this seems most unlikely.

Nothing in this programme for identifying the model's structure, moreover, would have suggested that the system, in reality, would have had anything other than an invariant structure populated with invariant parameters eventually identifiable as such, *once* the correct structure of the model had been revealed. That the parameters might have to be thought of as stochastic processes, *not* as random variables (i.e., not as unknown *constants*), was merely the means to an end. It followed from the interim measure of having to describe the behaviour of the system at an insufficiently fine scale of resolution, until the true invariant underlying structure could be discovered (at some more refined scale of description). Within the context of filtering theory it was perfectly natural for state variables to be masquerading as parameters in this way (Jazwinski, 1970), but only as a temporary expedient (Beck and Young, 1976).

In retrospect, this looks naïve. In the mesoscopic world of the environmental scientist, at least, it would seem we are bound to be using models that cannot capture all the fine-grained flutter and detail of the real thing. Pursuit of a successive unfolding of the coarser-scale parameters into more finely resolved states and parameters may not in truth have a stopping point, except in the absurdity of the model becoming indistinguishable from the system. Viewing parameters as stochastic processes has intrinsic merits; merits, indeed, that have very rarely been exploited. In one of the few such studies – on competitive growth among algal populations – Kremer (1983) observed that "... real causal mechanisms operating below an arbitrary level of resolution would be perceived as rapid, 'stochastic' changes in the [rate constants]". In others, more recently (Nielsen, 1994, 1995; Jørgensen, 1999), temporal variations in the parameters of a "structural dynamic" model of a shallow lake have been acknowledged as integral to the design of that model. Building on a concept first set out in Straškraba (1979), parametric variation is determined therein *not* as an empirical function of the deviation between theory and observation, but as a function of finding those values of the parameters maximising a hypothetical goal function for the system, for example, its exergy, at each successive step in time. In much grander form, but driven by the same need to understand the seeming idiosyncrasies of phytoplankton population dynamics, the notion that model parameters may have to vary with time appears to have become a legitimate topic of enquiry (Woods, 1999):

> This raises an interesting point about the difference between physics and biology. Physics was established at the Big Bang, and according to Einstein has not changed since then; our task is to discover its laws and the equations and parameters that best describe them... Biology, too, has Laws to be discovered, which can be described mathematically. But the biological equations *and in particular the parameters* are not all fixed once and for all. Many are continually evolving; some quite rapidly. The task of marine biologists is to discover the laws by laboratory experiment and provide first guess values for the parameters. Our task as ecologists is to include in the model the rules for the parameters to adapt by mutation and natural selection, as they do in Nature.
>
> Ecological models are primitive equation models with the physical parameters determined by experiment and the biological parameters free to adapt by mutation and natural selection.

That this "freedom" might be the gate to a fiddler's paradise was probably not intended.

The concept of a structure appearing to evolve with time, or actually so evolving, was born of the increasingly frequent appearance of the phrase "structural change" in Economic Theory (for example, Freeman, 1990). If the parameters of our models are always bound to be capable of change with time, then what may appear to us to be radical dislocations in the behaviour of the system's state variables could in fact be the consequence of smooth, subtle change in the model's parameters. And what we should therefore seek, as we contemplate the possibility of such a structural shift, are the "rules" (or a dynamic model) of this apparent parametric change, as a surrogate for knowledge of the real thing. For this – knowledge of the real thing – will *always* be lacking. A stochastic process, with given mean and variance for a given sampling interval in time, would be the simplest such surrogate. In other words, a model for forecasting in the presence of the potential for structural change would have a fixed network of interactions among its state variables, but with parameters presumed to be stochastic processes, the elementary features of which latter might be used for extrapolation.

In comparison with some ambitions this is modest stuff. What would happen if, as Allen (1990) has suggested, we were to seek not just the rules governing parametric change in a pre-specified, invariant structure, but the rules for predicting decline, disappearance, death, birth, growth, and emergence in the nodes and branches of the network of interactions? Any concern that the problems motivating this monograph may be trans-scientific – as questions that may be asked of science, but which science cannot answer (Weinberg, 1972) – must shrink in comparison with the programme Allen has set before us (Allen, 1990):

> [I]f the world is viewed as some kind of 'machine' made up of component parts which influence each other through causal connections, then instead of simply asking how it 'works', evolutionary theory is concerned with how it got to be as it is.
>
> The Newtonian paradigm was not about this. It was about mechanical systems either just running, or just running down.
>
> The key issue is centred on the passage between detailed microscopic complexity of the real world, which clearly can evolve, and any aggregate macroscopic 'model' of this.
>
> In economics... [t]wo time-scales are supposed. A very short one, for the approach to price equilibrium where all markets clear, and a longer one which describes the displacement over time of this equilibrium as a result of changing 'parameters'. Change is always then exogenous to the model, being driven by imposed changes of the relevant parameters. In other words, this corresponds merely to a 'description' of change (and not an accurate one), mechanically impacting on a system of fixed structure, [through] imposed changes in parameter values. Indeed, calibrating any such model becomes simply a task of finding changing values of parameters such that it reproduces the observed time variations of variables. And this amounts to a 'curve fitting' exercise with no real content.
>
> The central question which arises is that in order even to think about reality, to invent words and concepts with which to discuss it, we are forced to reduce its complexity. We cannot think of the trillions of molecules, living cells, organisms, individuals and events that surround us, each in its own place and with its own history. We must first make a taxonomic classification, and we must also make a spatial aggregation. This is shown in Figure [3.1]. On the left we represent the cloudy, confused complexity of the real world. Each part is special, each point unique. On the right is a 'model' of that reality, in terms of 'typical' elements of the system, where classifications and spatial aggregation have been carried out. But the point is that however good the choice of variables, parameters and interaction mechanisms may be, these only concern average behaviour. If we compare reality with the predictions of our model, then we shall necessarily find that variables and parameters 'fluctuate' around average values, and also that there is much greater microscopic diversity than that considered at the level of the macroscopic model.
>
> [I]f, in addition to our basic taxonomic and spatial aggregations, we assume that only average elements make up each category, and that only the most probable events actually occur, then our model reduces to a 'machine' which represents the system in terms of a set of differential equations governing its variables.
>
> But such a 'machine' is only capable of 'functioning', not of evolving. It cannot restructure itself or insert new cogs and wheels, while reality can!

Adopting such a programme, which in our framework would imply the discrete event of the birth of a new state equation in the model, would go well beyond the scope of the monograph; it is a challenge beyond even that which we have set ourselves.

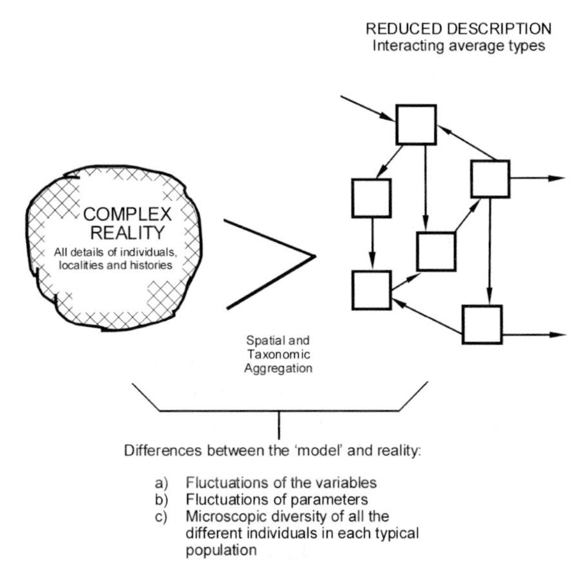

Fig. 3.1. "Modelling, and even thinking about a complex system, necessitates a simplification into categories, which constitute the system. We make a 'mechanical' replica of present reality. But evolution concerns change in this structure – new boxes and arrows." (Allen, 1990) (redrawn from the original figure in Allen (1990)).

3.2 RECONCILING THEORY WITH OBSERVATION

Amidst the flux of this evolutionary outlook the pillar of a truly invariant structure for the model of a system's behaviour seems an anachronism – and perhaps it now is. For belief in reaching this pillar of constancy has been relinquished, slowly but surely, and over many years (Beck, 1994). With it must be shed too the much cherished notion of finding a uniquely best set of values to be assigned to the model's parameters. It is not that these goals are necessarily inappropriate; rather, like parameters that need no calibration, they are useful illusions.

An age of uncertainty, whose origins (at least heuristically) lie precisely in the classical programme of modelling designed to reduce uncertainty, has settled upon us. In a seminal paper, O'Neill (1973) noted that the error in the predictions from a model should decrease with a decreasing degree of model aggregation (in his case the aggregation of behaviour of several species of organisms into a single state variable, for example). However, he also noted that precisely this increasing refinement of detail – more complex kinetic expressions, more state variables – would tend to increase the prediction errors resulting from the necessarily increasing number of model parameters with uncertain values. Errors of 10% in the parameter values were found to yield errors of prediction greater than 100% (expressed as a coefficient of variation).

It was but a small step from this to note that success in reducing predictive uncertainty should follow from success in reducing, *inter alia*, parametric uncertainty and that success in this latter would be governed by success in reconciling the behaviour of the model with the observed historical record. What is more, upon the issue of uncertainty could be constructed an argument that would expose still further the limitations of the classical programme of modelling, as well as those of the paradigm of "small is beautiful". There was a dilemma (Beck, 1981; 1983). With a small model, containing just a few parameters well identified from the record of the past, one might predict an entirely "incorrect" future and, worse still, attach great confidence to this erroneous forecast. This had always been readily appreciated. But with a large model containing many poorly identified parameters, it might be possible to forecast a "correct" future; yet one would attach little confidence to this prediction. This, though it seems trivially obvious in retrospect, had not been widely appreciated.

No-one should take comfort from this dilemma. The ubiquity of a lack of model identifiability is troubling, for the not-so-large (Kleissen et al., 1990) just as much as for the quite-large-enough (Beck, 1987; Beck et al., 1990; Beven and Binley, 1992). Of course, none of this should matter, at least not in the pragmatic context of making decisions, in which one might need to demonstrate merely that the best course of action is robust in the face of such ambiguity in the interpretation of past behaviour. Neither would it matter if the parameters of our (environmental) models were either universal constants or somehow independently measurable in the field; free, as we have already said in Chapter 2, of any presumption about the nature of the relationship between the observable quantities, in which presumed *model* the parameter appears. It would not matter if the parameter could be so "measured" at a scale identical with that of the model, or if the parameter possessed a value independent of any scale. That is to say, we would like to have a coefficient for the fluid-mechanical dispersion of a contaminant, for example, that has the same value irrespective of the volume of fluid over which it is evaluated (in some manner). But we cannot have these things (Beven, 1989). In short, we might as well assume that the problem of a lack of model identifiability *is* more or less universal. For we are asking too much of too little: to determine too many unknowns, the parameters, from too few knowns, the observable quantities. And if this is so, the contemporary epistemological scene ought to appear more troubled than it is. For we are saying, in effect, that our models cannot be reconciled with the available observations at this level of detail and things may only get worse. The goal of model validation is being substituted by the goal of mere data assimilation.

In an article on interactive computing as a teaching aid MacFarlane (1990) presented a three-element characterisation of knowledge. According to the American philosopher Lewis these three elements are (as reported by MacFarlane):

(i) the given data;

(ii) a set of concepts; and

(iii) acts which interpret data in terms of concepts.

It is readily apparent that the problem of system identification (the derivation of a model whose behaviour bears the closest possible resemblance to the observed behaviour of the actual system) is covered exactly by the third of the three elements. What now are the prospects for success in these "acts which interpret data in terms of concepts", that is, for success in reconciling the candidate model with the given data? After all, the power of the classical experiments of laboratory science lay presumably in promoting the possibility of "acts which interpret data in terms of concepts" by reducing the "set of concepts" under scrutiny to as small a set as possible and by maximising the scope for acquiring a large volume of the "given data".

Our common experience tells us that the scope and resolution of both the "given data" and the "set of concepts" necessary for understanding the behaviour of environmental systems are expanding at an increasing rate. We know too that this rate of expansion, if anything, is greater in respect of the set of concepts (the model). The General Circulation Models (GCMs) of climatology and meteorology are of suitably massive proportions, with typically seven or more state variables (wind velocities, air temperature, and so on) to be accounted for at some 10^6 spatial locations – and doubtless already still more (Watterson, 1993). The computing capacity now available for realising models of the behaviour of a system offers us a truly staggering, expanding universe of possibilities. Indeed, in the popular scientific press this potential was neatly captured as long as a decade ago in headlines such as "*Is It Real, Or Is It A Cray?*" (Pool, 1989) and "*Speculating in Precious Computronium*" (Amato, 1991). Not only do we have the possibility of designing hard, mechanical objects, such as aluminium cans, in the virtual world of the computer, but now we have the "... science of Virtual Ecology, which will change the way we learn how life in the sea varies both geographically and in time" (Woods, 1999).

Technical support for manipulating the logical consequences of our "set of concepts" is thus assured. Technical support for acquiring the "given data" is likewise assured, although it might always be argued to be (relatively) inadequate. For example, in the area of biological oceanography Woods (1999) has calculated that the most elaborate of today's monitoring exercises, the Joint Global Ocean Flux Study (JGOFS) of the IGBP and Scientific Committee on Oceanic Research (SCOR), was approaching a yield of one million four-byte words of data. And herein lies part of the problem: this is less than a mere one-millionth of the four-byte words required to characterise an elementary experiment in Woods' world of Virtual Ecology. In classical terms this is surely asking too much of too little. But then let us consider what these numbers mean for system identification, those "acts which interpret data in terms of concepts". Indeed, how does this "interpretation" actually come about? It is a result of juggling with, and sifting through, a unique assortment of disparate facts and figures assembled by the individual, upon which some kind of order is eventually imposed. It is a subjective mental process. Yet imagine juggling with 10^6 four-byte words juxtaposed with 10^{12} four-byte words in order to arrive at an understanding of the object under scrutiny. How does one, in particular, reconcile a large-scale geophysical model of global deglaciation with (reconstructed) relative sea level observations at 392 sites spanning a period of some 15,000 years (Tushingham and Peltier, 1992)? More specifically, which constituents of the very large and very complex assembly of micro-scale theory is at fault when the model fails – as inevitably it does – to match the relatively macroscopic historical observations?

That there will be significant developments in the technical support necessary for engaging the model in a meaningful interpretation of the data, is by no means assured. News of advances in computational capacity is abundant; news of advances in the technology of instrumentation and remote sensing is commonplace; news of the *increasing* capacity of the brain to juggle with disparate facts and concepts is non-existent.

Consider our predicament, then. For many systems we possess a wealth of concepts, which increasingly we shall be unable to evaluate against *in situ* field observations in the conventional, *classical* sense. Were we to forsake this classical notion of "verification" it would be somewhat akin to dismantling the signposts of scientific procedure, with profoundly unsettling implications. Yet what should we make of the following account of how the models of Virtual Ecology are to be verified, with its mixture of the old and the new in scientific enquiry? Why should we expect, as does Woods (1999), that the primitive equation models of ecology should behave as well as weather forecasting models? For this is his reasoning about model verification (Woods, 1999):

> Experience with primitive equation models in meteorology, tells us that the simulations are realistic (though not necessarily correct in detail) if the basic equations are sound.
>
> Even without verification the Virtual Ecosystems are likely to do quite well. Verification is important to get the *details* right. The philosophy of model verification is [as follows]. We start by identifying the very few features that have been reliably observed, and then we:
>
> [a] simulate those features under the conditions of the observations
> [b] compare observation and simulation
> [c] measure that matching error
> [d] adjust the model equations, often by adding new plankton guilds
> [e] iterate stages a–e until the matching error is minimized.
>
> As we seek to simulate each new set of observations made by another great experiment in biological oceanography, our model tends to become more elaborate; it includes more of the complexity of Nature, more biodiversity. As usual we follow the principle of Oc[c]am's razor, starting with a minimum model and only adding new equations when forced to do so by verification against good data.
>
> ...I expect discoveries in Virtual Ecology to start contributing to Theoretical Ecology... [Virtual Ecology] promises to convert biological oceanography to a branch of experimental science in which research cruises are designed to test theories by collecting data for comparison with phenomena encountered in Virtual Ecosystems.

Lewis's three-element characterisation of knowledge may no longer suffice for guidance, since, as MacFarlane (1990) has asserted (and as the foregoing account of Virtual Ecology tends to confirm):

> Modern scientists and engineers no longer work only in terms of theory and experiment. Their attack on the problems of describing Nature and on creating useful artefacts now has three fronts:
>
> - experiment
> - theory
> - computation
>
> Computation is opening up vast new continents in Popper's World 3.[2]

In other words, understanding – that is, assimilation of material into an appropriate mental structure (or mental model) – will derive increasingly from the *belief* that the virtual computational world has been founded upon true and correctly applied micro-scale theories and does not generate broad, macroscopic, qualitative predictions in obvious, absurd discord with whatever can be observed of the real thing in the physical world.

3.3 ERASING THE DIVIDE BETWEEN PAST AND FUTURE

Hawking's summary of the fate of the universe (Hawking, 1993) (Figure 2.1 in the preceding chapter) – epitomises our problem of extrapolation beyond the observed record. It does not have "time" as its abscissa, but it has the dots of the observations and the curves of the candidate models. And as we move to the right-hand edge of the span of observations, all manner of diverging curves are unleashed, each with a profoundly different implication for the future of the universe. In particular, the minuscule differences in the shape of the curve between the last two pairs of observations bear the seeds of all the prospective variety. If we could only push our capacity to observe the apparent brightness (or magnitude) a little further to the right of the figure, this would surely allow us to disentangle on which of the courses the universe might be embarked.

As we stand on the threshold of the future, we draw there our ordinate, the vertical line of the present that separates the emptiness to the right from the dots of the observations of the past falling away behind us (Figure 3.2(a)). We are indeed most strongly accustomed to the idea of performance – that which is observed over time of the features of interest about the system – being specified in terms of a time-series of observations. Each dot has been acquired through an instrument of sampling and analysis. The more remote such observation has become from the

---

[2] Popper's pluralistic approach to knowledge (Popper, 1972) views the world as consisting of three sub-worlds that exist quite separately: World 1, which is the physical world, or world of physical states; World 2, which is the mental world, or the world of mental states; and World 3, which is the world of all possible objects of thought, existing outside of and independent of any individual.

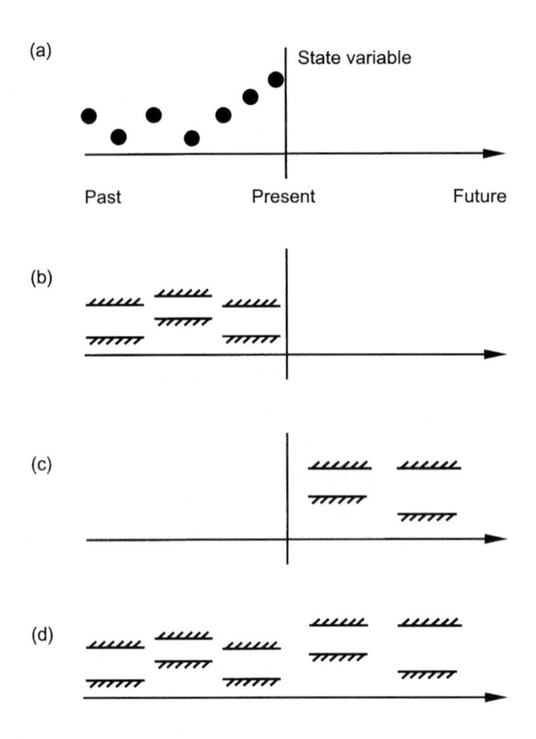

Fig. 3.2. Representation of the empirical knowledge of the system's behaviour: (a) conventional observations of past behaviour as a time-series; (b) empirical categorisation of "qualitatively" observed past behaviour; (c) imagined domains of possible future behaviour; (d) domains of past and possible future behaviour.

immediacy of our own common senses, the stronger has become the dividing line of the present, between the observed past and the unobservable future.

Our minds have been set: the nature of our observations of the past is fundamentally other than anything we can state of the future. So firmly set were they becoming in the 1970s that it was really very difficult to take a casual, personal observation seriously. Time-series analysis, forecasting, and control demanded the abstraction of a crisp, electronically encoded number, divorced from the taint of personal observation. Few things, moreover, could be more tainted than the smell of sewage being treated. Yet the treatment plant manager would make this observation in order to formulate his actions for controlling the performance of that system (Tong et al., 1980), and quite successfully so.[3] Here we were, at the same time, however, building an upturned pyramid of procedure for model structure identi-

[3] That this was actually the case was not recorded in the journal article – it did not seem reputable at the time – and when later there was a proposal to develop an "electronic nose" this struck one (from the ignorance of inter-disciplinary distance) as really quite an object of fun.

fication upon the mightily restrictive assumption of the availability of comprehensive time-series observations (Beck, 1982, 1986). This flew in the face not only of the experience of the wastewater treatment plant manager, but also of the common experience in which the majority of environmental issues would present themselves, with at most just sparse and rather messy data. It is one of those curiosities that nothing was done to resolve this contradiction until what seemed at the time to have been quite late in the day, with the work of Hornberger, Spear and Young on a problem of eutrophication in the Peel-Harvey Inlet of Western Australia (Young et al., 1978; Hornberger and Spear, 1980; Spear and Hornberger, 1980).

Thus was the mould broken. For what the Peel Inlet study showed was how casual, personal, observation in the field could be utilised in a meaningful and respectable manner: to distinguish those constituents of a richly composed, possibly high-order model, that appeared to be key, from those apparently redundant to the task of matching the performance of the model with past behaviour thus observed – and in the presence of gross uncertainty. This was neither model structure identification nor a reconciliation of theory with observation in the classical sense. But it was an idea whose time had come.

If we may now describe the empirical basis of the past in this linguistic, subjective fashion, why should we not express what we fear or desire of the future in the same terms? For this is what we all do, as scientists, non-scientists, members of the public, or policy-makers. Management and control are founded on the very principle of specifying a goal that is to be attained in the future. This is the behaviour we desire to observe, or fear may come to pass, in the future. Inasmuch as the dots on the left-hand side of the page can be replaced by "domains" in which the behaviour of the system was sensed by the field scientist to have resided in the past (Figure 3.2(b)), so we can populate the right-hand side of the page with domains in which behaviour in the future should (should not) come to reside (Figure 3.2(c)). With this the ordinate of the present may be erased, leaving us simply with a set of domains through which the system may pass on its course from the past to the future (Figure 3.2(d)). And when our models are reconciled with this sequence of domains, the nature of the curvature, or the subtle flexure in the model's parameter values towards the end of the record of the past, may be identified from its course into the future – as we imagine, believe, or fear it to be.

3.4 A KNOT OF CONSTRAINTS, OPPORTUNITIES AND CONTRADICTIONS

Here, then, is where we have come in our reflections. First, whereas we strive to assume, or demonstrate, invariance in the structure of the relationships among the state variables of a system, the reaching of such may not be possible in practice. Some, notably Allen (1990), would argue beyond this, that the presumption of invariance in a model's structure is to deny the possibility of systems capable of evolving; and our everyday experience is of a world replete with systems that *are* evolving. Second, while we may informally possess a wealth of knowledge about what is believed to be the structure underlying the observed behaviour of a system (that relating to the structure of food webs in lake ecosystems is an especially good example; Porter, 1996; Porter et al., 1996), its translation into the formal framework of a mathematical model is unlikely to sustain attempts at corroboration or refutation in a conventional sense. Third, and last, if we permit common-sense experience of the past behaviour of a system to be juxtaposed with a formally prepared expression of the assembly of constituent hypotheses about the causes of that behaviour, i.e., the conventional mathematical model, then so too might we allow the useful juxtaposition of this same, quantitative model with what is informally imagined of the possible future behaviour of the system.

Such opportunities and constraints, jostling together in what we have just recounted, may not have been apparent to many. That they hide a looming contradiction gives us cause for dismay. For how could we maintain a position of acknowledging the near impossibility of reconciling models with crisp, quantitative and (in some cases) plentiful observations, yet aspire to identifying significant *changes* (not mere constancy) in a model's parameter values on the basis, moreover, of just sparse, qualitative, casual observations and feared images of the future? Our purpose has been to provide a candid account – from the collection of a modest number of personal experiences – of the way things are: of the tightening knot of problems at the core of extrapolation in the Environmental Sciences. We now need to untie this knot and begin to resolve the contradictions, by attempting something different in our approach to exploring environmental change.

REFERENCES

Allen, P.M., 1990. Evolution, innovation, and economics. In: *Technical Change and Economic Theory* (G. Dosi, C. Freeman, R. Nelson, G. Silverberg and L. Soete, eds.). Pinter, London, pp. 95–119.

Amato, I., 1991. Speculating in precious computronium. *Science*, **253**, 23 August, 856–857.

Åström, K.J. and Eykhoff, P., 1971. System identification – a survey. *Automatica*, **7**, 123–162.

Beck, M.B., 1973. The application of control and systems theory to problems of river pollution, PhD Thesis, Department of Engineering, University of Cambridge.

Beck, M.B., 1978. Random signal analysis in an environmental sciences problem. *Appl. Math. Modelling*, **2**, 23–29.

Beck, M.B., 1979. Model structure identification from experimental data. In: *Theoretical Systems Ecology* (E. Halfon, ed.). Academic, New York, pp. 259–289.

Beck, M.B., 1981. Hard or soft environmental systems? *Ecol. Modelling*, **11**, 237–251.

Beck, M.B., 1982. Identifying models of environmental systems' behaviour. *Math. Modelling*, **3**, 467–480.

Beck, M.B., 1983. Uncertainty, system identification and the prediction of water quality. In: *Uncertainty and Forecasting of Water Quality* (M.B. Beck and G. van Straten, eds.). Springer, Berlin, pp. 3–68.

Beck, M.B., 1986. The selection of structure in models of environmental systems. *The Statistician*, **35**, 151–161.

Beck, M.B., 1987. Water quality modeling: a review of the analysis of uncertainty. *Water Resour. Res.*, **23**(8), 1393–1442.

Beck, M.B., 1994. Understanding uncertain environmental systems. In: *Predictability and Nonlinear Modelling in Natural Sciences and Economics* (J. Grasman and G. van Straten, eds.). Kluwer, Dordrecht, pp. 294–311.

Beck, M.B. and Young, P.C., 1976. Systematic identification of DO-BOD model structure. *Proc. Am. Soc. Civil Eng., J. Environ. Eng. Div.*, **102**(EE5), 909–927.

Beck, M.B., Kleissen, F.M. and Wheater, H.S., 1990. Identifying flow paths in models of surface water acidification. *Rev. Geophys.*, **28**(2), 207–230.

Beven, K.J., 1989. Changing ideas in hydrology – the case of physically-based models. *J. Hydrol.*, **105**, 157–172.

Beven, K.J. and Binley, A.M., 1992. The future of distributed models: model calibration and predictive uncertainty. *Hydrol. Processes*, **6**, 279–298.

Box, G.E.P and Jenkins, G.M., 1970. *Time Series Analysis, Forecasting and Control.* Holden Day, San Francisco, California.

Freeman, C., 1990. Introduction. In: *Technical Change and Economic Theory* (G. Dosi, C. Freeman, R. Nelson, G. Silverberg and L. Soete, eds.). Pinter, London, pp, 1–8.

Funtowicz, S.O. and Ravetz, J.R., 1990. *Uncertainty and Quality in Science for Policy*. Kluwer, Dordrecht.

Hawking, S., 1993. The future of the universe. In: *Predicting the Future* (L. Howe and A. Wain, eds.). Cambridge University Press, Cambridge, pp. 8–23.

Holling, C.S. (ed.), 1978. *Adaptive Environmental Assessment and Management.* Wiley, Chichester.

Hornberger, G.M. and Spear, R.C., 1980. Eutrophication in Peel Inlet. I, Problem-defining behaviour and a mathematical model for the phosphorus scenario. *Water Res.*, **14**, 29–42.

Jazwinski, A.H., 1970. *Stochastic Processes and Filtering Theory*. Academic, New York.

Jørgensen, S.E., 1999. State-of-the-art of ecological modelling with emphasis on development of structural dynamic models. *Ecol. Modelling*, **120**, 75–96.

Kalman, R.E., 1960. A new approach to linear filtering and prediction problems. *Am. Soc. Mech. Eng. Trans., J. Basic Eng.*, **83D**, 95–108.

Kleissen, F.M., Beck, M.B. and Wheater, H.S., 1990. The identifiability of conceptual hydrochemical models. *Water Resour. Res.*, **26**(12), 2979–2992.

Kremer, J.N., 1983. Ecological implications of parameter uncertainty in stochastic simulation. *Ecol. Modelling*, **18**, 187–207.

MacFarlane, A.G.J., 1990. Interactive computing: a revolutionary medium for teaching and design. *Comput. Control Eng. J.*, **1**(4), 149–158.

Nielsen, S.N., 1994. Modelling structural dynamic changes in a Danish shallow lake. *Ecol. Modelling*, **73**, 13–30.

Nielsen, S.N., 1995. Optimization of exergy in a structural dynamic model. *Ecol. Modelling*, **77**, 111–122.

O'Neill, R.V., 1973. Error analysis of ecological models. In: *Radionuclides in Ecosystems*, Conf 710501. National Technical Information Service, Springfield, Virginia, pp. 898–908.

Pool, R., 1989. Is it real, or is it a Cray. *Science*, **244**, 23 June, 1438–1440.

Popper, K.R., 1972. *Objective Knowledge*. Oxford University Press, Oxford.

Porter, K.G., 1996. Integrating the microbial loop and the classic food chain into a realistic planktonic food web. In: *Food Webs: Integration of Patterns and Dynamics* (G.A. Polis and K. Winemiller, eds.). Chapman and Hall, New York, pp. 51–59.

Porter, K.G., Saunders, P.A., Haberyan, K.A., Macubbin, A.E., Jacobsen, T.R. and Hodson,

R.E., 1996. Annual cycle of autotrophic and heterotrophic production in a small, monomictic Piedmont lake (Lake Oglethorpe): analog for the effects of climatic warming on dimictic lakes. *Limnol. Oceanogr.*, **41**(5), 1041–1051.

Spear, R.C. and Hornberger, G.M., 1980. Eutrophication in Peel Inlet. II, Identification of critical uncertainties via generalised sensitivity analysis. *Water Res.*, **14**, 43–49.

Straškraba, M., 1979. Natural control mechanisms in models of aquatic ecosystems. *Ecol. Modelling*, **6**, 305–321.

Tong, R.M., Beck, M.B. and Latten, A., 1980. Fuzzy control of the activated sludge wastewater treatment process. *Automatica*, **16**, 695–701.

Tushingham, A.M. and Peltier, W.R., 1992. Validation of the ICE-3G Model of Würm-Wisconsin deglaciation using a global data base of relative sea level histories. *J. Geophys. Res.*, **97**(B3), 3285–3304.

Watterson, I.G., 1993. Global climate modelling. In: *Modelling Change in Environmental Systems* (A.J. Jakeman, M.B. Beck and M.J. McAleer, eds.). Wiley, Chichester, pp. 343–366.

Webb, J.K., Murphy, M.T., Flambaum, V.V., Dzuba, V.A., Barrow, J.D., Churchill, C.W., Prochaska, J.X. and Wolfe, A.M., 2001. Further evidence for cosmological evolution of the fine structure constant. *Physical Letter Reviews*, 87(9), 091301-1–091301-4, 27 August.

Weinberg, A.M., 1972. Science and trans-science. *Minerva*, **X**(2), 209–222.

Woods, J.D., 1999. Virtual ecology. In: *Highlights in Environmental Research* (B.J. Mason, ed.). Imperial College Press, London.

Young, P.C., 1982. Personal communication.

Young, P.C., 1984. *Recursive Estimation and Time-series Analysis*. Springer, Berlin.

Young, P.C., 1998. Data-based mechanistic modelling of environmental, ecological, economic, and engineering systems. *Environ. Modelling and Software*, **13**(1), 105–122.

Young, P.C., 1999. Nonstationary time series analysis and forecasting. *Progr. Environ. Sci.*, **1**, 3–48.

Young, P.C., Shellswell, S.H. and Neethling, C.G., 1971. A recursive approach to time-series analysis. Report CUED/B-Control/TR16, Department of Engineering, University of Cambridge.

Young, P.C., Hornberger, G.M. and Spear, R.C., 1978. Modelling badly defined systems – some further thoughts. In: *Proceedings SIMSIG Conference*, Australian National University, Canberra, pp. 24–32.

Environmental Foresight and Models: A Manifesto
M.B. Beck (editor)

CHAPTER 4

Structural Change: A Definition

M.B. Beck

4.1 A METAPHOR AND SOME ACCOMPANYING CONCEPTUAL APPARATUS

What, more precisely, is this thing called "structural change"?

Let us assume that at some level of resolution we imagine behaviour to be described by a seamless web of interactions among the state variables, such as that depicted in Figure 4.1(a). We may note that this is already a "simplification into categories" in the sense of Allen (1990), whereby a "mechanical" replica of present complex reality has been constructed in our mind's eye (as in Figure 3.1 of the preceding chapter). In Figure 4.1(a) the nodes represent state variables and the branches causal interactions among them, with the arrows signalling the direction of influence – of cause provoking an effect. The nature of these interactions can in general be nonlinear and, equally generally, they will normally be described mathematically by expressions involving both the state variables ($\boldsymbol{x}$) engaged in the given interaction and parameters ($\boldsymbol{\alpha}$). We shall use this pictorial metaphor to define what is meant by structural change. In order to do so, however, it will be important to retain this figurative attachment of $\boldsymbol{x}$ to the nodes of the network and of $\boldsymbol{\alpha}$ to its branches.

The system under investigation – that piece of the real world in which one has a special interest – is as yet undefined in the web of Figure 4.1(a), for the web itself is simply an expression of a universe of interactions. To provide an identity for the system it is conventional to attach the labels "input" ($\boldsymbol{u}$) and "output" ($\boldsymbol{y}$) to those nodes of the network (state variables) that may be observed as stimulus (cause) and response (effect) respectively. Motivated by the problem at hand we wish to explore the manner in which variations in $\boldsymbol{u}$ are related to variations in $\boldsymbol{x}$ as observed through $\boldsymbol{y}$.

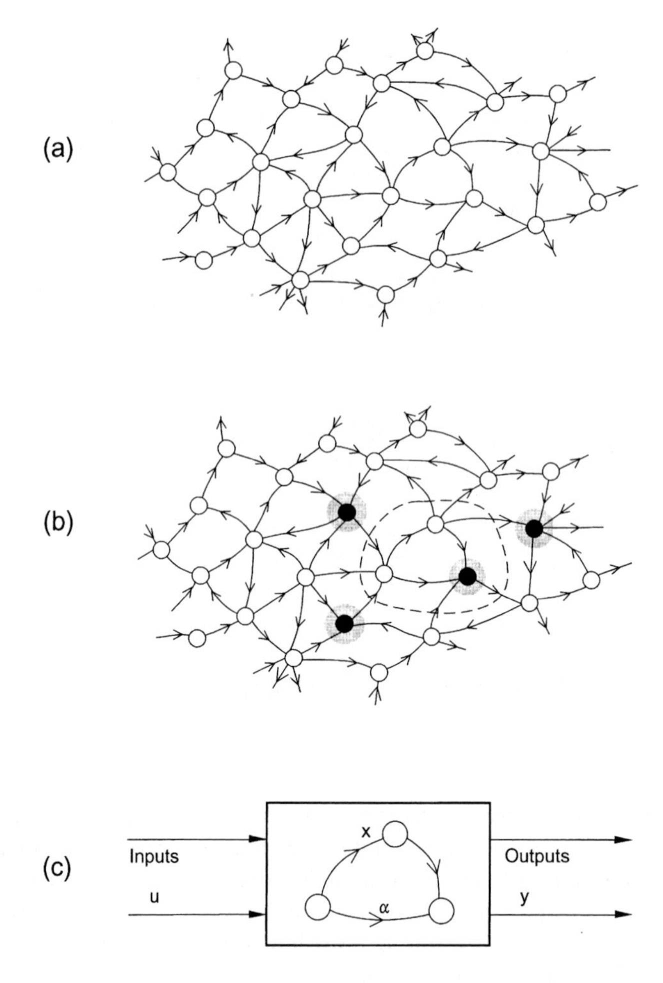

Fig. 4.1. Conceiving of the structure of the system's behaviour: (a) web of universe of interactions (cause and effect relationships); (b) abstraction of a particular piece of the web of interactions, i.e., identification of the system to be investigated, with observable inputs (***u***), states (***x***), and outputs (***y***) being denoted by emboldened nodes; and (c) further abstraction into the conventional representation of a system with inputs and outputs.

Thus, in Figure 4.1(b), for example, behaviour of the system is to be summarised by a model of the interactions among three state variables, two inputs, and two outputs, with causal influences flowing along the connecting branches from the input to output nodes. In this instance, and for the sake of emphasising the point, one of the observable outputs is identical with one of the model's state variables; the other is not.

Further abstraction allows the system, together with its inputs and outputs, to be cut out from the seamless web of interactions and placed in the laboratory world of computer simulation for more careful scrutiny, in the conventional guise of Figure 4.1(c). The branches between the inputs and state variables and between the state variables and outputs have been omitted simply for convenience. It may also be helpful to note that, especially in the case of the outputs, the arrows emanating from the system's block will in general indicate flows of information (as opposed to flows of material).[1] In addition, we should acknowledge that the metaphor of the branch–node network within the block of the system in Figure 4.1(c) does not entirely capture the mathematical nature of the state-variable dynamics. Some of the forces acting on a given state, thereby causing its value to vary with time, may be expressions involving merely that state and some parameters, which would require representation in Figure 4.1(c) by a branch starting and ending at the same node (as would be apparent in more formal graph representations). Our metaphor is intended to explain the nature of a problem, of structural change. It is not a device for any formal or quantitative analysis of that problem, although we shall show how it might take on such a more formal role in Chapters 11 and 15.

Let us assume, then, that the model of the environmental system can be defined by the following (lumped-parameter) representation of the state variable dynamics,

$$\mathrm{d}\boldsymbol{x}(t)/\mathrm{d}t = \boldsymbol{f}\{\boldsymbol{x},\boldsymbol{u},\boldsymbol{\alpha},t\} + \boldsymbol{\xi}(t) \tag{4.1a}$$

with the observed outputs being defined as follows,

$$\boldsymbol{y}(t) = \boldsymbol{h}\{\boldsymbol{x},\boldsymbol{\alpha},t\} + \boldsymbol{\eta}(t) \tag{4.1b}$$

in which $\boldsymbol{f}$ and $\boldsymbol{h}$ are vectors of nonlinear functions, $\boldsymbol{\xi}$ and $\boldsymbol{\eta}$ are notional representations, respectively, of those attributes of behaviour and output observation that are not to be included in the model in specific form, and t is continuous time. Should it be necessary, spatial variability of the system's state can be assumed to be accounted for by, for example, the use of several state variables of the same attribute of interest at the several defined locations.

Equation 4.1(b) caters for the possibility that the output $\boldsymbol{y}$ may not be simply, as noted above, a direct but error-corrupted measurement of a state variable. In spite of the obvious fact that all observation involves a more or less complicated measuring system (containing its own network of interacting state variables), we tend nevertheless to conceive of "direct" observation of the state variables as much more the norm in practice. In the illustration of Figure 4.1(b) the function $\boldsymbol{h}$ of equation 4.1(b) would not, however, collapse down to $\boldsymbol{x}$ alone for the "indirect" output located outside the dashed line defining the system of interest; and the parameters $\boldsymbol{\alpha}$ appearing in $\boldsymbol{h}$ would be those associated with the interaction between this output

---

[1] In a body of water, for example, the outputs of the system may be observed attributes of the water at points within its body, not necessarily observed attributes of a flux of water passing out of the system across one of its physical (and conceptual) boundaries.

and the state variable inside the system. It is not hard to think of a more complex (indirect) output of this kind. For example, the model inside the dashed line of Figure 4.1(b) might be intended to describe wind-induced sediment resuspension in a shallow lake, such that the state variable in the model, i.e., that connected to the output outside the dashed line, must represent strictly the concentration of suspended solids due only to this feature. If the output were a measure of total suspended solids at some point in the water column, however, it would doubtless encompass solid matter present at that location but arising from features quite other than those included in the model (Beck, 1985). This could cover particulate matter from the growth of suspended algal biomass, such that output $\boldsymbol{y}$ is influenced by state variables other than those within the dashed line enclosing that part of the system of interest (and the network of interactions influencing the indirect output in Figure 4.1(b) exhibits this type of possibility for the purposes of illustration).[2]

In very broad terms, then, the choices of $[\boldsymbol{x},\boldsymbol{\alpha},\boldsymbol{f},\boldsymbol{h}]$ signify that which we presume (or wish) to know of the system's behaviour, relative to the purpose of the model, while $[\boldsymbol{\xi},\boldsymbol{\eta}]$ acknowledge in some form that which falls outside the scope, or below the resolving power, of the model. Much, of course, must be subsumed under the definitions of $\boldsymbol{\xi}$ and $\boldsymbol{\eta}$. We may have chosen to exclude from the model some of that which was known beforehand (but which is judged not to be significant); there may be features for which there are no clear hypotheses (and therefore no clear mathematical expressions), other than that these may in part be stochastic processes with presumably quantifiable statistical characteristics; there may be yet other features of conceivable relevance, but of which we are simply ignorant; and, as is most familiar, there may be factors affecting the process of observation such that we are unable to have uncorrupted, perfect access to knowledge of the values of the state variables (the inputs and outputs, as depicted in Figure 4.1(b)).[3] Indeed, much of potential significance may have been lost in proceeding from Figure 4.1(b) to the abstraction of Figure 4.1(c) and its mathematical expression in the form of equation 4.1. In particular, careful inspection of Figure 4.1(b) reveals input disturbances of two of the model's state variables that are not the subject of observation (a feature usually subsumed under the definition of $\boldsymbol{\xi}$) and feedback loops whereby variations in the model's state variables are capable of inducing variations in the inputs (an omission not of great prominence before the recent and current discussion of feedbacks between the land surface, the oceans, and the atmosphere).

At some level of resolution, cutting out the representation of the system of interest and abstracting it away from its context in the seamless web of interactions in

---

[2] A more subtle example of a complex output, where the observation itself may need to be acknowledged and represented as a system in its own right, with an interacting set of state variables, can be found in Beck (1979).

[3] There is an asymmetry about the manner in which these last are accommodated in the model of equation 4.1. Typically $\boldsymbol{\eta}$ will be assumed to account for errors in observing the outputs, whereas the errors in observing the inputs ($\boldsymbol{u}$) must be incorporated into composite specifications of $\boldsymbol{\xi}$.

which it sits, requires a line to be drawn between that which we know (and wish to know), the {presumed known}, and that of which we know little or nothing, the {acknowledged unknown}. In this, however, the phrase "at some level of resolution" is not at all as innocuous as it may seem. For the resolving power of our models must always be bound to be macroscopic relative to reality. Except in the absolutely limiting case of a representation of behaviour purporting to describe the interactions among the elementary particles of the universe, the web of interactions of Figure 4.1(a) cannot represent reality resolved at its finest degree.

To make further progress in defining the nature of our problem we must rotate our conceptual apparatus somewhat. Imagine thus the web of interactions of Figure 4.1(a) to be set out on a horizontal plane: the upper plane of Figure 4.2. It is in the very nature of a model that a more refined description of the behaviour of the system can be supposed to exist at some more fundamental level, such as the more fine-grained web cast in the lower plane of Figure 4.2. At this (higher) level of resolution a relatively more microscopic description of the system's behaviour, with more state variables and more interactions, frequently embracing more parameters, would be apparent. Magnification of the object of our enquiry will have been achieved as the focus of the lens of our model is moved from the upper to the lower plane of Figure 4.2.

All models, as we have said, are bound to be approximative in some way (and strongly so in the study of environmental systems). That the parameters of the model

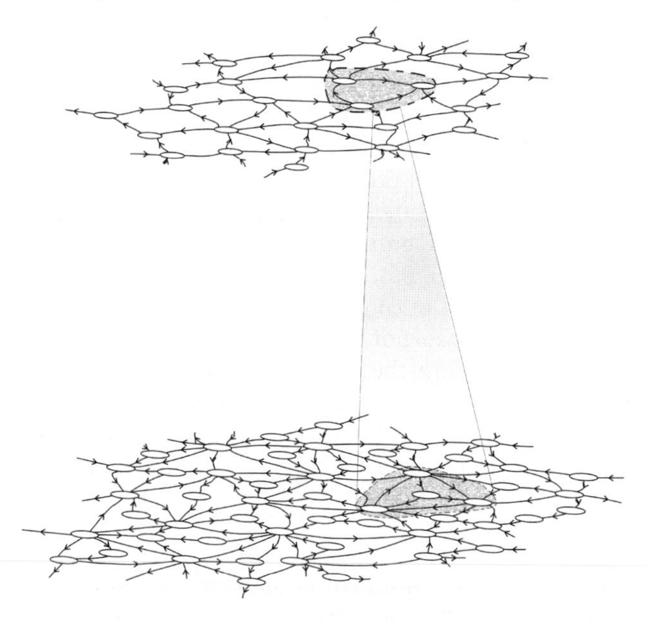

Fig. 4.2. Conceiving of the system's behaviour in terms of models with greater or lesser powers of resolution: upper plane is a relatively macroscopic description of the web of interactions (relatively crude resolving power); lower plane is a relatively microscopic description of the web of interactions (relatively refined resolving power).

should reflect the seat of this approximation, and that changes in the structure of the system's behaviour should take place in the parameter space of the model, are matters that we have set out as a cardinal point of this monograph. Further, we argue that one principal means of access from the upper plane of Figure 4.2 to the more refined scale of representation in the lower plane is via the parameters of the more macroscopic model. This is precisely the unfolding of the coarser-scale parts of some of the model's parameters, into a set of more refined state variables with interactions incorporating a more refined set of (invariant) parameters, with which we opened the discussion of Chapter 2. Viewed from a different perspective, the parameters in the representation of the upper plane of Figure 4.2, which we might denote as the parameters $\boldsymbol{\alpha}^0$ in a model of the form of equation 4.1, subsume under their definition a set of state variables and parameters $[\boldsymbol{x}^1,\boldsymbol{\alpha}^1]$ at the next (higher) level of resolution (in the lower plane of Figure 4.2). In turn, the parameters $\boldsymbol{\alpha}^1$ subsume under their definition a set of state variables and parameters $[\boldsymbol{x}^2,\boldsymbol{\alpha}^2]$ at some still more refined scale of representation, and so on (Figure 4.3).

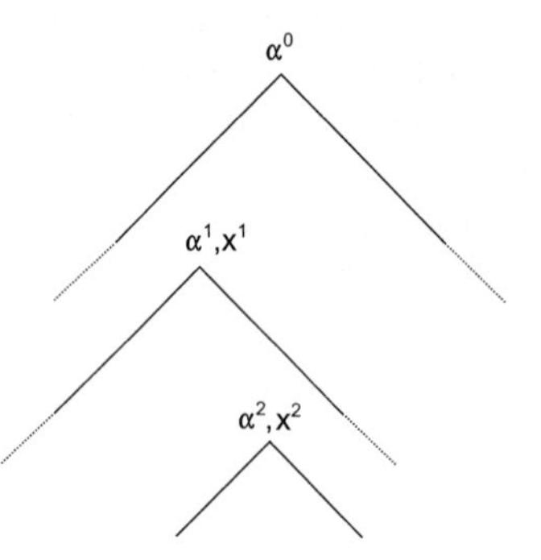

Fig. 4.3. Hierarchy of system descriptions: parameters of the more macroscopic representation ($\boldsymbol{\alpha}^0$) subsume parameters and state variables of the more microscopic representation ($\boldsymbol{\alpha}^1, \boldsymbol{x}^1$), and so on.

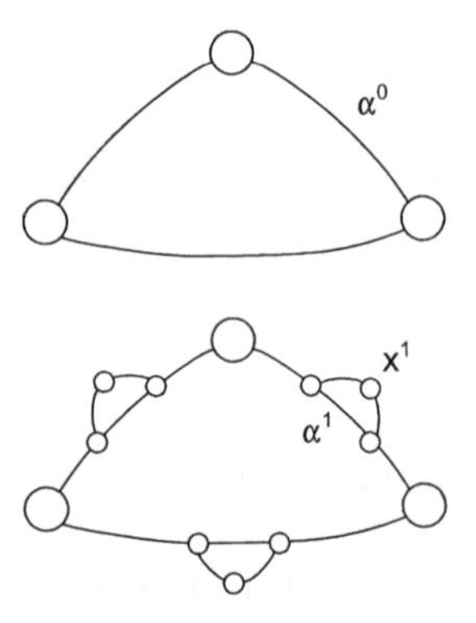

Fig. 4.4. Branch-node expression of the hierarchy of system descriptions: parameters of the more macroscopic representation ($\boldsymbol{\alpha}^0$) subsume parameters and state variables of the more microscopic representation ($\boldsymbol{\alpha}^1, \boldsymbol{x}^1$).

With this insight, let us now return to the stylised representation of the model in Figure 4.1(c). There, each of the three branches of this structure, parameterised through $\boldsymbol{\alpha}^0$, might be perceived at some higher level of resolution as in fact comprising other (more finely resolved) state variables and parameters $[\boldsymbol{x}^1,\boldsymbol{\alpha}^1]$, and so on (Figure 4.4).

4.2 AN EXAMPLE

In order to make this general argument more specific, consider the case of microbially mediated decomposition of a pollutant (x_1) in an aquatic environment. It might be appropriate to postulate that decomposition occurs simply as a matter of first-order chemical kinetics, with a rate constant α_1 (in fact such a model has been used successfully for most of the past century in directing investments in the restoration of acceptable water quality in river systems). To be more precise, let us denote these attributes as $[x_1^0, \alpha_1^0]$ in order to attach an index (0) to the resolving power of the model. Further, for the time being, let us drop $[\boldsymbol{\xi},\boldsymbol{\eta}]$ from our discussion, these being the nominal acknowledgements in equation 4.1 of all that has not been included in our enquiry in explicit hypothetical and mathematical form. Furthermore, since its presence does not materially affect our argument, we shall not engage any form of equation 4.1(b) in this example.

At some level of resolution, then, our model of the real thing – of pollutant degradation in a body of water – may be expressed as follows:

Level 0 description

$$\dot{x}_1^0(t) = [\text{mass transport terms}] - \alpha_1^0 x_1^0(t) \tag{4.2}$$

in which the dot notation in $\dot{x}_1^0(t)$ denotes differentiation with respect to time t. At a more refined level of resolution, however, we are fortunate enough to know in this instance that pollutant degradation is in reality mediated by a bacterial biomass, which we might denote as a second state variable (x_2). Thus, at this second level, with its attached index of superscript 1, the model would instead be as follows:

Level 1 description

$$\dot{x}_1^1(t) = [\text{mass transport terms}] - \alpha_1^1 x_1^1(t) x_2^1(t) / [\alpha_2^1 + x_1^1(t)] \tag{4.3a}$$

$$\dot{x}_2^1(t) = [\text{mass transport terms}] - \alpha_3^1 x_1^1(t) x_2^1(t) / [\alpha_2^1 + x_1^1(t)] \tag{4.3b}$$

where the nature of the interaction between the two states (pollutant and biomass respectively) has been expressed as a nonlinear Monod function.

Comparison of the structures of equations 4.2 and 4.3(a) now reveals the manner in which the parameter in the more macroscopic model (in the upper plane of Figure

4.2) provides access to the more microscopic model (in the lower plane of Figure 4.2). In particular, we find the following equivalence

$$\alpha_1^0 \equiv \alpha_1^1 x_2^1(t)/[\alpha_2^1 + x_1^1(t)] \tag{4.4}$$

In other words, some of the state variables of the more refined representation, i.e. $[x_1^1(t), x_2^1(t)]$, are in effect masquerading as the parameter $[\alpha_1^0]$ of the more macroscopic model. In general, however, an explicit form of equivalence, such as that in equation 4.4 would not be at all apparent, so that we could at most conclude that

$$[\alpha_1^0] \subset [x_1^1(t), x_2^1(t); \alpha_1^1, \alpha_2^1] \tag{4.5}$$

intended here in the sense of $[\alpha_1^0]$ subsuming $[x_1^1(t), x_2^1(t); \alpha_1^1, \alpha_2^1]$. Indeed, knowing beforehand that, in reality, neither the biomass nor the pollutant can be considered truly as singular entities, since they comprise many species of microorganisms and chemicals respectively, all interacting with one another in a complex foodweb, this same attribute – of a parameter at the relatively macroscopic level subsuming states and parameters at a more microscopic level of resolution – would repeat itself many times as the description of the system's behaviour is successively unfolded into its more elementary parts.

4.3 THE NUB OF THE PROBLEM

Herein lies the nub of the issue, of defining what is to be understood as *structural change*. While we may imagine the "truth" of the matter to be as shown by our metaphor according to Figure 4.5(a), at some very refined level of resolution, the resolving power of our models must always be bound to be macroscopic relative to this picture of reality. The structure of the behaviour of the system, as captured in our current models, might therefore appear as in Figure 4.5(b), relative to Figure 4.5(a). There will *always*, no matter how refined our descriptions, be attributes of the system's behaviour lying below the resolving power of our models; and very often these are subsumed under the definition of the model's parameters, as illustrated in the foregoing example. In general, and essentially for all practical purposes in working with models of environmental systems, we may suppose that

$$[\boldsymbol{\alpha}^i] \subset [\boldsymbol{x}^{i+1}(t); \boldsymbol{\alpha}^{i+1}] \tag{4.6}$$

irrespective of how high (i) may be the resolving power of the model. Making this presumption, however, does not require any supposition about how we imagine the truth of Figure 4.5(a) to be organised. In particular, we do not have to suppose an image thereof that is fixed, or evolutionary (as in the sense of Allen (1990)), or anything else (other than that its description will involve quantities that we can comfortably conceive of as state variables and parameters).

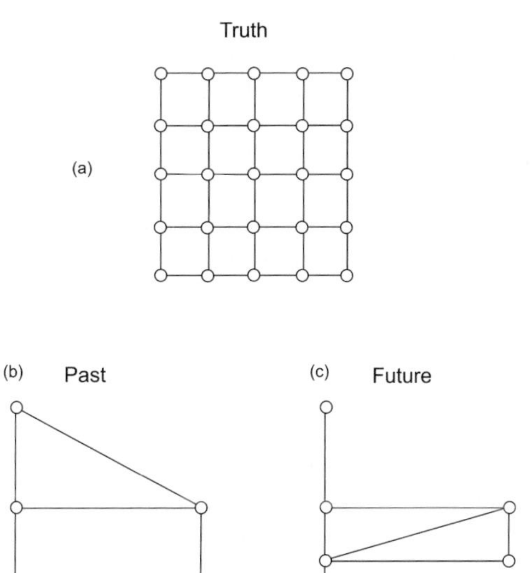

Fig. 4.5. Apparent structural change in the behaviour of the system as a consequence of inevitably incomplete understanding: (a) idealised representation of the structure of the truth; (b) apparent structure of the dominant modes of the system's past behaviour; and (c) different apparent structure of the modes that may come to dominate behaviour in the future.

As we stand, then, on the threshold of the future, with an obligation or desire to forecast the behaviour of the system, unexpected or even unimagined attributes of the behaviour of the system – attributes below the resolving power of the model (features that may in truth have been present but not observed or detected as such) – may come to dominate future behaviour. To us, with our inevitably macroscopic models, it would appear that this behaviour might be governed by quite a different structure, such as that of Figure 4.5(c). In the most difficult of these problems (as envisaged in the work of Allen (1990)), there is the possibility of new state variables seeming to emerge from below the resolving power of the model and of others seeming to disappear below it. The assembly of nodes, and therefore the branches, in the structure of the model will appear to have changed with time (a kind of metamorphosis), as is the case in the evolution from the *past* description of Figure 4.5(b) to that which would be required as the *future* description of Figure 4.5(c). It is as though variations in the number of differential equations in the mathematical model must be forecast ahead of time.

Yet without reference to knowledge of the truth (Figure 4.5(a)), how are we to achieve this and, of particular importance to this monograph, how might we detect the seeds of this change, buried – as they may be – in the observed record of behaviour in the past?

A simple-minded insight into possible ways of answering this latter question is the notion that branches in our metaphor for the structure of the model will disappear when the associated parameter (α) declines from a non-zero value to zero (and emerge under the converse condition). And given the argument of the preceding illustrative example of pollutant degradation in a river we would have every reason to believe that the parameters of a model will in truth rarely, if ever, be invariant with time. The state variables $[x_1^1(t), x_2^1(t)]$ subsumed under $[\alpha_1^0]$ in equation 4.5, albeit varying relatively slowly with time, give palpable substance to this supposition, as does equation 4.6 for the general case. In short, for the more general case, we may suppose that changes over time of the parameters of our models will be the norm, i.e., that

$$\dot{\boldsymbol{\alpha}}(t) \neq \mathbf{0} \tag{4.7}$$

The nub, if not the whole, of our charge is to find ways of detecting this feature, of diagnosing its cause, describing it for the purposes of extrapolation, and forecasting and managing the behaviour of the system in the express knowledge thereof.

REFERENCES

Allen, P.M., 1990. Evolution, innovation, and economics. In: *Technical Change and Economic Theory* (G. Dosi, C. Freeman, R. Nelson, G. Silverberg and L. Soete, eds.). Pinter, London, pp. 95–119.

Beck, M.B., 1979. Model structure identification from experimental data. In: *Theoretical Systems Ecology* (E Halfon, ed.). Academic, New York, pp. 259–289.

Beck, M.B., 1985. Lake eutrophication: identification of tributary nutrient loading and sediment resuspension dynamics. *Appl. Math. Computat.*, **17**, 433–458.

Environmental Foresight and Models: A Manifesto
M.B. Beck (editor)

CHAPTER 5

The Manifesto

M.B. Beck

5.1 THE CHALLENGE

Our challenge is this: to discern whether any "surprising" dislocations in the behaviour of the environment are imminent, especially as imagined by the community of stakeholders; to encode these scenarios in a crudely quantitative form; to seek to associate the possible realisation of these dislocations with key components in the ill understood and highly complex, seamless web of interacting mechanisms that is our current, best quantitative representation of the system's behaviour; and to implement field studies and to develop regulatory policy, designed respectively to clarify the seeds of the potential change and to cope with the consequences of their further propagation.

Our talents, as authors of this monograph, reside predominantly in reconciling knowledge in one form (theory) with knowledge in another, *maximally independent* form (observation). We have declared ourselves most familiar with solving problems of system identification (or model evaluation) and in coping with the inevitable uncertainty surrounding this task. Were we to have access to a form of knowledge capable of substituting for what we would really need, i.e., behaviour as observed in the future, we should then be able to bring some well tried methods to bear on the issue of structural change (as defined in Figures 4.5(b) and 4.5(c) of the foregoing chapter). But what form of knowledge, we must ask, would allow us to erase the divide between past and future in this manner, yet, crucially, satisfy the requirement of being "maximally independent" of current theory?

5.2 BELIEF NETWORKS: GENERATING THE FEARED DISLOCATIONS

Let us begin by looking more closely at the hierarchy of knowledge attaching to the resolving power of the model in Figures 4.2 and 4.3 and picturing respectively stacks of planes and pyramids set both above and below the illustrative few shown in these conceptual representations. It is not hard to imagine that something akin to the heterotic superstring theory for expressing the interactions among the elementary particles of the universe, would rest on the very lowermost plane (of Figure 4.2), at the base of the pyramid (of Figure 4.3). Traversing the lens of the microscope in the opposite direction, however, would bring us up to knowledge in the form of a set of personal, public, policy-maker, or lay-stakeholder beliefs in the uppermost plane, towards the apex of the pyramid of knowledge representation. Indeed, the less these beliefs have been prejudiced by those responsible for developing the models of environmental science, so much the better. For in this context, of seeking maximal independence of the forms of knowledge that are to be reconciled, the "very obscurity of the sums" in which we engage (in Schaffer's (1993) terms) is an advantage. That the public might fear the extinction of striped bass in Lake Erie, or be jolly pleased to see the zebra mussel population disappear – in either case without knowing of the possible significance of a Coriolis force or a microbial loop in the foodweb – would therefore be no bad thing. The goal is to maximise the separation of the "deep" knowledge, encoded within the framework of a familiar and conventional mathematical model, from the impressions – the respectable "shallow" knowledge (*not* intended in the pejorative sense) – of what the state of the environment may become.

However imagined or deduced, the end-point of such impressions of the feared (or desired) dislocations of behaviour would be a quantification of the domains of the state space into which the state of the system should not (or should) enter in the future (as suggested in Figure 3.2(c) or 3.2(d) of Chapter 3). Put simply, for example, these domains might be specified as

$$\boldsymbol{x}^{l}_{RD} \leq \boldsymbol{x}(t^{+}) \leq \boldsymbol{x}^{u}_{RD} \tag{5.1}$$

in which the superscripts l and u denote lower and upper bounds respectively, and the subscript RD connotes, say, *radically different* behaviour at some point in future time (indicated by t^{+}). Specifications richer than equation 5.1 are both possible and desirable. For instance, rather than a single end domain to be avoided or entered in a finite time, a sequence of intermediate domains defining a path to the end state might be specified (an idea reminiscent of some of the earlier work of Clark (1986)). In the present context richness of behaviour specification will tend to be more strongly associated with the number of constraints, of the form of equation (5.1), as opposed to the narrowness of the domain bounds.

We know now what we need – equation 5.1, that is – but not how to obtain it. How in fact would anyone start to conceive of future possibilities, the more outlandish the better? For almost all of us, one suspects, the answer would be quite intuitive and utterly familiar: in terms of causal chains of elemental "If"–"Then" rules. One need only turn to the pages of documents such as those dealing with performance assessment of proposed facilities for the storage of radionuclide wastes to find strings of these linguistic rules. For example, these have been extracted but little changed from the text of Davis et al. (1990):

> "If" renewed glaciation "or" "if" increased atmospheric CO_2 "then" climate change.
> "If" climate change "then" wetter conditions.
> "If" wetter conditions "and" "if" more dense vegetation "then" increased infiltration "and" decreased runoff.
> ...

Alternatively, the "causal dendogram" of Figure 5.1 expresses much the same kind of reasoning, accompanied (as it is) by the following text (DeAngelis and Cushman, 1990):

> [A] climate change will, presumably, lead to a warming of stream, lake, estuarine, and marine waters (2→3 [in Figure 5.1]) and to changes in streamflows, currents and ocean upwellings (2→4). Temperature increases could decrease the physical mixing of water in lakes and estuaries by increasing thermocline strength and, therefore, vertical stability (3→4), thus decreasing nutrient movement from lower levels to surface waters...

Even though we may presume to have more detailed knowledge of the cogs and wheels turning within the system, at some more refined level of reasoning (precisely as DeAngelis and Cushman (1990) proceed to demonstrate), our point is *not* to emulate it. In order for our analysis to be of any use the risk of falling into the trap of reconciling the same knowledge base with itself must be minimised. Generating scenarios for changes over the millennia in the future environment of a nuclear waste facility, for example, is a most earnest business and one for which there have been several attempts at a more systematic organisation of the opinions of the *experts*. The opinions of the experts are just what we may *not* need, however, if we are to achieve maximal independence of the two sources of knowledge to be reconciled. It may well be essential to shun the temptation to proceed to some deeper basis of knowledge in composing a set of impressions of the form of equation 5.1. That which is serendipitous and not entirely logical, perhaps even bizarre in the light of the current, scientific knowledge base, could be a virtue.

But where should we search for the cultivation of impressions such as this? Thompson (1997) argues there are four forms of social solidarity (or outlooks on the man–environment relationship), which map one-to-one onto Holling's (1986) myths of Nature. We are here probably not looking for those (the Individualists) subscribing to the myth of "Nature benign", since this seems essentially a peril-free outlook in which the environment is thought capable of absorbing all insult and injury before returning assuredly to a pre-disturbance equilibrium position. We may find something of what we need in those (the Hierarchists) subscribing to the myth of "Nature perverse but tolerant". In this perspective, restoration to the pre-disturbance equilibrium will be the norm, for as long as the environment is not placed under

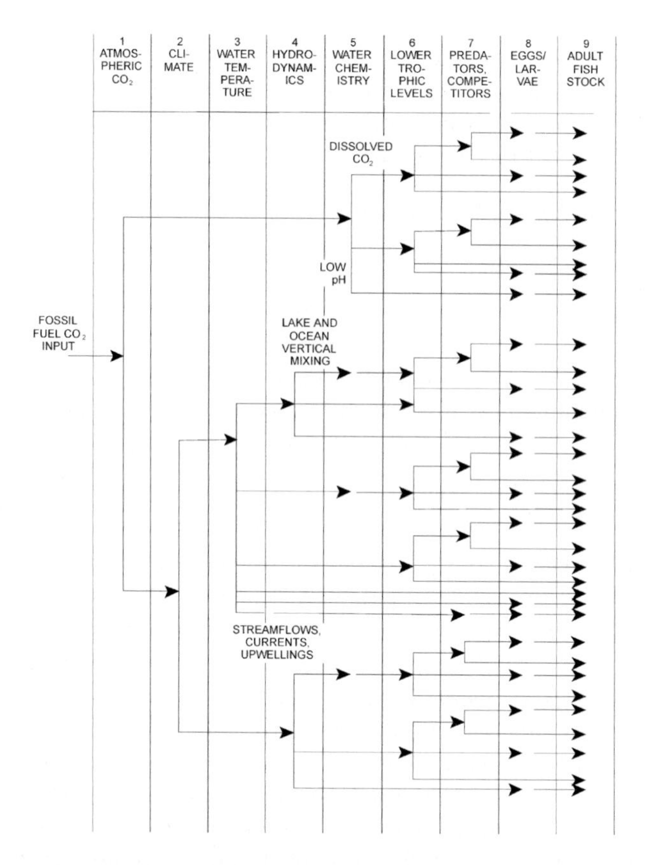

Fig. 5.1. Envirogram showing a variety of causal chains linking fossil fuel CO_2 input to the atmosphere and thence to a fish stock of interest (after DeAngelis and Cushman (1990)).

excessive stress, in which case excursions into the largely unfamiliar and undesirable will ensue. "Nature ephemeral", being the myth to which an Egalitarian solidarity should attach, may be what we are really in search of, however: any disturbance, no matter how small, may plunge the system into the wholly undesirable and unfamiliar.[1] For it is from this perspective, for instance, as articulated by scientifically well qualified environmental activists, that the greatest creativity about what might go wrong in the future may be brought to bear.

We have a sense, then, of the kind of thinking that is to be encouraged, even deliberately cultivated, but not of how to encourage or stimulate it, i.e., through what organised process. In the case study of Lake Lanier, Georgia (USA), which we shall discuss in Chapter 11, we sought to elicit the (feared) target futures directly from the

[1] The fourth solidarity, that of the Fatalists, who believe the myth of "Nature capricious", is hard to incorporate, since it fails to recognise "man" as requiring any relationship with the "environment".

lay stakeholders, as "untainted" by the interventions of professional scientists as possible, first by means of a survey and subsequently through a facilitated "foresight" workshop. Other alternatives, perhaps inferior options, could have been to manipulate computationally encoded forms of the stakeholders' mental models, using fuzzy logic or qualitative simulation, for example. In fact, we shall settle later (in Chapter 10) upon the choice of belief networks (Pearl, 1988; Varis, 1995) in order to illustrate how such manipulation might have been accomplished. In this context the word "belief" signals the probability – the degree of belief – that some statement of cause, effect, or (significantly) the relationship between the two, is likely to be true. We could thus take any form of knowledge representation, with all the relationships among the inputs, states, parameters, and outputs assembled in a network, and propagate probabilities of these features through a corresponding network of beliefs. In the relationships of Figure 5.1, for instance, the crudely quantified attributes in the state space – {warming of lake waters}, {decreased vertical currents}, {greater thermocline stability}, {decreased upward fluxes of nutrients} and so on – would thus form the nodes of the knowledge network, while its branches would comprise assertions such as [{warming of lake waters} causes {decreased vertical currents}]. Yet we might very well want to strike out against the normal tendency to build consensus from an initial set of heterogeneous beliefs (Varis and Kuikka, 1999). Our goal is not to stifle the vitality and imaginative diversity of the Egalitarian solidarity. A rogue element of randomness might need to be injected into the network of knowledge representation. In places where the constituent assertions are tenuous, for example, in the sense that the influence of cause over effect may be negative or positive, the structure of the network might be altered by rupturing or replacing some of its elemental "If"–"Then" rules (as reported in Rosenhead (1989) on the early work of Sandberg (1976)). We might even consider reversing the logic of what is believed to be cause and what effect.

The imagined future outcomes – feared or desired states of the environment – could thus be generated from barely believable future (input) circumstances fed through a network of tenuous causal connections. Whether we conduct such computational manipulations of belief networks, the basis of which is set out in Chapter 10, or consult directly the hopes and fears articulated by Thompson's Egalitarian solidarity, the principal results we seek are statements about future domains of behaviour in the form of equation 5.1. If we can get to this end-point, we shall have constructed a problem of system identification. Now we must examine the prospects for success in solving it.

5.3 HIGH-ORDER MODELS: RANDOM SEARCH AND THE REACHABILITY OF TARGET FUTURES

Our goal is to explore the extent to which the fears of a community – about the future behaviour of an environmental system – may come to pass; and, if their occurrence is to be expected, to identify to what aspects of the theories underpinning our current understanding of these systems such expectations are most sensitive. The shallow knowledge of the imagined futures is to be pitted against the more conventional deep knowledge of environmental science, as typically represented in mathematical models. The latter, one might say, is to be reconciled with the former, a maximally independent source of knowledge (but admittedly not a customary form of observation).

Stretched apart, towards opposite ends of the spectrum of knowledge representation, the target behaviour to be matched is of a roughly expressed lower order (in the terms of equation 5.1), while the model to be identified can be expected to be of a high order, i.e., with high-order state and parameter vectors $[\boldsymbol{x},\boldsymbol{\alpha}]$. In principle, the deeper the knowledge encapsulated in the model, so the higher the order of these vectors, and equally so the greater the distinction between the knowledge bases tapped for the specification of behaviour and the mechanisation of theory. The parameters of the model may be assumed to be unknown constants, i.e., to be random variables, most simply assumed further to be distributed uniformly between given upper ($\boldsymbol{\alpha}^u$) and lower ($\boldsymbol{\alpha}^l$) bounds,

$$\boldsymbol{\alpha}^l \leq \boldsymbol{\alpha} \leq \boldsymbol{\alpha}^u \tag{5.2}$$

The problem of identification, as always, is to find those values of $\boldsymbol{\alpha}$ enabling the simulated behaviour of the model to match the target behaviour, subject to the constraints of equation 5.2. In our case, however, this is behaviour as imagined, *not* as observed.

Consider the following procedure. It is nothing more than the customary procedure of what Hornberger, Spear, and Young have called a Regionalised Sensitivity Analysis (RSA) (Young et al., 1978; Hornberger and Spear, 1980; Spear and Hornberger, 1980), albeit cast in the domain of an imagined future behaviour. Candidate values $\boldsymbol{\alpha}^i$ for the parameters may be drawn at random from their parent probability distributions (equation 5.2) and substituted into the model in order to generate candidate trajectories of the states of the system, $\boldsymbol{x}^i(t^+)$, into the future. Any such trajectory can be classified according to whether $\boldsymbol{x}^i(t^+)$ matches the target behaviour, i.e., satisfies the constraints of the form of equation 5.1. The associated candidate parameterisations of the model ($\boldsymbol{\alpha}^i$) may thus be classified as either giving the defined behaviour or not. Given a sufficiently large number (i) of possible parameterisations, with sufficiently large samples in both the "behaviour-giving" and "not-behaviour-giving" categories, tests may be made of the extent to which the (posterior) distributions of the two categories of parameterisations may differ from each other. As Hornberger and colleagues would then argue, those elements of the parameter vector, $\boldsymbol{\alpha}_j$, for which the two distributions of "behaviour-giving" and "not-behaviour-giving" are significantly different, are deemed to be *key* to discriminating between whether or not the target behaviour has been matched. Other elements of the parameter vector, for which there is no such difference, are considered *redundant*, in the sense that it seems to matter not at all what values for these parameters are drawn from the distributions of equation 5.2. Any old value for such parameters may, or may not, give the defined behaviour.[2] The outcome of the RSA is therefore a cleaving of the parameter vector $\boldsymbol{\alpha}$ into two sets, $\{\boldsymbol{\alpha}^K\}$ and $\{\boldsymbol{\alpha}^R\}$, representing respectively those parameters that are key and those that are redundant in reaching, or entering into, the target domains of the state space defined by equation 5.1.

We have thus a method for identifying on which of the constituent hypotheses – within the web of interactions encapsulated in the model of the system – may turn the reaching of a feared pattern of future behaviour for the system. Such a result has intrinsic merit. For a start, if there are very few candidate parameterisations enabling the target future to be reached (out of a large number of trial attempts), one might conclude that, within our current understanding, with all its uncertainties and imperfections, the public's fears are not well founded. Alternatively, if the feared future is reachable with a significantly non-zero probability, if this appears to hinge upon knowledge of a relatively small number of interactions (parameters, hypotheses), and if we assume that further scientific enquiry cannot be sustained on all conceivably relevant fronts, then we have an indication of where to direct our limited resources of enquiry. Or yet again, if a succinct parameterisation of the system's inputs is possible, as an elementary form of random walk, for example,

$$u(t) = g\{\alpha_j;\omega(t)\} \tag{5.3}$$

wherein $\omega(t)$ is, say, a stochastic process, we could then attempt to answer the question of what variations in the annual cycle of the thermocline position in Lake Erie, for instance, would lead to a feared future with no striped bass population. This is tantamount to asking: "on what changes of climate, in concert with what key unknowns in our present understanding of northern, temperate, dimictic lake ecosystems, might the reachability of a future with the loss of a valued species turn"? Indeed, if such questions and target behaviours reflect well the fears of stakeholders, one could observe that here is quite clearly a case of the public interest directing the torch-light of science.

But this does not address the issue of structural change, of change over time in that which is dominating the behaviour of the system. For this we would need to examine whether there is the possibility of change in the dominant modes of the system's behaviour as it migrates from the observed past to the imagined future (through the domains of Figure 3.2(d) in Chapter 3). In the simplest statement of this problem a second set of constraints of the form of equation 5.1 could be specified in order to bound the domain of the state space in which behaviour, denoted $\boldsymbol{x}(t^-)$, has been observed to reside in the past (t^-). In just the same manner as the key and

[2] Things are not quite this simple, of course, since we are dealing here with a high-order parameter space in which it may very well be that specific *combinations* of candidate values for several of the parameters are key to discriminating between whether the behaviour is given or not. The results of the analysis ought ideally to be guided by a knowledge of the nature of such correlations; the difficulty is that the task of exploring their character is by no means trivial (Spear et al., 1994; Osidele and Beck, 2001).

redundant parameters have been distinguished above, with respect to the giving or not of a target *future*, so key and redundant parameters can be identified with regard to discriminating between whether *past* observed behaviour has been matched or not. More precisely, we would have the following diagnostic evidence with which to work: $\{\boldsymbol{\alpha}^K(t^+)\}$ and $\{\boldsymbol{\alpha}^R(t^+)\}$ for the future; and $\{\boldsymbol{\alpha}^K(t^-)\}$ and $\{\boldsymbol{\alpha}^R(t^-)\}$ for the past. If the memberships of the respective key and redundant sets differ markedly between those relative to the future and relative to the past (or, alternatively, if their memberships are essentially identical), what does this imply about the potential for a structural shift in the behaviour of the system? If a quite different set of parameters is expected to be the key to behaviour in the future, is this a specific realisation of evolution in the structure of the system's behaviour, in the sense of the rise to significance of currently minor modes of behaviour (and the fall from significance of the currently dominant modes of behaviour)?

Such questions may be given greater immediacy by setting them in the specific context of a case capturing much recent attention. Figure 5.2 is taken from an article (Weaver, 1995) on the issue of how the behaviour of the North Atlantic, which acts as a large-scale conveyor of heat from low to high latitudes, might be affected by climate change. Point 1 – defined rather precisely – specifies the domain of current, i.e., recent past, behaviour, while Point X marks equally precisely a domain of the state space into which one might well not want behaviour to enter in the future. It is referred to ominously as the "point of no return", a phrase undoubtedly capable of striking fear into the minds of the public. What we could imagine finding, then, from a pitting of the deep knowledge of a very high-order geophysical model against the

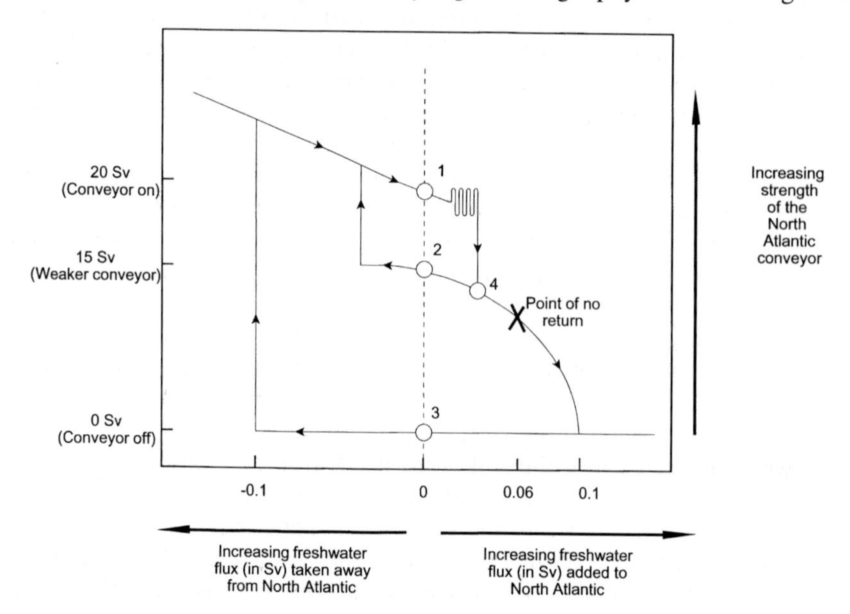

Fig. 5.2. State–space representation of possible trajectories in the behaviour of the ocean conveyor-belt in the North Atlantic (after Weaver (1995)).

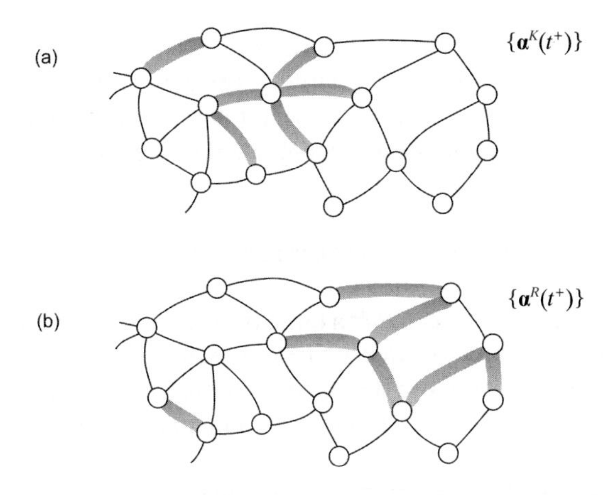

Fig. 5.3. Key (branches emphasised; $\boldsymbol{\alpha}^K$) and redundant parameters in (a) times past (t^-) and (b) times future (t^+).

shallow knowledge of Points 1 and X on Figure 5.2, might look something like Figure 5.3 – wherein the past and future hinge on different branches of the web of deep knowledge. This can indeed be the case, as we shall see later in Chapter 11.

There is always the possibility, of course, that the results of such an analysis will be rendered impotent and lacking in insight, in part through the gross uncertainty surrounding both equations 5.1 and 5.2, but also in no small part because the model has *not* been composed in a manner designed to answer questions of this kind. The custom is to construct the model, especially a high-order model, so that it reflects our best judgements about what is believed to be significant in governing the system's behaviour; or it incorporates a *consensus* set of mechanisms enabling the widest possible applicability, to many specific instances of a generic type of system. Our purpose herein, however, is to determine what may be the most important unknowns – including those speculative constituent hypotheses towards the *fringes* of contemporary understanding – that are worthy of further sustained investigation, because they appear to be key in reaching a feared (desired) pattern of future behaviour.

Subtle might this change of emphasis be in the design of the model. Yet crucial too may it be in increasing the chance of success in our present endeavour: of being able to discriminate more clearly the critical from the not-so-critical and the redundant. There are those who can imagine "... a dozen reasons [a dozen target futures] why – in the face of the stakes involved with the global-warming issue – societies, and economies, have cause for grave concern" (Leggett, 1996). There are others who can surely enumerate the several candidate (positive) feedback loops within the mechanism of the earth system on which the reachability of cataclysmic change may turn (Schellnhuber, 1999). Were we able to compose a coupled model of the atmosphere, oceans, and land surface, it could be designed expressly for discovery of which among these loops may come to dominate the reaching of the greatest variety of imagined future threats. And towards these few (of the many) candidate unknowns, should the scarce resources of scientific enquiry be directed. No prediction of actual *future behaviour* would have been made; instead a plan for *current action* could have been fashioned.

5.4 EVOLVING CLUSTERS OF CANDIDATE PARAMETERISATIONS

Proximity to a uniquely best parameterisation of the model – or a uniquely unambiguous interpretation of past observed behaviour – has no meaning in a Regionalised Sensitivity Analysis (RSA). Many would now concede that uniqueness and optimality in these terms are in any case fanciful notions (Beven, 1989, 1993, 1996). The original intent of the analysis, however, was not to undermine the trust we have traditionally placed in being able to approach the goal of a single, optimal parameterisation of the model. Much as in assessing the reachability of target futures, its purpose was rather one of identifying those (hopefully) few key hypotheses onto which further experimental effort should be concentrated in historically data-impoverished problem contexts. That much of the motivation in forsaking the long-cherished objective of locating a uniquely best set of estimates for a model's parameters should stem from the experience of hydrology, may therefore seem curious. After all, in hydrology, with its archetypal struggle to characterise the rainfall-runoff process mathematically, there are the greatest of all opportunities to maximise the number of available observations $[\boldsymbol{u},\boldsymbol{y}]$ relative to the number of internal attributes $[\boldsymbol{x},\boldsymbol{\alpha}]$ of the model that are to be estimated. Hydrology is a data-rich member among the sub-disciplines of environmental science. Yet the record of experience is littered with attempts at optimal estimation of the model's parameters that have failed because of a lack of identifiability (Beck, 1987, 1994). That is to say, many combinations of estimates for the values of the model's parameters were found to yield more or less equally good matches of the observed behaviour. We now know, from practical experience (for example, Hornberger et al., 1985; Jakeman and Hornberger, 1993; Beven, 1993; and Young and Beven, 1994) as much as from more theoretical analysis (Beck et al., 1990; Kleissen et al., 1990) and philosophical conjecture (Beven, 1989, 1993), that the bounds on what is possible are close by. At most just a handful of parameters can be recovered unambiguously, for example, from a single input–output pair monitored with a high frequency.

The crux of the problem of a lack of model identifiability, or over-parameterisation, is that what one would like to know about the internal description of the system $[\boldsymbol{x},\boldsymbol{\alpha}]$ is of a substantially higher order than that which can be observed of its external description $[\boldsymbol{u},\boldsymbol{y}]$. In less abstract terms, the model may be said to contain surplus or even spurious content (Young, 1978).

An RSA seems almost to celebrate the absence of identifiability. It has two defining elements: (i) random search over a parameter space, as realised most simply through the framework of Monte Carlo simulation; and (ii), much more importantly for what now follows, juxtaposition of the properties of one subset of candidate parameterisations against those of another (with a view to identifying what determines similarity and what difference between the two subsets). Whether or not behaviour has been matched is the simple determinant of classification into one or the other subset. In its original form, no candidate parameterisation of the model ($\boldsymbol{\alpha}^i$) was deemed any more likely than any other to have given rise to the behaviour. Each candidate parameterisation, it could be argued, would be equally worthy and fit for the purposes of forecasting future behaviour (Fedra et al., 1981). Such elementary notions, whose supreme advantages derive from their very simplicity, are capable of refinement.

If we believe that the point values of the observed states, i.e., the time-series of outputs $\boldsymbol{y}(t_k)$, are indicative of the past trajectory of the system's behaviour (their uncertainty notwithstanding), classical measures of the mismatch between observed and estimated behaviour may be used to introduce preferences and ranking among the candidate parameterisations, as in Beven and Binley (1992). Whatever the measure might be (it is typically a sum of the squared mismatches between estimates and observations), let us denote it as V. A sample of candidate pairs $\{V^i,\boldsymbol{\alpha}^i\}$ can be generated and $\boldsymbol{\alpha}^i$ ranked according to the order of V^i. This ordering of $\boldsymbol{\alpha}^i$ may then be used to achieve two results: (i) rejection of all the candidate parameterisations falling beyond a (subjective) threshold value of the measure of mismatch, as being "unworthy" in some sense; and (ii) a transformation such that the remaining most inferior parameterisation has the lowest non-zero probability and the most superior the highest probability of having given rise to the observed behaviour (Beven and Binley, 1992). This latter population of acceptable candidate parameterisations can now be cleaved into two subsets in a slightly different dimension. What is it, we might ask, that is essential in distinguishing membership in the subset of, say, the best 10% from that of the worst 10%. This is not a crisp juxtaposition of what "does" with what "does not" match the specified behaviour, but it is a quite logical extension of the principle.

The outcome of this more refined analysis, conducted under conditions rather more favourable than those of *gross* uncertainty and with access to more conventional characterisations of what constitutes behaviour, should yield insights of a character more subtle than those derived from the unrefined, basic RSA (as in Chapter 11). The occurrence of fire in a watershed can wreak substantial change in the dynamics of water fluxes through that system and into the associated streams. One would expect that lower-frequency parametric shift should be detectable from the interpretation of windows of high-frequency, rainfall-runoff observations before and after the event, perhaps even for a hydrological model with a high resolving power. Reconciliation of the same (candidate) fixed structure of the model (herein that of Beven et al. (1995)) with pre- and post-fire observed behaviour can indeed reveal snapshots, as it were, of the implied longer-term parametric changes. The storage of water in the root zone, considered as a parameter, is found to be much reduced just after the fire, but subsequently returns towards the domain of pre-event values. Cleaved and juxtaposed thus, window by window of high-frequency observations, we shall find in Chapter 12 clear evidence both of structural change, i.e., of $\dot{\boldsymbol{\alpha}}(t) \neq 0$, and of those *constituent* parameters in which this feature may be lodged. Parametric change can indeed be detected; it is discernible, all the uncertainty notwithstanding.

But projections of structural change onto just one dimension, that is, the identification of change with time in the central tendency of the distribution of each parameter isolated in turn from the whole of the high-dimensional vector ($\boldsymbol{\alpha}$) in which it sits, is not consistent with the premise that significance lies in the indivisibility of the candidate *combination* of values chosen for all the elements in the entire parameter vector (Beven, 1993; Spear et al., 1994; Osidele and Beck, 2001). In this higher-dimensional space of change, how are the changes in all the elements of the parameter vector orchestrated over time into some organised, but evolving, ensemble? Rules for the evolving statistical properties of the clusters of acceptable candidate parameterisations will be needed, their variance–covariance structure (pairwise correlation coefficients, for instance) at the very least. At present, we make extrapolations into the future by default according to the lowest form of any such rule: that the single variance–covariance structure crystallized out from interpretation of the record of the past will obtain for the future, without change in time (Fedra et al., 1981). The ensemble of parametric correlations simply remains fixed for all time.

5.5 SIMPLICITY OUT OF COMPLEXITY

Understanding of something is achieved when information about this thing is assimilated into an appropriate mental model (MacFarlane, 1990). We believe we have understood the behaviour of Lake Erie, for example, when what is observed to happen can be faithfully reflected in terms of the simple, every-day, household concepts of beakers with stirrers, warm fluid rising towards the surface, switches that prevent the beaker's contents from mixing, and food for the growth of organisms. We believe we have understood the circulation of heat within the ocean when provided with insights through the image of simple switching mechanisms along the path of a conveyor-belt (Weaver, 1995). Policy-makers and the public understand the implications of the increased release of carbon dioxide into the atmosphere and the associated possibility of a change in climate because of the analogy of the humble, but most familiar, greenhouse (familiar at least to those from northern temperate climates). It is through appeals to the commonplace, the mundane, and the familiar that we all have access to understanding and insight across the disciplines of science.

Of course, we know that things are not quite this simple. But until we have arrived at such a distillation of the essence of the matter – at this macroscopic, conceptual level (towards the apex of the pyramid of knowledge representation) – we are unable to make sense of all the information available. Yet what should we do if this "making sense" is achieved through a defective mental model? After all, there are some well-known, perhaps even infamous, cases where sense was manufactured on the basis of a defective mental model, William Bowie's rejection of Alfred Wegener's theory of continental drift being one of them (Oreskes, 1999). What should we do, for example, if the intuitive appeal of the ocean conveyor-belt – whose realisation has itself been enabled as much by the availability of very high-order ocean circulation models as by interpretation of the empirical observations – is misleading? How might we acquire any sense that the analogy of the conveyer-belt is wrong? How indeed would we provide evidence to support the generation of an alternative candidate high-level conceptual model?

Macroscopic conceptual understanding and basic, microscopic theory are set in very different planes; on these our tentative descriptions of the "truth" are resolved at very different levels. It is as though the scope of our enquiry is focused at very different powers of magnification. Whereas there may be observed current fields against which to match the behaviour of a high-order computational model of ocean circulation, we cannot literally observe the quantities of heat at given positions on the conveyor-belt or the times at which the belt was switched on, slowed down, switched off, and so forth. We cannot *directly* reconcile the conceptual insight – vital though it is to understanding – with any empirical evidence, any more than we can easily come up with an insight very different from the ocean conveyor-belt. But we might be able to undertake some useful form of evaluation, provided interpretations of theory and observation can be derived at more or less compatible powers of resolution.

More crisply stated, let us suppose we have a high-order computational model (HOM), with a very high resolving power, i.e., with many state variables and parameters, denoted $[\boldsymbol{x}^{+n},\boldsymbol{\alpha}^{+n}]$, yet we can only identify from the field observations models of a low order (LOM), denoted $[\boldsymbol{x}^{-n},\boldsymbol{\alpha}^{-n}]$, in which n is large, for the sake of dramatic contrast. In other words, the HOM and LOM are symmetrically reflected about the nominal model of equation 4.1 used in Chapter 4 to help define structural change. It is clear we cannot embellish the latter in order to achieve the detail of the former. But we could seek to reduce the order of the former in order to compare a distillate of its essence with the latter. This, of course, will readily be recognised as a problem of sensitivity analysis, akin to the ideas of "meta-modelling" discussed in Kleijnen (1987), for example. In Chapter 13, then, such an analysis of sensitivity is referred to as a dominant mode analysis: the goal is to approximate the dynamic behaviour of the given high-order model by identifying the small number of dynamic modes that appear to dominate its response to perturbations in the input variables. There, the LOM (or "meta-model") is in the form of an input–output time-series model, a type of model easily identifiable from the given data, but not customarily imbued with insights of an immediate conceptual, "mechanistic" nature. Having gone thus to such an abstract representation, the manner of then re-extracting some kind of quasi-conceptual interpretation from the so identified LOM is discussed at length in Young (1998), as well as Chapter 13. To summarise, our need is to know the nature of the mapping between $\boldsymbol{\alpha}^{-n}$ (in the LOM) and $\boldsymbol{\alpha}^{+n}$ (in the HOM). In particular, we wish to know how the elements of $\boldsymbol{\alpha}^{-n}$ that are identifiable from the observed record, i.e., from $[\boldsymbol{u},\boldsymbol{y}]$, might map onto those elements of $\boldsymbol{\alpha}^{+n}$ key to discriminating between whether this behaviour is given or not. When, as we have said in Chapter 3, a large-scale, geophysical model of global deglaciation fails to match the relatively macroscopic observations of variations in sea level, which constituents of the very large and very complex assembly of micro-scale theory are at fault? Should the ocean conveyor-belt be shown to be a less than perfect conceptual insight, on which defective elements of oceanographic theory should the blame be placed?

The movement of carbon-bearing substances in the global environment can be represented by a set of twenty-six ordinary differential equations of the form of equation 4.1(a) (Enting and Lassey, 1993). Although not all that complicated, this must serve the role of a high-order computational model for present purposes. Time-series observations are available over the period 1840–1990 for the single input of carbon-bearing fossil fuel substances (u) and the single output of atmospheric CO_2 concentration (y). In simple terms, it appears that the relationship between u and y can be succinctly identified from the empirical record as two continuously stirred tank reactors (CSTRs) in series, one with a time constant of about 4.5 years, the other of some 115 years (Chapter 13). That is to say, for this *data-based* LOM, whose structure is parameterised in this manner, say $\hat{\boldsymbol{\alpha}}^{-n}(o)$ (i.e., two compartments in series with given time-constants and steady-state gains), a notional (unit) impulse of fossil fuel carbon from anthropogenic sources would provoke something of a classical bell-shaped transient response in the atmospheric CO_2 concentration before returning (over a matter of centuries) to its pre-perturbation level. If the same notional input perturbation ($\hat{u}$) is applied to the nonlinear HOM (at some equilibrium condition) and the resulting simulated response in atmospheric CO_2 recorded (as $\hat{y}$), an alternative *theory-based* LOM can similarly be identified from $[\hat{u}, \hat{y}]$ and parameterised as, say $\hat{\boldsymbol{\alpha}}^{-n}(m)$. This latter is found to be a linear system of four compartments (CSTRs) acting in parallel, one of which is a pure integrator, the others having time constants of roughly 3.5, 16, and 459 years. When distilled down to its essence in these particular terms, then, it appears the original, nonlinear model of the global carbon cycle exhibits linear properties, but the structure of this behaviour is fundamentally *not* in agreement with the conceptual structure identifiable from the observed record. What is more, the atmospheric concentration of CO_2 simulated by the HOM does not return to its pre-disturbance level once perturbed by a notional unit impulsive input of fossil fuel carbon; it remains, for ever, at a higher steady state (Chapter 13).

When set on much the *same* plane, in a relatively coarse-grained representation of the system's behaviour, the two "interpretations" of this case study, i.e., the high-level conceptual descriptions (the one based on the observations, $\hat{\boldsymbol{\alpha}}^{-n}(o)$, the other on theory, $\hat{\boldsymbol{\alpha}}^{-n}(m)$), are inconsistent. That giving four compartments acting in parallel, with four parallel paths from the emission ultimately to the atmosphere, is in fact ambiguous. The identified behaviour could equally well have been expressed as other structural arrangements, of compartments in series, in parallel, and with feedbacks (Chapter 13). Such results are thought-provoking, not least when they confront us with somewhat surprising possibilities for conceptual interpretation. In this study of the global carbon cycle, the interpretation of the observed record (in contrast to that of the basic nonlinear simulation model) suggests that once converted from its fossil form and emitted into the atmosphere, carbon does not behave in a conservative manner. In other examples, the observed correlation of

40.5-month periodicities between the atmospheric CO_2 concentration and sea-surface temperature anomaly can be represented as an almost autonomous feedback system driven solely by a purely stochastic process, i.e., white noise (Young et al., 1991). Still more curious, if it could not be reduced to an absurdity, would be to conjure up some "physical" reasoning to explain why the identified relationship between rainfall and runoff in a catchment may have a feedback loop in it (Young and Lees, 1993).

The analysis of the global carbon cycle has established a principle: a means of mapping the dominant modes of behaviour of a high-order model and the strictly identifiable modes of observed behaviour onto a common plane in which – through the lens of our enquiry – both can be brought into sharp, compatible focus. And from there, a constructive discourse might flow; one that has the possibility of corroborating or refuting the candidate high-level conceptual description (the mental model). Such discourse might perhaps provoke conception of an alternative: that the system may even be evolving (in the sense of Allen, 1990), from one arrangement of the boxes, switches and belts to another.

5.6 PARAMETERISING PARAMETRIC CHANGE

Any observed record of behaviour embraces a spectrum of fluctuations with different characteristic time constants and frequencies of oscillation. In classical time-series analysis these are usually separated by convention into the categories of high-frequency *noise*, a mid-frequency *signal* (conveying the system's predominant dynamic response), lower-frequency *seasonal fluctuations*, and still lower-frequency *trends* (Young, 1999). Quintessentially, the noise is regarded as spurious: not capable of meaningful interpretation, other than as the action of pure chance. All the power of the analysis is directed at recovery and interpretation of the mid-frequency signal.[3] Yet for our problem, this goal is not the priority. We shall not seek to subordinate extraction of the high- and low-frequency fluctuations in order to recover the mid-frequency signal. Rather, our goal is to subordinate extraction of the *signal* – and discrimination against the spurious effects of the noise – in order to explore the properties of the low-frequency components of recorded behaviour. Are we able to detect the seeds of structural change, as manifest in the low-frequency fluctuations of the parameters ($\hat{\boldsymbol{\alpha}}^{-n}(o)$) in the model of the mid-frequency signal? Could we then construct a model of these parametric fluctuations, using a set of parameters at a slightly more refined scale of resolution, i.e., $\boldsymbol{\alpha}^{-n+1}$? What is more, could we use this model of the parametric variations for extrapolation of the potential changes of structure that may occur in the future? In short, would it be possible to discern how

[3] Of course, for systems whose behaviour is believed to be chaotic the classical presumption of what is the "signal" and what the "noise" is not necessarily self-evident. As expressed by Stewart (1993): "[o]ne person's noise may be another person's signal ..."

the boxes, belts and switches of the system appear to be being re-arranged and to forecast how they might continue to re-arrange themselves in the future?

Daily, even hourly, precipitation and streamflow have been recorded over many years in some hydrological catchments. When all the sophistication is stripped away, the results of analysing these data can be distilled down to a consistent conceptual interpretation of the system's mid-frequency signal and seasonal oscillation (Jakeman and Hornberger, 1993; Jakeman et al., 1994; Young and Beven, 1994; Chapter 14). Nonlinearity of behaviour can be subsumed under the computation of an effective rainfall. This effective rainfall provokes a response in streamflow that is linear in character and comprises behaviour equivalent to a pair of parallel flowpaths passing through two stores (tanks) in the catchment. One of these tanks is of small volume, thus giving rise to what is known as "quick" flow; the other is larger and is the source of "slow" flow. When combined, the two account for the mid-frequency signal of streamflow response, so long the focus of hydrological study. A third tank, whose time constant is presumed to vary with time as a function of observed temperature, provides for the nonlinear transformation of rainfall into effective rainfall. Conceptually, this tank simulates the manner in which only a fraction of the water incident on the land surface as rainfall is destined to emerge with the streamflow. Some of it will be returned to the atmosphere via evapotranspiration, which is strongly governed by variations in temperature. Thus, to summarise, this high-level conceptual description of a catchment's behaviour has the structure of a tank followed by two tanks acting in parallel. Relatively high-frequency fluctuations attach to the element of quick flow; these are faster than the bundle of fluctuations associated with the slow flow component; and these latter are in turn fluctuations with a frequency somewhat higher than that of the seasonal oscillation in evapotranspiration. The template of the model, as it were, is not over-parameterised ($\boldsymbol{\alpha}^{-n}$), so as to suffer from a lack of identifiability. Nor is it too poorly specified to fail to "snag and snare" upon the more significant of those minor variations ($\boldsymbol{\alpha}^{-n+1}$) – on the one basic theme – that capture the richness and diversity of catchment hydrology.

As with the occurrence of a fire, other changes of vegetation cover, most obviously as a consequence of deforestation and afforestation (and more subtly as a consequence of a changing climate), will have a substantial impact on the structure of the relationship between rainfall and streamflow. We might indeed suppose that much of this impact would be focused on the first of the three tanks of our conceptual template, wherein reside all the features of evapotranspiration associated with the vegetation cover. In particular, such features are subsumed under the definitions of the model's parameters ($\boldsymbol{\alpha}^{-n}$). And these, assuming we can discriminate against the unwanted corruption of the high-frequency noise, reflect collectively the expected fluctuations in behaviour down to those of the annual cycle of seasons. Any lower-frequency change, as a consequence of changes of vegetation cover, for example, would then have to be revealed as a change in the estimates of these parameters from one year to the next. That is to say, if we were to channel the interpretation of the observed record through the "rigid" template of our high-level conceptual model, a change of structure would have to be expressed as a change in the estimates of the model's parameters $\hat{\boldsymbol{\alpha}}^{-n}(t_k)$ from one year (denoted k) to the next. Unscrambling the parts of the observed record of interest would therefore *not* be conducted entirely within the space of the observed quantities $[\boldsymbol{u},\boldsymbol{y}]$ (as it is in classical time-series analysis), but also in the model's parameter space ($\boldsymbol{\alpha}$).

In short, $\hat{\boldsymbol{\alpha}}^{-n}(t_k)$ would differ significantly from $\hat{\boldsymbol{\alpha}}^{-n}(t_{k+1})$. But beyond this, our hope is that the sequence of estimates $\{\hat{\boldsymbol{\alpha}}^{-n}(t_k)\}$, for $k = 0, 1, ..., N$, would exhibit a pattern capable of description in terms of a set of truly invariant parameters ($\boldsymbol{\alpha}^{-n+1}$) at a more refined scale of resolution, i.e.,

$$\hat{\boldsymbol{\alpha}}^{-n}(t) = g\{\boldsymbol{\alpha}^{-n+1}; \boldsymbol{u}(t); \boldsymbol{y}(t)\} \tag{5.4}$$

in which $[\boldsymbol{u},\boldsymbol{y}]$ represent observed states of the system.[4] We would thus have achieved a parameterisation of the *low-frequency* parametric variations associated with structural change.

In the Picaninny catchment in Victoria, Australia, the low-frequency impact of deforestation can be revealed – from underneath the clutter of all the higher-frequency fluctuations in the observed record – as first a contraction and then an expansion in the buffering capacity of the first tank in the model (Jakeman et al., 1994). Something of the same is apparent in the post-deforestation behaviour of the Jarrah catchment in Western Australia (Figure 5.4). Here, however, buffering capacity is shunted over the years away from the first soil-moisture and vegetation tank into the secondary parallel-path tank associated with slow flow (Jakeman et al., 1994). A pattern is emerging in both cases. What we can visualise in the abstract terms of the model's structure is akin to the more familiar crumpling, deformation, and distortion of an engineering structure (as in a building subjected to a seismic disturbance or a vehicle absorbing the shock of mechanical impact). We would like to be able to find the rules for propagating this low-frequency evolution, arrested incomplete in the present with all its ramifications still to be played out into the future.

5.7 ELASTO-PLASTIC DEFORMATION OF THE STRUCTURE

Deformation (or evolution) of the model's structure occurs because of the excessive stresses imposed upon it as attempts are made to reconcile the model – as the less refined structure in the upper plane of Figure 4.2 (Chapter 4) – with observations of the more richly composed subtleties of the real thing (set, for the purposes of illustration, in the lower plane of Figure 4.2).

[4] We note that equation 5.4 closely resembles equation 5.3. Formally, there is no difference between the way in which one would attempt to identify a model of parametric variation (as here) or reconstruct the pattern of an unknown input disturbance (as in reconstructing global patterns of greenhouse gas sources; Mulquiney et al., 1998).

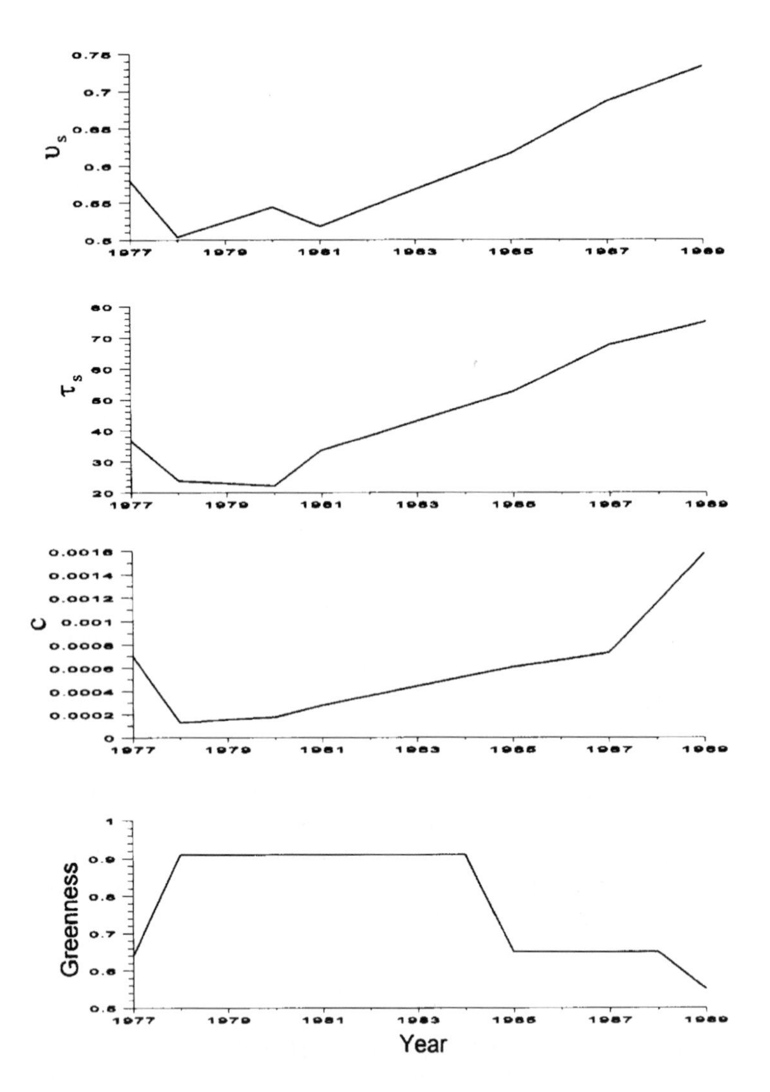

Fig. 5.4. Model parameter values and greenness index for calibration periods in Western Australia Jarrah catchment (reproduced with permission from "From data and theory to environmental model: the case of rainfall runoff", by A.J. Jakeman, D.A. Post, and M.B. Beck, *Environmetrics*, **5**(3), 297–314. © 1994 John Wiley & Sons Limited).

In particular, suppose that the "truth" of the matter is as stylised in Figure 5.5(a), a structure with five state variables and five parameters, but that our model has just four states and four parameters (Figure 5.5(b)). Further, let us suppose that significant variations with time in the state x_5^1 (of the "truth") have been caught in the observed record of past behaviour. And finally, let us imagine that the model has been embedded within a filtering algorithm, or some other scheme for tracking the predominant patterns in the observed record. The task is to reconcile the template of

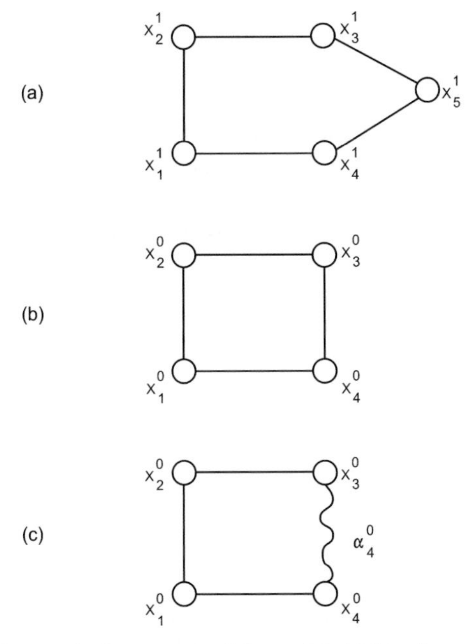

Fig. 5.5. Detection of inadequacy in the structure of the model: (a) "true" structure of behaviour (level of resolution 1); (b) approximative model structure (level of resolution 0); and (c) deformation of the approximative model's structure in respect of the constituent hypothesis associated with parameter α_4^0.

the model's *relatively* crude structure with that of the observed behaviour, which unbeknown to the analyst embraces a more richly specified structure. The rigidity of the model resides in its arrangement of the nodes, one to another. Elasticity, and thus flexibility of the structure, is vested in its branches. The capacity for the recursive estimates of the model's parameter values to be changed with time (as the structure is twisted and turned in pursuit of the trajectory of observed behaviour) ought to reveal when the template of the model has snagged upon the more significant of the subtleties of the "truth". In principle, moreover, the particular member(s) of the structure exhibiting such deformation ought to signal the precise point(s) of access to greater refinement in the resolving power of the model. In our hypothetical example, omission of the interactions of state x_5^1 with x_3^1 and x_4^1 ought to be revealed through the pattern of flexure in the estimate of parameter α_4^0 (stylised thus as the oscillatory branch of Figure 5.5(c)).

We can discern, then, how the seeds of subtle change in the structure of the model might be sought, at least in principle. For we have yet to formulate this search in a more systematic manner. We have indeed yet to explore how our model (equation 4.1),

$$\mathrm{d}\boldsymbol{x}(t)/\mathrm{d}t = \boldsymbol{f}\{\boldsymbol{x},\boldsymbol{u},\boldsymbol{\alpha},t\} + \boldsymbol{\xi}(t)$$

$$\boldsymbol{y}(t) = \boldsymbol{h}\{\boldsymbol{x},\boldsymbol{\alpha},t\} + \boldsymbol{\eta}(t)$$

might be designed with this express purpose in mind. To begin to do so, let us first recall the fundamental division between $[\boldsymbol{x},\boldsymbol{\alpha},\boldsymbol{f},\boldsymbol{h}]$, denoting that which we presume to know of the system's behaviour, and $[\boldsymbol{\xi},\boldsymbol{\eta}]$, the acknowledgement of all that which falls outside the scope, or below the resolving power, of the model. We can measure $[\boldsymbol{u},\boldsymbol{y}]$, but are not free to measure $\mathrm{d}\boldsymbol{x}(t)/\mathrm{d}t$. We can quantify the mismatch between what we think is happening and what is observed to happen, at the focal point of the difference, $\boldsymbol{\varepsilon}$, between $\boldsymbol{y}$ and $\hat{\boldsymbol{y}}$ (respectively the observed and model-based estimate of the output quantities). What we cannot know are $[\boldsymbol{\xi},\boldsymbol{\eta}]$, since these gauge the distance between the inevitable approximation of the model and the unknowable "truth" (*not* the knowable observation thereof).

Within the context of filtering theory, however, the roles of $[\boldsymbol{\xi},\boldsymbol{\eta}]$ can be approximately substituted by $\boldsymbol{\varepsilon}$ in a specific "predictive" arrangement of the model, given as (Ljung, 1979, 1987; Chapter 15),

$$\mathrm{d}\boldsymbol{x}(t)/\mathrm{d}t = \boldsymbol{f}\{\boldsymbol{x},\boldsymbol{u},\boldsymbol{\alpha},t\} + \boldsymbol{K}\boldsymbol{\varepsilon}(t) \tag{5.5a}$$

$$\boldsymbol{y}(t) = \boldsymbol{h}\{\boldsymbol{x},\boldsymbol{\alpha},t\} + \boldsymbol{\varepsilon}(t) \tag{5.5b}$$

where $\boldsymbol{h}\{\boldsymbol{x},\boldsymbol{\alpha};t\}$ is in effect the predicted output $\hat{\boldsymbol{y}}$.[5] Given the measurable quantities $[\boldsymbol{u},\boldsymbol{y}]$ and given the presumed relationships $[\boldsymbol{f},\boldsymbol{h}]$ among these and the unknown, but computable, quantities $[\boldsymbol{\varepsilon},\boldsymbol{\alpha},\boldsymbol{K}]$, the latter can duly be computed. More specifically, given $\hat{\boldsymbol{y}}$ and $\boldsymbol{y}$, $\boldsymbol{\varepsilon}$ can be determined from equation 5.5(b) and substituted into equation 5.5(a), from which a new value of $\boldsymbol{x}$, and thence $\hat{\boldsymbol{y}}$ in equation 5.5(b), can be generated for comparison with a new observation $\boldsymbol{y}$, as time proceeds. Estimates of $\boldsymbol{\alpha}$ and $\boldsymbol{K}$ can be derived as by-products of this primary calculation.

The divide between that which we presume to know and that of which we acknowledge we know little or nothing, can be drawn between the first and second terms on the right-hand sides of each of equations 5.5(a) and 5.5(b). In equation 5.5(a), in particular, $\boldsymbol{\alpha}$ and $\boldsymbol{K}$ fulfil the role of labels, capable of tagging (or tracing) the source of any inadequacy in our representation of what we believe we understand of the system's behaviour. The one ($\boldsymbol{\alpha}$) provides a trace to the source of mis-approximation in that which has been presumed to be correctly and *expressly* known (the {presumed known}). The other ($\boldsymbol{K}$) provides a trace to sources of mis-approximation arising from that which has been omitted (the {acknowledged unknown}). In the simplest of terms, one could think of this as follows. Equation 5.5 provides a mechanism of continuous prediction of the output over a short span of time ahead into the future. Adaptation and "learning" in this process proceed along two parallel paths: first, through the application of mid-course corrections, fed back

[5] The notation, in particular, for the arguments of the variables in equation 5.5, is not a strictly correct representation of the actual computational procedure that could be employed in solving this problem (see, for example, Stigter and Beck, 1994, 1995). Nevertheless, equation 5.5 serves the immediate purpose of explaining the strategic principles of an approach to solving the problem; much more will be said of this in Chapter 15.

from the continuous, high-frequency comparison of what is predicted with what is observed to occur (via ε); and second, through lower-frequency changes to the estimated values of $\boldsymbol{\alpha}$ and $\boldsymbol{K}$. In particular, as the estimates of $\boldsymbol{K}$ veer from 0.0 to 1.0, say, they signal that prediction has shifted from a reliance on what is presumed "known" to a reliance on the mismatch (ε) with what was most recently observed to have occurred. In effect, the shift identifies a move towards reliance of the predictive instrument on the "unknown" aspects of behaviour (observed, though not in accord with the theory).

But like so many things of interest to us in this monograph, such refinements of procedure – quite apart from the delicate elegance of the algorithms available and the smallness of scope of the prototype problems on which they have been set to work – are fine-sounding in principle, quite something else in practice (as we shall indeed see in Chapter 15). The likelihood of achieving success in tracing the potential cause of deformation in the candidate model's structure, as revealed in the estimated trajectories of $\boldsymbol{\alpha}$ *and* $\boldsymbol{K}$, remains to be seen (even on problems that are well known to us). Our experience suggests that access to $\boldsymbol{K}$ brings a sharper contrast to when the candidate structure succeeds and when it fails to track what has been observed, but more in the sense of crystallisation around a plastic deformation of the structure (Stigter and Beck, 1994, 1995). This is not without interest in its own right. For it gives specific substance to what is, in effect, a Popperian programme of falsifying the boldly stated, constituent hypotheses assembled in the model (Beck, 1994). Yet our present need is to detect – in the patterns of distortion (in $\hat{\boldsymbol{\alpha}}(t)$, in particular) – the rules for evolution in the structure of the model, as it catches on the more richly composed subtleties of the behaviour of the system itself.

5.8 DETECTING AND FORECASTING GROWTH IN THE SEEDS OF STRUCTURAL CHANGE

A common and readily apparent change of structure is that which occurs when a population of organisms (such as phytoplankton or bacteria), surviving at a concentration more or less below the lowest levels of detection of our observing instruments, is stimulated into rapid and substantial growth. Perhaps more probable still, traces of the behaviour of an unmonitored population – quite outside the purview of any experimental design – may burst in upon the scene of those states that *have* been selected for observation. We know, only too well, that this is precisely what is believed to have happened in the observed behaviour of the River Cam in 1972. For the first 35 days of the record the dynamics of the dissolved oxygen (DO) and biochemical oxygen demand (BOD) concentrations in the river water (the two macroscopic observed states) exhibit no hint of what is about to occur. Indeed after over twenty years of poring over these data we had come to expect just this, nothing more. And that, in many ways, epitomises the challenge of this monograph. Here we have it in a microcosm, in the case study of the Cam: is it possible to discern the unexpected (yet imminent) dislocation in the behaviour of the system; divine its causes; predict its propagation into the future; and do all of this before the event?

Scientific enquiry is motivated by dogged pursuit of the invariant. It is the unending quest for an *invariant* structure in the relationships – populated by *invariant* parameters – among quantities that essentially are expected to *vary* with time. This pillar of constancy remains a useful illusion, as the ever-receding object of the unending search. We presume the structure underlying the observed behaviour of the system is invariant and we seek the model best mimicking this invariance. Exposing deformation of the candidate model's structure, through the enforced twists and turns of its parameters (as in the foregoing), is a procedural device to be applied successively in our approach to that which is invariant. Yet for all this, the pillar of constancy is an illusion nonetheless. It has *not* been the custom to presume that our models, no matter how high their resolving power, are macroscopic approximations whose structures may be *expected* to change with time. Structural error, hence the potential for apparent structural change, hence the recognition that stasis may not prevail, are features only recently recognised and acknowledged more generally (for example, Oreskes and Belitz, 2001). We are therefore ill-equipped to shed light on the evolving trajectories of the model's parameters, $\boldsymbol{\alpha}^0(t)$, *by design*. Somewhere, between the most probable, mean pattern of behaviour and the improbable, possibly absurd, interpretation of the outliers, may reside the observed consequences of these trajectories. What we seek is evidence of actual behaviour diverging from the consensus, yet not straying into something of an absurdity. We seek insights at the *fringes* of what is reasonable.

Consider, then, Figure 5.6(a), where the "truth" has been cast as rather more complex than that of Figure 5.5(a). Perception of this truth in the observed record, however, is not invariant with time: as time passes (from period t^0 through t^1 and into t^2) we suppose in Figure 5.6(b) that that which was perhaps dominant, e.g., α_1, falls from significance, while that which was buried in the clutter of the noise, e.g., α_3, rises to a position of dominance, as reflected in the accompanying trajectories of the parameters.[6] Behaviour associated with the central frame of the structure, so to speak, is manifestly present throughout the observed record; that generated by its section to the left withers, whereas that to the right grows in importance. We would be drawn, by convention, to identifying a model structure that does not vary with time, yet is capable of capturing to the maximum extent possible the mean, most probable, or consensus pattern of behaviour *throughout* t^0, t^1, and t^2. We would be drawn, in effect, to identifying the structure of Figure 5.6(c).

What is important here are not so much small values of $|\varepsilon|$ in the engine of prediction (equation 5.5), since these suggest that observed behaviour is diverging but little from the consensus, as encapsulated in the presumed theory of $[\boldsymbol{x}, \boldsymbol{\alpha}, \boldsymbol{f}, \boldsymbol{h}]$. Nor are large values of $|\varepsilon|$ especially significant, since these imply a spurious absurdity in the data. The greatest weight should instead be placed on values of $|\varepsilon|$

[6] The fact that these parameters of the "truth" may vary with time suggests, of course, that the structure of Figure 5.6(a) is itself merely an approximation that still fails to get to the bottom of the matter; the truth in fact lies below the resolving power of what is still just an approximation to it.

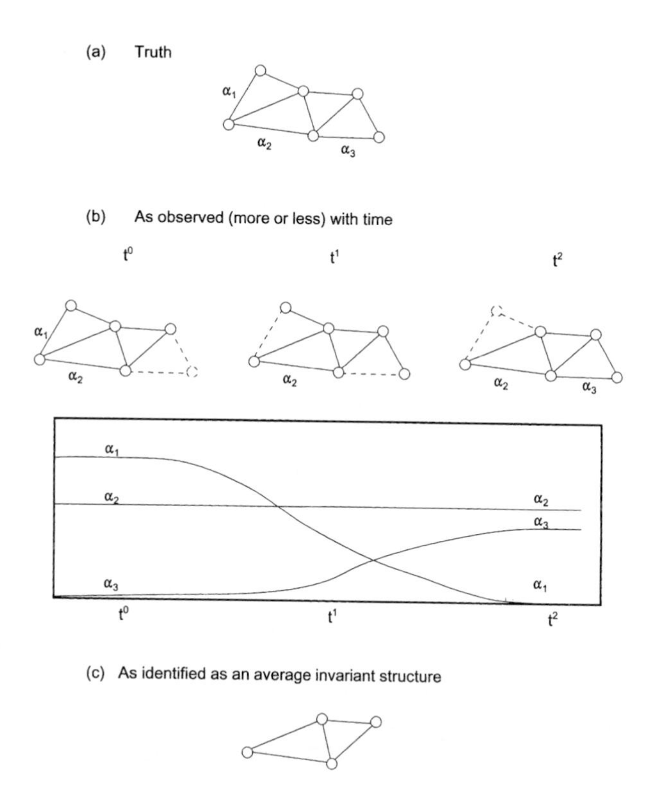

Fig. 5.6. Apparent "evolution" in the structure of the system's behaviour: (a) "true" structure of behaviour; (b) the structure as identified over several spans of time (t^0, t^1, and t^2) (dashed branches denote poorly identified parameters, as reflected in the temporal trajectories of the parameter estimates); and (c) the structure as it might be identified as an average over all the spans of time.

that are neither small nor large, in order thus to magnify patterns in the distortions of $\boldsymbol{\alpha}$ (the {presumed known}) and $\boldsymbol{K}$ (the {acknowledged unknown}) as the structure of the approximative model snags on the subtle, but significantly divergent elements of observed behaviour. Such weighting would clearly be easy to achieve, provided one had prior knowledge of the distribution of $\boldsymbol{\varepsilon}$. Errors ($\boldsymbol{\varepsilon}$) lying between one ($\boldsymbol{\sigma}_\varepsilon$) and two ($2\boldsymbol{\sigma}_\varepsilon$) standard deviations from the mean might be given the greatest weight, for example.

Suppose the most probable (consensus) interpretation of the system's observed behaviour is as portrayed in the trajectory of $\bar{y}(t)$ of Figure 5.7(a). The structure of the model underlying this reference trajectory is assumed to be invariant with time and has the form of the identical structures shown in Figures 5.7(b), (c), and (d). In principle, trajectory $\bar{y}(t)$ ought to reflect a path through the mode of the corridor of scattered observations $\boldsymbol{y}(t)$, so that we can now engage in what will be tantamount to a sensitivity analysis of the RSA of Young et al. (1978), Hornberger and Spear (1980), and Spear and Hornberger (1980). To reiterate, the issue here is one of establishing

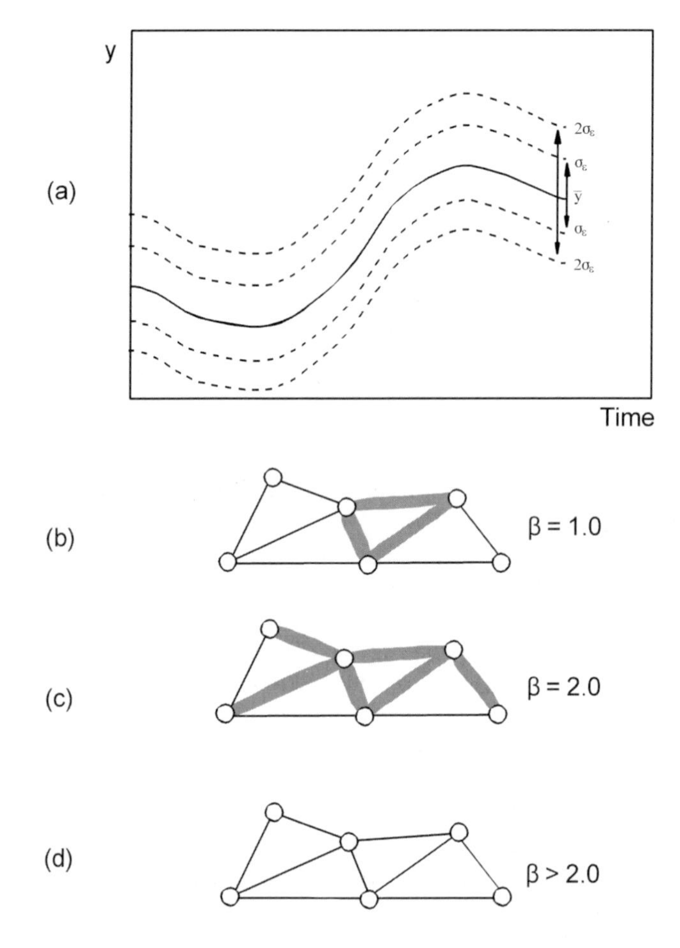

Fig. 5.7. Probing possible determinants of the structure of the system's behaviour towards the fringes of the observations: (a) "mean" and one and two standard-deviation bounds on observed behaviour; (b) key parameters (branches emphasised) when $\beta = 1.0$ (in equation 5.6); (c) key parameters (branches emphasised) when $\beta = 2.0$; and (d) key parameters (branches emphasised) when $\beta > 2.0$.

which elements of the model's structure can be identified as being key to the generation of observed behaviour that is distant from the consensus yet not too remotely positioned beyond the fringes of current knowledge. In other words, as the bounds defining the domain of "past behaviour" are expanded outwards from $\bar{y}(t)$ – for instance, as a function of the variance of the errors of mismatch between $\boldsymbol{y}$ and $\bar{\boldsymbol{y}}$ (say, $\boldsymbol{\sigma}_\varepsilon$) – candidate trajectories ($\hat{y}$) should satisfy constraints of the following form,

$$\bar{y}(t) - \beta\boldsymbol{\sigma}_\varepsilon \leq \hat{y}(t) \leq \bar{y}(t) + \beta\boldsymbol{\sigma}_\varepsilon \tag{5.6}$$

in which the breadth of the envelope of acceptable trajectories can be controlled by β. In particular, we wish to know which of the model's parameters enter into the set

$\{\boldsymbol{\alpha}^K(\beta)\}$, and which depart from it, as β passes between 1.0 and 2.0, say, and beyond (Figure 5.7(a)). According to the hypothetical results shown in Figures 5.7(b), (c), and (d), $\{\boldsymbol{\alpha}^K(b)\}$ might undergo change from $\{\alpha_5, \alpha_8, \alpha_9\}$ through $\{\alpha_4, \alpha_5, \alpha_6, \alpha_7, \alpha_8\}$ and on to the empty set $\{\cdot\}$ as β covers in turn behaviour in the vicinity of the consensus ($\beta = 1.0$), towards the fringes of the interestingly unexpected ($\beta = 2.0$), and out into the realms of the absurd ($\beta >> 2.0$). From this we might deduce that the seeds of structural change are located within whatever are the knowledge bases underpinning the elements $\{\alpha_4, \alpha_6, \alpha_7\}$ of the consensus model structure.

Whether we should set up this kind of test for different periods of time (t^0, t^1, and so on) is a refinement to which consideration might be given once the basic principle can be seen to have real potential. Whether we should, in the spirit of Beven and Binley (1992), juxtapose the properties of the subset of unworthy candidate parameterisations with those of the subset of the worst 10% of the worthy – and enquire what is it that redeems the "absurd" by bringing it back into the fold of the "interestingly unexpected" – is a further alternative. In the microcosm of the Cam case study, where the problems are recalcitrant and the procedure of analysis often lacking in elegance, whatever may work in this forensic science – of finding clues to the seeds of structural change – will be of value. There is no orthodoxy, other than that variety of approach and perspective is essential.

Phytoplankton, when in "bloom", are hardly a *subtle* feature in the dynamics of an ecological system, although their trajectory towards the start of such a condition undoubtedly is. Before the event, we can discern parametric drift in the consensus model structure; it is apparent that whatever is happening is affecting both of the observed state variables; the drift is correlated with changing patterns of streamflow and temperature, but *not* with the pattern of sunlight; there is *not* "nothing new under the sun", and the something there is can be parameterised and incorporated formally into the model; better still, as this characterisation of the fault is propagated into the future, it affords superior prediction of the imminent dislocation in actually observed behaviour. All of this will be revealed in Chapter 16. After more than twenty years of arguably excessive scrutiny of the Cam data, such was not expected.

5.9 PROBING THE SHORES OF IGNORANCE

So much, then, of that to which we aspire in this monograph lies well beyond our current capabilities: of seemingly elegant yet fragile methods that can be shown to work with some success on problems of but small proportions. And these successes have been hard won, even in the microcosm of the Cam case study.

For the host of environmental problems of any reasonable size, no such methods are available and the culture of research required to engage in developing better methods seems not to be in the ascendant (witness Woods, 1999). The prevailing lines of enquiry in these fields – of constructing "virtual worlds" of environmental systems – are headed in other directions. They presume that an ever-expanding capacity for computation will enable that which we know, i.e., $[\boldsymbol{f},\boldsymbol{h}]$ in our instruments of prediction (such as equation 5.5), to colonise more of the territory of that which we do not, [ε]. We may therefore render negligible the capacity of the unknown to contain within it any seeds of significant structural change. The edifice of our science base would then be secure, with no scope for play or vibration in its constituent members and sufficient robustness to remain untoppled by any seismic disturbances originating from the unexpected.

This is the goal we all seek. Yet it is a distant, if not ever-receding, goal. In the virtual worlds of supercomputers and teraflops it is all too easy to think of a model as just the vessel to be laden passively with all that we know. This is not what we want. Our models – including those of a very high order $[\boldsymbol{x}^{+n},\boldsymbol{\alpha}^{+n}]$ – should be designed expressly to maximise the possibility of discovering that of which we are ignorant. We wish to relegate the passive archiving of what is known and promote the active apprehension of that which is not known. We acknowledge, of course, that this task will be made vastly more difficult the more we presume to know, not least because that which remains to be discovered would *seem* to be of vanishing significance. But we do not believe our ignorance has yet been reduced to insignificance. Another look at Figure 2.2 (from Chapter 2) will remind us of the wavering science base of atmospheric chemistry (and this indeed is the supposedly "simpler" stratospheric component thereof). The resolving power of our models has not reduced our ignorance to insignificance and design for its discovery is still a most worthy exercise, towards which we shall attempt to make further progress in Chapter 15. If we are fortunate enough to have detected the peculiarity of correlated fluctuations in the atmospheric CO_2 concentration and Pacific Ocean sea-surface temperature anomaly, with a counter-intuitive period of about 40.5 months and a miniscule amplitude (Young et al., 1991), how are we to explain this? How should we assess whether this peculiarity contains the seeds of low-frequency structural change that may come to dominate the system's behaviour in the future?

In the troposphere daily variations in ozone concentration were first understood as a balance between ozone release in the photolysis of nitric oxide (NO_2) to nitrous oxide (NO) and its consumption in the reverse process of oxidising NO to NO_2. The emission of volatile organic chemicals (VOCs) from urban sources upset the equilibrium of this understanding. For these chemicals, given the prerequisite of a hydroxyl radical, could also influence the "forward" conversion of NO to NO_2, which could then exacerbate the unwanted excessive presence of ozone, through the "reverse" conversion of NO_2 to NO. And thus the presence of VOCs could turn ever faster the cycle – from NO to NO_2 and back to NO – spinning off a higher net balance of unhealthy ozone (at least on sunny days). This high-level conceptual description was in turn toppled by the perception that peroxides were driven forward to hydroxyl radicals, themselves the prerequisites of the VOC accelerator, by the forward conversion of NO to NO_2. The VOC radicals reduced in the same forward conversion were then found to be capable of oxidation back to a peroxide form, which could in turn participate in inducing further revolutions in the $NO \rightarrow NO_2 \rightarrow NO$ cycle. The hydroxyl radicals could be taken out of the system by themselves taking out NO_2, terminating it in the form of nitric acid and applying the brake to the cycle out of which the ozone was being spun. It is as though we were to conceive of all this as a grand clockwork mechanism in which we sought to control the position of the hour hand with observed knowledge of neither the speeds of rotation of the minute and second hands nor the evolving (and nonlinear) gearings among the spinning wheels of the seconds, minutes and hours.

How, then, could we maximise the probability of realising that the "hour-hand" of the system is not quite where it is expected to be or, more probably (in the case of tropospheric ozone concentration), that it is in the correct position but kept there by incorrectly conceived inner cycles turning at the wrong speed? How could we determine which elements of the structure of the mechanism are the source of this anomaly? Could we determine where teeth are shearing away from the minute- and second-hand wheels, so to speak, causing a progressive slippage in the ratios between the cogs in the mechanism? And how would we tell whether this slippage is growing in amplitude? In short, how could we tell whether an innocuous chemical species in the real system, masquerading as a kinetic rate parameter in the counterpart virtual system, is in truth not that insignificant, or may inadvertently be made very significant as a result of actions taken to change the position of the ozone concentration?

We can observe the real thing, but only in macroscopic terms, at the apex of the hierarchy of descriptions in Figure 5.8(a). We can observe the behaviour of its virtual

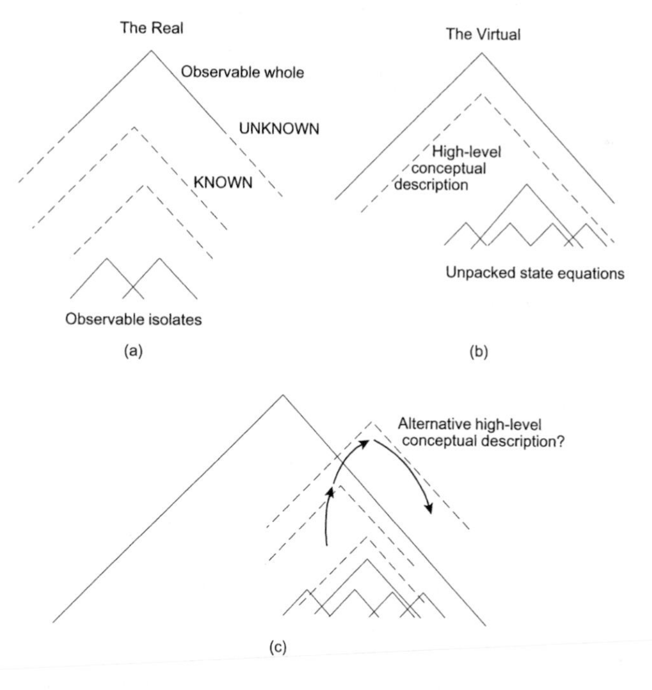

Fig. 5.8. Hierarchies of system description: (a) the observed behaviour of the prototype – what is "unknown" lies outside the pyramid, what is "known" lies within (what is "observable" is drawn as a continuous line and what is not, as a dashed line; (b) the simulated virtual reality; and (c) unpacking of state variable equations may provoke an alternative high-level conceptual description.

counterpart and we can unpack this into its myriad microscopic, component *parts* (Figure 5.8(b)); yet these parts of the virtual will still not be as fine-grained as may be the significant parts of the real. We can observe the behaviour of some of these parts of the real thing – cut away and isolated from the web of interactions in which they sit – but not therefore as the part would behave within the whole. In smog-chamber experiments, for instance, models of the behaviour of the chemical interactions at the air-wall interface must be proposed in order to reconstruct a description of the chemistry of the free atmosphere. Yet we are presumably no more sure of the one than the other. Understanding either the real or the virtual *whole* is achieved when information about these entities is assimilated into an appropriate mental model, a high-level conceptual description, towards the apex of the virtual system (Figure 5.8(b)). We do not comprehend the behaviour of environmental systems, nor apprehend the imminence of their possible departure from the expected, through the minutiae of their elemental parts. Somehow, the focus of our lens of enquiry and the juxtaposition of the resulting images must be articulated around and within the two (the real and the virtual of Figures 5.8(a) and (b)).

If the parts of the equations of state of the virtual system are unpacked from the whole, for example, one can perceive that removal of the hydroxyl radicals from the $NO \rightarrow NO_2 \rightarrow NO$ cycle, as *observed* in the real world through the concentration of nitric acid (HNO_3), is not the predominant (*virtual*) source of HNO_3 (Jang et al., 1995a,b; and Jeffries, 1995).[7] Contrary to expectations, as much as half the *observed* HNO_3 concentration derives from nitrate radicals in the *virtual* world of the model. Interpretation of the "dial" of nitric acid concentration as a gauge of the "speed" of hydroxyl "rotation" around the $NO \rightarrow NO_2 \rightarrow NO$ cycle may therefore be quite misleading. We have begun thus to grasp the essence of tropospheric ozone production through the macroscopic, conceptual device of the position, velocity, and acceleration of a set of coupled chemical wheels (Chapter 17). We have gained essential insight into the circulation of heat in the ocean through the persuasive image of a conveyor-belt. We can appreciate the implications of climate change through the concept of the greenhouse. These are crude analogs without the filigree of microscopic detail; and they are successful precisely because of this. They will doubtless inspire the formulation and testing of further hypotheses. They may determine the future direction of the associated science. Yet they may also become mind-sets: marvellously simple concepts that are increasingly difficult to dislodge in the search for the unexpected.

We must ask not merely, then, to what defective, microscopic part is the failure of the macroscopic whole due? We have also to ask: is the "grand clockwork mechanism" itself plain inappropriate? Towards what alternative high-level conceptual description, perhaps outside the current image, could we re-assemble the

---

[7] This device, of opening up and examining one to another the trajectories of the several constituent "forces" acting on the balance of the given, single state variable, has been used before, in studies of both Lake Erie (Di Toro and Connolly, 1980) and Lake Huron (Di Toro and Matystik, 1980).

unpacked knowledge of the virtual, microscopic parts – as in Figure 5.8(c)? And having done so, if we reflect back down the new thinking released by this alternative simplified, conceptual insight, would this imply an unpacking of our knowledge into a novel set of microscopic parts? Our challenge lies not just in interpreting observed behaviour through the vehicle of the virtual counterpart in order to generate understanding and explanation of the past. It lies instead in probing the real *and* the virtual to divine what might happen in the future.

5.10 VISUALISATION AND LEARNING

We are, of course, tilting at immensely complex and largely intractable problems. Enormous volumes of numbers, representing the simulated [$\boldsymbol{x}$] and observed [$\boldsymbol{y}$] states of the environment over substantial tracts of space and time, must be assimilated, with the express purpose of provoking the apprehension that something is not quite as it was expected to be. The eye travels over the *relative* juxtaposition of $\boldsymbol{x}$ and $\boldsymbol{y}$ in a graph. It will spot convergence and divergence between the two. As it scans across the many elements of these vectors it may even spot the illusion of a convergence that is in truth the cancellation of correlated divergences elsewhere. But without details of the magnitude of any divergence *relative* to measures of the uncertainty attaching to both $\boldsymbol{x}$ and $\boldsymbol{y}$ no conclusion on the significance of the anomaly may be drawn. More numbers must be available – at the least on the variances of the uncertainties attaching to $\boldsymbol{x}$ and $\boldsymbol{y}$ – before we may proclaim the existence of an anomaly. But anomalies, once identified, are not rectified in the state (output) space. Their cause lies within the body of the model and there, in the parameter space [$\boldsymbol{\alpha}$], still more numbers will have to be available: over possibly equally large tracts of space and time and of means *relative* to variances.

Thirty years ago we could have pasted the walls and ceiling of our office space with sheets of computer graph paper, the condensate of many of these numbers; we could have spun our heads around this space; then, somehow, absorbing the black and white patterns of the graphical forms, have come to a conclusion on the anomaly; and perhaps too, have glimpsed a hint of the corner in which the source of the anomaly might lie. Today, we may keep our heads still and move the information around us, as it were. We need only look at the screen of the computer and observe an animated film of the flexure, stresses, and strains wrought in the structure of the model as it seeks to pass through the hoops and hurdles of the observations of the past and the hopes and fears for the future (such as those of Figure 3.2, from Chapter 3). We can exploit movement and colour. The complex state-parameter network of the model of Figure 4.1 (from Chapter 4) might be composed in the form of a three-dimensional structure – as the analog of either a building or an organic molecule – and the colour and shape of its nodes and branches designed to change with the evolving patterns of divergence, means, and variances of all the host of numbers attaching to the attributes of $\boldsymbol{x}$, $\boldsymbol{y}$, and $\boldsymbol{\alpha}$. And thus, we might conclude, reaching some feared pattern of behaviour in a lake may turn on the red vibrating distortions in the branches of the model's structure surrounding the sector where insect larvae (*Chaoborus*) prey upon copepods in our best – yet still defective – contemporary understanding of a pelagic foodweb.

5.11 FORESIGHT FOR ACTION

If we were to be successful in moving towards our goal, what should we do with all the foresight thus generated? If we had the means of deriving the feared dislocations, of establishing their plausibility, of identifying an imminent change of structure, or of discovering what we know we do not know – at the earliest possible juncture – what actions should we take? If one had this wealth of foresight, one must surely then be obliged to propose a course of action, *not* further contemplation. Should we not have come up with some novel framework in which to formulate and implement policy?

Not necessarily, for it is hard to better the concept of adaptive control. We have an understanding of the behaviour of our environment, yet recognise this understanding is bound to be incomplete. We can contemplate the outcomes of our actions and sense thus the direction in which things might be headed, notably with the assistance of the computer. Actions imply reactions. Since we do not have complete understanding, yet the observation of action and reaction implies an opportunity for learning, management may be designed not only to direct the actual but also to explore the space of possible futures for a system's behaviour. As in the adaptive control of engineering systems (Åström and Wittenmark, 1989), or procedures for the design of robust decisions (Rosenhead, 1989), or adaptive environmental assessment and management (Holling, 1978), we appreciate the *dual* function of pragmatic actions: to probe that which is unknown and to control that which is observable. We appreciate too the *recursive* nature of adaptive management: of action; observation; adaptation of action; further observation; further adaptation of contemplated action; and so on. This is so eminently sensible and commonplace. It is very hard, in fact, to better the concept of adaptive control.

What, then, will we have achieved, in terms of taking practical steps into the future? Perhaps the answer will be little more than a change of outlook, or a subtle change of emphasis: from a focus on "action" within adaptive management to one on "apprehension" – that something is amiss, not just with the system itself but with our understanding thereof. How must the science base change, we ask, not merely our policies, in order for all our imagined futures to be made reachable? We suggest it does not have to be the case that all of our science and interpretation of the record of the past should be funnelled into a singular, rigid, crystalline, immutable structure of the model, from which predictions of future consequences are drawn, albeit covered in an equally crystalline pattern of uncertainty. Stasis does not rule supreme. We acknowledge that changes in the structures of our models (the science base) will occur; that we shall have to reorient our computational analyses to accommodate this; and accept that contemplation of the future may indeed point back to re-interpretation of the path by which the observed past has led up to the present. Good actions will be ones that are robust in the face of this absence of stasis – which does not imply scientific anarchy, or some form of "freestyle science" – and capable of probing where, in particular, the forecast of future behaviour may be bumping up against our ignorance.

REFERENCES

Allen, P.M., 1990. Evolution, innovation, and economics. In: *Technical Change and Economic Theory* (G. Dosi, C. Freeman, R. Nelson, G. Silverberg and L. Soete, eds.). Pinter, London, pp. 95–119.

Åström, K.J. and Wittenmark, B., 1989. *Adaptive Control*. Addison Wesley, New York.

Beck, M.B., 1987. Water quality modeling: a review of the analysis of uncertainty. *Water Resour. Res.*, **23**(8), 1393–1442.

Beck, M.B., 1994. Understanding uncertain environmental systems. In: *Predictability and Nonlinear Modelling in Natural Sciences and Economics* (J. Grasman and G. van Straten, eds.). Kluwer, Dordrecht, pp. 294–311.

Beck, M.B., Kleissen, F.M. and Wheater, H.S., 1990. Identifying flow paths in models of surface water acidification. *Rev. Geophys.*, **28**(2), 207–230.

Beven, K.J., 1989. Changing ideas in hydrology – the case of physically-based models. *J. Hydrol.*, **105**, 157–172.

Beven, K.J., 1993. Prophecy, reality and uncertainty in distributed hydrological modelling. *Adv. Water Resour.*, **16**, 41–51.

Beven, K.J., 1996. Equifinality and uncertainty in geomorphological modelling. In: *The Scientific Nature of Geomorphology* (B.L. Rhoades and C.E. Thorn, eds.). Wiley, Chichester, pp. 289–313.

Beven, K.J. and Binley, A.M., 1992. The future of distributed models: model calibration and predictive uncertainty. *Hydrol. Processes*, **6**, 279–298.

Beven, K.J., Lamb, R., Quinn, P.F., Romanowicz, R. and Freer, J., 1995. TOPMODEL. In: *Computer Models of Watershed Hydrology* (V.P. Singh, ed.). Water Resources Publications, Highlands Ranch, Colorado, pp. 627–668.

Clark, W.C., 1986. Sustainable development of the biosphere: themes for a research program. In: *Sustainable Development of the Biosphere* (W.C. Clark and R.E. Munn, eds.). Cambridge University Press, Cambridge, pp. 5–48.

Davis, P.A., Price, L.L., Wahi, K.K., Goodrich, M.T., Gallegos, D.P., Bonano, E.J. and Guzowksi, R.V., 1990. Components of an overall performance assessment methodology, Report NUREG/CR-5256, SAND88-3020, Sandia National Laboratories, Albuquerque, New Mexico.

DeAngelis, D.L. and Cushman, R.M., 1990. Potential application of models in forecasting the effects of climate changes on fisheries. *Trans. Am. Fish. Soc.*, **119**, 224–239.

Di Toro, D.M. and Connolly, J.P. (1980), Mathematical models of water quality in large lakes, Part 2: Lake Erie, Report EPA-600/3-80-065, US Environmental Protection Agency, Environmental Research Laboratory, Duluth, Minnesota.

Di Toro, D.M. and Matystik, W.F., 1980. Mathematical models of water quality in large lakes, Part 1: Lake Huron and Saginaw Bay, Report EPA-600/3-80-056, US Environmental Protection Agency, Environmental Research Laboratory, Duluth, Minnesota.

Enting, I.G. and Lassey, K.R., 1993. Projections of future CO_2, Technical Paper 27, Division of Atmospheric Research, CSIRO, Melbourne, Australia.

Fedra, K., van Straten, G. and Beck, M.B., 1981. Uncertainty and arbitrariness in ecosystems modelling: a lake modelling example. *Ecol. Modelling*, **13**, 87–110.

Holling, C.S. (ed.), 1978. *Adaptive Environmental Assessment and Management*. Wiley, Chichester.

Holling, C.S., 1986. The resilience of terrestrial ecosystems: local surprise and global change. In: *Sustainable Development of the Biosphere* (W.C. Clark and R.E. Munn, eds.). Cambridge University Press, Cambridge, pp. 292–320.

Hornberger, G.M. and Spear, R.C., 1980. Eutrophication in Peel Inlet, I, Problem-defining behaviour and a mathematical model for the phosphorus scenario. *Water Res.*, **14**, 29–42.

Hornberger, G.M., Beven, K.J., Cosby, B.J. and Sappington, D.E., 1985. Shenandoah watershed study: calibration of a topography-based, variable contributing area model to a small forested catchment. *Water Resour. Res.*, **21**, 1841–1850.

Jakeman, A.J. and Hornberger, G.M., 1993. How much complexity is warranted in a rainfall-runoff model? *Water Resour. Res.*, **29**(8), 2637–2649.

Jakeman, A.J., Post, D.A. and Beck, M.B., 1994. From data and theory to environmental model: the case of rainfall runoff. *Environmetrics*, **5**(3), 297–314.

Jang, J.C., Jeffries, H.E. and Tonnesen, G.S., 1995a. Sensitivity of ozone to model grid resolution – II. Detailed process analysis for ozone chemistry. *Atmos. Environ.*, **29**, 3101–3114.

Jang, J.C., Jeffries, H.E., Byun, D. and Pleim, J.E., 1995b. Sensitivity of ozone to model grid resolution – I. Application of high-resolution regional acid deposition model. *Atmos. Environ.*, **29**, 3085–3100.

Jeffries, H.E., 1995. Photochemical air pollution. In: *Composition, Chemistry, and Climate of the Atmosphere* (H.B. Singh, ed.). van Nostrand-Reinhold, New York, pp. 308–348.

Kleijnen, J.P.C., 1987. *Statistical Tools for Simulation Practitioners*. Marcel Dekker, New York.

Kleissen, F.M., Beck, M.B. and Wheater, H.S., 1990. The identifiability of conceptual hydrochemical models. *Water Resour. Res.*, **26**(12), 2979–2992.

Leggett, J.K., 1996. The threats of climate change: a dozen reasons for concern. In: *Climate Change and the Financial Sector: The Emerging Threat, the Solar Solution* (J.K. Leggett, ed.). Gerling Akademie Verlag, Munich, pp. 27–57.

Ljung, L., 1979. Asymptotic behaviour of the extended Kalman filter as a parameter estimator. *IEEE Trans. Automatic Control*, **24**, 36–50.

Ljung, L., 1987. *System Identification: Theory for the User*. Prentice-Hall, Englewood Cliffs, New Jersey.

MacFarlane, A.G.J., 1990. Interactive computing: a revolutionary medium for teaching and design. *Comput. Control Eng. J.*, **1**(4), 149–158.

Mulquiney, J.E., Taylor, J.A., Jakeman, A.J., Norton, J.P. and Prinn, R.G., 1998. A new inverse method for trace gas flux estimation: 2. Application to tropospheric $CFCL_3$ fluxes. *J. Geophys. Res.*, **103**, 1429–1442.

Oreskes, N., 1999. *The Rejection of Continental Drift: Theory and Method in American Earth Science*. Oxford University Press, New York.

Oreskes, N. and Belitz, K., 2001. Philosophical issues in model assessment. In: *Model Validation: Perspectives in Hydrological Science* (M.G. Anderson and P.D. Bates, eds.). Wiley, Chichester, pp. 23–41.

Osidele, O.O. and Beck, M.B., 2001. Analysis of uncertainty in model predictions for Lake Lanier, Georgia. In: *Proceedings AWRA Annual Spring Specialty Conference* (J.J. Warwick, ed.). TPS-01-1, American Water Resources Association, Middleburg, Virginia, pp. 133–137.

Pearl, J., 1988. *Probabilistic Reasoning in Intelligent Systems: Networks of Plausible Inference*. Morgan-Kaufmann, San Mateo, California.

Rosenhead, J., 1989. Robustness analysis: keeping your options open. In: *Rational Analysis for a Problematic World* (J. Rosenhead, ed.). Wiley, Chichester, pp 193–218.

Sandberg, A., 1976. *The Limits to Democratic Planning*. Liberforlag, Stockholm.

Schaffer, S., 1993. Comets and the world's end. In: *Predicting the Future* (L. Howe and A. Wain, eds.). Cambridge University Press, Cambridge, pp. 52–76.

Schellnhuber, H.J., 1999. 'Earth System' analysis and the Second Copernican Revolution. *Nature*, **402** (Supplement, 2 December), C19–C23.

Spear, R.C. and Hornberger, G.M., 1980. Eutrophication in Peel Inlet, II, Identification of critical uncertainties via generalised sensitivity analysis. *Water Res.*, **14**, 43–49.

Spear, R.C., Grieb, T.M. and Shang, N., 1994. Parameter uncertainty and interaction in complex environmental models. *Water Resour. Res.*, **30**(11), 3159–3169.

Stewart, I., 1993. Chaos. In: *Predicting the Future* (L. Howe and A. Wain, eds.). Cambridge University Press, Cambridge, pp. 24–51.

Stigter, J.D. and Beck, M.B., 1994. A new approach to the identification of model structure. *Environmetrics*, **5**(3), 315–333.

Stigter, J.D. and Beck, M.B., 1995. Model structure identification: development and assessment of a recursive prediction error algorithm, Working Paper WP-95-105, International Institute for Applied Systems Analysis, Laxenburg, Austria.

Thompson, M., 1997. Cultural theory and integrated assessment. *Environ. Modeling Assess.*, **2**, 139–150.

Varis, O., 1995. Belief networks for modelling and assessment of environmental change. *Environmetrics*, **6**, 439–444.

Varis, O. and Kuikka, S., 1999. Learning Bayesian decision analysis by doing: lessons from environmental and natural resources management. *Ecol. Modelling*, **119**, 177–195.

Young, P.C., 1978. General theory of modelling badly defined systems. In: *Modelling, Identification and Control in Environmental Systems* (G.C. Vansteenkiste, ed.). North-Holland, Amsterdam, pp. 103–135.

Weaver, A.J., 1995. Driving the ocean conveyor. *Nature*, **378** (9 November), 135–136.

Woods, J.D., 1999. Virtual ecology. In: *Highlights in Environmental Research* (B.J. Mason, ed.), Imperial College Press, London.

Young, P.C., 1998. Data-based mechanistic modelling of environmental, ecological, economic, and engineering systems. *Environ. Modelling and Software*, **13**, 105–122.

Young, P.C., 1999. Nonstationary time series analysis and forecasting. *Prog. Environ. Sci.*, **1**, 3–48.

Young, P.C. and Beven, K.J., 1994. Data-based mechanistic modelling and the rainfall-flow non-linearity. *Environmetrics*, **5**(3), pp. 335–363.

Young, P.C. and Lees, M., 1993. The active mixing volume: a new concept in modelling environmental systems. In: *Statistics and the Environment* (V. Barnett and R. Turkman, eds.). Wiley, Chichester, UK, pp. 3–43.

Young, P.C., Hornberger, G.M. and Spear, R.C., 1978. Modelling badly defined systems – some further thoughts. In: *Proceedings SIMSIG Conference*, Australian National University, Canberra, pp. 24–32.

Young, P.C., Ng, C.N., Lane, K. and Parker, D., 1991. Recursive forecasting, smoothing and seasonal adjustment of non-stationary environmental data. *J. Forecasting*, **10**(1&2), 57–89.

Environmental Foresight and Models: A Manifesto
M.B. Beck (editor)

CHAPTER 6

Epilogue

M.B. Beck

6.1 AN EVOLUTIONARY APPROACH IN FORM

Approaches must by definition lead somewhere. Yet we are all dressed up, it might seem, with nowhere to go. We have no all-encompassing case study by which to demonstrate the superiority of the *whole* of what we are proposing. We might be able to conceive of something that would fit this bill: of a piece of the environment in which those holding a stake have a legitimate fear for its future; where, within our lifetimes, changes may occur that will threaten the integrity of the system we cherish; and a sense that such changes may seem to turn more upon the imperfections of our science base than on what we can well imagine – from experience of the workings of our societies, industries and economies – will induce stress and damage within that environment. To be sure, the *parts* of our approach have been born of practical experience: from work on the big and grand – the global carbon cycle – to study of the small and humble, such as the idiosyncrasies of algal population dynamics in freshwater river systems.

In retrospect, what has motivated our enquiry – or, in truth, what has been beyond our grasp, somewhere creatively within the domain of Weinberg's trans-science (Weinberg, 1972) – has been this (again in the words of Allen (1990)):

> The key issue is centred on the passage between detailed microscopic complexity of the real world, which clearly can evolve, and any aggregate macroscopic 'model' of this.
>
> But such a 'machine' [model] is only capable of 'functioning', not of evolving. It cannot restructure itself or insert new cogs and wheels, while reality can!

In the beginning (in the mid-1960s) there was much underlying theory available and waiting to be articulated through the instrument of a computational model, a backlog as it were, of classical fluid mechanics and chemistry. Navier, Stokes, St Venant, Darcy, Richardson, Manning, Taylor are the names of scientists past – before the modern era of the computer – whose contributions mark this corpus of theory. The "computerised" model was a passive device for mobilising theory in the service of generating predictions. These devices have become ever more faithful mimics of the underlying theory. Theory, however, hardly stands still. Changing elements in both mathematics and computational power enable new articulations of nascent theories, themselves having fewer restrictive assumptions than the idealisations that have gone before (as in the application of topological ideas to fluid mechanics; Ricca and Berger, 1996). Thus, as the theoretical front moves outwards, covering more of the unknown ever more succinctly, computational models will follow, some distance behind. From time to time a uniquely useful visualisation of the results from the trailing computational front will have the potential to accelerate the outward expansion of the leading front of theory (perhaps in the manner of the foregoing discussion of Figure 5.8 from Chapter 5). Emulation of this principle is exemplified across large tracts of Geophysics. On occasion its impact is dramatic, as in the insight of the ocean behaving as a conveyor-belt – an elementary concept formed from finally clearing the backlog of theory (waiting to be articulated) in the classical fluid mechanics of physical oceanography. The predominant image of how the programme proceeds is therefore that of a "mediating model" (Morton, 1993) chasing after the outward migration of theory, *not* observation.

As theory climbs up from the physics of the Earth's environment (through its chemistry and biology) so to reach its ecology, there must be compromise. If for no other reason, our vehicles of articulation simply cannot hold all this theory within them. Approximations must be parameterised and the action of chance invoked. Chance, in any case, seems intrinsic to adaptation over time of the biological (model) parameters that summarise a sub-structure of behaviour in which random mutation and natural selection are believed to be dominant (Woods, 1999). Yet this has not been the principal route by which uncertainty has leaked into our otherwise deterministic view of the environment. Our common experience has been of models failing to mimic not theory but the observed behaviour of the real thing. In the knowledge that compromise has left the model with parameters that are not precisely knowable quantities – not strictly deducible *a priori* from the universal constants of Nature – it has become a commonplace to view them not as constants but as random variables. The rigidity of determinism can usefully be replaced by model structures that are subject to random vibrational play in the parametric links between the fixed nodes of the state variables. We know only too well that our observations are sparse and uncertain; that the concepts articulated within our models are only approximately correct; that reconciliation of the one with the other will leave the model with a characteristic form of random vibrational play in its structure; and that this retrospective signature of uncertainty should be carried forward with predictions, into the making of decisions wherein the risk of failure has replaced the certainties of costs and benefits of successful action alone. Yet in such a programme we do not seek to explain, or enquire into the nature of, the random vibrational play in the model's structure.

In preparing the monograph we have laboured to re-define some of the problems of forecasting environmental change and thence to approach them from yet another angle. What we see now is this. The action of chance, when integrated up and over the various levels of the fine-grained sub-structure of behaviour to the relatively crude, macroscopic approximations of the model, may seem more like a partly determined, partly erratic low-frequency drift than an entirely random structural "vibration". We shall put aside belief in the parameters being imprecisely known yet "in truth" constant. We expect them *not* to be as random variables but as stochastic processes: as random walks, or perhaps more figuratively, random crawls. The orientation of the nodes (the state variables) of the model's structure, one to another, is still fixed; but now we expect this structure to evolve through varying shapes in time. And our principal task, if we – the public, the policy-makers, the stakeholders – can conjure up what may happen in the future, is to impose that imagined shape on the structure of the model and *quintessentially* to discern on what elements of parametric drift this change from the retrospective shape of the past to the prospective shape of the future may hinge. What is more, in the vicinity of the present, we wish to know whether there are even the faintest of indications of the potential for this structural change. This is not "evolutionary" in the sense to which Allen's programme aspires. The change of structure is imposed by our hands – through our imagination of the future – not by the metaphorical hand of God. There is no semblance in our programme of the model evolving, of its restructuring itself or, of its own accord, inserting new cogs and wheels.

6.2 PARAMETRIC CHANGE AS THE AGENT OF CONTROL

Hope in the restoration of our environment to some "equilibrium", away from which we believe it has been pushed in the past, is instinctive, no matter how ephemeral the concept of a fixed equilibrium may now seem. The remedy has been to remove the stress that induced the strain in the first place; and we have rather presumed this prescription to be quite predictable in its effect.

In contrast, if we can imagine some feared future, which has not been encountered in the past (and is perhaps therefore feared for this very reason), what must happen – specifically within the terms of the defective conceptual and theoretical bases encoded in our models – for this feared future to come to pass in the long term? What *trajectories* of parametric change with time ($\boldsymbol{\alpha}(t)$), we might ask, would be required to transfer the system from its recent past and present states to some target future state? This, however, is but a slightly eccentric reformulation of the central question of control theory: of establishing what sequence of input control actions ($\boldsymbol{u}(t)$) would be required to transfer the system from a present to a future state. What would happen, then, if we were to transpose the roles of $\boldsymbol{u}$ and $\boldsymbol{\alpha}$ in the conventional problem of control system synthesis? In other words, could systematic solutions be derived from the following formal statement of our problem, of determining the temporal course of changes in the model parameters that will transfer the system from its past and present observed states ($\bar{x}_p(t^-)$) to some specified future state, say $\bar{x}_f$ at time t^+? That is to say (and we shall begin to explore this possibility in Chapter 18):

Given:

$$\mathrm{d}\boldsymbol{x}(t)/\mathrm{d}t = \boldsymbol{f}\{\boldsymbol{x},\boldsymbol{u},\boldsymbol{\alpha},t\} + \xi(t) \qquad (6.1)$$

and given:

$$\boldsymbol{x}(t) = \bar{\boldsymbol{x}}_p(t^-) \text{ for } t = t^-; \boldsymbol{x}(t) = \bar{\boldsymbol{x}}_f(t^+) \text{ for } t = t^+; \boldsymbol{u}(t) \text{ for all } t$$

Determine what $\boldsymbol{\alpha}(t)$ for $t^- \leq t \leq t^+$ will transfer $\boldsymbol{x}(t)$ from $\bar{\boldsymbol{x}}_p(t^-)$ to $\bar{\boldsymbol{x}}_f(t^+)$.

Here, as previously, t^- and t^+ denote somewhat liberally interpreted blocks of time in the past and future respectively. More specifically, we might suppose that over the span of the past and the future the state of the system must pass through a sequence of prescribed domains of the state space (that is, as observed in the past and as feared for the future, much as in Figure 3.2, from Chapter 3).

Instead of finding, for example, the sequence of changes in input fossil fuel flux ($\boldsymbol{u}$) necessary for stabilising atmospheric carbon dioxide concentration by the year 2300 (as in Figure 2.3, from Chapter 2), we seek those changes in the parameterisation ($\boldsymbol{\alpha}$) of the associated global carbon cycle model that would achieve much the same targeted response. In particular, our search is for clues as to whether, for instance, the necessary parametric change involves a modest upturn, as opposed to a modest downturn, in the values of the parameters in the vicinity of the present and recent past. Does it appear, for example, that the coefficient of downward vertical diffusion of organic detrital carbon into the bottom-most layers of the ocean must increase, or decrease, for the purposes of reaching a stable atmospheric CO_2 concentration? If we could secure our projection of the required structural change sufficiently far into the future, this ought to define what subtle flexure in the model's parameter values would be necessary – in the here and now – for departure on the path towards that target future. Small differences in current flexure may determine the imminence, or otherwise, of any potential dislocations in the behaviour of the system.

In the limit, imagining a complete transposition in the roles of $\boldsymbol{u}$ and $\boldsymbol{\alpha}$ would bring the sharpest of contrasts to this discussion: between the classical purpose of control system synthesis and the questions that are now of interest to us. For we are concerned primarily, albeit not exclusively, with *low*-frequency fluctuations associated in some way with the model's parameters ($\boldsymbol{\alpha}$) in the presence of any suitable set of assumptions about the *high*-frequency flutter in the inputs ($\boldsymbol{u}$). The vagaries of economic and regulatory cycles of our manipulations of the environment – rather like those over seven centuries in the Davos valley of Switzerland (as chronicled in Price and Thompson, 1997) – might then be seen as mere random vibrational play towards the margins of the strategic sweep of potential change in the structure of our science base. In turn, as the model's parameters describe a smooth, slow trajectory of change, its state variables ($\boldsymbol{x}$) may exhibit behaviour that slips, in a sequence of dislocations, through qualitatively quite different ensembles of relatively high-frequency fluctuations (such as, for example, those arising from the multiple attractors, catastrophes, subharmonics, and chaos of the relatively simple model system of nutrients, phytoplankton, zooplankton, and fish described by Doveri et al. (1993)).

Smooth, slow, dynamical change in the model's parameters has thus been made the focus of our problem's formulation, while account of the faster flutter in the input variables has been relegated to the sidelines.

6.3 INCLINED TO SURVIVE (OR OTHERWISE)

As on so many occasions throughout this discussion, we have imagined ourselves to be as though standing on the threshold of the future, contemplating what salient trajectory of parametric change might bring the environment to a target domain of the state space. We have imagined too what might be the system's final destination, in order to explore those changes of model structure required to reach it, hence to obtain specific preferences in searching through the empirical evidence of the past for the seeds of these possibly imminent changes. This, we know, has been ambitious. There is a simpler and more basic issue to be resolved, perhaps, before embarking on such sophisticated lines of enquiry. Given our typically pitifully narrow empirical basis – as if we had observed the nature of but a single tile within the mosaic of the whole – is the reachability of the feared future more probable, as a consequence of the conditioning of this observed experience, than it would seem otherwise, in the maximally uninformed condition of having no such experience to which to make reference? To what extent does the angle at which the observed record of the past is set predispose the predicted hypothetical trajectories of the system's future behaviour to enter the target domain? In a sense, if we could generate the path of the system's movement through the state space from past to future in the absence of any observations, by how much would behaviour projected into the future be deflected from this path by the impact of the observed record on the otherwise hypothetical course through the past?

For this one last time we shall again presume that the feared future patterns of the system's behaviour can be derived in some manner, thereby cutting the state space apart into a (feared/not feared) binary classification. We further presume that an observed record of input and output variables $[\boldsymbol{u}, \boldsymbol{y}]$ is available for a short span of time in the past (t^-) relative to the horizon of the forecast into the future (t^+). Given these assumptions, we wish to develop methods of analysis that will allow exploration of the extent to which reconciliation of the model with the observations of the past can be shown to make the entry of future behaviour, $\boldsymbol{x}(t^+)$, into a defined domain of the state space, more or less likely than for the maximally uninformed case of having no access to an empirical record of past behaviour. For convenience we shall refer to this as a problem of *comparative forecasting*, thereby emphasizing that we are not seeking an absolute probability of some feared future behaviour being attained. On the contrary, outcomes of the analysis would be of a relative form: that the loss of a biological species from an ecosystem, given sparse past observations, is more probable than the forecast of such an occurrence on the basis of the model alone (in which no reference is made to past observed behaviour). If this were in fact the outcome of the comparative forecast, it would imply there is evidence in the empirical record of the first signs of a trajectory towards an undesired future pattern of behaviour.

Yet again, then, one would be drawn to investigate the properties of the record of the past conditioned upon prior assessment of the attainability of some target future behaviour. And what now follows is how we might resume our attack on this central problem, from yet another angle.

In the long run an environmental system may be assumed to have just two radically different futures, denoted "survival" and "collapse" for brevity. The essential task is to detect – *now*, and relative to the available empirical evidence – towards which of these two futures behaviour may be tending. We assume that the system operates over an infinite time horizon composed of relatively short intervals (days, months, or years), indexed by k, $k+1$, and so on. Further, we make the assumption that if the system's behaviour in consecutive periods is quantified (by observation or prediction), it is possible to determine categorically whether behaviour has deteriorated ($s_k = 1$) or improved ($s_k = 0$) from k to $k+1$. Here deterioration or improvement signals movement of the system's state ($\boldsymbol{x}$) along a trajectory towards, or away from, respectively, the condition of eventual "collapse". We have, therefore, an infinite sequence of (0/1) state transitions characterising the evolution of the system's behaviour, an encoding procedure used elsewhere for describing the dynamics of natural systems (Kryazhimskii et al., 1996). This is the essential element in our concept of comparative forecasting, which we shall introduce more fully in Chapter 19.

Let us suppose, for instance, we have a model of the conventional form of equation 6.1 for a phytoplankton-fish system, such as that of Doveri et al. (1993). For the sake of argument, let us further suppose that the several parameters of this model (maximum specific growth rate and predation rate "constants", for example) are to be treated not as random variables but as stochastic processes, $\boldsymbol{\alpha}(t)$, as in Finney et al. (1982) and Kremer (1983). These parameters may, however, be assumed to be subject to an upper bound on their rate of temporal variation. A feared future, comprising collapse (C), may also be defined: as a domain of the state space into which one would not wish to have the system enter (a combination of, say, an annual peak phytoplankton biomass above a critical level and an annual peak fish biomass below a critical level). The behaviour of the system into the future is simulated, for a sample of admissible candidate parameter trajectories and appropriate assumptions – brought in from the sidelines, as we have put it – about future input disturbances of the system (including, possibly, policy actions). For simplicity, we could assume that over the relatively short interval k the parameters are held constant, as $\boldsymbol{\alpha}_k$, but that they vary from one interval to the next in some manner, i.e., we have the sequence of parametric variations $(\ldots,\boldsymbol{\alpha}_k,\ \boldsymbol{\alpha}_{k+1},\ldots)$. Should any of the candidate future state trajectories enter the feared domain of collapse, the time at which they first do so is noted and this is the shortest time to collapse from the given period. More formally, if the current time is denoted as k, the shortest time to collapse from this period, over all admissible sequences of $(\boldsymbol{\alpha}_{k+1}, \boldsymbol{\alpha}_{k+2},...)$, can be denoted r_k. Should the time to collapse from the next period, r_{k+1}, computed in the same manner, be less than r_k, the transition from the state of the system at k to the state at $k+1$ will be deemed to be "bad" (1), and vice versa for "good" (0). Thus we shall have generated, in principle, our sequence $\boldsymbol{s}$, i.e., $(...,s_k,s_{k+1},...)$, of (0/1) state transitions characterising the evolution of the system's behaviour.

Now we assert that a strong domination of bad transitions (of deterioration) in a relatively long sequence of consecutive transitions will cause the system to undergo *only* bad transitions in the future, as it subsequently moves *irreversibly* towards the target state of collapse. It is possible, therefore, to introduce and define a parameter representing a threshold in the degree of domination of bad transitions (within the total set of transitions), such that when the proportion of bad transitions in the sequence exceeds this threshold the system becomes locked on to an irreversible path towards collapse.[1] Conversely, degrees of domination by good transitions, and the accompanying introduction of a complementary threshold parameter, can be defined for the case in which the system may eventually lock on to a path towards survival. Thus, as soon as either of these thresholds is passed, the behaviour of the system ceases to be stochastic; there is only deterministic change towards collapse or survival.

Our concept of comparative forecasting can thus be put to work as follows. If we were to make no reference to the record of the past, that is, to $[\boldsymbol{u}, \boldsymbol{y}]$ over t^-, except for an assumed point of departure (or initial condition for the computation, back in the past), an unconditional probability of collapse, $P(C)$, can be generated. The label of "unconditional" (or "prior") connotes here a property of the maximally uninformed bundle of hypothetical trajectories of the system's behaviour through the state space from past to future. If an empirical record of the system's past transitions, $\boldsymbol{s}(t^-)$, say, has been observed, and the model then reconciled with this record (giving a different bundle of trajectories in the past) the conditional probability of collapse, $P\{C\,|\,\boldsymbol{s}(t^-)\}$, can be found. We can therefore identify the "tendency to collapse" of the system, if the observed record increases the probability of collapse relative to the unconditional probability, i.e., $P\{C\,|\,\boldsymbol{s}(t^-)\} > P(C)$, where no reference is made in the latter to the system's past observed behaviour. Otherwise, "no tendency to collapse" is identified; and likewise the status of the survival-oriented forecasts can be classified (a condition of both a "tendency to collapse" and a "tendency to survival" is not

[1] This conception of our problem has been motivated by the work of Arthur et al. (1987, 1988), who used the device of an urn scheme in their study of path-dependent stochastic processes in relation to technological innovation and "lock-on" (urn-type path-dependent processes have also been investigated in the context of evolutionary games; Smale (1980), and Fudenberg and Kreps (1993)). Technically, the proposed use of the concept herein may require analysis to be applied to a sequence of state transitions of finite length, in fact a changing memory of transitions.

possible in the concept of comparative forecasting). Most importantly, in these comparative forecasts the precise values of the probabilities are not important, merely that which is the greater will be of significance.

We may have glimpsed thus the manner in which the observed record of the past will have impinged upon our understanding of the behaviour of environmental systems – encapsulated in a model – deflecting the course of its otherwise purely hypothetical projection into the future, as we imagine, believe, or fear it to be. Above all, perhaps, our next step should neither lower the probability of reaching some desired future several steps ahead nor raise the probability of seeing our fears come to pass. What is more, with each passing step these hopes and fears about the ever-receding horizon are themselves evolving.

REFERENCES

Allen, P.M., 1990. Evolution, innovation, and economics. In: *Technical Change and Economic Theory* (G. Dosi, C. Freeman, R. Nelson, G. Silverberg and L. Soete, eds.). Pinter, London, pp. 95–119.

Arthur, W.B., Ermoliev, Y.M. and Kaniovski, Y.M., 1987. Adaptive growth processes modeled by urn schemes. *Kibernetika*, **6**, 49–57 (in Russian); *Cybernetics*, **23**, 779–789 (English translation).

Arthur, W.B., Ermoliev, Y.M. and Kaniovski, Y.M., 1988. Nonlinear adaptive processes of growth with general increments: attainable and unattainable components of terminal sets, Working Paper WP-88-86, International Institute for Applied Systems Analysis, Laxenburg, Austria.

Doveri, F., Scheffer, M., Rinaldi, S., Muratori, S. and Kuznetsov, Y., 1993. Seasonality and chaos in plankton-fish model. *Theoret. Popul. Biol.*, **43**(2), 159–183.

Finney, B.A., Bowles, D.S. and Windham, M.P., 1982. Random differential equations in river water quality modeling. *Water Resour. Res.*, **18**(1), 122–134.

Fudenberg, D. and Kreps, D., 1993. Learning mixed equilibria. *Games and Econ. Behav.*, **5**, 320–367.

Kremer, J.N., 1983. Ecological implications of parameter uncertainty in stochastic simulation. *Ecol. Modelling*, **18**, 187–207.

Kryazhimskii, A.V., Maksimov, V.I., Solovyov, A.A. and Chentsov, A.G., 1996. On a probabilistic approach to a quantitative description of dynamics of natural processes. *Problems of Control and Informatics*, **1**(2), 23–41 (in Russian).

Morton, A., 1993. Mathematical models: questions of trustworthiness. *Br. J. Philos. Sci.*, **44**, 659–674.

Price, M.F. and Thompson, M., 1997. The complex life: human land uses in mountain ecosystems. *Global Ecol. Biogeogr. Lett.*, **6**, 77–90.

Ricca, R.L. and Berger, M.A., 1996. Topological ideas and fluid mechanics. *Phys. Today*, **49**(12), December, 28–34.

Smale, S., 1980. The prisoner's dilemma and dynamical systems associated with non-cooperative games. *Econometrica*, **48**(7), 1617–1633.

Weinberg, A.M., 1972. Science and trans-science. *Minerva*, **X**(2), 209–222.

Woods, J.D., 1999. Virtual ecology. In: *Highlights in Environmental Research* (B.J. Mason, ed.). Imperial College Press, London.

PART II
CASE HISTORIES

Environmental Foresight and Models: A Manifesto
M.B. Beck (editor)

CHAPTER 7

Lake Erie and Evolving Issues of the Quality of its Water

W.M. Schertzer and D.C.L. Lam

7.1 INTRODUCTION

Lake Erie has been the subject of intensive investigation for over four decades, from which inevitably a wealth of physical, chemical, biological, and socio-economic data has accumulated. At least four key issues of concern to water quality, some overlapping with others in time, can be identified. Between 1960 and 1970 investigations concentrated on understanding processes and impacts related to eutrophication (the basic elements of which have already been recounted in Chapter 2). From 1970 onwards, through to the early 1980s, research dealt with the development of anoxia in the hypolimnion of the central basin, as a function of both eutrophication and variations in weather patterns. During the 1980s and 1990s interest grew in these latter, and their association with changes of climate, so that they became the focus of attention, although not exclusively, for there was also considerable concern at the introduction of exotic species into the lake's ecosystem. From a rather different perspective – indeed arguing that we have been largely unaware of this particular perspective – it has been said that Erie exhibits a classical form of "two-attractor catastrophe cusp" behaviour (Kay et al., 1999). A pelagic attractor prevails when the majority of the energy of solar irradiance is utilised in the warm upper waters of the epilimnion, while a benthic attractor describes behaviour when the majority of this incoming energy is utilised towards the bottom of the lake (in the hypolimnion and sediments). In the longer sweep of things, these authors suggest that Lake Erie is at present in an intermediate state, as its behaviour begins to "flip" from the pelagic to the benthic attractor. We, from our perspective, will be equally strategic in outlook. Reflection on the extent to which the use of models and forecasting have helped to clarify understanding, inform policy, diagnose the apparent failure of policy, and identify issues perhaps just beyond the horizon, is thus chronicled in this chapter, albeit with the benefit of very substantial hindsight.

7.2 EUTROPHICATION

To begin with the facts, Lake Erie (Figure 7.1) is an international water resource shared by Canada and the USA. Compared with the other Laurentian Great Lakes, Erie is the shallowest with a mean depth of 18.7 m. It has a surface area of 25,320 km^2 and a volume of 470 km^3. The lake has three distinct basins. The western basin is 10 m deep, the central basin is 25 m deep and the eastern basin is 64 m deep. Major hydrological inflow comes from the Detroit River (mean monthly inflow of 5320 m^3/s) and the major outflow is through the Niagara River (mean monthly outflow of 5740 m^3/s) with a water residence time of 7.2 years. Climate and hydrological characteristics are reviewed in Schertzer and Croley (1999). Located within a densely populated region of the Great Lakes, Erie receives chemical loadings from municipal, industrial, agricultural and atmospheric sources, all of which affect the lake's water quality. These issues of water quality have been a predominant area of scientific research on Lake Erie for over 40 years.

7.2.1 Emerging Perception of a Problem

As early as 1916, the International Joint Commission (IJC) had expressed concern over worsening water quality in Lake Erie. Direct discharge of raw sewage had been identified as a major source of pollution in the Detroit and Niagara Rivers and had generated concern over municipal water supplies. Oxygen deficiency was first

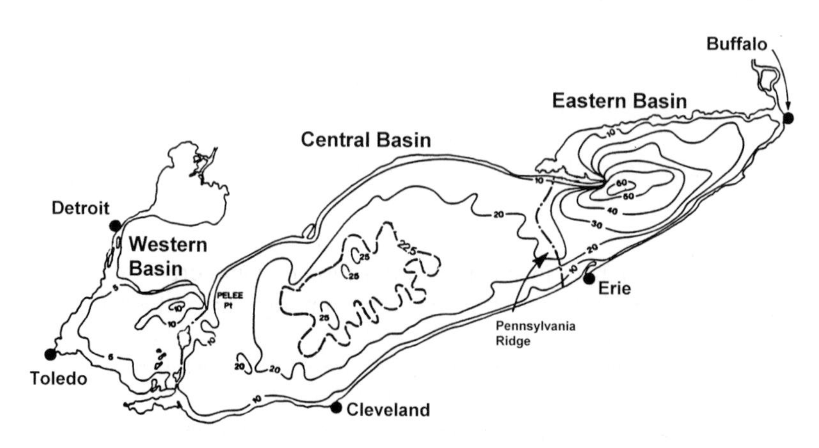

Fig. 7.1. Lake Erie bathymetry and major basins.

observed in the central basin of Lake Erie in 1929 (Fish et al., 1960), gradually increasing during the period 1950–1970 over the central basin hypolimnion (Carr, 1962; Dobson and Gilbertson, 1972). A critical oxygen depletion rate of 3.0 mg/L/month appeared to determine whether much of the central basin hypolimnion would become deoxygenated before fall overturn; Erie had achieved such a value by about 1960. The extent of anoxia in the central basin can be significant. For example, a hypolimnion oxygen deficiency of 2 mg/L or less was estimated to extend over an area of 6735 km^2 (FWPCA, 1968). Assuming negligible oxygen entering the bottom water after thermal stratification and an average hypolimnion depth of about 4 metres, the central basin of Lake Erie became mesotrophic around 1940 and was becoming increasingly eutrophic (Dobson and Gilbertson, 1972).

As already outlined in Chapter 2, accelerated eutrophication can present serious difficulties for the preservation of ecological well-being. Large amounts of organic matter produced by algae in the (upper) epilimnion settles to the hypolimnion (colder, lower water) layer where decomposition depletes oxygen reserves. Under severely eutrophied conditions, as have occurred in Lake Erie, hypolimnion anoxia (dissolved oxygen lower than a specified threshold concentration, e.g., less than 1 mg/L) can occur and have serious ramifications. For example, depletion of summer-time oxygen reserves can adversely affect lake biota and the health of fisheries. It can lead to odorous and unpalatable water unsuitable for potable water supply. In addition, decaying masses of algae and heavy cladophora growth can foul bathing beaches, block water intakes and reduce shoreline property values. It is also well known that anoxia may cause the release of nutrients from the sediments into the water column, possibly resulting in a cyclical process of self-fertilisation of the lake (Mortimer, 1941).

7.2.2 Delayed Response: A Public Call to Arms

While these early water quality observations (before the 1960s) identified some major sources of pollution and the oxygen deficiency in the central basin, public concern over worsening water quality seemed to lag behind the initial scientific findings. In 1964, the IJC was requested by the scientific community to assess the extent and causes of the pollution problem. Based on limited scientific research largely conducted in the 1960s, the IJC reported in 1969 that there was advanced eutrophication in the western part of Lake Erie (IJC, 1969). It was postulated that worsening water quality was the result of increased nutrient inputs, especially phosphorus. Trophic status was also observed to be changing in the other basins of Lake Erie. Dramatic changes in species composition of commercial fish catches and bacterial pollution of nearshore water was noted among other deleterious effects.

7.2.3 Understanding the System: the {Known} **versus** *Our {Ignorance}*

By the early 1970s the scientific community had begun to investigate eutrophication in greater detail, implementing basic research to increase knowledge and to establish the factors responsible for the eutrophic conditions and observed anoxia within the lake. Project Hypo (Burns and Ross, 1972a), was a co-ordinated Canada-USA investigation of the oxygen depletion problem in the central basin of Lake Erie. A number of important findings were made, creating a firmer basis on which eventually to construct policies, controls and targets for Erie, not to mention early developments in modelling the system.

Physical limnological studies provided information on fundamental thermal and circulation characteristics of the lake. For example, Blanton and Winklhofer (1972) described the characteristics of the central basin thermocline and noted that in upwelling zones nutrient-enriched anoxic hypolimnion water became mixed with epilimnion water and largely accounted for profuse algal blooms. Upwelling, intrusion of east basin water, and other advection and vertical entrainment mechanisms were suggested as playing a role in volume changes in the central basin hypolimnion. Sediment oxygen demand (SOD) experiments tended to confirm the long-term trend of increasing rate of depletion, indicating that SOD may be an important factor in the measured hypolimnion oxygen depletion rate. Murthy (1972) suggested that a strong thermocline 2–3 m thick with a steep temperature gradient could effectively act as an almost impenetrable barrier for upward vertical mixing.

Chemical and biological measurements were also seen as integral to understanding eutrophication and the anoxia problem. For example, Lucas and Thomas (1972) found that oxygen produced by the photosynthetic activity of algae on the lake bottom offset the SOD during part of the day. Furthermore, algae and chlorophyta deposited on the bottom were the major source of organic carbon utilized in the consumption of oxygen as a result of bacterial activity at the water–sediment interface. Oxygen demand from physical and chemical phenomena was found to be quite strong in comparison. Menon et al. (1972) found that intense bacterial decomposition of organic matter at the sediment–water interface resulted in the formation of reduced compounds which also depleted oxygen in the hypolimnion. These compounds were subsequently oxidized by chemo-autotrophic bacteria with further oxygen losses, pointing thus to bacteria as principal factors in depleting oxygen in the hypolimnion of the central basin of Lake Erie.

A simple mass balance model was developed by Burns and Ross (1972b), called the Sequential Mesolimnion Erosion Model, to investigate chemical budgets under conditions of changing hypolimnion volume. The model did not explicitly include complex thermocline dynamics determined from the physical limnological investigations; rather, hypolimnion volume and heat transfers estimated between lake-wide surveys made it possible to demonstrate the probable chemical pathways. The study found that approximately 88% of the oxygen uptake was due to bacterial degradation of algal sedimentation with 12% of the oxygen being taken up in the oxidation of reduced metallic species. In 1970, approximately 33% of the hypolimnion was anoxic by August 25, and completely anoxic conditions were present by September 23. A crucial hypothesis from this study was that a reduction of nutrients (phosphorus) would lead to a corresponding decrease in the oxygen depletion rate and that the lake would soon return to an acceptable state, if phosphorus inputs to the lake were decreased such that oxygenated conditions were maintained all year in the lake.

Project Hypo concluded then with the following:

> Phosphorus input to Lake Erie must be reduced immediately; if this is done, a quick improvement of the lake can be expected; if it is not done, the rate of deterioration of the lake will be much greater than it has been in recent years.

The basis for this conclusion was a reliance on an intensive investigation of Erie's physical and biochemical observations. The investigation was by no means exhaustive but it allowed a fundamental understanding of some of the mechanisms involved in eutrophication of the lake.

7.2.4 Synthesis: A Low-Order Model (LOM)

Vollenweider (1968) was the first to attempt to unify all of the nutrient budget data available in the literature. Based on phosphorus budgets for a number of lakes, plots of areal total phosphorus loading against mean depth demonstrated that lakes could be categorized according to their degree of eutrophy into oligotrophic, mesotrophic or eutrophic lakes (Figure 7.2). An arbitrary set of "permissible" and "critical" loading levels provided a guide for determining "dangerous" levels leading to eutrophic conditions. Vollenweider cautioned that the relationship was approximate and that other parameters such as flushing rate, internal loading and lake dimensions may also be important factors to consider (Dillon, 1974).

Vollenweider's empirical loading-depth relationship considered the loading characteristic for a lake with a measure of its dimension. This lower-order model

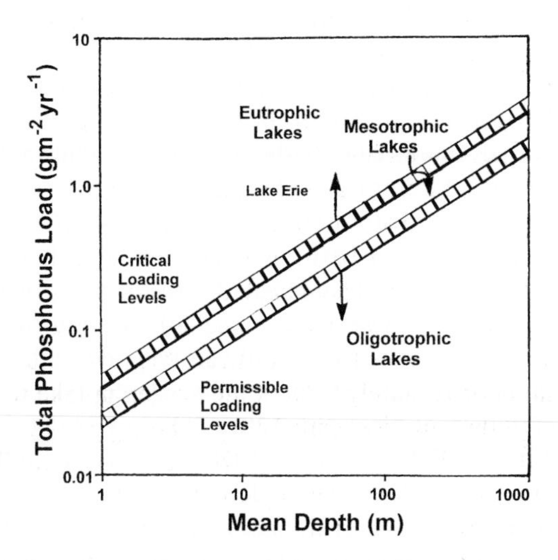

Fig. 7.2. Total phosphorus *versus* mean depth relationship (modified from Vollenweider, 1968).

(LOM) – in the context of the monograph's discussion – provided a rationale for the removal of phosphorus from sewage treatment effluents in the Great Lakes basin. It was recognized that growth, reproduction, biomass or yield of a community could be limited if phosphorus were in short supply (Thomas et al., 1980). Compared with carbon, nitrogen, oxygen, hydrogen and sulphur, phosphorus was considered by far the most common limiting factor in the Great Lakes. Without explicit description of the system dynamics, this LOM allowed scientists to "forecast" that a reduction in phosphorus loading would have a measurable impact on the reduction of the eutrophic conditions in a lake.

7.2.5 Formulating Policy: Great Lakes Water Quality Agreement (GLWQA)

Although the complete system dynamics and ecosystem interrelationships were not well understood, the measurements of the 1960s and early 1970s, and in particular those from Project Hypo, provided a good understanding of some of the physical, chemical and biological interrelationships pertinent to Erie's eutrophication issue. These observations, combined with Vollenweider's simple loading-depth relationships, the limiting hypotheses based on phosphorus, and the "forecast" that reducing phosphorus loading should reverse the trend to eutrophic conditions, all led to the introduction of policies to reduce the effect of eutrophication by controlling the phosphorus loading.

The IJC advised the Governments of Canada and USA in 1970 to enter into an agreement on an integrated program of phosphorus control to include: (1) a reduction of phosphorus in detergents; (2) reduction of phosphorus in municipal and industrial waste effluents, and (3) reductions in phosphorus discharges from agriculture. Recommendations included developing a timetable for reductions and the regulation of any new uses of phosphorus which could result in appreciable additions to the lakes.

The 1972 Great Lakes Water Quality Agreement (GLWQA) set target phosphorus loads based on the available information at that time. They were designed to represent the loads that would exist if the apparently achievable goal of 1 mg/L phosphorus in effluents of treatment plants in excess of 1 mgd were met. An important objective was the restoration of year-round aerobic conditions in the hypolimnetic water of the central basin of Lake Erie. Another policy goal was to restrict the growth of nuisance algae through raising the dissolved oxygen concentration in the central basin (Thomas et al., 1980). Phosphorus and chlorophyll-a values in the eastern and western basins would correspond to the loads necessary to achieve the desired central basin water quality. Based (significantly) on model output, a maximum allowable total phosphorus load of 11,000 MT/yr was recommended to maintain summertime hypolimnetic aerobic conditions. An average hypolimnion dissolved oxygen concentration of 4 mg/L was recommended to maintain aerobic conditions in the central basin hypolimnion and to reduce the areal extent of hypolimnion anoxia. Maintenance of aerobic conditions was postulated as being sufficient to prevent the large flux of phosphorus from the sediment.

7.2.6 And the Auditing of Policy

Evaluation of the phosphorus control policies included monitoring point source phosphorus loads and measurement of in-lake conditions using lake-wide surveillance cruises. By 1976 the IJC reported that Lake Erie had begun to show signs of recovery (Sugarman, 1980). However, target loads were not achieved and near-shore areas had not recovered as expected. There was a realisation that target loads were probably not achievable with the then current programs. Surveillance of the lake indicated that even with reduced phosphorus loadings, the problem of central basin hypolimnion anoxia was still present in most years. Given the non-achievement of the target loads and following re-evaluation of initial policy measures, continued monitoring (surveillance) of water quality was clearly warranted. From a scientific viewpoint, increased research on the system dynamics was required to understand the mechanisms making Erie's central basin continue to experience hypolimnion anoxia in some years, even when phosphorus loads were decreasing.

Other major studies, for example, the Upper Lakes Reference Group Studies (UGLRS) and Pollution from Land Use Reference Group (PLUARG) raised the question of the importance of available phosphorus *versus* its unavailable forms. It was estimated that only 40% of the particulate phosphorus in tributary discharges was of the available form. This was significant because the early models had erred in considering *all* phosphorus to be equally responsible for primary production. If it were, this would suggest that municipal sewage treatment plant phosphorus would have to be reduced to zero in order to effect satisfactory lake-wide phosphorus and oxygen conditions. We see here then a good example of the potential problems of setting policy without adequate knowledge of the system dynamics.

7.2.7 Back to the Bench

Considerable effort was thus made to predict the degree of eutrophication in lakes, but largely using LOMs. Dillon (1974) reported cases in which lakes (with high flushing rates) did not fit Vollenweider's original phosphorus loading-mean depth categorisation, which had been used as a rationale for removal of phosphorus from sewage treatment plant effluents. In response Vollenweider (1975) modified his simple phosphorus loading (L_p) *versus* mean depth (z) relationship to include the mean residence time of water (τ_w) (Figure 7.3), a refinement having a more realistic representation with a basis in theoretical considerations of nutrient modelling. The refined model (L_p *versus* z/τ_w) was more precise and allowed for predicting the trophic status that would result following both changes in population and reduction of phosphorus in sewage effluent (Dillon, 1974). This re-evaluation demonstrated the need for model verification over a large set of data to ensure the applicability and validity of forecasts.

Vollenweider's loading plot was based on steady-state solutions of a simplified mass balance model for a mixed reactor. The limitation of such models is that estimates of in-lake conditions can be derived only after equilibrium has been

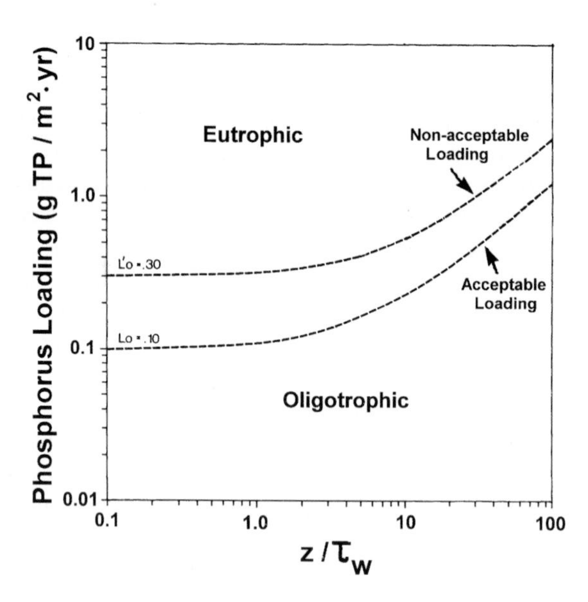

Fig. 7.3. Total phosphorus load *versus* mean depth/water residence time (modified from Vollenweider, 1976).

reached. They could not provide information on system response times to a change in phosphorus load (Bierman, 1980). A mass balance model involving a dynamic calculation for total phosphorus concentration with time, given phosphorus load, volume, depth and hydraulic retention time was developed by Chapra (1977), which indeed allowed computation of system response times to a change in phosphorus loads. Chapra reported that to conform to the policy of restoring year-round aerobic conditions, phosphorus concentrations must be controlled so that the amount of organic matter produced and settling to the bottom, with subsequent decomposition, should not be so large as to consume the oxygen supply of the bottom waters. Chapra's (1979) model indicated that a springtime average phosphorus concentration not exceeding 10 μg/L under normal meteorological and limnological conditions, should result in no appreciable area of the central basin becoming anaerobic during summer stratification.

7.2.8 Towards Higher-Order Models (HOMs)

By the late 1970s and early 1980s, the basic concern in water quality policy for Lake Erie was this: even with a significant reduction in phosphorus loading approaching the target loads, anoxia in the central basin hypolimnion was still being observed. The scientific community responded with increased research on the system's dynamics and the development of dynamic state–space models. These more sophisticated models were used to forecast the response of lake total phosphorus concentrations with reduced loadings and the consequent effects on selected lake system components. In effect, they were used to test and forecast lake response to policies imposed to alleviate water quality problems. Table 7.1 provides a summary of the principal characteristics of several of the Great Lakes eutrophication models. We provide here only a brief account of some of the important findings from three of these models, followed below by a more in-depth discussion of the model of Lam et al. (1987a).

HOM development undoubtedly played an important role in building an understanding of the mechanisms contributing to observed eutrophication and the anoxic

Table 7.1

Summary of principal model characteristics of several Great Lakes eutrophication models (modified from Bierman et al., 1980)

Model characteristics	Model[1]					
	1	2	3	4	5	6
Model Type						
Deterministic	–	–	*	*	*	*
Empirical	*	*	–	–	–	–
Time Dependence						
Dynamic	–	*	*	*	*	*
Steady-state	*	–	–	–	–	–
Spatial Segmentation						
None	*	–	–	–	–	–
Horizontal	–	*	*	*	*	*
Vertical	–	–	*	*	*	*
Input Requirements						
External loads	*	*	*	*	*	*
Depth	*	*	*	*	*	*
Volume	*	*	*	*	*	*
Hydraulic detention time	*	*	–	–	*	*
Temperature (in lake)	–	–	*	*	*	*
Light	–	–	*	*	*	*
Water circulation rates	*	*	*	–	*	*
Sediment nutrient release rates	–	–	–	–	*	*
Primary Variables						
Phosphorus	*	*	*	*	*	*
Nitrogen	–	–	*	*	–	–
Silicon	–	–	–	*	–	–
Total forms only	*	*	–	–	–	–
Available/unavailable forms	–	–	*	*	*	*
Secondary Variables						
Chlorophyll	*	*	*	–	–	–
Diatom/non-diatom chlorophyll	–	–	–	*	–	–
Multi-class biomass	–	–	–	–	–	–
Zooplankton	–	–	*	*	–	–
Dissolved oxygen	–	–	–	*	–	*

[1] Model: (1) Vollenweider (1975); (2) Chapra (1977); (3) Thomann et al. (1975); (4) Di Toro and Connolly (1980); (5) Simons et al. (1979); (6) Lam et al. (1987a)

conditions. For example, Thomann (1976) developed a two-layered, horizontally well-mixed model for Lake Ontario for simulating phytoplankton and nutrient distribution. This model demonstrated that spring growth and peak chlorophyll-a concentrations were related primarily to increased light and temperature and that phosphorus limitation essentially controlled the spring peak (Thomann and Segna, 1980). Di Toro (1980) extended Thomann's (1976) model to predict the impact of reducing phosphorus loads on the annual cycle of phytoplankton production, its relation to the supply of nutrients, and the effect on dissolved oxygen concentrations in Lake Erie. Results indicated that a reduction in total phosphorus loading of between 9,500 and 12,000 MT/year would be sufficient to eliminate anoxia in the central basin of Lake Erie – the range attributed to uncertainty surrounding the long-term sediment response. Simons et al. (1979), evaluated the ability of water quality models to simulate seasonal variations, long-term trends and spatial variations with a two-component phosphorus model of Lake Ontario, which included only organic phosphorus and soluble reactive phosphorus. Equally satisfactory simulations were obtained with a variety of parameterisations, regardless of conditions of annual periodicity imposed on the solution. Simons and Lam (1980) demonstrated that seasonal verification studies of dynamic models by themselves are not sufficient to diagnose the utility of such models for predicting long-term trends. They concluded that the uncertainty surrounding the formulation of sedimentation and nutrient regeneration, in conjunction with the sensitivity of models to assumptions regarding dynamic balance between lake concentrations and nutrient loadings, undermined the long-term predictive capability of these HOMs.

7.2.9 The Keys to Insight and Understanding

Unfortunately, in spite of the significant reduction in phosphorus loading, which was approaching target values, in the range of 10,000–12,000 MT/yr by 1982 in Figure 7.4 (Fraser, 1987), anoxia was still observed to occur in the central basin hypolimnion in most years and the spatial extent was observed to vary from year to year. One hypothesis for the continued occurrence of central basin hypolimnion anoxia was that the weather would have a controlling influence on the lake's thermal structure (Schertzer et al., 1987) and a shallow hypolimnion appeared to be associated with the development of anoxia (Schertzer and Lam, 1991). Although there was consideration of some of the system's physical dynamics in the development of the earlier water quality models (e.g., diffusion across a fixed-level thermocline), emphasis on the biochemical aspects resulted in no detailed attempt to model the dynamical changes in the lake's thermal structure, even though early investigations (e.g. Blanton and Winklhofer, 1972; Burns and Ross, 1972b) had clearly implicated the importance of stratification in the occurrence of anoxic conditions.

Lam et al. (1987a) set out, therefore, to investigate this hypothesis by developing a modelling framework incorporating the influence of weather variability into the simulation of water quality. Their model has three basic sub-models, to specify the source terms (heat fluxes and hydrology), vertical temperature structure, and

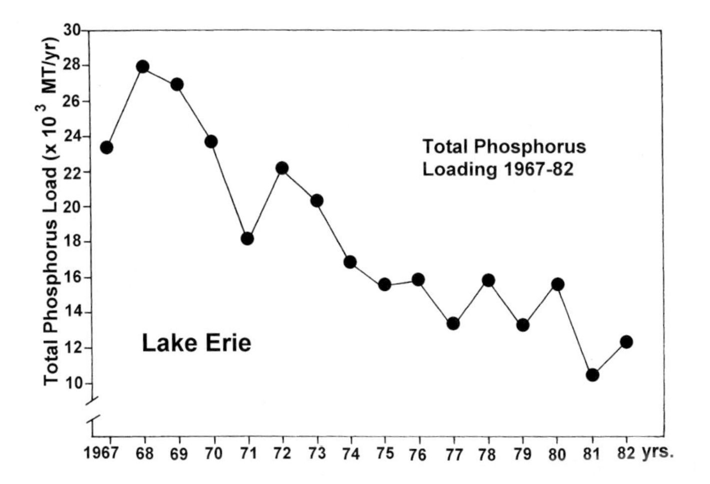

Fig. 7.4. Total phosphorus loading to Lake Erie (modified from Fraser, 1987).

biochemical behaviour. It utilized the two-component phosphorus dynamic model developed by Simons et al. (1979) and included an additional sub-model to simulate dissolved oxygen concentrations. Figures 7.5 and 7.6 illustrate physical processes within the National Water Research Institute (NWRI) 9-box water quality model. The framework and application are briefly discussed below.

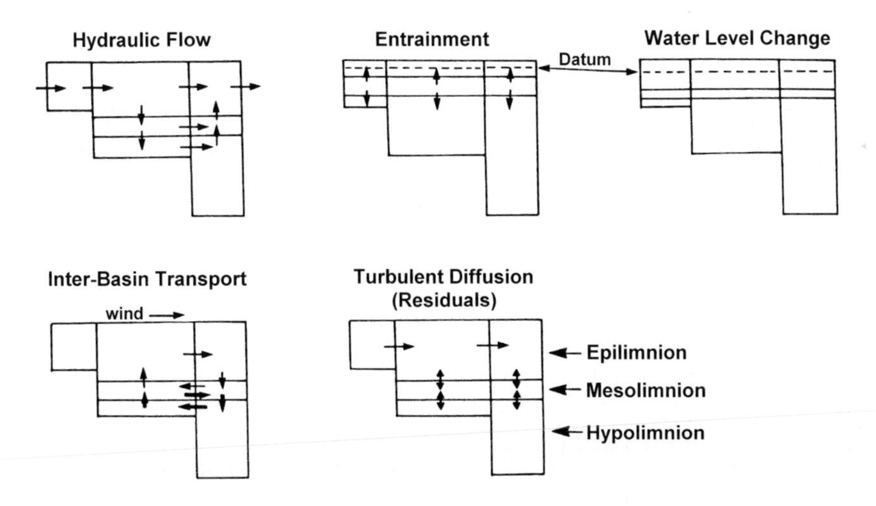

Fig. 7.5. Major physical components of the NWRI 9-box water quality model framework (adapted from Lam et al., 1987a).

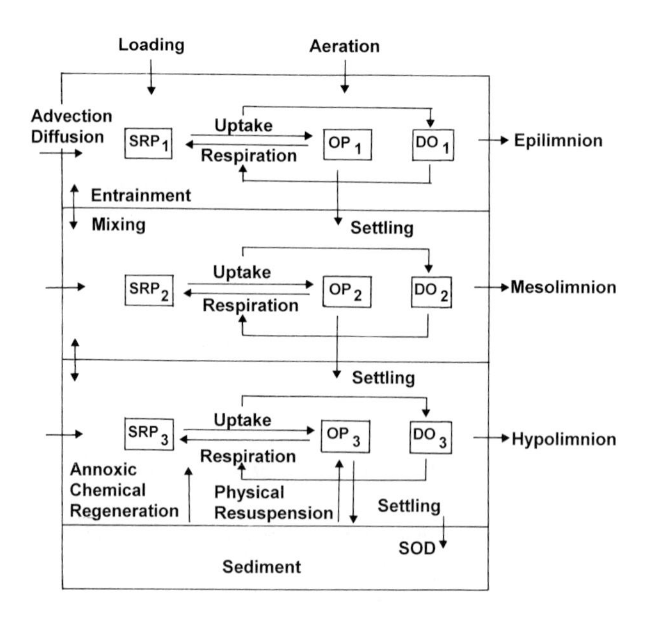

Fig. 7.6. Schematic of the biochemical sub-model of the NWRI 9-box water quality model (adapted from Lam et al., 1987a).

Model Framework

The model considers three basins, straightforwardly conceptualised according to the bathymetric characteristics of Figure 7.1. Vertical resolution into three layers represents a maximum thermal stratification into an upper (epilimnion), transition (mesolimnion) and lower (hypolimnion) layer. The model framework differs from other models applied to Lake Erie in that the thermal layers are dynamic and respond to meteorological conditions through considerations of the surface heat flux, wind mixing, and hydrological balances (Schertzer, 1987). A 1-D thermocline model is used as a link with the surface meteorological forcing.

Lake Stratification (1-D Thermocline Model)

There are a number of one-dimensional thermocline models (e.g. Kraus and Turner, 1967; Mellor and Durban, 1975; Walters et al., 1978), each with its own physical basis and domain of application. However, few are designed for areally-averaged, whole-lake or basin applications. A 1-D model similar to Walters et al. (1978) was developed and modified for the Lake Erie case (Lam and Schertzer, 1987). It specifies the daily basin-wide vertical temperature structure and thus defines the vertical compartments of the model. The vertical temperature model reflects the direct link with surface weather conditions, underscoring the fact that weather effects were believed to be important in the development of hypolimnetic anoxia in Lake Erie.

Physical Processes

Physical processes were recognized as an important feature in the horizontal and vertical dynamics of the lake (see Burns and Ross, 1972b). The model includes important processes (Figures 7.5 and 7.6), having meteorological and hydrological linkages, which were not explicitly included in previous analyses. Exchange between basins was represented by the hydraulic flow deriving from the Detroit River inflow and the Niagara River outflow. Vertical thermal interface displacements were parameterized in terms of an entrainment required to conform with observations (i.e. Blanton and Winklhofer, 1972; Ivey and Boyce, 1982). Such inclusion of inter-basin transport recognized the importance of the exchange occurring between the central and eastern basins involving the channel at the Pennsylvania Ridge (Blanton and Winklhofer, 1972), which has the important potential to introduce oxygenated water from the east basin into the central basin and in some circumstances directly into the central basin hypolimnion. Since the lake volume is not static, daily water level change was also incorporated into its physical representation. Turbulent diffusion residuals were accounted for as an additional exchange mechanism across areal interfaces in which the nutrient concentrations in both affected layers are mutually disturbed by this exchange. Consideration of both entrainment and vertical diffusion mechanisms sets this model apart, then, from others, which consider only vertical diffusion.

Phosphorus–Oxygen Submodel

A set of biological and chemical processes takes place within each model compartment, represented by a three-state model of soluble reactive phosphorus (SRP), organic phosphorus (OP) and dissolved oxygen (DO) (Lam et al., 1987a). The biochemical kinetics, rate formulations and constants relate to phosphorus uptake and respiration, settling velocity and the light factor (Simons and Lam, 1980). With respect to oxygen dynamics, oxygen is produced by plankton photosynthesis in the photic zones and by reaeration at the air–water interface. Most of the time oxygen is saturated or even supersaturated in the upper layer, with the saturation being a function of the water temperature calculated using the thermocline model. In the mesolimnion and hypolimnion, oxygen can be produced by photosynthesis, since these layers may be within the photic zone, particularly during the early part of the stratification period.

The Sediment Oxygen Demand (SOD)

SOD, as we have seen, was implicated as one of the major factors responsible for removing oxygen from the overlying waters in the central basin hypolimnion (Burns and Ross, 1972c). Estimates of SOD varied widely, ranging from 0.08–0.88 g O_2 m^{-2} d^{-1} (e.g., Lucas and Thomas, 1972; Charlton, 1980; Herdendorf, 1980; Di Toro and Connolly, 1980; Mathias and Barica, 1980; Snodgrass, 1987) and were not easy to characterise in mathematical form. The NWRI 9-box model incorporates Monod kinetics in its SOD submodel, which not only produced essentially the same results as the most complex SOD model (Snodgrass, 1987), but also successfully simulated the biological sediment oxygen demand as well as the water oxygen demand. The chemical sediment oxygen demand was found to be relatively small and was described by first-order kinetics.

Simulation Results: Weather Effects as a Major Controlling Influence on Lake Erie Central Basin Water Quality and Hypolimnetic Anoxia

The model revealed that decreased phosphorus loadings resulted in a declining trend in lake total phosphorus concentrations, particularly in the western basin. Irregularities in the downward trend were influenced by processes such as loading pulses, wind-wave resuspension and anoxic regeneration. A significant finding was that the effect of the phosphorus removal program on dissolved oxygen concentration in the central basin hypolimnion was masked by large seasonal variations in this variable. Inclusion of the dynamic, physical factors within the model framework successfully demonstrated that weather controlled the thermal layer thickness, water temperature, interbasin transport, vertical entrainment and vertical diffusion, which in turn dominated the seasonal variations of lake temperature and dissolved oxygen concentration.

Three main meteorologically induced thermal responses of the lake were identified (Figure 7.7): the "normal", "shallow-hypolimnion", and "entrainment reversal" types. A comparison between computed and observed total phosphorus, soluble

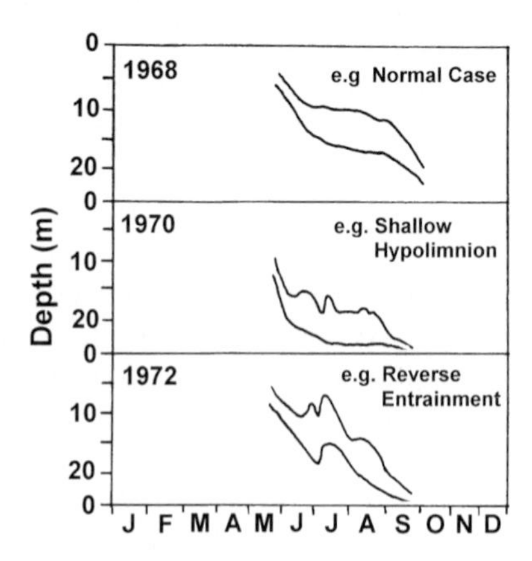

Fig. 7.7. Examples of the three main meteorologically induced thermal responses of Lake Erie characterising "normal", "shallow hypolimnion", and "entrainment reversal" types. Results are based on daily vertical temperature simulations, the top line representing the position of the epilimnion–mesolimnion interface and the lower line the position of the mesolimnion–hypolimnion interface depth (based on Lam and Schertzer, 1987).

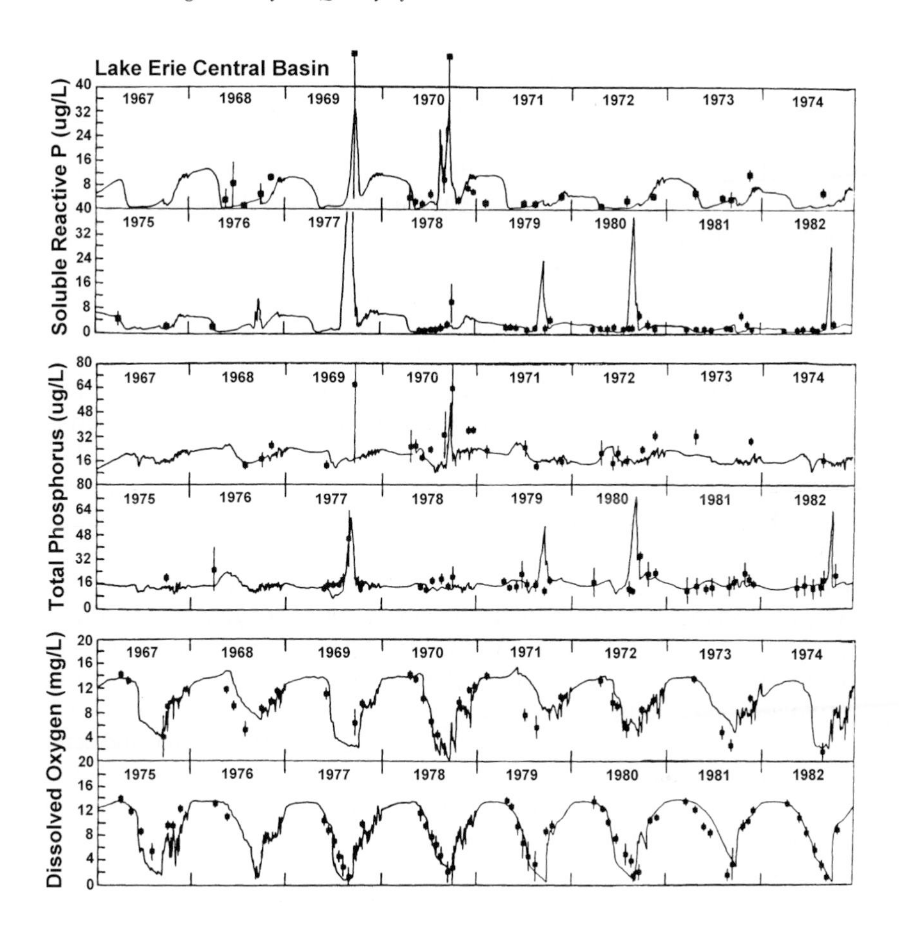

Fig. 7.8. Computed and observed total phosphorus, soluble reactive phosphorus, and dissolved oxygen concentrations for the central basin of Lake Erie (adapted from Lam et al., 1987a).

reactive phosphorus and dissolved oxygen concentrations for the central basin hypolimnion of Lake Erie from 1967 to 1982 is shown in Figure 7.8. The simulations demonstrated that both nutrient loading effects and thermocline dynamics were responsible for observed changes in phosphorus concentrations. For the years 1970, 1977, 1978, 1980 and 1982 in which the hypolimnion thickness was shallow and the vertical diffusion was small (Lam and Schertzer, 1987), the computed oxygen concentration decreased to less than 1.5 mg/L and was in agreement with observed anoxic DO concentrations. Conversely, when the hypolimnion thickness was not so shallow (as in 1971 and 1972) or when the diffusion was relatively large (e.g. 1975 and 1981), anoxia was not attained so readily. It was found that the shallow hypolimnion layer thickness and small vertical diffusion are highly correlated ($r = 0.83$) with the probability of anoxic occurrences (Lam et al., 1987b).

To summarise, our HOM illuminated the following. Only when seasonal meteorological changes were accurately accounted for (as reflected in the thermocline dynamics) could the long-term effect of phosphorus loading reduction on the hypolimnion oxygen depletion in the central basin be estimated. Anoxia of the hypolimnion in the central basin had a tendency to occur in response to the shallow-hypolimnion type and would not likely occur under entrainment reversal episodes. These two response types approximated the upper and lower bounds of meteorological influences. In short, the NWRI 9-box model simulations essentially negated one of the main premises of the phosphorus control strategy: that reduction of total phosphorus to the target load level would result in an eradication of anoxic conditions in the central basin. In fact, further investigations (Schertzer and Sawchuk, 1990), which examined the response of Erie under a warm year, determined that even with a thicker hypolimnion, weather conditions which lengthen the stratified period by several weeks can also be significant in allowing the development of anoxia in its central basin.

While this model was primarily used to investigate the dynamics of the phosphorus and dissolved oxygen questions, it is interesting to note that it was also extended to postulate various strategies to "adapt" or to "engineer" possible ways to avert anoxia occurrence. One suggested such adaptation was that anoxia could potentially be averted by the careful regulation of lake water level – under normal meteorological conditions. It was recognised, however, that regulation of water level carried with it the possibility of significant shoreline property damage and ecological effects for wetlands resulting from adverse weather events. The significance of this "adaptive" investigation is that although additional environmental stresses such as climate warming were relatively "unforeseen" problems for Lake Erie, there was evidence that the model structure could be modified to investigate the potential responses of the lake to other stresses and developing options for managing them.

7.3 "UNFORESEEN" STRESSES OF THE LATE 1980s AND 1990s

By the late 1980s two further overlapping concerns for the health of Lake Erie had arisen: the potential impacts on the lake's physical, chemical and biological regime as a result of climate warming and the potential impacts on the lake's ecology brought about by the inadvertent introduction of exotic species (e.g., zebra mussels), which occurred in about 1988.

7.3.1 Looking Beyond the Horizon: Climate Change

While the NWRI 9-box model had demonstrated the importance of weather on the occurrence of central basin hypolimnion anoxia, it did not address the question of to what extent climatic change might compromise recovery from anoxia. Only during the late 1980s (and up to the present) has there been a concerted effort to investigate the potential effects of climate warming on Erie's physical, chemical and biological regime, with these being interwoven with socio-economic concerns (Mortsch et al., 1998; Hofmann et al., 1998).

A review of the current state of climate change research on the Laurentian Great Lakes system (see Lam and Schertzer, 1999) has now been completed. This has included discussion of the modelling of climate and hydrological components (Schertzer and Croley, 1999), lake thermodynamics (McCormick and Lam, 1999), large-scale circulation (Beletsky et al., 1999), wind-waves (Liu, 1999), ice cover (Assel, 1999) and water quality (Atkinson et al., 1999). The main finding is that modelling techniques for the simulation of climatic change effects on hydrodynamics and water quality of large lake systems are available, but key uncertainties surround climatic scenarios from which to derive definitive statements on causal climatic (warming) effects (Schertzer and Lam, 2000). Future meteorological conditions are clearly uncertain. In the face of this the most common methodology has been to rely on modification of deterministic models largely developed for current climatic conditions and then to impose climatic scenarios on key meteorological variables driving these models. Several "deterministic" modelling approaches have been applied to Lake Erie to examine the potential effects of climate warming (see Schertzer and Croley, 1999). We provide an example of one such approach which demonstrates the adaptability of currently developed "deterministic" modelling approaches to address this largely unforeseen problem in the context of water quality.

Atkinson et al. (1999) have provided preliminary results of climate change effects on the lower Great Lakes related to water quality, although there is recognition of the limited information on input and boundary conditions to predict a definitive outcome, especially when spatial and temporal interactions of the hydrodynamic and biological processes are involved. For example, a critical element is the uncertainty of future hydrodynamical circulations, which influence the reliability of spatial distributions of phosphorus concentrations in the lake. Consequently, one is forced to examine the issue by hypothesis and scenario testing. Atkinson et al. (1999) used $1 \times CO_2$ and $2 \times CO_2$ climate scenarios based on the Canada Climate Centre Global Circulation Model (CCC-GCM-II). Meteorological data were used as input to a fully mixed two-dimensional hydrodynamical model (Simons, 1974) to produce a simulated circulation pattern, with the computed currents then used in a transport model with a primary production sub-model (Simons and Lam, 1980; Lam et al., 1987a) to generate spatial distributions of water quality variables. Water quality simulations for the $1 \times CO_2$ and the $2 \times CO_2$ climate scenarios for Lake Erie (Figure 7.9) indicate that under a climate warming scenario, the particulate phosphorus (which includes algae) shows an increase in primary (biomass) productivity with pronounced changes in the central and eastern basins. The results suggest that under the climate warming scenario water temperature rises relatively more quickly in these basins, since they are shallower, and this contributes to more nutrient uptake and algal production. Atkinson et al. (1999) caution that the GCM scenarios used are only two of the many potential climate scenarios that could occur (see Schertzer and Croley, 1999). This hypothetical assessment indicates that hydrodynamic and

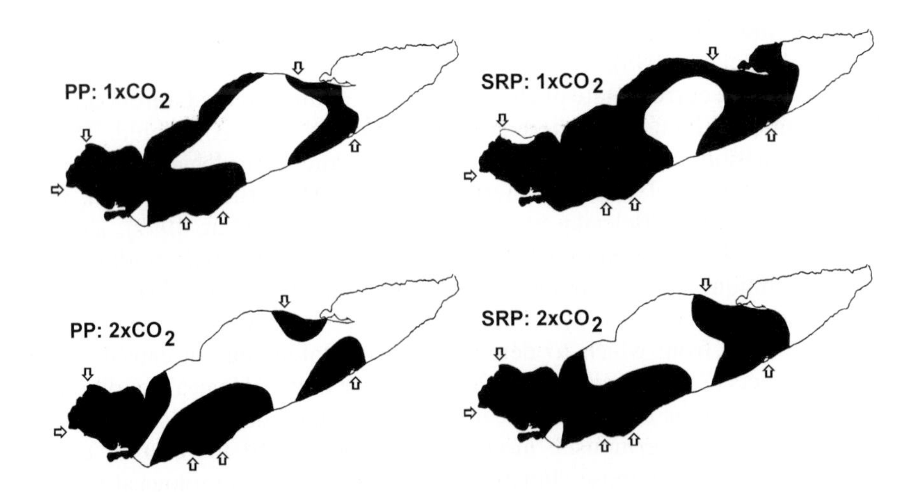

Fig. 7.9. Comparison of simulated results (PP = particulate phosphorus; SRP = soluble reactive phosphorus) for $1\times CO_2$ and $2\times CO_2$ scenarios for Lake Erie: the progression of pollution concentrations (dark regions), having received river inputs (arrows) for 100 days from April 1 in a typical year, is slower in the $2\times CO_2$ case due to projected calmer winds and reduced river flows, in spite of possible increases in nutrient uptake due to projected warmer water temperature in the lake.

thermal changes in the lake combined with changes in nutrient loading may make some water quality conditions, such as oxygen depletion, more acute.

Unlike the problem of eutrophication, investigation of the potential impacts of climate warming on the lake has not benefited from decades of past research on the system. Current research is only at the initial stages of developing models and methodologies with which to assess possible impacts. For the time being the "deterministic" model approach must rely on "best" estimates of future climate scenarios. Most researchers warn that GCMs are still evolving and that any climate scenarios used may be further updated as the GCMs are improved. In fact, several alternatives to GCM-generated climate scenarios have been employed to provide inferences regarding physical, chemical, biological and water quality responses in lakes. These include, for example, transposition climate scenarios (Croley 1990; Schertzer and Croley, 1999), responses of hypothetical lakes over other climate regimes (e.g., Meyer et al., 1994), and observations as analogues of climate warming (Schertzer and Sawchuk, 1990). Not surprisingly, therefore, assessment of the consequences of climate warming assessment still has many unknowns. Indeed, precisely because of this, a range of results from several methodologies will be required in order to minimize costly mal-adaptations (Mortsch, 1993; Schertzer and Leung, 1994). Such prototype explorations with models have been important, since they provide initial assessments of potential ecosystem problems from which a range of possible mitigation and/or adaptive strategies could be drafted for a limited number of issues.

7.3.2 Exotic Species

Around 1988 scientists were alerted to the presence of the zebra mussel in the Great Lakes and its possibly adverse impacts on water quality and the health of the lake's entire ecosystem. Being filter feeders, these mussels are able to extract available phosphorus from the water column. Since existing policies for managing water quality require further reductions in the amount of phosphorus discharged to the lake to 9,000–10,000 MT/year (compare with Figure 7.4), additional "removal" by exotic filter feeding organisms would threaten the availability of nutrients for other species in the ecosystem. Current policy may, then, inadvertently lead to previously unexpected stresses on fisheries, with thus socio-economic impacts. Zebra mussels now form 90–99% of the benthic biomass in the lake, numbers of large-sized phytoplankton (diatoms) have declined, the amphipod population of *Diporeia* (which fed on detrital diatom matter) has disappeared, and the diet of (exotic) rainbow trout smelt has shifted from large cladocerans to smaller copepods (Dermott and Munawar, 2002). Whereas the commercial catch of these smelt remained above 4 million kg per year between 1961 and 1989, it has dropped since 1993 to less than 2 million kg.

There is good reason, therefore, to explore the possible consequences of the unforeseen introduction of the zebra mussels, indeed using models. Given the interest in the effect of the mussels on the phosphorus cycle in Erie, it is obvious the earlier NWRI 9-box model could be brought back into service (once more). However, this would require significant changes to the structure of the model, in particular, more detailed characterisation of the sediment behaviour, as well as increasing the number of physical segments (boxes) into which the lake is conceptually divided, providing, for example, for the distinction of significantly differing behaviour in the near-shore and mid-lake zones (Lam et al., 2000). As in the discussion of the "ozone problem" in Chapter 9 (and again in Chapter 17), we sense our work is bumping up against the borders of what we know we do not know, our ignorance. And equally, as much as in the exploration there of the ozone problem, it will be essential here for the development of models to proceed in very close harmony with the maintenance of field monitoring programmes, even the further development of instrumentation technology for acquiring better observations. The formulation of effective policies should be enhanced by the properly orchestrated interplay between field observation and computational model development.

7.4 LESSONS LEARNED

Application of Vollenweider's input–output, steady-state mass balance model, was the first instance – for Lake Erie (and the Great Lakes) – of a model being used to bring scientific understanding into the policy arena. It led to a multi-million dollar programme of pollution abatement. The gravity of the eutrophication problem and its political importance, in terms of the large-scale effort of banning the use of phosphorus in detergents, and the subsequent massive overhaul of municipal waste-water infrastructure, undoubtedly attracted the attention of the public, policy makers, and scientists. The phosphorus problem, and therefore the associated building of models, was the centrepiece of the Great Lakes Water Quality Agreement of the early 1970s (IJC, 1978).

***Lesson 1**: Computational models, even in their simplest form, when applied correctly, are capable of influencing the development and modification of policies.*

Public concern over worsening water quality prompted politicians to fund more research to better understand the problem, ultimately for sounder policy development. The scientific community began to implement observational programs and data collection, ranging from a single focus on hypolimnetic processes in the central basin of Lake Erie (Project Hypo) to the full investigation of all forms of phosphorus and nutrients and their relationships with ecosystem components. At one time, the IJC Great Lakes Modelling Task Force (Sonzogni et al., 1986) counted over 70 models developed following the use of Vollenweider's original model: from input–output box models to fully 3-dimensional time-dependent hydrodynamic models incorporating chemical and biological details. These latter offered the best and the worst of the higher-order modelling approach: the best in the sense they synthesized in an orderly and systematic manner the full knowledge base of the day; the worst because they required evaluation, or estimation, of so many inputs and model coefficients for computational realisation, and produced too many output variables over a finely resolved space–time field. There were not enough data for model calibration, let alone model confirmation. The IJC Phosphorus Management Task Force (Rast, 1980) concluded that the results of five complex models basically agreed with those of Vollenweider's in terms of forecasting the target load for phosphorus.

***Lesson 2**: Complex models are useful scientific research tools, but in this case, are often no better, in terms of predictive capability, than simple models tailored to the specific problem.*

***Lesson 3**: As models become more complex, structural indeterminacy may become an increasingly serious problem.*

***Lesson 4**: While models developed early in a process may be found to require revision as we learn more, the early models are important to help define data collection, further areas of research, and initial policy actions. Thus, analysis of complex environmental problems needs to be iterative and include evaluation phases.*

By the mid-1970s, and into 1980s, the phosphorus problem for Lake Erie appeared to have been solved, until continued monitoring revealed that, while the phosphorus concentration in the lake had attained its target, oxygen levels in the hypolimnion of the central basin were, at irregular times, still very low. The developers of the models returned to the drawing board, so to speak, and re-discovered previous work (Burns and Ross, 1972a; Charlton, 1987) on sediment oxygen demand and its relationship with the physical phenomenon of thermal stratification. With a one-dimensional vertical diffusion model coupled to phosphorus and oxygen processes, Lam et al. (1987a) were able to show that, besides the controllable source of phosphorus loading, there was an uncontrollable element intervening – that of weather (solar heating, wind, cloudiness, and so on) – which affected the thermal structure of the lake and therefore oxygen concentrations.

Lesson 5: *There is always an element of the unknown or uncertainty in environmental forecasting. Models incorporate only as much as is known. Improvements to a model come from monitoring, detecting and understanding these unknown elements (our {ignorance}). It is possible that the unknown elements may be the most important source of uncertainty.*

Lesson 6: *Narrowly defined or directed research (intended, for example, to answer only what the policy defines) can cause problems for new issues, since the process may ignore or miss important, or broader system aspects of the problem.*

Lesson 7: *As knowledge increases, the complexity of models increases over time. Since the simple (LOM) models are not replaced as the more complex (HOM) models are added, the store of available models is also increased.*

The "mechanistic" approach to modelling employed in the investigation of Erie – the playing out of what has been defined as Programme (1) in the discussion of Chapter 1 – requires a rather good understanding of the relevant constituent processes *a priori*. Yet uncertainties will arise from input data, model coefficients, and the structure of the model, so that we may be obliged to adopt elements of the procedure summarised under Programme (2) of Chapter 1 (and set out elsewhere, for example, in Beck and van Straten (1983), van Straten (1985) or Beck (1987), as well as in Chapter 12). Other areas of uncertainty are more difficult to define – the unforeseen event, for example, of invasion by the zebra mussels, or whether climate change will eventually perpetuate the anoxia in the central basin in the future.

Lesson 8: *One can make the case for or against using high-order models, depending on one's expectation, but it is clear that treating even these large-scale models in a stochastic framework cannot cover all the facets of uncertainty encountered now or in the future. In the problem settings of the case studies of this monograph, integrated modelling approaches (including the use of neural networks, as illustrated in Lam et al., 1998, or the dominant mode analysis of Chapter 13) may have considerable potential, as lower-order models for exploring and generating extreme conditions or the "feared" futures of Chapter 5.*

ACKNOWLEDGEMENTS

The authors acknowledge constructive comments on this manuscript provided by Bruce Beck, Robin Dennis and Olli Varis, as well as useful discussions with members of the International Task Force on Forecasting Environmental Change. We also acknowledge Isaac Wong (NWRI) for his assistance in preliminary examination of neural network applications of Lam et al. (1998) and Oskar Resler for formulating some of the figures. Research included in this chapter has been supported by many agencies over the years. In particular, we acknowledge support from Environment Canada's Great Lakes 2000 Climate Program, the US Environmental Protection Agency, and the International Joint Commission.

REFERENCES

Atkinson, J.F., DePinto, J.V. and Lam, D.C.L., 1999. Water Quality. In: *Potential Climate Change Effects on Great Lakes Hydrodynamics and Water Quality* (D.C.L. Lam and W.M. Schertzer, eds.). American Society of Civil Engineers, ASCE Press, Reston, Virginia, pp. 180–217.

Assel, R.A., 1999. Great Lakes ice cover. In: *Potential Climate Change Effects on Great Lakes Hydrodynamics and Water Quality* (D.C.L. Lam and W.M. Schertzer, eds.). American Society of Civil Engineers, ASCE Press, Reston, Virginia, pp. 159–179.

Beck, M.B., 1987. Water quality modeling: a review of the analysis of uncertainty. *Water Resour. Res.*, **23**(8), 1393–1442.

Beck, M.B. and van Straten, G., 1983. *Uncertainty and Forecasting of Water Quality*. Berlin, Springer Verlag.

Beletsky, D., Lee, K.K. and Schwab, D.J., 1999. Large lake circulation. In: *Potential Climate Change Effects on Great Lakes Hydrodynamics and Water Quality* (D.C.L. Lam and W.M. Schertzer, eds.). American Society of Civil Engineers, ASCE Press, Reston, Virginia, pp. 99–140.

Bierman, V.J., Jr., 1980. A comparison of models developed for phosphorus management in the Great Lakes. In: *Phosphorus Management Strategies for Lakes* (R.C. Loehr, C.S. Martin and W. Rast, eds.). Ann Arbor Science, Ann Arbor, Michigan, pp. 235–255.

Blanton, J.O. and Winklhofer, A.R., 1972. Physical processes affecting the hypolimnion of the central basin of Lake Erie. In: *Project Hypo, an Intensive Study of the Lake Erie Central Basin Hypolimnion and Related Surface Water Phenomenon* (N.M. Burns and C. Ross, eds.). Canada Centre for Inland Waters, Paper No. 6, Burlington, Ontario, pp. 9–38.

Burns, N.M. and Ross, C., 1972a. *Project Hypo, an Intensive Study of the Lake Erie Central Basin Hypolimnion and Related Surface Water Phenomenon*. Canada Centre for Inland Waters, Paper No. 6, Burlington, Ontario, 182 pp.

Burns, N.M. and Ross, C., 1972b. Oxygen–nutrient relationships within the Central Basin of Lake Erie. In: *Project Hypo, An Intensive Study of the Lake Erie Central Basin Hypolimnion and Related Surface Water Phenomenon* (N.M. Burns and C. Ross, eds.). Canada Centre for Inland Waters, Paper No. 6, Burlington, Ontario, pp. 85–119.

Burns, N.M. and Ross, C., 1972c. Project Hypo – discussion of findings. In: *Project Hypo, an Intensive Study of the Lake Erie Central Basin Hypolimnion and Related Surface Water Phenomenon* (N.M. Burns and C. Ross, eds.). Canada Centre for Inland Waters, Paper No. 6, Burlington, Ontario, pp. 120–126.

Carr, J.F., 1962. Dissolved oxygen in Lake Erie, past and present. In: *Proc. 5th Conf. on Great Lakes Research*, Great Lakes Research Division, Pub. No. 9, University of Michigan, pp. 1–14.

Chapra, S.C., 1977. Total phosphorus model for the Great Lakes. *Proc. Am. Soc. Civ. Engrs., J. Env. Engineering*, **103** (EE2), 147–161.

Chapra, S.C., 1979. Applying the phosphorus loading concept to the simulation and management of Great Lakes water quality. Prepared for Conference on Phosphorus Management Strategies for the Great Lakes, 11th Annual Cornell University Conference, Rochester, NY, April 17–20, 1979.

Charlton, M.N., 1980. Oxygen depletion in Lake Erie: Has there been any change? *Can. J. Fish. Aquat. Sci.*, **37**, 72–81.

Charlton, M.N., 1987. Lake Erie oxygen revisited. *J. Great Lakes Res.*, **13**(4), 697–708.

Croley, T.E., II, 1990. Laurentian Great Lakes double – CO_2 climate change impacts. *Climate Change*, **17**, 27–47.

Davis, C.C., 1964. Evidence for the eutrophication of Lake Erie from phytoplankton records. *Limnol. Oceanogr.*, **9**(3), 275–283.

Dermott, R. and Munawar, M. (2002), Structural changes in Lake Erie food-web due to biological invasions, *Verhand. Internat. Verein. Limnol.,* in press.

Dillon, P.J., 1974. The application of the phosphorus loading concept to eutrophication research, Pub. No. NRCC 13690, National Research Council of Canada, Ottawa, Canada, 42 pp.

Di Toro, D.M., 1980. The effect of phosphorus loadings on dissolved oxygen in Lake Erie. In: *Phosphorus Management Strategies for Lakes* (R.C. Loehr, C.S. Martin and W. Rast, eds.). Ann Arbor Science, Ann Arbor, Michigan, pp. 153–205.

Di Toro, D.M. and Connolly, J.P., 1980. Mathematical models of water quality in large lakes, Part 2: Lake Erie. Report EPA-600/3-80-065, U.S. EPA Environmental Assessment Laboratory, Duluth, Minn., 231 pp.

Dobson, H.H. and Gilbertson, M., 1972. Oxygen depletion in the hypolimnion of the Central Basin of Lake Erie, 1929 to 1970. In: *Project Hypo, an Intensive Study of the Lake Erie Central Basin Hypolimnion and Related Surface Water Phenomenon* (N.M. Burns and C. Ross, eds.). Canada Centre for Inland Waters, Paper No. 6, Burlington, Ontario, pp. 3–8.

Fish, C.J. and Associates, 1960. Limnological Survey of Eastern and Central Lake Erie, 1928–1929. U.S. Fish and Wildlife Service, Special Scientific Report – Fisheries No. 334, 198 pp.

Fraser, A.S., 1987. Tributary and point source total phosphorus loading to Lake Erie. *J. Great Lakes Res.*, **13**(4), 659–666.

FWPCA (Federal Water Pollution Control Administration), 1968. *Lake Erie Environmental Summary: 1963–1964.* US Department of the Interior, Great Lakes Region Publication 170 pp.

Herdendorf, C.E., 1980. Lake Erie Nutrient Control Program – An Assessment of its Effectiveness on Controlling Lake Eutrophication. Report EPA-600/3-80-062, U.S. EPA, Environmental Research Laboratory, Duluth, Minn.

Hofmann, N., Mortsch, L., Donner, S., Duncan, K., Kreutzwiser, R., Kulshreshtha, S., Piggott, A., Schellenberg, S., Schertzer, W.M. and Slivitsky, M., 1998. Climate change and variability: impacts on Canadian water. In: *The Canada Country Study: Climate Impacts and Adaptation, National Sectoral Issues, Vol. VII* (G. Koshida and W. Avis, eds.). EARG, Downsview, Ont., Canada, pp. 1–121.

IFYGL, 1981. *The International Field Year for the Great Lakes* (E.J. Aubert and T.L. Richards, eds.). National Oceanic and Atmospheric Administration, Great Lakes Environmental Research Laboratory, U.S. Department of Commerce, Ann Arbor, Michigan, 410 pp.

International Joint Commission (IJC), 1969. Report to the International Joint Commission on the Pollution of Lake Erie, Lake Ontario and the International Section of the St. Lawrence River, Vol. 2, 316 pp.

International Joint Commission (IJC), 1978. Great Lakes Water Quality Agreement Between the United States of America and Canada, International Joint Commission, Windsor, Ontario, 56 pp.

Ivey, G.N. and Boyce, F.M., 1982. Entrainment by bottom currents in Lake Erie. *Limnol. Oceanogr.*, **27**, 1029–1034.

Kay, J.J., Regier, H.A., Boyle, M. and Francis, G., 1999. An ecosystem approach to sustainability: addressing the challenge of complexity. *Futures*, **31**, 721–742.

Kraus, E.G. and Turner, J.S., 1967. A one-dimensional model of the seasonal thermocline: II. The general theory and its consequences. *Tellus*, **19**, 98–105.

Lam, D.C.L. and Jacquet, J.M., 1976. Computations of physical transport and regeneration of phosphorus in Lake Erie, Fall 1970. *J. Fish. Res. Bd. Can.*, **33**, 550–563.

Lam, D.C.L. and Schertzer, W.M., 1987. Lake Erie thermocline model results: comparison with 1967–1982 data and relation to anoxic occurrences. *J. Great Lakes Res.*, **13**, 757–769.

Lam, D.C.L. and Schertzer, W.M. (eds.), 1999. *Potential Climate Change Effects on Great Lakes Hydrodynamics and Water Quality*. American Society of Civil Engineers, ASCE Press, Reston, Virginia, 227 pp.

Lam, D.C.L., Schertzer, W.M. and Fraser, A.S., 1987a. Oxygen depletion in Lake Erie: Modelling the physical, chemical and biological interactions, 1972 and 1979. *J. Great Lakes Res.*, **13**, 770–781.

Lam, D.C.L., Schertzer, W.M. and Fraser, A.S., 1987b. A post-audit analysis of the NWRI nine-box water quality model for Lake Erie. *J. Great Lakes Res.*, **13**, 782–800.

Lam, D.C.L., Schertzer, W.M. and McCrimmon, R.C., 2000. Modelling changes in phosphorus and dissolved oxygen pre- and post- zebra mussel arrival in Lake Erie, NWRI Report, 48 pp.

Lam, D.C.L., Wong, I. and Schertzer, W.M., 1998. Neural network and expert system techniques: machine learning tools for environmental forecasting and modelling. Paper presented at the 3rd Workshop: Task Force on Forecasting Environmental Change, International Institute for Applied Systems Analysis, Laxenburg, Austria.

Liu, P.C., 1999. Wind-waves on large lakes. In: *Potential Climate Change Effects on Great Lakes Hydrodynamics and Water Quality* (D.C.L. Lam and W.M. Schertzer, eds.). American Society of Civil Engineers, ASCE Press, Reston, Virginia, pp. 141–158.

Lucas, A.M. and Thomas, N.A., 1972. Sediment oxygen demand in Lake Erie's Central Basin, 1970. In: *Project Hypo, an Intensive Study of the Lake Erie Central Basin Hypolimnion and Related Surface Water Phenomenon* (N.M. Burns and C. Ross, eds.). Canada Centre for Inland Waters, Paper No. 6, Burlington, Ontario, pp. 39–44.

Mathias, J.A. and Barica, J., 1980. Factors controlling oxygen depletion in ice covered lakes. *Can. J. Fish. Aquat. Sci.*, **37**, 185–194.

McCormick, M.J. and Lam, D.C.L., 1999. Lake thermodynamics. In: *Potential Climate Change Effects on Great Lakes Hydrodynamics and Water Quality* (D.C.L. Lam and W.M. Schertzer, eds.). American Society of Civil Engineers, ASCE Press, Reston, Virginia, pp. 79–98.

Mellor, G.L. and Durban, P.A., 1975. The structure of and dynamics of the ocean surface mixed layer. *J. Phys. Oceanogr.*, **5**, 718–728.

Menon, A.S., Marion, C.V. and Miller, A.N., 1972. Microbiological studies related to oxygen depletion and nutrient regeneration processes in the Lake Erie Central Basin. In: *Project Hypo, an Intensive Study of the Lake Erie Central Basin Hypolimnion and Related Surface Water Phenomenon* (N.M. Burns and C. Ross, eds.). Canada Centre for Inland Waters, Paper No. 6, Burlington, Ontario, pp. 71–84.

Meyer, G., Masliev, I. and Somlyody, L., 1994. Impact of Climate Change on Global Sensitivity of Lake Stratification, Working Paper WP-94-28, International Institute for Applied Systems Analysis, Laxenburg, Austria, 56 pp.

Mortimer, C.H., 1941. The exchange of dissolved substances between mud and water. *J. Ecol.*, **29**, 280(1941); **30**, 147(1942).

Mortsch, L., Koshida, G. and Tavares, D. (eds.), 1993. Great Lakes – St. Lawrence Basin Project: Adapting to the Impacts of Climate Change and Variability. Proc. of the Great Lakes – St. Lawrence Basin Project Workshop, February 9–11, 1993, Quebec City, Quebec, Atmospheric Environment Service, Downsview, Ontario, 88 pp.

Mortsch, L., Quon, S., Craig, L., Mills, B. and Wrenn, B. (eds.), 1998. Adapting to Climate Change and Variability in the Great Lakes – St. Lawrence Basin. Proc. of a Binational Symposium on Adapting to Climate Change and Variability in the Great Lakes – St. Lawrence Basin, May 13–15, 1997, Toronto, Environmental Adaptation Research Group, Atmosphere Environment Service, Downsview, Ontario, 193 pp.

Murthy, C.R., 1972. An investigation of diffusion characteristics of the hypolimnion of Lake Erie. In: *Project Hypo, an Intensive Study of the Lake Erie Central Basin Hypolimnion and Related Surface Water Phenomenon* (N.M. Burns and C. Ross, eds.). Canada Centre for Inland Waters, Paper No. 6, Burlington, Ontario, pp. 39–44.

Rast, W. (ed.), 1980. *Proceedings of the Conference on Phosphorus Management Strategies for Lakes*. Ann Arbor Science, Ann Arbor, MI.

Reckhow, K.H., 1979. The use of a simple model and uncertainty analysis in lake management. *Water Resources Bull.*, **15**(3), 601–611.

Reckhow, K.H. and Chapra, S.C., 1983. *Engineering Approaches for Lake Management, Vol. 1: Data Analysis and Empirical Modelling.* Butterworth, Boston.

Schertzer, W.M., 1987. Heat balance and heat storage estimates for Lake Erie, 1967 to 1982. *J. Great Lakes Res.*, **13**, 454–467.

Schertzer, W.M. and Croley, T.E., II, 1999. Climate and lake responses. In: *Potential Climate Change Effects on Great Lakes Hydrodynamics and Water Quality* (D.C.L. Lam and W.M. Schertzer, eds.). American Society of Civil Engineers, ASCE Press, Reston, Virginia, pp. 5–78.

Schertzer, W.M. and Lam, D.C.L., 1991. Modeling Lake Erie water quality – A case study. In: *Water Quality Modeling: Volume IV, Decision Support Techniques for Lakes and Reservoirs* (B. Henderson-Sellers, ed.). CRC Press, Boca Raton, FL, pp. 27–69.

Schertzer, W.M. and Lam, D.C.L., 2000. Modeling of climate-change impacts on large lakes and basins with consideration of effects in related sectors, Contributions to the International Hydrological Program (IHP-V) by Canadian Experts. *IHP Tech. Doc. in Hydrol.*, **33**, 127–155.

Schertzer, W.M. and Sawchuk, A.M., 1990. Thermal structure of the lower Great Lakes in a warm year: Implications for the occurrence of hypolimnion anoxia. *Trans. Amer. Fish. Soc.*, **119**(2), 195–209.

Schertzer, W.M., Saylor, J.H., Boyce, F.M., Robertson, D.G. and Rosa, F., 1987. Seasonal thermal cycle of Lake Erie. *J. Great Lakes Res.*, **13**, 468–486.

Simons, T.J., 1974. Verification of numerical models of Lake Ontario: Part 1. Circulation in

Spring and early Summer. *J. Phys. Oceanogr.*, **4**, 507–523.

Simons, T.J. and Lam, D.C.L., 1980. Some limitations of water quality models for large lakes: a case study of Lake Ontario. *Water Resour. Res.*, **16**, 105–116.

Simons, T.J. and Schertzer, W.M., 1987. Stratification, currents and upwelling in Lake Ontario, summer 1982. *Can. J. Fish. Aquat. Sci.*, **44**, 2047–2058.

Simons, T.J., Boyce, F.M., Fraser, A.S., Halfon, E., Hyde, D., Lam, D.C.L., Schertzer, W.M., Shaarawi, A.H., Willson, K. and Warry, D., 1979. *Assessment of Water Quality Simulation Capability for Lake Ontario*. Environment Canada, Inland Waters Directorate, Sci. Ser. No. 111, Ottawa, Ontario, 220 pp.

Snodgrass, W.J., 1987. Analysis of models and measurements for sediment oxygen demand in Lake Erie. *J. Great Lakes Res.*, **13**, 738–756.

Sonzogni, W.C., 1987. Large lake models – uses, abuses and future. *J. Great Lakes Res.*, **13**(3), 387–396.

Sonzogni, W.C., Canale, R.P., Lam, D.C.L., Lick, W., MacKay, D., Minns, C.K., Richardson, W.L., Scavia, D., Smith, V. and Strachan, W.M.J., 1986. Uses, abuses and future of great lakes modeling. Report to the Great Lakes Science Advisory Board, February 1986.

Sugarman, R.J., 1980. Conference objectives. In: *Phosphorus Management Strategies for Lakes* (R.C. Loehr, C.S. Martin and W. Rast, eds.). Ann Arbor Science, Ann Arbor, Michigan, pp. 7–10.

Thomann, R.V. and Segna, J.S., 1980. Dynamic phytoplankton-phosphorus model of Lake Ontario: Ten-year verification and simulations. In: *Phosphorus Management Strategies for Lakes* (R.C. Loehr, C.S. Martin and W. Rast, eds.). Ann Arbor Sci, Ann Arbor, Michigan, pp. 153–205.

Thomann, R.V., Winfield, R.P. and Di Toro, D.M., 1975. Modelling of Phytoplankton in Lake Ontario 1. Development and Verification, Report EPA-660/3-75-005, Environmental Research Laboratory, U.S. Environmental Protection Agency, Corvallis, Oregon.

Thomann, R.V., Winfield, R.P. and Di Toro, D.M., 1976. Modelling of Phytoplankton in Lake Ontario 2. Simulations using Lake 1 Model, Report EPA-660/3-76-065, Environmental Research Laboratory, U.S. Environmental Protection Agency, Duluth, Minnesota.

Thomas, N.A., Robertson, A. and Sonzogni, W.C., 1980. Review of control objectives: new target loads and input controls. In: *Phosphorus Management Strategies for Lakes* (R.C. Loehr, C.S. Martin and W. Rast, eds.). Ann Arbor Science, Ann Arbor, Michigan, pp. 61–90.

van Straten, G., 1985. Analytical methods for parameter–space delimitation and application to shallow-lake phytoplankton-dynamics modelling. *Appl. Math. and Comp.*, **17**, 459–482.

Vollenweider, R.A., 1968. Scientific fundamentals of the eutrophication of lakes and flowing waters with particular reference to nitrogen and phosphorus as factors in eutrophication. Organization for Economic Co-Operation and Development – DAS/CSI/68.27.

Vollenweider, R.A., 1975. Input–output models with special reference to the phosphorus loading concept in limnology. *Swiss J. Hydrol.*, **37**, 53–84.

Vollenweider, R.A., 1976. Advances in defining critical loading levels for phosphorus in lake eutrophication. *Mem. 1st. Ital. Idrobiol.* Dott Marco de Marchi Pallanza Italy, **33**, 53–83.

Vollenweider, R.A., Rast, W. and Kerekes, J., 1980. The phosphorus loading concept and Great Lakes eutrophication. Proceedings of the Conf. Phosphorus Management Strategies for Lakes. Ann Arbor Science, Michigan, pp. 207–234.

Walters, R.T., Carey, G.F. and Winter, D.F., 1978. Temperature computation for temperate lakes. *Appl. Math. Modelling*, **2**, 41–48.

Environmental Foresight and Models: A Manifesto
M.B. Beck (editor)

CHAPTER 8

Impacts of Acidic Atmospheric Deposition on the Chemical Composition of Stream Water and Soil Water

G.M. Hornberger

8.1 INTRODUCTION

The problem of forecasting the effects of deposition of acidic chemicals from the atmosphere on the chemical composition of stream water and soil water has several attributes that make it quite difficult. These effects are likely to become manifest on time scales of decades, leading to limitations on direct experimental testing. Descriptions of coupled processes are complex. This complexity, in conjunction with the spatial heterogeneity of natural soils and transients induced by weather events, renders problematic a truly mechanistic approach to model building (Beven, 1993), the virtues of Chapter 17 notwithstanding. The lack of a thoroughgoing mechanistic framework leads to the use of conceptual models, ones in which different investigators select different box-and-arrow representations of the system. Such models must be calibrated by matching output to observations. Monitoring data for problems such as acidification are not plentiful, however, and where they do exist, their use for testing alternative explanations is limited because of natural variability and the long time constants associated with the processes.

The difficulties of making long-range forecasts of environmental change have led to a fair amount of pessimism regarding the utility of models. Sarewitz et al. (1999) suggest that scientists must articulate clearly the uncertainties in any forecast, even if the large error bounds lead decision makers to dismiss the use of models for setting policy. They include models of effects of acid rain among those that, at least in their view, have essentially no credibility. Unfortunately, it is not clear what alternative Sarewitz et al. (1999) envision. As Rastetter (1996), in a discussion of ecological models of response to global change, points out: "... these models are likely to remain a vital part of any evaluation of the responses to changes in global climate and carbon dioxide because the alternatives are worse. It would be foolhardy to plunge blindly into the future without some attempt to evaluate the global consequence of human activities."

The forecasting problem for acid rain is important for both ecological protection and economic competitiveness. Acidification of surface waters has been implicated as a cause of loss of fisheries, while soil acidification has been implicated as a cause of forest decline. To assess the need for measures to protect ecological resources, it is clear that a way to apportion blame to the effects of acid rain is required. Given that policies to mitigate the effects of atmospheric acidic deposition can affect large segments of the economies of the world, it is equally clear that useful forecasting tools are desired.

Although forecasting effects of atmospheric acidic deposition on streams and soils is a daunting task, several models have been proposed and exercised. Throughout the decades of the 1980s and the 1990s, these models have been analyzed, criticized, defended, and compared. In this chapter, the way in which models were constructed for forecasting the effects of acid rain is examined. It is important to see how, over the years of study, comparison of the model output with data from experimental catchments and from synoptic surveys of surface waters led to certain structural changes in the models.

8.2 BACKGROUND

Scientists reported the potentially adverse impacts of acid rain in the decades of the 1960s and 1970s (e.g., see Likens et al., 1979). The growing weight of evidence that acid rain was having an effect in Scandinavia, Canada, and the northeastern US led to a burgeoning research effort in field, laboratory, and modelling exercises aimed at understanding the problem. The mainstream view was that sulphate deposited at the land surface by the atmosphere was a mobile anion and would move through soils to streams and lakes. The sulphate ion had to be balanced by some cation and, to the extent that the balance was achieved by mobilizing acidifying ions (hydrogen ion or aluminium), surface waters would be acidified. There was disagreement on the nature of the problem and on how forecasting future change could be accomplished. Evidence from time-series of measured water quality variables was viewed by some as convincing enough to establish the cause-effect relationship between acid rain and changes in the quality of surface waters and soils, but what was viewed as a *prima facie* case to some was challenged by others (e.g., Krug and Frink, 1983; Rosenqvist, 1978). One challenge revolved around the role of organic acids in buffering the soil solution and in aluminium speciation. Some people argued that atmospherically deposited sulphate would be rendered essentially harmless by natural organic material.

Historical time-series of sufficient length for detailed statistical proof were lacking and experimentation was not viewed as the complete answer because time scales for the occurrence of effects are likely to be on the order of decades (e.g., see Cosby et al., 1985a). All of the ingredients for a difficult forecasting problem were at hand. The hydrological and biogeochemical processes that operate to produce soil and stream waters of a given quality from the precipitation delivered to the ground surface are complex. The spatial variability in material properties affecting the incompletely known processes is very large. The signal of the response of a water-quality variable to changes in atmospheric deposition may be small with respect to the natural variability due to seasonal changes or to the occurrence of meteorological events. Given the nature of the problem, it is not surprising that several different approaches to forecasting the aquatic effects of atmospheric acidic deposition were taken – from trend analysis of data to simple highly aggregated models to rather complex, less aggregated models.

8.2.1 Direct Use of Data

In the face of the large uncertainties associated with modelling effects of acid rain, one argument is that a strictly data-driven analysis is appropriate. Indeed, there are now records of streamflow and chemical composition of stream water spanning several decades for a number of intensively studied catchments around the world. Trend analysis of water quality might appear to be the best way to decide on how to make sensible forecasts of future change.

The difficulties associated with discerning trends in long-term data and then associating them unambiguously with a specific cause are well known. One example is given by Reynolds et al. (1997) who concluded that in one of two neighbouring catchments in Wales, sulphate in stream water showed a decreasing trend over a sixteen-year period but that the other did not. Furthermore, they found no evidence of a trend in the atmospheric deposition of sulphate to the catchments. In general, analyses of trends in data are used to suggest hypotheses that may explain them. For example, concentrations of base cations at the Hubbard Brook have decreased over the past few decades. The mobile-anion hypothesis attributes these decreases to the drop in the amount of sulphate deposited by the atmosphere, leading to decreases in concentrations of base cations that partially balance the acidic anions. Driscoll et al. (1989) point out that atmospheric deposition of base cations themselves have decreased over the period, however, and this decrease in supply could equally well explain concentration decreases in streamflow (a discussion of competing hypotheses is given in Church, 1997). The conclusion of such analyses is often that an even longer record is needed to discriminate among various hypotheses. In the meantime, models provide an alternative for examining different hypotheses.

8.2.2 Conceptual Nature of Acidification Models

Faced with the challenge of making adequate (if not totally accurate) forecasts, a traditional scientific response is to use the best (mechanistic) theory available to create a model. For environmental systems, such models generally describe three spatial dimensions, temporal change, and are replete with nonlinear relationships among the state variables and parameters. Although there are those who debate the wisdom of the reliance on complex models, this traditional approach generally is taken as the gold standard by scientists. It seeks to emulate the apparent security of physics, as we have argued in Chapter 2.

Despite the scientific allure of developing models based on rigorous physical laws, models of the effects of acid rain on aquatic ecosystems fall short of the ideal in many respects. Mechanistic modelling of only the hydrological response of upland catchments to precipitation events, let alone their biogeochemical response, is a task that has been argued to be essentially impossible (Grayson et al., 1992), primarily because the details of spatially variable hydraulic properties and the variation of properties with time are unknowable. Although there is not universal agreement that the analysis of Grayson et al. (1992) is correct (e.g., see Smith et al., 1994), the fact remains that no mechanistic, spatially distributed model for the acid-rain problem has been described. That the information needed to specify detailed initial and boundary conditions, as well as values of the parameters, in a three-dimensional mechanistic model of a catchment's hydrochemistry will be readily available in the near future, must be subject to grave doubt. Hence the utility of such models for the purpose of forecasting must also be seriously doubted. Nonetheless, such models are important scientifically; they provide a useful laboratory for performing numerical experiments to sharpen the focus of observations, field experiments, and formulation of simpler models. Work on mechanistic models has continued apace in recent years (e.g., see Ewen et al., 2000) and these models are being used to explore process dynamics on complex landscapes.

One might argue, therefore, that the *intent* of the more complex models that have been developed for the acidification problem is to take a rigorous mechanistic approach. Yet these models simplify process representations and space–time resolution to such an extent that they are not even close to being suitable for mechanistic forecasting in the way that, say, weather forecasting is achieved. In essence, despite any suggested appearance of greater physical reality, models of the effects of acidic deposition on streams and soils are lumped conceptual models in the terminology generally applied in hydrology (for example, Anderson and Burt, 1985). Water is routed from one storage reservoir to another. Functional forms are assumed to direct flows and chemical reactions within each store. The fundamental question of how one evaluates such models is vexed (Beck et al., 1990).

8.3 THE CASE OF MAGIC: A TOOL FOR LONG-TERM FORECASTING

Rastetter (1996) suggests that four methods have been used to validate models of long-term change in environmental systems: (1) comparison with short-term data; (2) space-for-time substitutions; (3) reconstruction of the past; and (4) comparison with other models. To some extent all of these techniques have been applied to acidification models. And all are susceptible to Rastetter's main conclusion – that none of the methods can provide a crucial test. The use of the first of these methods – comparison with short-term data – is discussed below with respect to MAGIC.

The main point that Rastetter makes under this category is that comparison of model results with short-term data does not allow evaluation of important responses that may arise only over the long term. As one example, he cites the failure of relatively short-term data to constrain a major response that arises in the long term due to a feedback with a long time constant. Cosby et al. (1985a) show how recovery from a pulse of acidic deposition has a dominant time constant that is much longer than that for the response to the perturbation itself. The asymmetry arises because stripping base cations from exchange sites on the soil is a much faster process than replenishing them by primary mineral weathering. Thus, it is certainly possible that comparison with short-term data will fail to stress certain parts of the model that may be quite important to long-term behaviour.

The comparisons with short-term data in acidification modelling have been with regard to experiments performed on small catchments. The comparison with experiments is a very standard scientific approach to model testing. That is, a model is matched to historic data and a forecast is made of the result of an *experiment*. The experiment is performed and the model forecast is compared with the measured result. The outcome of such experiments most often is to expose shortcomings in the model structure used and to suggest avenues for further research.

Wright and Cosby (1987) and Wright et al. (1990) used data from two manipulated catchments in Norway to evaluate MAGIC forecasts of the catchment response to chemical amendment or removal. In one of the experiments, a 7200 m^2 catchment that had received little natural acidic deposition was artificially acidified by the addition of 70 to 100 meq m^{-2} y^{-1}. At a 1200 m^2 catchment in southern Norway, which had received decades of significant acidic deposition, a roof was built over the catchment. The natural (acidic) rain was excluded and clean precipitation was applied beneath the roof. Wright and Cosby (1987), in a preliminary test of MAGIC, concluded that the "The general agreement between predicted and observed alkalinity levels thus reflects the predictive strength of MAGIC". Differences between the model forecasts of changes in concentration of acid anions and the observed response led to modification of the model. The original version of MAGIC did not account for organic acids and part of the model deficiency in the case of the Norwegian experiments was attributed to this omission. Wright et al. (1990) revisited the application of a version of MAGIC modified to account for organic acids to the manipulated catchments in Norway. They used measurements of total organic carbon to produce a crude estimate of organic acidity. In this application, they note that "changes in both pH and aluminium concentrations are also well simulated by MAGIC" and reiterate their conclusion that the use of MAGIC as a forecasting tool is warranted. Cosby et al. (1995) used a different approach for handling organic acids in a subsequent application of MAGIC to the manipulated catchments. They used data from a survey of Norwegian lakes to estimate parameters in a model for organic acids presented by Driscoll et al. (1994). Cosby et al. (1995) also considered the model forecasts to be successful, although they note problems with simulating aluminium as judged by comparing simulated and observed values of various aluminium species.

Cosby et al. (1996) report a test of MAGIC for a paired-catchment experiment in Maine. Two neighbouring catchments, East and West Bear Brook, have been monitored since 1986. In 1989, West Bear Brook was manipulated chemically by the application of 300 equivalents of $(NH_4)_2SO_4$ per hectare on a bimonthly schedule. Some of the measured responses are increased levels of acid anions (e.g., sulphate, nitrate) and of base cations (e.g., calcium, potassium). Cosby et al. (1996) note that there are biases in certain predictions from the model (e.g., both the acid anions and the base cations were under-predicted). The biases cancelled in the calculation of acid neutralizing capacity (ANC), taken as the sum of base cations and acid anions, resulting in a prediction of ANC that was in remarkably good agreement with observations! The authors acknowledge that there are deficiencies in the model's ability to provide accurate forecasts but note that "in general, the dynamics of MAGIC are consistent with the general dynamic trends of the manipulated system." Cosby et al. (1996) argue that modelling comparisons such as theirs should be viewed as an effort to improve the conceptual, scientific understanding of the natural system as well as the knowledge of model performance and how to improve the model.

In the example applications of MAGIC above, we see a pattern that generalizes to the use of models in many environmental systems. The acidification experiments were designed to examine processes and, inasmuch as these processes are represented in a model, to test the model. Typically, the model is judged to capture the general trend of behaviour but fails to reproduce accurately the time series of all observed variables. In fact, in large field experiments, variables that are not even included in the model may be measured and judged by the field investigators to be quite important to knowledge about the system behaviour. The failure of the model leads to speculation about how the model can be improved, either by adding more processes or by changing functional forms embedded in the model. The model structure evolves by incorporating more and more complexity. The changes always improve the model results for the specific application (after all, they are suggested by specific failures of the model), but there is no guarantee that the model has been made better for general forecasting purposes, i.e., for forecasting acidification responses across a region rather than for one experimental catchment.

8.3.1 Application to Regional Assessments

For regional assessments, the application of a model to isolated, well-instrumented catchments is not one that is useful. There is no logical way to extrapolate results from one catchment to an entire region. An alternative is to apply a model to relatively sparse data collected synoptically. One example of the use of this approach to modelling effects of acid rain is the Direct-Delayed Response Project (DDRP), a large effort sponsored by the US Environmental Protection Agency (Church et al., 1990).

The Virginia team became a part of DDRP to apply MAGIC regionally. We brought to the project the argument that traditional calibration, against observed time-series from a specific catchment, was not a fruitful way to proceed. We had evolved the idea that data from surveys of sparse sampling across several catchments could be taken to be a statistical representation of water-quality status in the region. Given the statistical description of the region, we argued that the model could be used to fit the (statistical) distribution using the model in a Monte-Carlo simulation mode. The idea was that distributions of water-quality variables would be hindcast and forecast for the region, without ascribing a particular forecast to a particular catchment. The method produced the kinds of regional forecasts that we thought useful (Hornberger et al., 1986, 1989). The DDRP project team did not accept this approach, however, insisting that each catchment be modelled explicitly. In response, the Virginia team then developed a strategy for calibration that would preserve the notion that uncertainties in the model and in the data were such that a single optimum parameter set for a given catchment could not be defined. We developed a procedure designed to account partially for these uncertainties.

In the DDRP , MAGIC was calibrated to 126 catchments in the Northeastern US by adjusting a set of model parameters to minimize the sum of squared errors between model calculations and a set of surface-water and soil chemical measurements (Cosby et al., 1989). The chemical measurements at which calibration was directed are termed "target variables" in what follows. The selection of optimal model parameters is a standard (deterministic) practice and results in a single best parameter vector. The nature of both the data base and the incomplete knowledge embodied in the model itself suggests that such deterministic calibration would be misleading at best. Uncertainties were quantified in the MAGIC application by performing numerous of the standard optimisation runs to derive a range of equally likely optimisations. The method is a simplified version of the GLUE procedure of Beven, which will be set out and illustrated later in Chapter 12 (Zak et al.,1997, report on the application of GLUE to an acidification model). The procedure we used with the MAGIC model is elaborated below. Other applications of the procedure with MAGIC are given by Jenkins and Cosby (1989) and Ferrier et al. (1995).

Calibration of the model to a specific catchment is accomplished by: specifying the model forcing functions (inputs); setting the values of those parameters that can be measured ("fixed" parameters); and determining the values of the remaining parameters that cannot be measured ("adjustable" parameters) through an optimisation routine which adjusts those parameters to give the best agreement between observed and predicted values of simulated variables (target variables). The formal procedure for estimation of optimal values of the adjustable parameters (and their error variances) customarily ignores several sources of uncertainty in the calibration procedure. First, effects of the initial estimates of the adjustable parameters used in the optimisation algorithm are not considered (noisy response surface); second, effects of uncertainties in the values of the fixed parameters are not treated (noisy fixed parameters); third, effects of errors in the values of the measured variables used to evaluate the squared errors are not considered (noisy target variables); and fourth, effects of errors in the specified inputs used to drive the model are not considered (noisy inputs). The approach for calibrating MAGIC addressed the first three of these sources of uncertainty.

The specification of values of the fixed parameters for the models is based on measurements made in the DDRP (Church et al., 1990). These values are in many cases aggregated values meant to represent the average property over a whole catchment. The effect of fixed parameter uncertainty was evaluated for a catchment by performing multiple calibrations of the models using different randomly selected values of the fixed parameters for each calibration. The random selections for each parameter (for each calibration) were made from a distribution of possible values whose first two moments are the mean value and error (or spatial variance) of the measurements, respectively. Because all values of the fixed parameters chosen in this manner can be considered equally good estimates of the true parameter value, all acceptable optimisations achieved must be considered equally good calibrations of the model for the catchment in question. This approach defines an optimisation procedure using a distribution of fixed parameters – a series of calibrated models with a *range* of fixed parameter values (rather than single fixed values) is derived for each catchment.

The calibration procedure also took into account the fact that the target variables were uncertain. Due to the large spatial and temporal variability in the values of the target variables, sampling errors are assumed to be larger than measurement errors for the DDRP catchments. Thus, *ranges* of values rather than single values represented the target variables used in calibrating the models. In performing the optimisations, any simulation with values of the target variables falling within the specified ranges (or windows) is considered as good as any other calibration whose target variables also fall within those ranges – much as will be the case in Chapter 11 for analyses using the original form of Regionalised Sensitivity Analysis (Hornberger and Spear, 1980; Spear and Hornberger, 1980). In effect, the windows defined by the ranges of target variables set a minimum value of the sum of squared errors below which further adjustment of the parameters is not warranted (based on the noise in the target variables). This minimum value of the loss function defines an indifference region for the model (all parameterisations with lower values of the loss function), and the optimisation algorithm was stopped when the model entered this region rather than proceeding to some absolute minimum determined by single values of target variables.

The adjustable parameters in each model were optimized ten times for each catchment. Before the first optimisation, ranges of acceptable values were specified for (a) the initial values of the adjustable parameters, (b) the fixed parameters, and (c) the target variables (the ranges of acceptable values were based on measured data and the appropriate errors or uncertainties in those data). For each of the ten optimisations initial values of the adjustable parameters, and values of the fixed parameters and target variables were chosen from the specified ranges using a Latin Hypercube design (Iman and Conover, 1980). The optimisation algorithm was

Table 8.1

Results of calibration of MAGIC to catchment 1A2077, Northeastern U.S. All concentrations are given in μeq/L. Results are for 10 independent optimisations. (Adapted from Cosby et al., 1989)

Chemical variable	Observed value (1984)	Median of modelled value (1984)	Max–min modelled value (1984)	Max–min forecast value (2034)
Calcium	178.1	182.1	19.6	22.8
Magnesium	32.9	33.1	4.3	4.3
Sodium	43.5	43.6	12.4	11.7
Potassium	4.6	5.0	4.1	4.0
Chloride	6.5	5.5	1.8	1.8
Sulphate	121.2	120.3	41.0	34.7
Alkalinity	131.4	135.7	42.4	40.9
pH	6.95	7.01	0.14	0.13

implemented and those optimisations whose simulated values fell within the specified acceptable ranges were retained as successful calibrations. The final calibration of the models for each catchment consisted of the ensemble of successful calibrations. The final simulation results for each catchment are thus composed of a bundle of simulations all of which are considered equally good given the uncertainties in the available data. Uncertainty bands for *forecasts* are calculated by projecting each of the ten calibrated models into the future. For DDRP, one such forecast was for water quality in the year 2034 under an assumption that atmospheric sulphate deposition remained constant over the period from 1984 to 2034. A sample of the results from Cosby et al. (1989) indicate how this treatment of uncertainties yields a forecast of a range of values rather than a single, deterministic trace (Table 8.1).

The propagation of uncertainty into the future depends on the status of the catchment. For the particular case considered above – a poorly buffered catchment with future deposition continuing for 50 years into the future – uncertainties from the parameterisation of MAGIC do not grow with time. This is because the catchment is already close to a steady state with respect to the inputs in 1984; the changes in the catchment have already occurred and future trajectories are constrained by simple mass balance of inputs and outputs. In cases where processes represented in the model (sulphate sorption, base cation exchange, and so on) have not reached steady state, the uncertainties may grow into the future because storage and release mechanisms are active – the model is not so tightly constrained by an input–output mass balance.

It should be noted that the regional forecasts are simply that – the representation of possible futures given an acceptance of the model being correct. The procedure does not offer insight into structural inadequacies of the model or into inadequacies relative to assessing uncertainties. In an application of MAGIC to survey data in Norway, in which lakes sampled in 1974 were resampled in 1986, we found that forecasts of variances in regional data were smaller than those observed. "The larger variances of the observed changes *versus* the forecast changes suggest that the discrepancies between model and observations may be due to temporal or spatial stochastic variability that is not taken into account in the model. This points up the fact that the uncertainty bands calculated in our resampling procedure refer only to uncertainty in model parameters (e.g., weathering, selectivity coefficients, etc.) for a simulation sample size of 1000 and do not assess uncertainties in the form of the equations used in the model or in deposition sequences used to drive the model" (Hornberger et al., 1989; p. 2017). The evaluation of structural uncertainty can not be done within the framework of application of a single model to a regional assessment.

8.3.2 Uncertainty in Model Structure

The uncertainty associated with the parameters in a model of fixed structure cannot represent the true uncertainty associated with a forecast. Our conceptualisations of the interlinking of complex processes and our specification of a mathematical approach leads to a particular model *structure* (as in the visual metaphor of Chapter 4). Different model structures are advanced by different groups to represent basically the same overall environmental problem. And yet there are no useful rules for deciding the relative worth of competing structures for complex environmental models. What is the extent of this uncertainty in model structure? In fact, a key question addressed in this monograph is whether a new program of environmental modelling can be elaborated to provide a context for evaluating structural uncertainty. Again the case study for aquatic effects of acid rain provides some perspective.

Three watershed acidification models were used in the DDRP to forecast future changes in the quality of surface waters in the northeastern US due to acid rain (the sulphate impact of acid rain, that is). The models were the ILWAS model (Gherini et al., 1985), the MAGIC model (Cosby et al., 1985b) and the ETD model (Nikolaidis et al., 1989). The models have different levels of complexity and represent certain processes differently (as reviewed in Rose et al., 1991). Briefly, ILWAS is the most complex of the three models. It is a daily-time-step model that tracks 18 constituents affecting ANC in a series of sub-catchments that drain to a single lake. Within a sub-catchment, the canopy, snowpack, and five soil layers are represented explicitly. ETD is second in complexity. It is a daily-time-step model based on mass balances for sulphate and ANC. Three soil layers and a snowpack are represented. The DDRP version of MAGIC treats two soil compartments and uses an annual time step.

Part of the aim of the DDRP was to compare these three models of acidification, ILWAS, MAGIC, and ETD. The general conclusion was that all three gave fairly similar forecasts and that the robustness of conclusions about the likely effects of future acidification was therefore enhanced (Church et al., 1990). Rose et al. (1991) reported a more systematic comparison of these three models. The performance of the models was judged on the basis of forecasts of ANC. Absolute predictions (i.e., the actual values of ANC forecast by the models) and relative predictions (changes and variance of forecasts in response to presumed levels of acid rain in the future) were compared. For several of the situations considered by Rose et al. (1991), the models were seen to provide similar forecasts on a relative scale, but to result in substantial differences for some cases in terms of absolute forecasts. Rose et al. recognize that "the use of multiple alternative models to bound forecasts is a useful approach", but that "caution should be used in interpreting model predictions on an absolute scale".

An alternate view of the effects of model structure was presented in the "Workshop on Comparison of Forest–Soil–Atmosphere Models" (van Grinsven et al., 1995). The comparisons undertaken as part of this workshop were not as systematic as those of Rose et al. (1990), but they are instructive nevertheless. Eleven biogeochemical models were evaluated with regard to their calibration to water chemistry data at Solling, Germany (Kros and Warfvinge, 1995). A general conclusion from the study is: "Despite the failure of the models applied in the model comparison to reproduce details in the data set, the general trends and levels of ion concentrations and fluxes could be identified by the models." This conclusion is in agreement with most of the work that has been done on modelling the effects of sulphate deposition on long-term catchment response. The mechanisms of catchment acidification by sulphate seem to be captured in models to the extent that trends and rough levels of, say, ANC, can be forecast with some assurance.

The forecasting problem is far from solved, however! All tests of individual models and all comparisons of different models have pointed out deficiencies. As one example, consider the results of the comparison of the eleven different models used by Kros and Warfvinge (1995). They found that the models were not capable of modelling pH and aluminium concentrations simultaneously. More ominously, they found that modelling nitrate concentrations was not successful "despite several complex modelling approaches". As with all process-based models of environmental systems, deficiencies can be found with respect to the fit to data and this is generally interpreted to imply deficiencies in the scientific underpinnings of the model. The standard response to model deficiencies is to improve the model, and this always increases complexity.

8.4 STATUS AND FUTURE DIRECTIONS

As already noted, the DDRP was aimed at assessing the potential effects of the atmospheric deposition of *sulphate* as an acidifying agent. It was recognized that nitrate deposition also had the potential for acidifying surface waters, but the attention was focused on sulphate because it was thought to be the most important mobile anion. Nitrate is much more biologically active than sulphate, and is generally taken up by forests. That is, the first effect of nitrogen deposition on a forested catchment is as a fertilizer. The potential difficulty lies in nitrogen saturation. If a forest has nitrogen available in excess of the needs of growing plants, nitrate can leach from a catchment producing surface-water acidification. Because nitrogen species are so active in biological reactions, modelling nitrate breakthrough in a catchment is not easy. The results of Kros and Warfvinge (1995) indicate that none of the eleven models applied at Solling satisfactorily reproduced observed nitrate concentrations.

In any application of process-oriented models to environmental problems, deficiencies in the model will be noted. No model, regardless of complexity, can be expected to reproduce the complexity of the natural system. Because of this fact, scientists looking at how a given model is deficient will always want to add something to the model, for example, better descriptions of organic acids, more details in nutrient cycling, further disaggregation of hydrological processes, and so on. Conversely, regional assessments typically require relatively simple models, especially given the large uncertainties and limited data. Thus, models used to produce results to inform decision making are criticized for failing to represent important processes, while models developed to include a large number of processes that scientists think are important are criticized for being too uncertain. Scientists recognize the conflict. For example, Neal (1997) states: "... little trust can be placed in the predictive capacity of the current hydrochemical and environmental impact models associated with acidic deposition. Their use in determining environmental management policies may well be overemphasised, misplaced and counterproductive." A half paragraph later, Neal says: "The models serve an extremely valuable purpose of providing learning tools for examining the dynamics of water quality change both within the framework of hydrological events and of long term changes in catchment stores and evolving endmember compositions." Many scientists may well share this view currently: models are useful within science to inform the learning process, but not outside science where policy decisions are made. The problem with this view is that alternatives to using models to inform policy decisions, however imperfect the models may be, have not been articulated very clearly.

The conflict can lead to serious consequences in making policy decisions. In the US, the nitrogen saturation issue has reached the point where decisions on regulation of emissions will be made. Part of the debate has related to "scientific uncertainty" (Renner, 1995). The EPA's Scientific Advisory Board criticized the first modelling approach used by EPA because detailed biological processes were not included. In all likelihood, if and when the models are altered to include detailed biological processes, the criticism levelled will be that the uncertainties associated with such complex models renders them suspect.

Scepticism about various modelling activities notwithstanding, this case study in forecasting the effects of acid rain is basically a success story. Scientific understanding of the processes occurring in the field is certainly more sophisticated now than it was before the efforts to comprehend (and forecast) the effects. And the notion of critical loads has been embraced by European states as a policy tool, with the definition of these loads based at least in part on the results of model studies (e.g., Posch and de Vries, 1999). The uncertainties attending forecasts, including gross uncertainties related to model structure, have not been bounded or even brought under the reins of scientific understanding. Nor has the dim view of scientists about the accuracy of model forecasts been changed. Despite these limitations, however, scenario-based modelling has undoubtedly contributed to the policy debate about acid rain.

Scenario analyses will continue to be critical for informing decisions about environmental degradation. The Board on Sustainable Development, in grappling with implications of a broad array of environmental stresses, came to the conclusion that a suite of methods of scenario analysis are quite important to resolving the environmental sustainability problem (NRC, 1999). "At their best, however, these methods have helped determine efforts to probe the future implications of present trends, to identify the likely obstacles to sustainability, and to illuminate alternative options for moving toward specific sustainability goals. In doing so, they have helped us to learn a bit about what a transition to sustainability might actually entail. This learning is a process through which notable progress has been made using [various methods of scenario analysis]; that is a surprising and optimistic finding in itself" (NRC, 1999, pp. 159–160).

REFERENCES

Anderson, M.G. and Burt, T.P., 1985. Modelling strategies, in: *Hydrological Forecasting* (M.G. Anderson and T.P. Burt, eds.). John Wiley and Sons Ltd. Chichester, pp. 1–13.

Beck, M.B., Kleissen, F.M. and Wheater, H.S., 1990. Identifying flow paths in models of surface water acidification. *Rev. Geophys.*, **28**(2), 207–230.

Beven, K.J., 1993. Prophecy, reality and uncertainty in distributed hydrological modelling. *Adv. Water Resources*, **16**, 41–51.

Church, M.R., 1997. Hydrochemistry of forested catchments. *Ann. Rev. Earth Planet. Sci.*, **25**, 23–59.

Church, M.R., Thornton, K.W., Shaffer, P.W., Stevens, D.L., Rochelle, B.P., Holdren, G.R., Johnson, M.G., Lee, J.J., Turner, R.S., Cassell, D.L., Lammers, D.A., Campbell, W.G., Liff, C.I., Brandt, C.C., Liegel, L.H., Bishop, G.D., Mortenson, D.C., Pierson, S.M. and Schmoyer, D.D., 1990. Direct/delayed response project: Future effects of long-term sulphur deposition on surface water chemistry in the Northeast and Southern Blue Ridge Province, EPA/600/3-89/061, U.S. Environmental Protection Agency, Washington, DC.

Cosby, B.J., Hornberger, G.M., Galloway, J.N. and Wright, R.F., 1985a. Time scales of catchment acidification. *Env. Sci. Technol.*, **19**, 1144–1149.

Cosby, B.J., Hornberger, G.M., Wright, R.F. and Galloway, J.N. 1985b. Modelling the effects of acid deposition: assessment of a lumped-parameter model of soil-water and stream-water chemistry. *Water Resour. Res.*, **21**, 51–63.

Cosby, B.J., Wright, R.F., Hornberger G.M. and Galloway, J.N. 1985c. Modelling the effects of acid deposition: estimation of long-term water quality responses in a small forested catchment. *Water Resour. Res.*, **21**, 1591–1601.

Cosby, B.J., Hornberger, G.M., Ryan, P.F. and Wolock, D.M., 1989. MAGIC/DDRP Final Report. Project Completion Report, U.S. EPA Direct/Delayed Response Project, USEPA/CERL, Corvallis, OR.

Cosby, B.J., Wright, R.F. and Gjessing, E., 1995. An acidification model (MAGIC) with organic acids evaluated using whole-catchment manipulations in Norway. *J. Hydrology*, **170**, 101–122.

Cosby, B.J., Norton, S.A. and Kahl, J.S., 1996. Using a paired catchment manipulation experiment to evaluate a catchment-scale biogeochemical model. *Sci. Total Environ.*, **183**, 49–66.

Driscoll, C.T., Likens, G.E., Hedin, L.O., Eaton, J.S. and Bormann, F.H., 1989. Changes in the chemistry of surface waters. *Environ. Sci. Technol.*, **23**, 137–143.

Driscoll, C.T., Lehtinen, M.D. and Sullivan, T.S., 1994. Modeling the acid–base chemistry of organic solutes in Adirondack, New York, Lakes. *Water Resour. Res.*, **30**, 297–306.

Ewen, J., Parker, G. and O'Connell, P.E., 2000. SHETRAN: Distributed river basin flow and transport modeling system. *J. Hydrologic Eng.*, **5**, 250–258.

Ferrier, R.C., Wright, R.F., Cosby, B.J. and Jenkins, A., 1995. Application of the MAGIC model to the Norway spruce stand at Solling, Germany. *Ecol. Modelling*, **83**, 77–84.

Gherini, S.A., Mok, L., Hudson, R.J.M., Davis, G.F., Chen, C.W., and Goldstein, R.A., 1985. The ILWAS model: Formulation and application. *Water Air Soil Pollut.*, **26**, 425–459.

Grayson, R.B., Moore, I.D. and McMahon, T.A., 1992. Physically based hydrological modeling. 2. Is the concept realistic? *Water Resour. Res.*, **28**, 2659–2666.

Hornberger, G.M. and Spear, R.C., 1980. Eutrophication in Peel Inlet, I, Problem-defining behaviour and a mathematical model for the phosphorus scenario. *Water Res.*, **14**, 29–42.

Hornberger, G.M., Cosby, B.J. and Galloway, J.N., 1986. Modeling the effects of acid deposition: uncertainty and spatial variability in estimation of long-term responses of regions to atmospheric deposition of sulfate. *Water Resour. Res.*, **22**, 1293-1302.

Hornberger, G.M., Cosby, B.J. and Wright, R.F., 1989. Historical reconstructions and future forecasts of regional surface water acidification in southernmost Norway. *Water Resour. Res.*, **25**, 2009–2018.

Iman, R.L. and Conover, W.J., 1980. Small sample sensitivity analysis techniques for computer models, with an application to risk assessment. *Commun. Statist. Theor. Meth.*, **A9**, 1749–1842.

Jenkins, A. and Cosby, B.J., 1989. Modelling surface water acidification using one and two soil layers and simple flow routing. In: *Regional Acidification Models* (J. Kämäri, D.F. Braake, A. Jenkins, S.F. Norton and R.F. Wright, eds.). Springer-Verlag, New York, pp. 253–266.

Kros, H. and Warfvinge, P., 1995. Evaluation of model behaviour with respect to the biogeochemistry at the Solling spruce site. *Ecol. Modelling*, **83**, 255–262.

Krug, E.C. and Frink, C.R., 1983. Acid rain on acid soil: A new perspective. *Science*, **221**, 520–525.

Likens, G.E., Wright, R.F., Galloway, J.N. and Butler, T.J., 1979. Acid rain. *Sci. Am.*, **241**, 43–51.

Neal, C., 1997. A view of water quality from the Plynlimon watershed. *Hydrol. & Earth Systems Sci.*, **1**, 743–753.

Nikolaidis, N.P., Schnoor, J.L. and Georgakakos, K.P., 1989. Modeling of long-term lake alkalinity responses to acid deposition. *J. Water Pollut. Control Fed.*, **61**, 188–199.

NRC (National Research Council), 1999. *Our Common Journey: a Transition Toward Sustainability*. National Academy Press, Washington DC, 363 pp.

Posch, M. and de Vries, W., 1999. Derivation of critical loads by steady-state and dynamic soil models. In: Langan, S. (ed.), *The Impact of Nitrogen Deposition on Natural and Semi-natural Ecosystems*. Kluwer Academic Publishers, Norwell, MA, pp 213–234.

Rastetter, E.B., 1996. Validating models of ecosystem response to global change. *BioScience*, **46**, 190–198.

Renner, R., 1995. "Scientific uncertainty" scuttles new acid rain standard. *Environ. Sci. Technol.*, **29**, 464A–466A.

Reuss, J.O., 1980. Simulation of soil nutrient losses resulting from rainfall acidity. *Ecol. Modelling*, **11**, 15–38.

Reuss, J.O., 1983. Implications of the Ca–Al exchange system for the effect of acid precipitation on soils. *J. Environ. Qual.*, **12**, 591–595.

Reuss, J.O. and Johnson, D.W., 1985. Effect of soil processes on the acidification of water by acid deposition. *J. Environ. Qual.*, **14**, 26–31.

Reynolds, B., Renshaw, M., Sparks, T.H., Crane, S., Hughes, S., Brittain, S.A. and Kennedy, V.H., 1997. Trends and seasonality in stream water chemistry in two moorland catchments of the Upper River Wye, Plynlimon. *Hydrol. & Earth Systems Sci.*, **1**, 571–581.

Rose, K.A., Cook, R.B., Brenkert, A.L., Gardner, R.H. and Hettelingh, J.P., 1991. Systematic comparison of ILWAS, MAGIC, and ETD watershed acidification models. 1. Mapping among model inputs and deterministic results. *Water Resour. Res.*, **27**, 2577–2589.

Rose, K.A., Cook, R.B., Brenkert, A.L., Gardner, R.H. and Hettelingh, J.P., 1991. Systematic comparison of ILWAS, MAGIC, and ETD watershed acidification models. 2. Monte Carlo analysis under regional variability. *Water Resour. Res.*, **27**, 2591–2603.

Rosenqvist, I.I., 1978. Acid precipitation and other possible sources for acidification of rivers and lakes. *Sci. Total Environ.*, **10**, 39–49.

Sarewitz, D., Pielke, R. Jr., and Byerly, R., Jr., 1999. Prediction: a process, not a product. *Geotimes*, **44**(4), 29–31.

Smith, R.E., Goodrich, D.R., Woolhiser, D.A. and Simaton, J.R., 1994. Comment on: Physically based hydrological modeling. 2. Is the concept realistic? by R.B. Grayson et al. *Water Resour. Res.*, **30**, 851–854.

Spear, R.C. and Hornberger, G.M., 1980. Eutrophication in Peel Inlet, II, Identification of critical uncertainties via generalised sensitivity analysis. *Water Res.*, **14**, 43–49.

van Grinsven, H.J.M., Driscoll, C.T. and Tiktak, A., 1995. Workshop on comparison of forest–soil–atmosphere models: Preface. *Ecol. Modelling*, **83** 1–6.

Wright, R.F. and Cosby, B.J., 1987. Use of a process-oriented model to predict acidification at manipulated catchments in Norway. *Atmos. Environ.*, **21**, 727–730.

Wright, R.F., Cosby, B.J., Flaten, M.B. and Reuss, J.O., 1990. Evaluation of an acidification model with data from manipulated catchments in Norway. *Nature*, **343**, 53–55.

Zak, S.K., Beven, K.J. and Reynolds, B., 1997. Uncertainty in the estimation of critical loads: a practical methodology. *Water Air and Soil Pollution*, **98**, 297–316.

Environmental Foresight and Models: A Manifesto
M.B. Beck (editor)

CHAPTER 9

The Ozone Problem

R.L. Dennis

9.1 BACKGROUND AND CONTEXT

For complex models an examination of the series of steps undertaken over several years of study, to try to reach a desired future, contains many of the issues associated with trying to detect and avoid an undesired future. Both are confronted with predicting outside or beyond current conditions. Both are confronted with scientific ignorance and issues surrounding the correctness of the model's structure. Thus we shall embark on this, our third case history, with a view to illuminating some of the lessons to be learned of a more generic nature, beyond the specific details of the "ozone problem".

9.1.1 The Essential Problem

The original "problem" with ozone was that it was found to be the major oxidant in photochemical smog responsible for adverse human health effects and crop damage. Following seminal work in Los Angeles, California, a rudimentary scientific understanding of the source of ozone pollution was developed during the 1950s and early 1960s (Haagen-Smit, 1952; Leighton, 1961). Armed with this basic understanding it was felt that management strategies could be designed to mitigate the problem.

The United States promulgated National Ambient Air Quality Standards (NAAQS) in the 1970 Clean Air Act (CAA) for several pollutants – including ozone – termed criteria pollutants. But ozone was the only such criteria pollutant not *directly* emitted (as a primary pollutant), but rather formed in the atmosphere through chemical interactions (as a secondary pollutant). The standard for ozone was originally set at 80 ppb for a one-hour average, and then revised upward to 120 ppb for a one-hour average in 1977. The focus was on urban areas with high ozone levels. Management strategies were required by the 1970 CAA for the polluted urban areas, with ambitious deadlines to bring pollution levels down to the standard (attainment).

Emissions reduction programs were instituted in the 1970s and, while there was tangible success in reducing pollution levels in the 1970s for the primary pollutants, ozone was notable for the lack of associated success, even through the 1990s (Fiore et al., 1998). Modelling studies suggest that ozone concentration levels would have worsened had there been no emissions reduction program (Harley et al., 1997). Achieving the desired future has therefore proved to be a difficult task.

9.1.2 The Challenge for Management

The challenge for management stems from the fact that ozone is a secondary product of photochemistry produced by a nonlinear and self-buffering chemical system, in which two types of precursor pollutants are involved. These are the inorganic nitrogen oxides, or NO_X (= $NO + NO_2$), and the volatile organic compounds, VOCs. The relationship between the resulting ambient ozone concentrations and the amount of NO_X and VOC emissions can be illustrated by a simplified response surface of peak ozone generated in a box model as a function of a matrix of NO_X and VOC emissions (shown in Figure 9.1). The initial concentrations are scaled with the emissions and the axes labelled in terms of the initial concentrations. Values of the

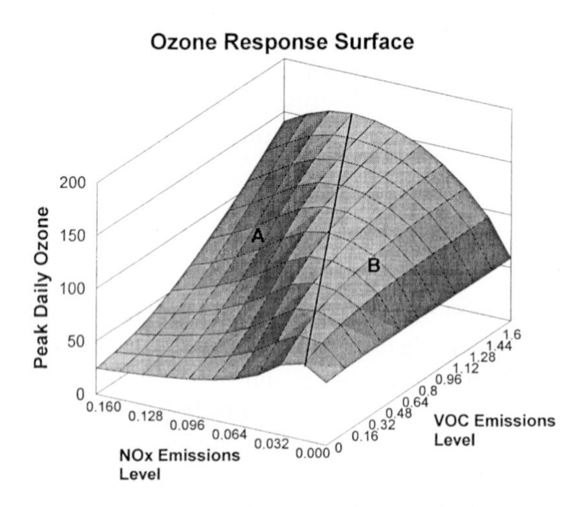

Fig. 9.1. Depiction of the ozone response surface through plotting daytime peak ozone predicted as a function of VOC and NO_X emissions levels for a wide range of combinations representative of both urban and rural conditions. The initial concentrations are scaled with the emissions and the axes labelled in terms of the initial concentrations.

peak ozone concentration predicted by the box-model form the ozone response surface. The grid shows lines along which NO_X emissions are held constant (left-to-right) or VOC emissions are held constant (front-to-back). The heavy line drawn down the slope at the point of maximum radial curvature is the ridgeline of maximum ozone production; it divides the response surface into two domains. At point A, to the left of the ridgeline, there is a low ratio of VOC/NO_X emissions, because concentrations of NO_X are relatively high, which is typical of major urban areas. Reducing NO_X emissions without likewise reducing VOC emissions will cause an increase in ozone concentrations. In this regime the system is radical-limited (or VOC-limited) and reductions in VOC emissions would be preferred as a means of reducing ozone. At point B, to the right of the ridgeline, there is a high ratio of VOC/NO_X, because concentrations of NO_X are low, this being typical of rural areas. In this regime the system is NO_X-limited and reductions in NO_X emissions would be preferred. Thus, there are two categories of precursor emissions to be managed and there are disbenefits to be dealt with for certain decisions.

9.1.3 The Nature of Management Cycles

The cycle of management for ozone was instituted by the 1970 CAA. It involves development of a strategy and formal plan for reductions in emissions of VOC, NO_X, or both, for a specific area. Execution of the plan is expected to result in a reduction of levels of ozone sufficient for the area to attain the NAAQS by a specified year. The first deadline for attainment was 1975 and plans had a purely urban focus. Attainment of the NAAQS for ozone was not achieved in this short time frame; in fact, little progress was achieved. The deadline was then deferred to 1980, and then to 1987. The difficulty of attaining the standard and the regional scale of the problem came to be recognized, so that in the 1990 CAA Amendments the deadlines for attainment were set at 2005 for severe areas and 2010 for extreme areas (and these are the deadlines currently obtaining). Several management cycles have been involved, therefore, over the past 30 years in trying to reach the desired future.

The use of complex, emissions-based models (higher order models; HOMs) for assisting the decision process has become increasingly important with each management cycle, especially during the last 20 years. The progression towards more complex modelling has coincided with improved understanding of the features and scale of the problem and the availability of greater computing power, as elaborated below. Models are expected to deal with complexity and provide a reliable basis for decision making. Over time the cost of the control programs for ozone has grown to more than \$1 billion (10^9) per year (US EPA, 1997).

9.2 EVOLUTION OF MODELS OF AIR QUALITY

To meet the demands of supporting environmental management, air quality models have had to address a number of challenges, one of the largest being to represent the complex associated chemistry in an abbreviated manner in order to fit this within computational constraints and to account for the chemistry faithfully, in spite of some key weaknesses in our knowledge base. Another substantial challenge has been to represent more reliably the evolution of an air mass and the interactions among different air masses, thus advancing air quality modelling from the paradigm of a stirred batch reactor to one of dynamic chemistry in a fluid flow. The impetus has always been to increase the sophistication of the models, as the modelling community tried to understand why the reductions in emissions did not seem to be as effective as expected, or as originally predicted.

9.2.1 Representation of Chemistry

A central feature of photochemistry, besides the breaking apart of key species by photons, is the attack on organic carbon species by the hydroxyl radical, OH. There are thousands of organic carbon species (VOCs) involved, with lifetimes from minutes to years. Yet the chemical mechanism used in the models must be condensed to only a few tens of species because of computer memory and power limitations. In addition, there are still major uncertainties and gaps in our knowledge of reaction mechanisms, products of important classes of organic compounds, and termination pathways in other classes (see US EPA, 1996, for a discussion). The size of the radical pool produced by daughter products, in dynamic interactions with radical initiation and termination, influences the cycling of OH and hence the production of ozone (Jeffries, 1995).

The inhibition of ozone production, and the creation of the so-called NO_X disbenefit in the radical-limited portion of the response surface to the left of the ridgeline, is controlled by the inorganic chemistry of NO_X. This chemistry is considered to be well understood and is represented relatively completely in the chemical mechanisms developed for the air quality models. There is little doubt about the general shape of the ozone response surface with its two regimes. It is with respect to VOC behaviour that the representation of the gas-phase chemistry is vulnerable to uncertainty.

The construction of photochemical mechanisms for use in the models is as much art and judgement as it is science. Smog chamber simulations, run at high concentrations typical of very polluted conditions, are used to constrain the parameterisation of the mechanisms and school expert intuition. There is thus a continual need for professional judgment about the structure and parameterisation of the chemical mechanisms. Understanding and representation of gas-phase chemical mechanisms have increased in size, from the order of 10 species to more than 50 species and 160 reactions (e.g., Dodge, 1977; Stockwell et al., 1990). Current gas-phase chemical mechanisms are typically responsible for 75% of the computer time used in air quality models that do not include particle (or heterogeneous) physics.

9.2.2 Representation of Transformation Coupled with Transport

Air quality models have changed considerably over the last 25 years regarding representation of mixing, transport and the interactions of air masses. The first computer model was a single box of variable height (including chemical mechanisms) designed to reflect the change in the mixing volume of the atmosphere as part of the diurnal cycle, termed OZIPR (Gery and Crouse, 1990). For the large urban areas it became clear that the magnitude of peak ozone, its location, and its response to controls, depended on the spatial details of the emissions and wind flow. A simple box description was inadequate. The use of Eulerian (fixed grid) urban photochemical models was allowed and then encouraged (U.S. EPA, 1986; U.S. EPA, 1991). Meteorology for input was interpolated. It then became clear that in the eastern United States regional transport was important; one city's downwind plume could set the upwind boundary conditions of an urban area in a neighbouring state. Regional models covering several thousand kilometres were developed to provide boundary conditions for the urban model (Lamb, 1983; US EPA, 1988). Meteorology was diagnosed from observations using equations to preserve mass consistency. Further study also showed that ozone itself was a key "player" in radical initiation and oxidant production in rural areas and that its regional background concentration was important and must therefore be included in considerations of management. The models have now advanced to the status of urban models being nested within regional models of the same structure and driven by meteorology produced by primitive equation meso-scale meteorological models. Air quality models are now simulating complex chemistry in a fluid flow – as a surrogate reality. The more advanced models include cloud processes (convective mixing, aqueous chemistry and wet deposition), air surface exchange (dry deposition), turbulent mixing, advection (transport), gas chemistry, aerosol chemistry and plume-in-grid treatments for very concentrated NO_X plumes from electric utility plants (Russell and Dennis, 2000).

Generation of emissions of NO_X and VOCs for the air quality models is a critical part of the environmental management cycle. Emissions are built up from many individual source categories. Typical diurnal and seasonal activity factors, together with hourly wind speed and temperature factors, are used to help create hourly emissions that are day-specific. VOC source profiles for the source categories containing hundreds of species are collapsed into the smaller number of groups of species for the chemical mechanisms. Many components of the emissions estimation procedure are based on engineering emission factors and activity factors, or surrogates, such as population. Two important components of emissions estimation, however – mobile source (passenger vehicles and trucks) and biogenic emissions – are themselves derived from models. For management scenarios one or more historical, several-day, high-ozone episodes are simulated to create a base case. The emissions are then changed to reflect a proposed emissions reduction strategy and the control scenario is simulated using the same meteorological period (the meteorology is held constant). The change in ozone, for example, peak ozone in the domain, between the base case and the control scenario, is used in the planning process.

9.3 THE FORECASTING PROBLEM: THE IMPORTANCE OF ISOPRENE EMISSIONS

The analysis of smog chamber data and early air quality data in support of the 1970 CAA led to the conclusion that reduction in VOC emissions would be the most effective strategy for reducing ozone concentrations in those urban areas with pollution problems. The first guidance for development of mitigation strategies, therefore, focused on reducing VOC emissions at the urban level. National reductions in automotive emissions of NO_X and VOC were also called for as a complement of the local controls. In retrospect, the focus on VOC emissions reductions is understandable, because smog chamber experiments had to be run at high levels of NO_X to avoid wall-effect contamination of the experimental results. Such levels were representative of a very polluted atmosphere, such as that of Los Angeles in the 1960s, and representative of the radical-limited regime to the left of the ridgeline in the ozone response surface. In effect all urban areas were assumed to be like Los Angeles.

The role of biogenic hydrocarbons in photochemical modelling, specifically isoprene, was the subject of considerable debate in the late 1970s (Dimitriades and Altshuller, 1977). The question was: "is isoprene important?" The answer at that time was "no". A scientific judgment was made (in the late 1970s and early 1980s) that isoprene emissions from plants were not important and did not need to be taken into account (Dimitriades, 1981). This judgment was based on very sparse data and analyses; a major degree of scientific intuition and judgment was thus required. In the analysis, empirical measurements were given primacy over model estimates. In effect, a decision about the structure of the chemistry incorporated in the models had been made. A parameterisation of isoprene oxidation that had been omitted in the chemical mechanisms continued to be left out of the mechanisms deliberately – as an explicit decision, that it was not important enough to include, given the computational constraints and other uncertainties.

The role of biogenic hydrocarbons was revisited with the advent of the regional models, where estimates of these emissions needed to be included for completeness, that is, to avoid the problem of incomplete model structure. The chemical mechanisms had to be quickly modified to incorporate the oxidation of isoprene and its products and calibrated against old and new smog chamber data. The structure of the emissions processing also had to be augmented to incorporate a model of isoprene emissions from plants, especially trees. New estimates for isoprene emissions (BEIS1) and modelling studies of their effects on regional ozone and control strategy results began circulating in the scientific community in the late 1980s (Roselle et al., 1991; Roselle, 1994). These studies indicated that it was important to include isoprene. They also pointed out the importance of the rural ozone background concentration on the boundary conditions for urban areas.

Concurrently, measurements in urban areas of the southeast United States and companion box-model analyses of the chemistry (to help interpret the measurements) concluded that consideration of biogenic hydrocarbons in ozone production in these cities, particularly isoprene, was important (Chameides et al., 1988). They also suggested that it was important to consider the effects of biogenic hydrocarbons in development of emissions reduction plans and that an exclusive focus on urban VOC reductions was erroneous guidance, thwarting effective planning in some urban areas. The debate over considering a more balanced approach, including control of NO_X emissions, reached a head with the report from the National Academy of Sciences (NAS) in 1991 (NRC, 1991).

Advancement in the science of biogenic emissions has led to new estimates of biogenic emissions (BEIS2) that are about five times higher than the earlier estimates (Geron et al., 1994). The scientific debate over the accuracy of these estimates continues, fuelled in part by the effect such higher estimates might have on emission reduction plans (Morris et al., 1997). Modelling studies continue to address the issue (Pierce et al., 1998).

9.4 FORECASTING IMPACTS IN PERSPECTIVE

A retrospective analysis can characterize in a consistent manner the impact on (ozone forecasts and) choice of control strategy resulting from different choices or expectations about the level of isoprene emissions. All other factors, such as differences in model structure, chemical mechanisms, initial and boundary conditions, other emissions and meteorological inputs, can be held constant. Such an analysis represents an idealized case for examining forecasts. It also allows us to examine the issue in a retrospective manner to ascertain how large the signals were that were being missed or ignored. For this analysis a multi-day high ozone episode – typical for the northeastern area of the US – was simulated with the Regional Acid Deposition Model (Chang et al., 1987). The horizontal resolution of the simulations was 20 km, which is expected to be small enough to capture the main features of the issue regarding the relative effectiveness of NO_X and VOC controls.

What is characterized is the change in base-case ozone resulting from simple, spatially uniform emissions reductions of 50% for either anthropogenic NO_X or VOCs. The uniform emissions reductions are simulated for three different cases: no biogenic emissions (No Beis), the first estimate of the magnitude of biogenic emissions (BEIS1), and the current estimate of biogenic emissions (BEIS2). The analysis is performed in this manner to highlight the issues. A reduction of 50% is typical for general analyses, although actual emissions reductions for State Implementation Planning (SIP) are spatially more complex and may be larger. However, use of spatially uniform emissions reductions is a standard interpretive approach, because it removes some extraneous features of the ozone response, and is the style of analysis used by the scientific community to illuminate the problem in the NAS report. Two of the high ozone days – July 30, 1988 and August 1, 1988 – are presented in the retrospective analysis. The purpose is to provide us with a sense of how large are the relative effectiveness changes from the no isoprene case, through the BEIS1 case, to the more current estimates obtained from BEIS2.

9.4.1 Retrospective Results for Spatially Uniform, Widespread Reductions

Percentage reductions for the New York area for the different cases of biogenic emissions are shown in Figure 9.2. In the urban core there are clear NO_X disbenefits, which diminish to near zero with BEIS2. There are clear benefits for the VOC emissions reductions in all cases; however, the percent effectiveness of VOC controls is cut in half between No Beis and BEIS2. For the urban peak downwind of the city, there are clear NO_X disbenefits for the No Beis case, but more mixed results for BEIS1 that nonetheless appear to favour VOC reductions. There are basically indistinguishable benefits for both NO_X and VOC emissions reductions for BEIS2.

Percentage reductions for the Detroit area for the different biogenic emissions cases are shown in Figures 9.3(a) and (b). Depending on the day simulated and the location, the choice of control benefits is either ambiguous, with a propensity towards favouring NO_X emissions reductions under BEIS2, while VOC controls are favoured in the area of the peak, regardless of the level of biogenic emissions.

Percentage reductions due to the widespread emissions reductions for the Philadelphia area for the different biogenic emissions cases are shown in Figure 9.4(a). Depending on the day simulated, for the No Beis case neither control approach is better in the urban core, yet reductions at the peak favour VOC control. For the BEIS1 case the separation towards favouring NO_X controls that is very evident for grid cells over the city is either muddled for the peak ozone prediction or accentuated. For the BEIS2 case, the separation between control effectiveness of the two choices is unambiguous, even for the peak. VOC controls are less than half as effective in BEIS2, as in No Beis, and NO_X controls are approximately 50% more effective in BEIS2 than in No Beis.

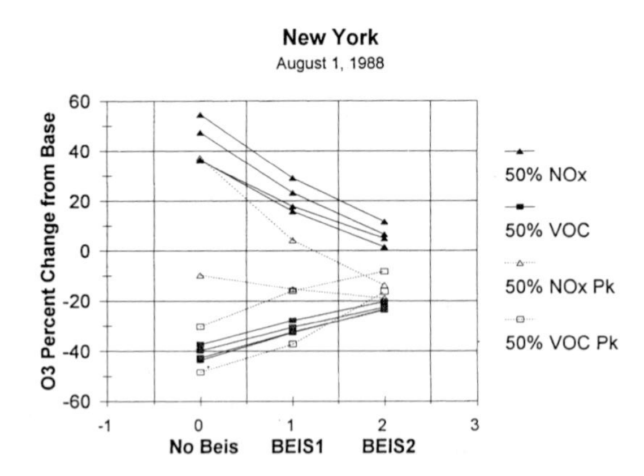

Fig. 9.2. The predicted percent change in daily peak O_3 from the base case due to region-wide 50% reductions in anthropogenic NO_X or VOC emissions plotted as a function of the level of biogenic emissions, from none (No Beis) to BEIS2, for New York City grid cells for the particular day of August 1, 1988. Solid lines are for grid cells in the city; dotted lines are for cells containing the ozone peak downwind and close to the city.

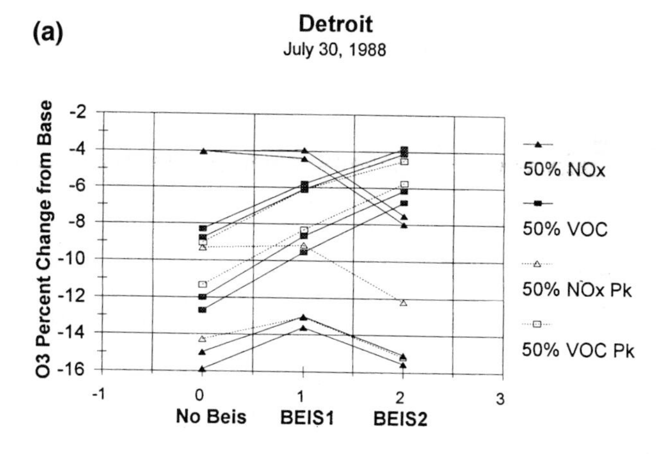

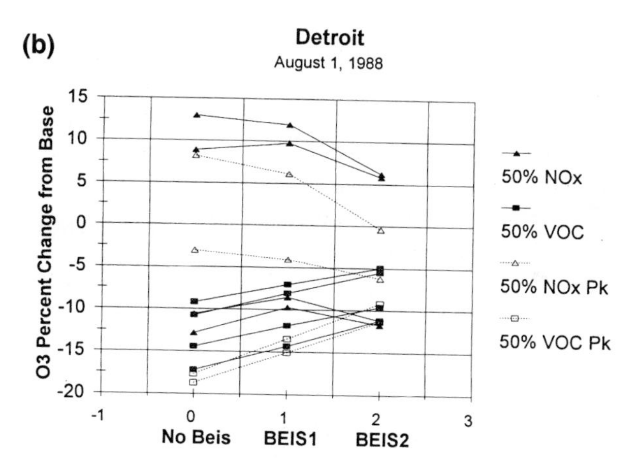

Fig. 9.3. Same construction as Figure 9.2 but for Detroit, Michigan, showing a city for which there is an ambiguous response: (a) for July 30, 1988; and (b) for August 1, 1988.

Percentage reductions for the Pittsburgh city and surrounding area for the different biogenic emissions cases are shown in Figures 9.4(b) and 9.5(a), respectively. The only time VOC controls show a slight advantage over widespread NO_X controls is for the No Beis scenario. With BEIS1 biogenic emissions, there is no doubt that widespread NO_X controls are more effective. The effectiveness of VOC controls is more than cut in half in BEIS2 compared to No Beis, and NO_X controls are twice as effective in BEIS2 relative to No Beis.

Percentage reductions for a rural site in central Pennsylvania (Scotia) for the different biogenic emissions cases are shown in Figure 9.5(b). As with Pittsburgh, once biogenic emissions reach the level of BEIS1, regional NO_X emissions reductions are clearly more effective.

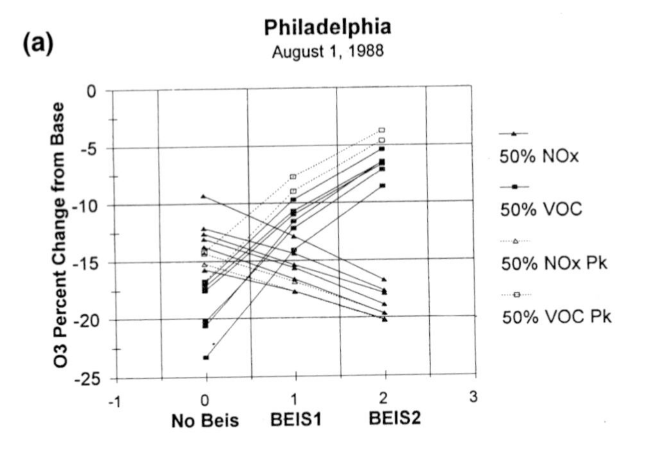

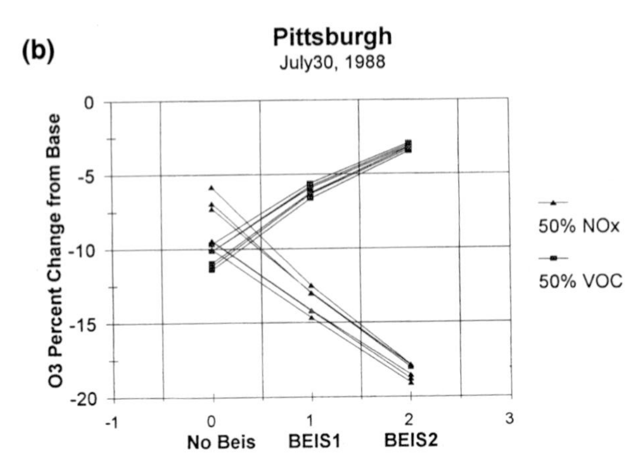

Fig. 9.4. Same construction as Figure 9.2 but for cities with a clear shift in control preference: (a) Philadelphia, Pennsylvania, for August 1, 1988; and (b) Pittsburgh, Pennsylvania, for July 30, 1988. For Pittsburgh city cells and peak cells coincide.

9.4.2 Retrospective Results for Urban-focused NO_X Reductions

The early modelling of the reductions only considered reductions in the urban emissions and estimated what would happen to boundary ozone coming into each urban area. Embedded in the results for all of the urban areas is the fact that background NO_X is reduced due to widespread regional NO_X reductions, hence reducing background ozone coming into the area. The question is, if only local NO_X controls were modelled, would the "obvious" signal noted above be diluted, i.e., that NO_X controls are very competitive and even out-compete VOC controls a majority of the time? Might a purely urban focus have missed a significant part of the NO_X

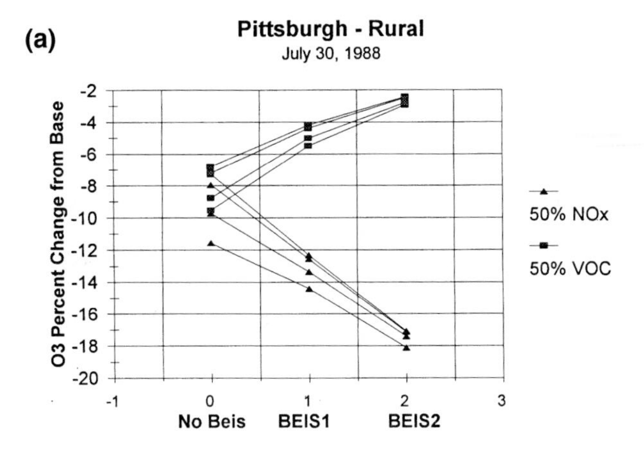

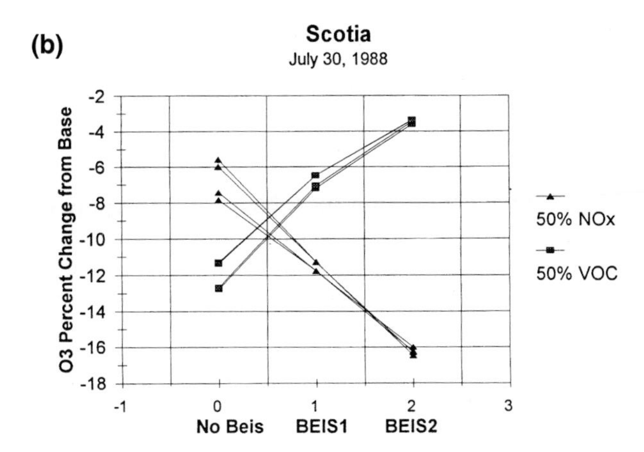

Fig. 9.5. Same construction as Figure 9.2 but for rural areas (separate considerations of peaks unnecessary): (a) 100–200 km downwind of Pittsburgh, Pennsylvania, for July 30, 1988; and (b) in the centre of rural Pennsylvania for July 30, 1988.

signal? To examine this, the above cases were simulated again, this time assuming only urban reductions of NO_X emissions in each of the cities at a relatively high control level of 75%.

The results of this urban NO_X scenario are compared to the widespread reduction results for July 30, 1988, in Figures 9.6 and 9.7. For New York (Figure 9.6), there is not much difference in the effectiveness of NO_X emissions reductions for either the urban-focused or widespread NO_X reductions. The disbenefit of NO_X remains high. For the other three cities (Figure 9.7), the competitiveness of NO_X emissions reductions is more than diluted; the competitiveness disappears when only local NO_X controls are implemented, even at a 75% reduction. VOC emissions controls are

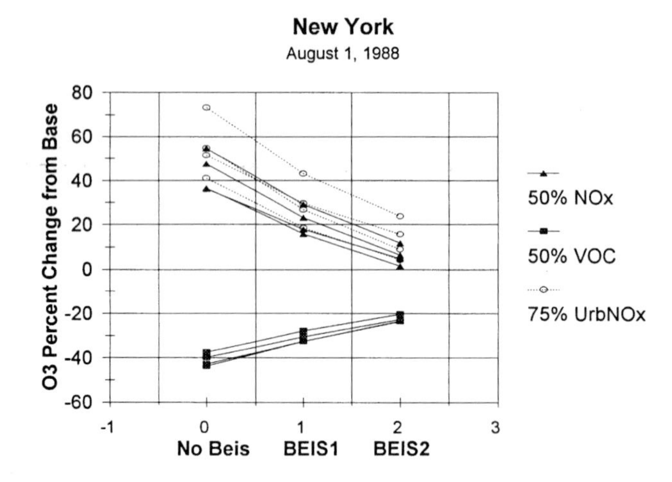

Fig. 9.6. The predicted percent change in daily peak O_3 from the base case due to region-wide 50% reductions in anthropogenic NO_X or VOC emissions, or 75% urban-only NO_X reductions, plotted as a function of the level of biogenic emissions, for New York City grid cells. Solid lines are for the 50% region-wide reductions; dotted lines are for the 75% urban-only NO_X reductions.

clearly favoured in all of the cities if only local NO_X reductions are considered. Only in Philadelphia for the BEIS2 case do local NO_X controls achieve any parity with VOC emissions controls. Since the actual controls that were modelled would focus on the peak ozone and assume some change in the boundary conditions, analysts were probably examining results that fell between the two sets of NO_X emissions reduction results developed here for the retrospective analysis. Nonetheless, it is clear that with a purely urban focus for these cities, there probably would not be a clear case for control of NO_X emissions.

9.4.3 In Summary

The regional analysis shows there is a clear NO_X disbenefit for New York. The VOC controls are effective throughout, although their effectiveness (magnitude of % reduction) is reduced. A mixed message is suggested for Detroit: VOC and NO_X controls are roughly of the same order, and the relative effectiveness depends on the meteorological day simulated. Results for Philadelphia and Pittsburgh suggest that NO_X control is definitely more beneficial. VOC controls are still effective, but their "power" has been cut in half in going from No Beis to BEIS1 and cut in half again in going to BEIS2. However, embedded in the results for all of the cities is the fact that background or regional NO_X is reduced and, hence, background ozone coming into the area is reduced. This is a critical factor. When NO_X reductions with a purely urban focus are examined, the apparent NO_X control advantage disappears.

A consistent modelling analysis shows there is a clear erosion of the power of the VOC controls as isoprene emissions come into play and it also shows that in many

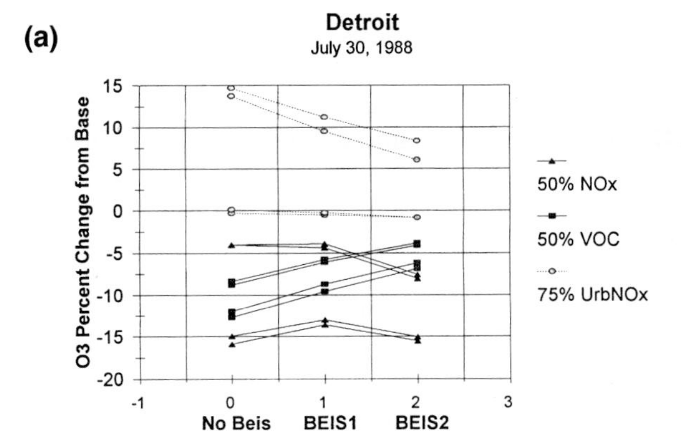

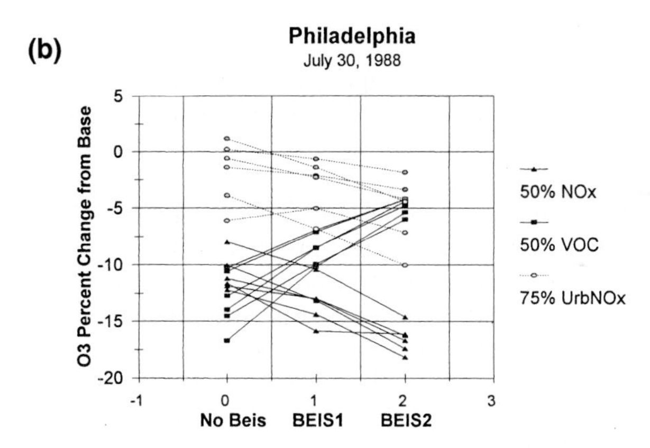

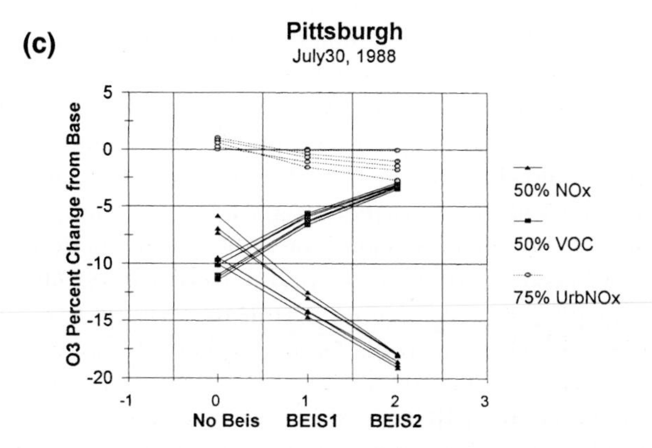

Fig. 9.7. Same construction as Figure 9.6 but for: (a) Detroit, Michigan, for July 30, 1988; (b) Philadelphia, Pennsylvania, for July 30, 1988; and (c) Pittsburgh, Pennsylvania, for July 30, 1988.

urban areas of moderate size NO_X controls should have been taken seriously, even with very low isoprene emissions. Yet it is the regional, not local, nature of the NO_X controls that is important to this conclusion. It is not absolutely clear that NO_X controls definitely would have been found to be more effective for any of the urban areas, had a full 3-D urban model analysis been carried out in the early stages. Note, the first attempt at modelling would have been using BEIS1.

9.5 DIFFICULTY DISCERNING THE TRUTH

While the effect of the isoprene misrepresentation looks relatively obvious in retrospect, it was not easy to discern at the time. Early on, and into the 1980s, nearly all of the SIP analyses used OZIP, the box-model approach, with a chemical mechanism that initially excluded isoprene. OZIP was used because it was easy and because of the severe limitation of computational power and very limited expertise to run the emerging 3-D models. So very optimistic predictions of the effectiveness of VOC controls would have resulted. The accompanying response surface of predicted peak ozone was effectively tuned in conjunction with monitoring data. This practice precluded the development of an awareness of any latent problems, because potential discrepancies between the model and measurements were rendered insignificant by calibration. Urban models today would not show such a rosy picture for VOC controls. It is expected that the problems of the early 3-D models in predicting high enough ozone, without isoprene emissions, would have raised questions. It is not known whether any comparisons have ever been made between earlier OZIP predictions and later 3-D model predictions.

The belief that it would be better to employ 3-D models (instead of OZIP), in areas with severe ozone problems, initiated a drive in the 1980s to begin using 3-D models for urban areas outside of Los Angeles, starting with New York. Thus, the urban region with the greatest amount of NO_X disbenefits and potentially the urban area in the eastern US most consistent with earlier OZIP thinking, was the first to be modelled there. This made it unlikely that the NO_X issue would dramatically or quickly emerge from these first 3-D model results. The urban focus of the early modelling would have resulted in only developing local NO_X controls and would have overlooked the regional signal. Coupled with this, simple assumptions about how boundary conditions would change were used. In retrospect, these assumptions were optimistic and, together with the incorrect urban-only focus, improperly attributed the source of changes in boundary ozone to the effectiveness of VOC controls.

Meanwhile, realisation of the more correct, full regional/urban modelling capability was a slow process. Regional models of ozone were being developed to allow the background influence of ozone from regional emissions to be simulated. The new regional models of the 1980s and early 1990s had difficulty producing enough background ozone with BEIS1, giving rise to distrust of their predictions. Moreover, there was poor coupling between regional boundary conditions and urban modelling. As a result, much judgement (or established intuition) continued to be used in setting boundary conditions for the urban models in order to represent future conditions; and these judgments continued to err on the side of being optimistic.

Use of models was (and is) very resource-intensive. The schedule for obtaining results and feeding them through the process was always very tight. This onerous schedule meant that no-one had time to look and learn. It was simply essential to complete the analyses.

For the urban 3-D model, the inputs, especially emissions, boundary conditions and meteorology, were judged by the entire modelling community to have large uncertainties. It was difficult to obtain acceptable predictions of ozone with the new 3-D models, as defined by EPA, and many simulations were made on the selected episodes until adequate performance was obtained. Thus, procedures to tune the model came into practice, and have actually been proposed as items to be codified into normal practice in Reynolds et al. (1996). This type of model tuning (a model application practice) masked model problems and biased the search for improvements using the 3-D model (Russell and Dennis, 2000). More recent simulations of peak ozone in the New York area are in broad agreement with the RADM results for downwind peaks presented above. When localized control strategies were examined, they showed less distinction between NO_X and VOC controls (when examining peak ozone). In addition, they show the confounding influence of meteorological uncertainty on the model prediction of control effectiveness (see Sistla et al., 1996).

9.6 THE JERKY EXCHANGE BETWEEN POLICY AND SCIENCE

The early choice in the 1970s to concentrate on VOC controls as the strategy of choice by EPA was based on economic and technical grounds and supported by early results from the models. While new results regarding isoprene were accumulating during the 1980s, EPA seemed to continue on its preordained path of VOC control. The onerous schedule set by Congress and EPA for carrying out the SIP process was not conducive to thoughtful reflection on the procedure. Decisions had to be made and had to be made “now”. Much attention was focused on the northeastern US cities along the eastern seaboard, where, given the urban focus and the large potential for ambiguity regarding relative effectiveness of the two control options, it would not be hard to find signs that VOC controls were continuing to appear effective.

Scientists were increasingly voicing concern and criticism of EPA’s VOC-only stance. The influential Chameides paper was published in *Science* in 1988.Yet the concerns of the scientific community were apparently not heeded until their criticism achieved formal status in a 1991 NAS report. Interestingly, many of the studies quoted in this report were performed by scientists with EPA’s Office of Research and Development (EPA/ORD), which was developing regional ozone models to address the regional-urban coupling and long-range transport of ozone. Through these models, with nominally an 18-km grid resolution, the importance of NO_X controls was evident, both regionally and for the smaller urban areas.

The response to the 1991 NAS report was to increase efforts in the realm of regional ozone studies and to increase attention on regional NO_X controls. EPA and the state air agencies made a strong commitment to address the regional character of the ozone problem and the importance of regional NO_X controls. The Ozone Transport Assessment Group (OTAG), a major consortium of state air directors in the eastern US, was formed to study the potential for additional regional NO_X controls. OTAG initiated a substantial exercise in modelling, indeed the largest ever. EPA responded to OTAG with the NO_X SIP call in 1998, using BEIS2, in which large reductions in utility NO_X emissions were called for. This "call" is still going through due process. While the VOC reductions remain in place, the policy focus has clearly shifted to NO_X. The 1990s seem thus to have witnessed the pendulum swinging from strong attention to VOCs to strong attention to NO_X. Actions, in fact, have the appearance of a swinging pendulum that is kicked in motion by political and technical decisions of a rather short number of years and then momentum takes over, leading to a jerky exchange between policy and science.

The process continues to evolve, as does the debate over emissions. A BEIS3 is expected very soon. Other parts of our ignorance are being discussed, including biogenic NO_X emissions and non-isoprene biogenic VOC emissions. Other parts of the chemistry are being tested and the possibility is being raised of particles influencing gas-phase chemistry (Cantrell et al., 1996). Issues relating to modelling scale and resolution are also surfacing: for example, analyses suggest there might be the possibility of oxidation production potential being a function of NO_X emissions density, suggesting that regional reductions from intense power plants may not be very effective (Ryerson et al., 1998). Most of these questions are being addressed piecemeal in a largely *ad hoc* manner within the scientific community, because resources have been shifted from the study of ozone towards that of particulates, the latest pollutant of concern (things, alas, do not change).

9.7 LESSONS TO BE LEARNED

There are several lessons we can learn from this case study, to be added to those already distilled from the case study of eutrophication (in Chapter 7).

***Lesson 1**: Given the limits of knowledge and of computational power, we are bound to be missing pieces of structure or represent them poorly in HOMs for the foreseeable future.*

The problems confronting simple low order models (LOMs), including questions of structure, also pertain to HOMs, their basis in the traditions of geophysical modelling notwithstanding. The ozone case study shows that the problem stemmed first from a decision on the structure of the model regarding the chemical processes to be included in it. Then the structural issues changed form and spread to additional parts of the system. These decisions on structure and their impacts moved then to the quality of the parameterisation of the isoprene chemistry (down a level in detail) and the structure of the emissions model required to produce the chemical species for input to the models (up and over a level of detail). The expertise for the emissions modelling was initially outside the traditional area of the scientists working on this problem (chemists imagining biology), although great strides were made in this area.

Lesson 2: *It is difficult to discover errors in the model or modelled processes affecting future predictions by examining state variables of current conditions. This is believed to be true for all models.*

Here, the study of current conditions was skewed by the monitoring stations being almost exclusively sited in urban areas, since that was where the issue was of concern; yet an urban area is where there is least sensitivity to the error (or is it that so many plausible explanations exist, the "real" truth is hard to discern?). It matters not where the problem might lie, its resolution is sought only in the space that happens to be illuminated.

Lesson 3: *It is especially difficult to discover errors affecting predictions in complex HOMs.*

This will be particularly true, if we only examine state variables predicted by the HOMs, because feedbacks and nonlinearities in the system can create compensating errors or self-buffering, so that what is really happening internally is masked (Dennis et al., 1999). This suggests the difficulty we will have with evaluating virtual systems (such as that of any "Virtual Ecology"; Woods, 1999).

Lesson 4: *There are many other imperatives and pressures arising from the formulation and implementation of policy and these can cause the searching for issues of structural change to be ad hoc and have therefore a high probability of failure.*

In this case the sensitivity to a signal of something possibly being wrong was masked by the practice of always tuning the models to meet a regulatory acceptance criterion on the state variable of concern, before use of the model could proceed. Similar masking arose from the slow transition to more adequate modelling of the full urban/regional scope of the problem. The commitment to an initial control path can also become entrenched due to political, cost, and equity factors. Thus, if we do not have a program actively seeking apprehension of structural change, and developing associated skills, procedures, and interests, we will become entirely absorbed in environmental management, miss the first sense of bumping up against our ignorance (the {acknowledged unknown}), and incur "expensive" costs and delays. Where the notion arises of designing a model for discovering what we know we do not know (in Chapters 5 and 15), it has been prompted by the experience recorded in this case study.

The ozone case study shows that when we are exploring the fringes of our scientific knowledge, where we expect to be looking for surprises, we will also be exploring the shores of our ignorance. Thus, the "message" that we obtain from the models about structural change is expected to have a significant degree of ambiguity. The message will not be sufficiently obvious for us to perceive it, especially if we did not happen to be looking for it. If we want to avoid the surprises before they are fully upon us, we need to be actively looking for them – an issue we shall take up again in Chapter 17.

REFERENCES

Cantrell, C.A., Shetter, R.E., Gilpin, T.M., Calvert, J.G., Eisele, F.L. and Tanner, D.J., 1996. Peroxy radical concentrations measured and calculated from trace gas measurements in the Mauna Loa Observatory Photochemistry Experiment 2. *J. Geophys. Res.*, **101**, 14,653–14,664.

Chameides, W.L., Lindsay, R.W., Richardson, J. and Kiang, C.S., 1988. The role of biogenic hydrocarbons in urban photochemical smog: Atlanta as a case study. *Science*, **241**, 1473–1475.

Dennis, R.L., Arnold, J.R., Tonnesen, G.S. and Li, Y., 1999. A new response surface approach for interpreting Eulerian air quality model sensitivities. *Computer Phys. Comm.*, **117**, 99–112.

Dimitriades, B. and Altshuller, A., 1979. International conference on oxidant problems: Analysis of the evidence/viewpoints presented, 1. Definition of key issues. *J. Air Pollut. Contr. Assoc.*, **27**, 299–307.

Dimitriades, B., 1981. The role of natural organics in photochemical air pollution: Issues and research needs. *J. Air Pollut. Contr. Assoc.*, **31**, 229–235.

Dodge, M.C., 1977. Combined use of modelling techniques and smog chamber data to derive ozone precursor relationships. In: *Proc. Int. Conf. On Photochemical Oxidant Pollution and Its Control* (B. Dimitriades, ed.). EPA-600/3-77-001b, U.S. Environmental Protection Agency, Office of Research and Development, Research Triangle Park, NC 27711, pp. 881–889.

Fiore, A.M., Jacob, D.J., Logan, J.A. and Yin, J.H., 1998. Long-term trends in ground level ozone over the contiguous United States, 1980–1995. *J. Geophys. Res.*, **103**, 1471–1480.

Geron, C., Guenther, A. and Pierce, T., 1994. An improved model for estimating emissions of volatile organic compounds from forests in the eastern United States. *J. Geophys. Res.*, **99**, 12,773–12,791.

Gery, M.W. and Crouse, R.R., 1990. User's Guide for Executing OZIPR, Order No. 9D2196NASA, U.S. Environmental Protection Agency, Office of Research and Development, Research Triangle Park, NC 27711.

Haagen-Smit, A.J., 1952. Chemistry and Physiology of Los Angeles Smog. *Ind. Eng. Chem.*, **44**, 1342–1346.

Harley, R.A., Sawyer, R.F. and Milford, J.B., 1997. Updated photochemical modelling for California's South Coast Air Basin: Comparison of chemical mechanisms and motor vehicle emission inventories. *Environ. Sci. Technol.*, **31**, 2829–2839.

Jeffries, H.E., 1995. Photochemical air pollution. In: *Composition, Chemistry and Climate of the Atmosphere* (H.B. Singh, ed.). Van Nostrand Reinhold, New York, pp. 308–348.

Lamb, R.G., 1983. Regional scale (1000 km) model of photochemical air pollution, Part 1. Theoretical formulation. Report No. EPA/600/3-83-035, U.S. Environmental Protection Agency, Office of Research and Development, Environmental Sciences Research

Laboratories, Research Triangle Park, NC.

Leighton, P.A., 1961. *Photochemistry of Air Pollution*. Academic Press, New York.

Pierce, T., Geron, C., Bender, L., Dennis, R., Tonnesen, G. and Guenther, A., 1998. Influence of increased isoprene emissions on regional ozone modelling. *J. Geophys. Res.*, **103**, 25,611–25,629.

Reynolds, S.D., Michaels, H., Roth, P., Tesche, T.W., McNally, D., Gardner, L. and Yarwood, G., 1996. Alternative base cases in photochemical modelling: Their construction, role and value. *Atmos. Environ.*, **30**, 1977–1988.

Roselle, S.J., Pierce, T.E. and Schere, K.L., 1991. The sensitivity of regional ozone modelling to biogenic hydrocarbons. *J. Geophys. Res.*, **96**, 7371–7394.

Roselle, S.J., 1994. Effects of biogenic emission uncertainties on regional photochemical modelling of control strategies. *Atmos. Environ.*, **28**, 1757–1772.

Russell, A.G. and Dennis, R.L., 2000. NARSTO Critical Review of photochemical models and modelling. *Atmos. Environ.*, **34**, 2283–2324.

Ryerson, T.B., Buhr, M.P., Frost, G.J., Goldan, P.D., Holloway, J.S., Hübler, G., Jobson, B.T., Kuster, W.C., McKeen, S.A., Parrish, D.D., Roberts, J.M., Sueper, D.T., Trainer, M., Williams, J. and Fehsenfeld, F.C., 1998. Emissions lifetimes and ozone formation in power plant plumes. *J. Geophys. Res.*, **103**, 22,569–22,583.

Sistla, G., Zhou, N., Hao, W., Ku, J.-Y., Rao, S.T., Bornstein, R., Freedman, F. and Thunis, P., 1996. Effects of uncertainties in meteorological inputs on urban airshed model predictions and ozone control strategies. *Atmos. Environ.*, **30**, 2011–2055.

Stockwell, W.R., Middleton, P. and Chang, J.S., 1990. The second generation regional acid deposition model chemical mechanism for regional air quality modelling. *J. Geophys. Res.*, **95** (D10), 16,343–16,367.

U.S. Environmental Protection Agency (U.S. EPA), 1986. Guidelines on air quality models. Report No. EPA/450/2-78-072R, U.S. Environmental Protection Agency, Office of Air Quality Planning and Standards, Research Triangle Park, NC.

U.S. Environmental Protection Agency (U.S. EPA), 1988. Regional Oxidant Model for Northeast Transport (ROMNET) Project, Alliance Technologies Corporations correspondence to Emissions Committee. Subject: Responses to the ROMNET Existing Control Questionnaires, October 14, 1988.

U.S. Environmental Protection Agency (U.S. EPA), 1991. Guideline for regulatory application of the Urban Airshed Model, Report No. EPA/450/4-91-013, U.S. Environmental Protection Agency, Office of Air Quality Planning and Standards, Research Triangle Park, NC.

U.S. Environmental Protection Agency (U.S. EPA), 1996. Air quality criteria for ozone and related photochemical oxidants, Report No. EPA/600/P-93/004aF, Volume 1, Office of Research and Development, Washington, D.C. 20460.

U.S. Environmental Protection Agency (U.S. EPA), 1997. Regulatory impact analyses for the particulate matter and ozone National Ambient Air Quality Standards and proposed regional haze rule, July 1997 Office of Air Quality Planning and Standards, Research Triangle Park, NC 27711.

Woods, J.D., 1999. Virtual ecology. In: *Highlights in Environmental Research* (B.J. Mason, ed.). Imperial College Press, London.

Williams, J., Roberts, J.M., Fehsenfeld, F.C., Bertman, S.B., Buhr, M.P., Goldan, P.D., Hübler, G., Kuster, W.C., Ryerson, T.B., Trainer, M. and Young, V., 1997. Regional ozone from biogenic hydrocarbons deduced from airborne measurements of PAN, PPN, and MPAN. *Geophys. Res. Lett.*, **24**, 1099–1102.

PART III
THE APPROACH

Environmental Foresight and Models: A Manifesto
M.B. Beck (editor)

CHAPTER 10

Belief Networks: Generating the Feared Dislocations

O. Varis

10.1 ON CERTAINTIES AND UNCERTAINTIES

To know something is a positive thing. The more one knows, the better – although the knowledge should not be overly one-sided. The associative ability to connect issues with one another, detecting analogies and controversies, is all the more important the more complex the issues one is dealing with (Rowe and Watkins, 1992).

Let us play with the idea of constructing a model for the future development of water resources on this planet, given the myriad interconnections of water with environment, poverty, urbanisation, food production, and other things that we can only put on paper. We are conscious of the underlying importance of the concept of sustainability, and understand that it means we should leave this planet to our children in a shape that gives them little reason to blame us. Then, after one generation they would do the same for their children, we hope. From this standpoint, it would be highly interesting to examine how this planet was one generation ago, and what it could be after one or two generations from now. We would select a few dozen of the most significant things that are undergoing change – such as population growth, urbanisation, climate, economy, globalisation, human resources, technology, and industry – and then try to define those affected by these changes: from ecology, to social issues, the economy, and so on. This passivity would soon start to bother us; we would brainstorm about how things could be driven in the direction we want, to make the world as good as possible after a few decades. We could also recall how policy makers, and in fact many other people, are enthusiastic about certain ideologies at certain times, and about others at other times. We would need to consider some policy tools and theoretical frames that would make our exercise a more active and attractive one.

It would be most interesting, then, to scrutinize all the various interconnections of these issues. Naturally, they all can be – and to a certain extent are – interconnected. To cover this, without losing track in handling so many interconnections at one and the same time, we need a systematic means of doing so. We should analyze each link systematically, whether it is negative or positive, whether it can be ignored, is small or large. To a certain extent, such features are universal, being similar everywhere, yet most of them are also specific to conditions with respect to climate, culture, nature, institutional set-up, and so on. The analysis should have a regionalised structure. So what we now need is region-specific knowledge on the states of the issues we want to study. But how should such diverse and uncertain things be presented in a way that could be used to address each of the issues in the same way? Perhaps a good idea would be to put aside quantification according to some absolute numerical scale, emphasizing instead the magnitudes of things merely as one relative to another, or as relative to themselves from one point in time to another. The present situation could be a good point of reference, from which to acquire some sense of whether an issue will grow (relatively speaking), decrease, or stay unchanged. Very few of us could state anything with certainty, however, preferring to say for a given item only that it is likely to grow, with a reasonable chance of not changing at all, or even declining. We could take probability distributions to describe all of this.

Now that we have written down what we know of the issues and their interconnections, it would be most rewarding to explore whether we have been at all consistent in our reasoning, what are the most critical issues with respect to some of the other issues, how would the other issues react, if we were to change something. We might even contemplate how we should advise our governments on making the world better for future generations. For that, we would need some sort of apparatus to help us with all the knowledge we have on the pieces of paper scattered across the desk. The available computer simulation models would help us quite a bit, of course, but rather as sources of reference information, not as final, usable products. For they contain only some of the interconnections, and they assume these are known precisely, which would make us suspicious. Such models tend also to reflect a view of the problems from a limited perspective, for example, from that of, say, water, or agriculture, or poverty, economics, population growth, or from some other point of view – but disregarding many of the issues we consider of importance. Such models appear too deterministic, missing many interesting things, especially associations between issues that do not belong to any single discipline.

At this point, somebody would perhaps say: why don't you just let the data tell us what will happen? But the data do not always tell all that much. For instance, the data on the arable land area of China differ with an error margin of 20–30% from one another (Smil, 1992). Another example of the many that could be taken is urban poverty in India: an issue that touches a nine-digit number of people each day (Figure 10.1).

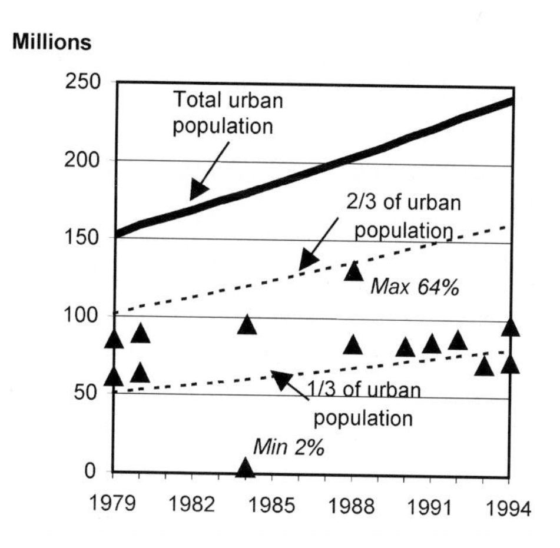

Fig. 10.1. Urban poverty in India. Triangles are estimates, which range from 2–64% of the urban population being defined as living in poverty (data collated by Amis (1997) from various sources).

What we could have achieved in this thought experiment would thus far have been accomplished essentially by brainstorming – using our own brains. Of course, we would have been forced to feed the brains with huge piles of books and reports and newspapers and whatever. For these kinds of extensive, tangled, barely quantifiable problems, we would need something to help us engage in the dull work of processing the melange of information, data, beliefs, threats, fears, opportunities, chances ... something, we argue, drawn from *Artificial Intelligence*.

From the position taken in this chapter, then (and it is rather different from that of most of the others), the nub of the challenge is to confess to all the gross, qualitative uncertainties, and then to do battle with them, specifically through the computational technique of Belief Networks (BNs). We shall examine case situations of consensus *versus* dispersion amongst the beliefs stakeholders have, and express the accompanying challenge: of needing new ways of exploiting awkward knowledge. Within the overall thread of the argument running through this part of the monograph, the ultimate goal is to illustrate a general computational means of generating impressions of the target futures, whose reachability might then be assessed with the procedure set out in the next chapter.

10.2 ARE WE OVERCONFIDENT IN OUR CONVENTIONAL NUMERICAL MODELS?

What will this planet look like after one or two generations from now? This is the toughest of questions, and nobody knows the answer with any certainty. The Population Project of the International Institute for Applied Systems Analysis (IIASA) listed all the certainties they were able to assemble on the basis of their population projections (Lutz and Prinz, 1994). They were:

> World population will continue to grow. By 2030, world population will increase by at least 50 percent and may even double in size. Short-term growth is inevitable: it is built into the age-structure of today's world population ...
>
> Developing countries will account for a greater share of the world's population. By 2030, today's developing countries will represent between 85 and 87 percent of the world's population – this is a very small margin of uncertainty resulting from extremely different scenarios ...
>
> All populations will become older ... The more rapidly fertility declines, the faster the population ages.

The rest is uncertain. And the forecasts differ quite a bit, although those of population are among the least uncertain of all global projections.

The 1990s indeed witnessed a boom in other forms of global forecasts, scenarios, projections, and assessments of the present and future states of natural resources, societies, environment, human development, and so forth. A series of UN Summit meetings was launched following the Earth Summit (held in Rio de Janeiro in 1992), together with some large-scale investigations of issues such as population, gender, food production, human settlements, and so on – all issues having complex interconnections with the evolution of our environment. In addition to the activities of the UN, innumerable other organisations have taken significant steps in the same direction, within their own specific spheres of interest. The various factors of concern, from the point of view of environmental change, can be grouped under the following five themes:

- Population growth
- Urbanisation and other patterns of migration
- Changes and variations in climate
- The economy and level of globalisation
- Human capital, technology, and industrialisation

What do these forecasts, assessments, projections, and scenarios tell us of the situation of our planet one or two generations hence? They imply very different futures, if compared one with another, as illustrated by Figure 10.2. The various respected institutions disagree strongly on future grain production, as well as regarding the amount of water used in the world. The global circulation models (GCMs) used to construct future scenarios for different parts of the world under a changed climate, cannot reproduce the present climate well enough to provide the future projections with much credibility. In the geographical regions noted in the lower set of 12 plots of Figure 10.2 – those with about one third of the world's population, 43% of the world's rural population, and two thirds of all the people judged to be living in absolute poverty – the GCMs cannot reproduce the present patterns of rainfall (the columns denoted as "Present") with any tolerable accuracy. Around 80% of the world's rural population depend almost entirely on monsoon rains, whose predictability is amongst the weakest elements of the GCMs. However,

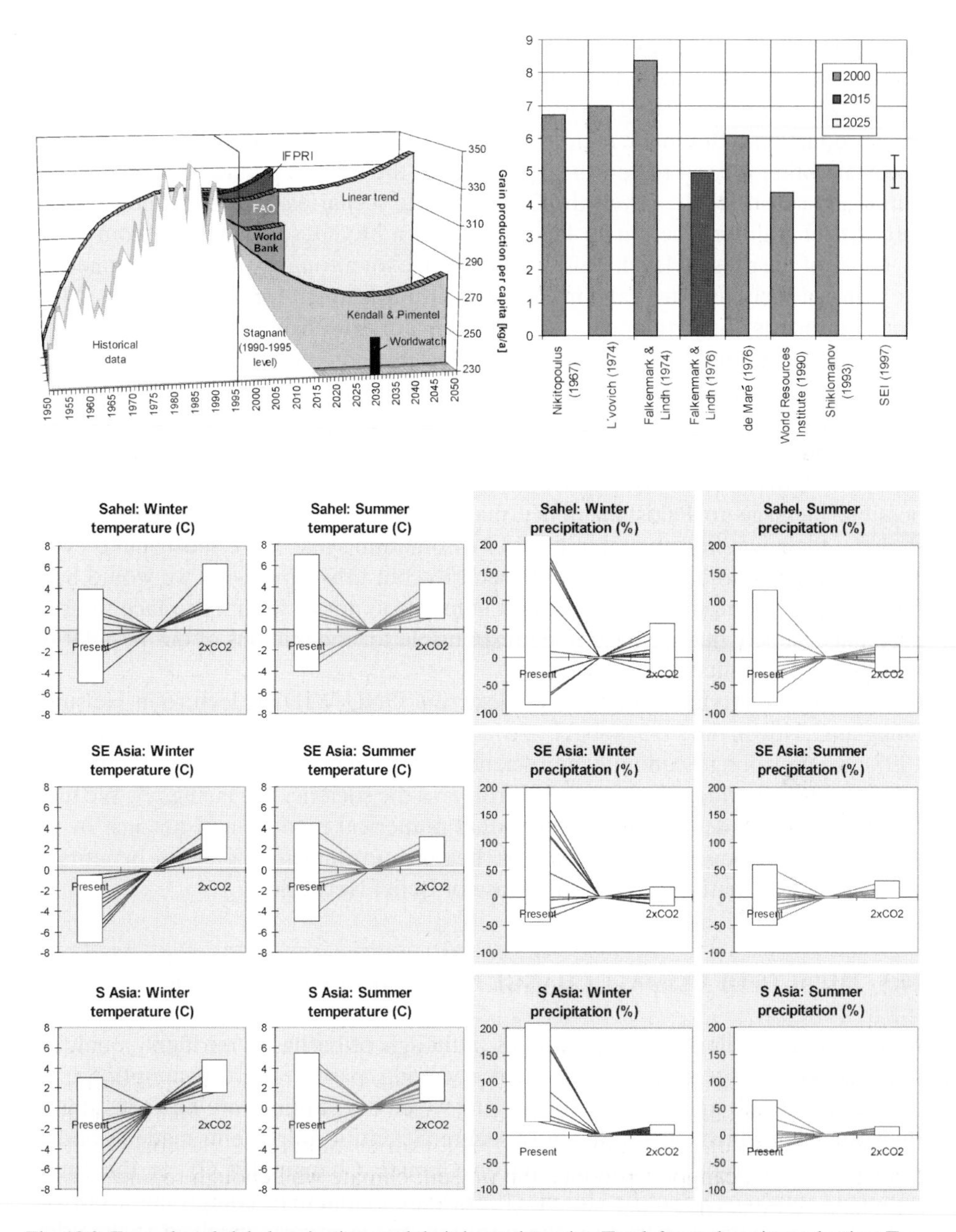

Fig. 10.2. Examples of global projections and their inconsistencies. Top left panel: grain production. Top right panel: water withdrawals (SEI, 1997). For these upper panels, references and a detailed analysis are given in Vakkilainen and Varis (1998, 1999). In the panels below are indications of climate change. The left-hand columns of each of these figures (marked "Present") represent the ranges of estimates of the current Global Circulation Models (GCMs) for current conditions. GCMs are used widely in climate change studies and for generating scenarios of the future; can we trust their predictions on this basis? (Data are from IPCC (1996) with detailed analysis given in Varis (1998d)).

we experts use these future predictions (the columns denoted as "$2\times CO_2$") for all sorts of scenario-generating purposes with seemingly no moral problems.

Elsewhere (Varis, 1998d) evidence has been presented on how projections of food production differ in the regions identified in Figure 10.2, along with production in certain other critical regions bearing most of the world's population. Projections differ very substantially. Given the GCM climate scenarios, the published range of forecasts of grain production for Pakistan under a $2\times CO_2$ scenario span from –60% to +60% of the present level. This is a huge range for a country that has remained on the brink of food insecurity and massive undernourishment for many decades. Most global forecasts, unfortunately, do not even acknowledge, let alone assess, the enormous uncertainties inherent in these projections. A user – a policy maker, scientist, journalist, layman – obtains merely a hint of the uncertainties spread across the results of the different studies; results that tend to diverge very much from one another, not only for global models, but also those cast at smaller scales. It is utterly clear, then, that such forecasts are enormously uncertain. So what might be the possible message to Pakistani policy makers of any of these individual studies? Perhaps their justified thoughts would be something like: "We must make every possible attempt to increase our food security, but this is just what we would have done anyway. Computer models do not help us too much; we have to face the real problems, namely that we have to be responsible for the millions of our population affected by our policy decisions".

In a press interview within the context of a UNU/WIDER lecture in Helsinki, Douglas C. North, the 1993 Nobel Prize laureate in Economics, put it this way (North 1997): "... prevailing Economics as a science has become so exact and mathematical, that it cannot provide solutions to the problems the societies are facing ...". We have become overconfident in our conventional numerical models, and not just in the domain of Economics. Let us therefore begin to explore the potential benefits of some of the alternatives, specifically those of Belief Networks.

10.3 IMPACTS OF CLIMATE CHANGE ON FINNISH WATERSHEDS

Models used in climate change studies, although often based on highly qualified scientific knowledge, nevertheless involve varying, partly expert, assumptions. No objective methods are available for judging which assumptions are more justifiable than others, or which GCM is the most secure. Scientific judgements made by bodies such as the Intergovernmental Panel on Climate Change (IPCC), or the International Council of Scientific Unions (ICSU), are bound to be compromises born of subjective evaluations, based on the intuition of the experts involved (Schneider, 1990). These panels and councils do not, however, use formal methods in the production of consensus summaries (Morgan and Keith, 1995). This is somewhat surprising for what is such a highly important step in using scientific climate-change knowledge, especially when the knowledge available in scientific journals is subject to strict scientific, peer review.

The need for intuitive knowledge in the evaluation of various information sources is due to the holistic nature of the problem. Although integrative methods, such as Geographical Information Systems (GIS) and decision support systems (DSS), are important for combining the various pieces of knowledge (Schackley and Wynne, 1995; Shlyakter et al., 1995), there is still a need for systematic procedures for evaluating subjective, expert knowledge. What is more, the method used must be interactive and sufficiently simple to allow the attainment of a consensus among experts, with yet an accompanying analysis of uncertainties. For example, Morgan and Keith (1995) used a structured interview, with elicitation methods, to evaluate the uncertainties in estimates made by climate change experts for single outcome variables. They did not, however, estimate overall uncertainty arising from knowledge of cause–effect relationships among the many variables of the problem as a whole. Their results indicate a higher degree of disagreement among the experts than reported in consensus documents, as we have already come to appreciate from the discussion of Chapter 2 (van der Sluijs et al., 1998). These results imply that the scientific uncertainties were underestimated. From the decision-making point of view, any underestimation of variance suggests that the need to account for risks has not been fully acknowledged, so that, in a sense, scientists may thus be overstepping the boundaries of their responsibilities.

Conversely, dissonance and disagreements among expert judgments are also an essential source of information when assessing the overall uncertainties attaching to a scenario. Indeed, Chapter 5 has argued that "fringe" opinion is to be nurtured, not least in the kinds of analysis illustrated in the following chapter. Within the IPCC system, for instance, scenarios are negotiated so as to yield an "amalgam" acceptable to each and every panel member. Yet a consensus-deriving procedure should keep careful track of the inconsistencies in information, in order to provide proper uncertainty estimates for the end user (as also argued in van der Sluijs et al., 1998). There is an evident risk that without such analytical procedures, some or much of this type of information is lost, and the scenarios appear more certain than they should be.

Let us examine, therefore, how we might go about preserving the uncertainties arising from divergences of opinion within the framework of a National Climate Change Research Programme (SILMU) for Finland – a comprehensive analysis conducted between 1990 and 1995 of the expected impacts of climate change on forests, agriculture, fisheries, waters, and human interactions (Kämäri, 1997). In the sub-programme on waters, the approach adopted was as follows:

> **The climate scenario used as the basis of all analyses.** The "best guess" scenario was constructed on the basis of outputs from various GCMs. The temperature scenario was the following: a linear growth over 1990–2100, yielding an (annual) mean temperature growth of 4°C, irrespective of the month under concern, by 2100. As it happens, the range of the original GCM scenarios varied well over 10°C (Räisänen, 1994), having even greater imprecision and higher uncertainty than those for the regions identified in

Figure 10.2. Precipitation was also assumed to grow linearly, summing to an additional annual increment of 60 mm by 2100. Low and high scenarios were also given. For both temperature and precipitation, changes in the low scenario were 25% of the best guess scenario and 150% in the high scenario. A stochastic scenario generator simulating daily cloud cover, temperature, and precipitation, given their observed variability, was also constructed.

The impact analyses on terrestrial and aquatic ecosystems and human interactions. These were made as simulations, by changing the temperature and precipitation conditions according to the scenarios described above. The results were predominantly presented as deterministic trajectories, or sets of trajectories, of model results (Figure 10.3).

Summary phase with popularisation of the results. In the final phase, emphasis in the investigations was shifted from specific impact studies towards integrated views. The goal was to produce material that would allow the public and policy makers to take advantage of the entire research programme.

Whether the scientists – who had been examining these issues with their computer models for six years – indeed believed that the trajectories, such as those in Figure 10.3, would reflect their perception of what would happen to the objects of study, is of great interest to us. The trajectories look so certain; and this appearance is not compensated adequately by a sentence or two in the accompanying report's conclusions to the effect that "the results are subject to various uncertainties". Seldom do scientists venture beyond such bland caveats. For example, elsewhere (Varis and Somlyódy, 1996) some 60 studies of climate change impacts on surface water quality have been reviewed. They indicate much the same: in most studies the research approach was as described above, including the "uncertainty excuse" in the conclusions.

In our case study of SILMU, then, a panel of climate and watershed experts was assembled from individuals who had been among the leading scientists within the

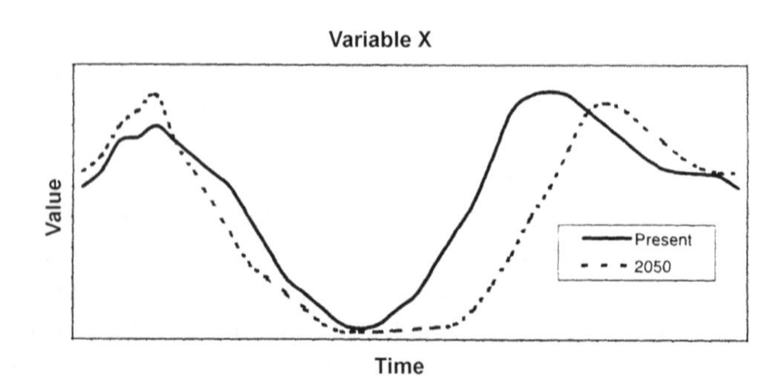

Fig. 10.3. Style of presentation typical of climate-change impact analyses.

programme. The perspective of each expert was weighted identically when processing the data. In addition to the basic analyses, comprehensive sensitivity analyses were performed on the information obtained. These included studies of the roles of the information obtained for the variables themselves (expressed as probability distributions), as well as the roles of the cause–effect relationships between the variables (expressed as links). Such sensitivity analyses were made separately for each of the variables, for climate, physical and biological properties of a watershed, and socially relevant issues and are reported fully in Kuikka and Varis (1997) and Varis and Kuikka (1997a). The following is a synopsis of this case study.

10.3.1 Methodology: Belief Networks

The concept of uncertainty has several facets in the present context. From a decision-theoretic view, uncertainty can be grouped into three clusters (Howard, 1968; Varis et al., 1994), involving:

(a) acquisition, presentation, and propagation of the information available;

(b) the expression of preferences and objectives for a given problem; and

(c) structural issues.

The basic standpoint in the development of the approach documented here is that all these types of uncertainty must be efficiently handled.

Pearl (1988) divides modern computational techniques for handling uncertain information into two groups: logic-based approaches (e.g., monotonous logic in rule-based systems) and probabilistic approaches (of Bayes, Dempster–Shafer, or fuzzy set theory, among others). Here, Bayesian calculus is used, because it is known to have a strong theoretical basis and to provide a unified approach to statistical and deterministic theories, as well as to questions of hypothesis testing and estimation (Howson and Urbach, 1991). Its essential mathematical elements are set out in the Appendix. Unlike its challengers, our Bayesian calculus is thus compatible with classical statistics, decision and risk analyses, expected utility theory, and so on.

Belief Networks belong therefore to the Bayesian family of modern computational techniques derived from research on artificial intelligence (AI). Characteristic of approaches in this group – known as belief, causal, Bayesian, qualitative Markov, or constraint networks, or influence diagrams – is the principle of networking nodes representing conditional, locally updated probabilities (Pearl, 1986, 1988; Horwitz et al., 1988; Oliver and Smith, 1990; Shafer, 1990; Szolovits and Pauker, 1993). The local updating principle allows construction of large, densely coupled networks without excessive growth of computation; moreover, the networks can easily be constructed to operate interactively. According to Bobrow (1993), a particularly successful technique has been the belief network approach of Pearl (1986, 1988), which was selected as the basis of the approach presented herein. In fact, Szolovits and Pauker (1993) have observed that "... Pearl's formulation has had a revolutionary impact on much of AI".

As is usual with such techniques, the entire model – the hypothesis space – is subjected to Bayesian analysis, not merely the parameter space (Gordon and Shortliffe, 1985; Pearl, 1986; Shenoy and Shafer, 1986). The basic idea of belief networks is to analyze the uncertainties of a system by using probability distributions for the variables and conditional probabilities for the (cause–effect) interdependencies between these variables, the interdependencies being known also as links or interconnections (Figure 10.4). In the present case the system consists of the climate–hydrology–limnology–interest/problem variables of watersheds under conditions in southern Finland (an area of around 200,000 km^2). Links indicate the strength of the interdependence between variables, i.e., the extent to which a change in one influences the states of the others, as is usual in conventional simulations. The more uncertain we are about the state of a variable, the more diffuse its distribution and the smaller its effect on the state of any other variables linked with it. Much as in the metaphor of Chapter 3, the model can be seen as a network (Figure 10.5), in which information is transformed from one variable to another, e.g., the effects of climate to water and to the interest and problem variables of relevance to society.

Our point of departure was a rating sheet with the most relevant variables describing the impacts of climate change on the inland waters of southern Finland by the year 2050, i.e., those variables indicated as nodes in Figure 10.5, and a set of outcomes (values) describing the possible future evolution of these variables. For each variable, a discrete probability distribution was used to describe the expected change in the variable, and each expert was required to assign the probabilities in this distribution. By using a cross-impact matrix, these experts were asked to judge the strength and character of the interdependencies between each pair of variables. A scale from –1 to 1 was used for this purpose, with –1 signifying an entirely negative interdependency (that the one variable would increase when the other would decrease), 0 no interdependency, and 1 complete positive interdependency. From a statistical viewpoint, this implies the proportion of the variance of each variable explained by the variance of the conditioning variable (analogous to Pearson's R^2), yet with *causal* interpretation of direction of influence. Accordingly, the sum of absolute values of all the link strengths heading to a variable may not exceed 1. The link strengths are given in two directions: from variable A to variable B and vice versa. Thus, for example, if the two variables under consideration are eutrophication and

Opposite page: Fig. 10.4. Examples of two variables, denoted as Nodes A and B. As in all Bayesian approaches they both have a prior and a posterior probability distribution. The conditional probabilities between the two nodes are expressed through links. (a) Here, in the uppermost pair of panels, there are no causal links between the two variables in the model – strictly speaking, their links have a strength of 0 – so that priors and posteriors are identical for each variable. (b) The link has a strength $h = 0.5$, with causal direction A to B, so that the posterior probabilities of variable B can accordingly be updated as a function of the new information deriving from variable A. (c) The link has a strength $h = 0.7$, so that here the posteriors of A differ from its priors. (d) Links in both directions are present. Computational details of these manipulations can be found in the Appendix.

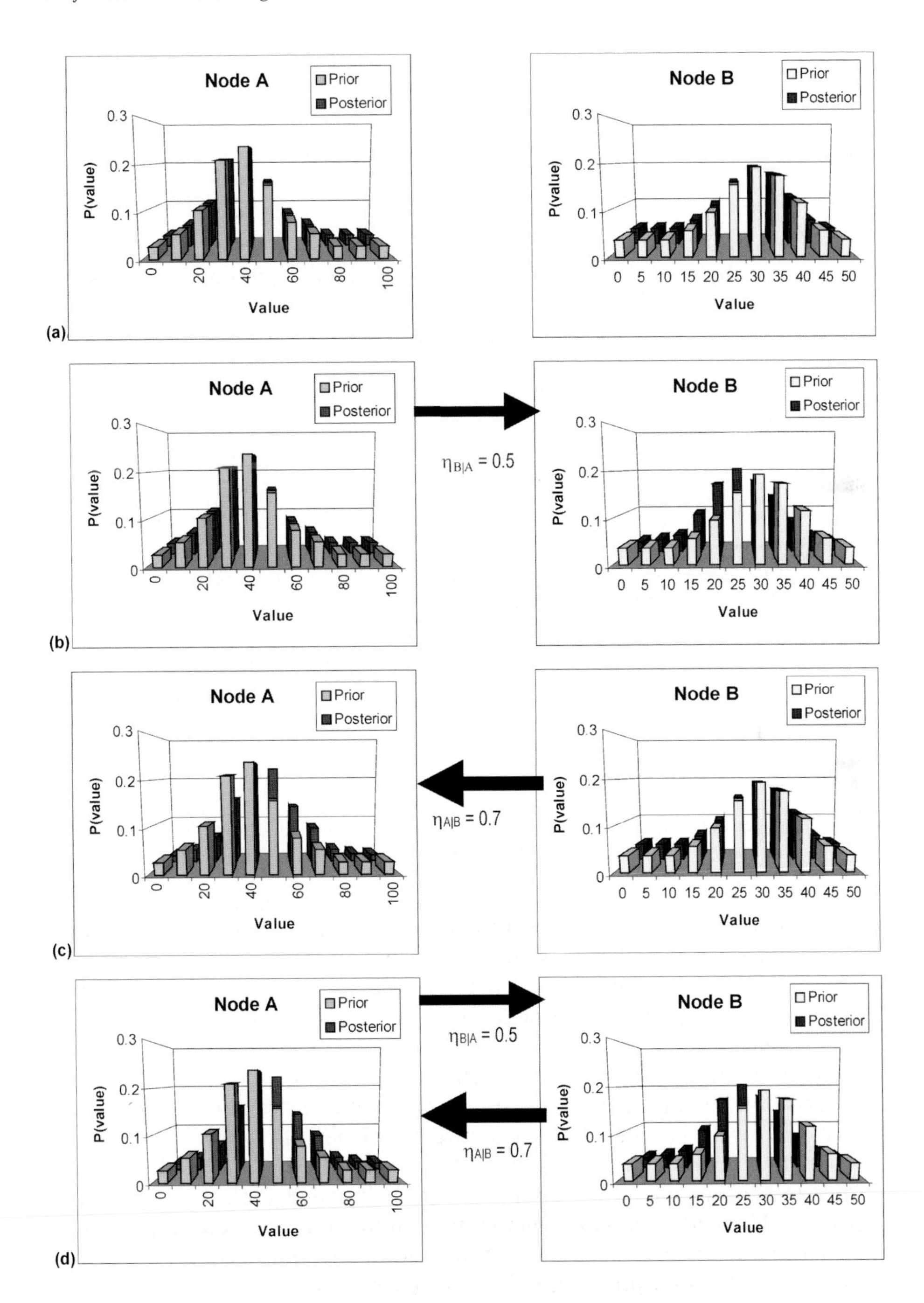

Fig. 10.4 (caption opposite).

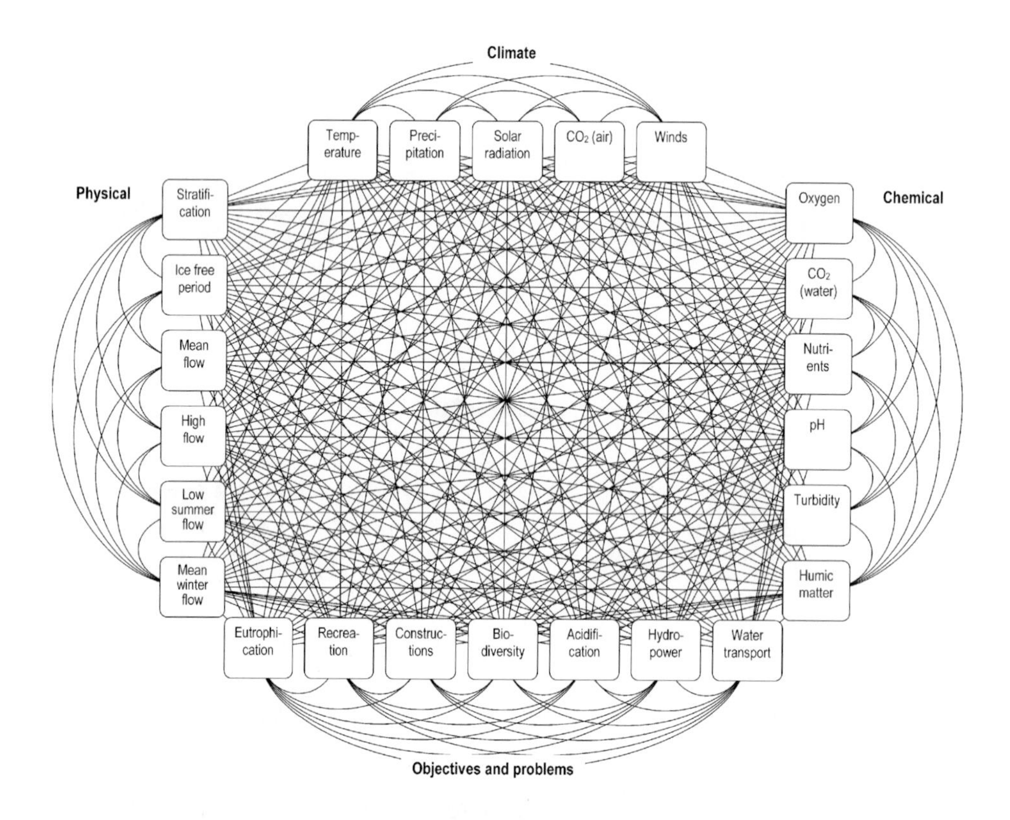

Fig. 10.5. The model structure and variables. Each line between two nodes (or variables) expresses a two-directional link (or conditional probability) between these nodes.

temperature, the link strength from eutrophication to temperature is apparently 0, but the inverse link strength may be, say, 0.5 (Figure 10.4).

The probability distributions and link strengths of the impact matrix form a Bayesian network – a network of nodes and state variable interactions, the latter frequently being referred to as "links". This network allows one to update the prior distributions inserted by the expert for each of the variables, using information from other parts of the network (model), i.e., the probability distributions for the other variables and for the interdependencies between each pair of variables, with the flow of information being regulated by the links. The updated posterior distributions incorporate features, therefore, of both the prior knowledge on the states of the interest variables and information on the states of the other variables controlled by the causal knowledge given by the links. The posterior probability should not diverge too much from the prior probability. If it does (as often occurs) this might indicate inconsistency in the assessment. That is to say, the states of the other variables and their connections do not fully accord with the beliefs underlying the state of the given variable. A sharper posterior distribution indicates that the information content of the entire model allows a decrease in the uncertainty of the specific variable under consideration.

10.3.2 Interviews: Collection of the Data

Eight scientists from the SILMU Programme participated in our case study; three were educated in hydrology and five in limnology. Self-evidently they represent the best available scientific knowledge of the aquatic impacts of climate change at the watershed scale in the given circumstances. They had, after all, been assessing climate change impacts on Finnish watersheds using various models for over five years. Definition of the impact matrix, i.e., selection of the variables included in the analysis, was designed with the aid of these experts, taking as its starting point the cross-impact matrices for the climate-change impact assessment for lakes and rivers of Varis and Somlyódy (1996). Five different possible outcomes were employed for the discrete probability distributions attaching to the expected direction and estimated magnitude of change: a large decrease, a small decrease, no change, a small increase, and a large increase. This non-numerical scaling had a subjective element to it and helped the experts express degrees of belief in the changes. In the event that an expert was very uncertain about the state of one variable, he/she inserted a flat distribution (compared with the other variables).

During the assessment phase, each expert worked for about six to eight hours to construct the model. The interviewer first described the methodology, explained the assessment task, and defined the content of each variable – a process usually requiring almost an hour. The expert was asked to state the prior probabilities required in order to express his/her personal views on the state of climate and watersheds in southern Finland by about the year 2050. Any factors excluded from the study, but which might influence the state of the variables (for example, use of fertilizers in agriculture), were asked to be left out of the evaluation. The given prior distribution described the expected direction and magnitude of changes caused by climate change only and the links were given as a general, causal macro-scale description relevant to watersheds in southern Finland. In practice, the experts took each variable in sequence and chose the other variables directly affecting it, using an arbitrary scale for the link strengths. They later estimated the proportion of the effects arising from variables excluded from the model (a form of structural error, denoted here as p), so that the link strengths heading to the variable were scaled so as to sum to $1 - p$.

Since the interviewed individuals were not climate experts, an additional data set was collected from five climate experts on the possible changes in climate-associated variables. These results were made available to the watershed experts during the assessment, with their subsequent use depending on how each watershed expert used his/her own scaling. The belief network analysis was implemented in spreadsheet-based software (Varis, 1998b) and enabled interactive use of graphics facilities to illustrate the changes in distributions during construction of the model.

10.3.3 Sample Results: Scientists are in Private Much More Uncertain than Their Scientific Papers Would Suggest

The prior probabilities assigned by the experts showed no consensus for most variables, not even with respect to the expected direction of change. For example, in Figure 10.6 there are appreciable differences amongst perceptions of the outcomes of some of the interest/problem variables, such as those for acidification and recreation. This must be considered a factor of uncertainty when using these results for policy making or advising on future research priorities. It must also be noted that some of the experts systematically used wider distributions than the others. The differences in personal scaling cause some differences in the results too, for example, in the definition (threshold) of a change. But the basic issue – of what is the direction of change and with what probability – turns out to be largely insensitive to these shortcomings.

The link structures of the models assessed by the experts differed to some extent. By way of illustration, Figure 10.7 shows the links between the climate variables and the physical watershed variables, for the models of six randomly selected experts. The strengths of most links are fairly consistent, yet some experts favour markedly more densely coupled (complex) models than others; for instance, the upper-left model has far less links than the lower-right one.

In addition to the analysis of the raw data, which reveals directly any differing judgements among the experts, an "averaged" model was constructed on the basis of all eight individual models, and the attaching outcome probabilities shown in Figure 10.8. Our experts were most confident about increases in the length of the ice-free

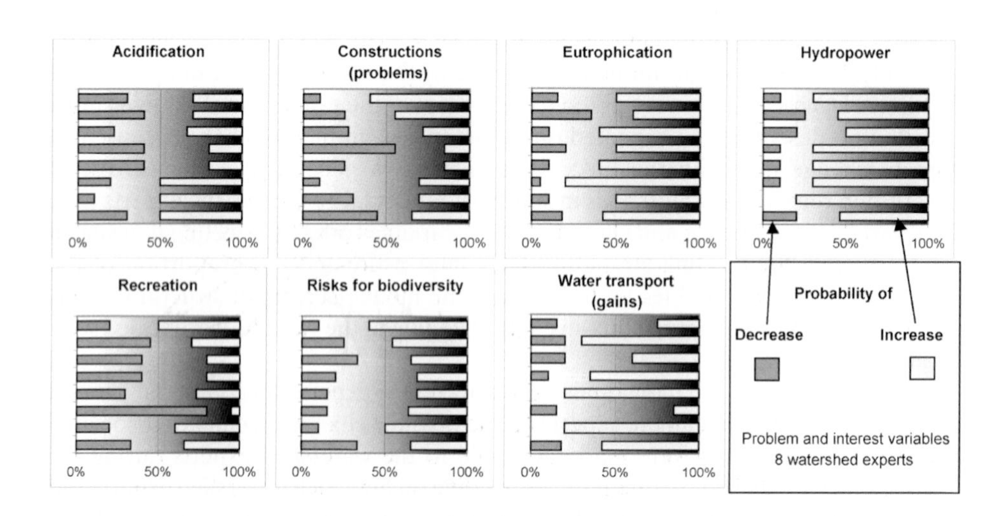

Fig. 10.6. Prior probabilities assessed by eight watershed experts for the evolution of the problem and variables of concern.

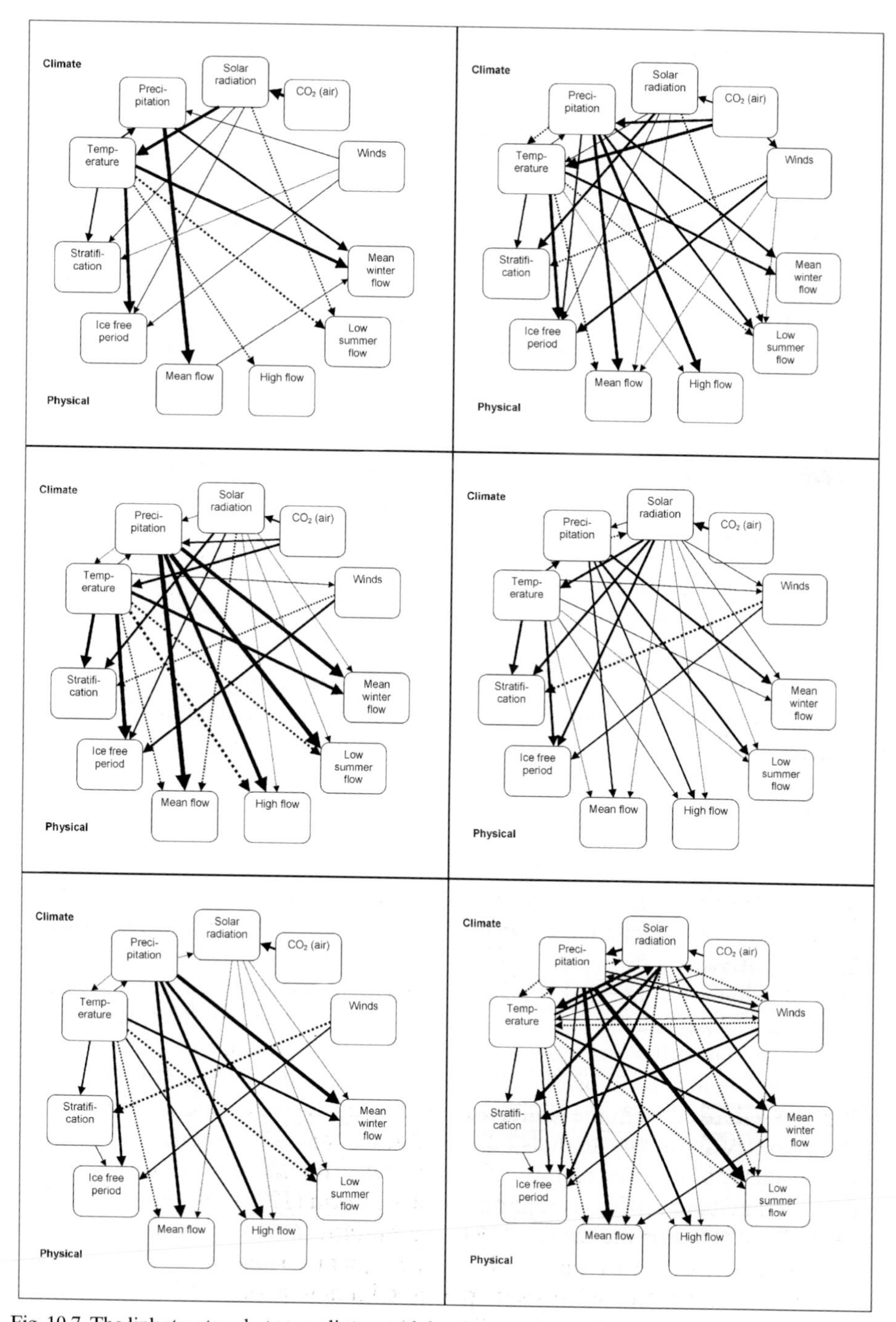

Fig. 10.7. The link structure between climate and the physical properties of the watersheds, as assessed by six of the experts. The weight of the line denotes the strength of the associated link. Grey lines indicate links having a negative value for their strength.

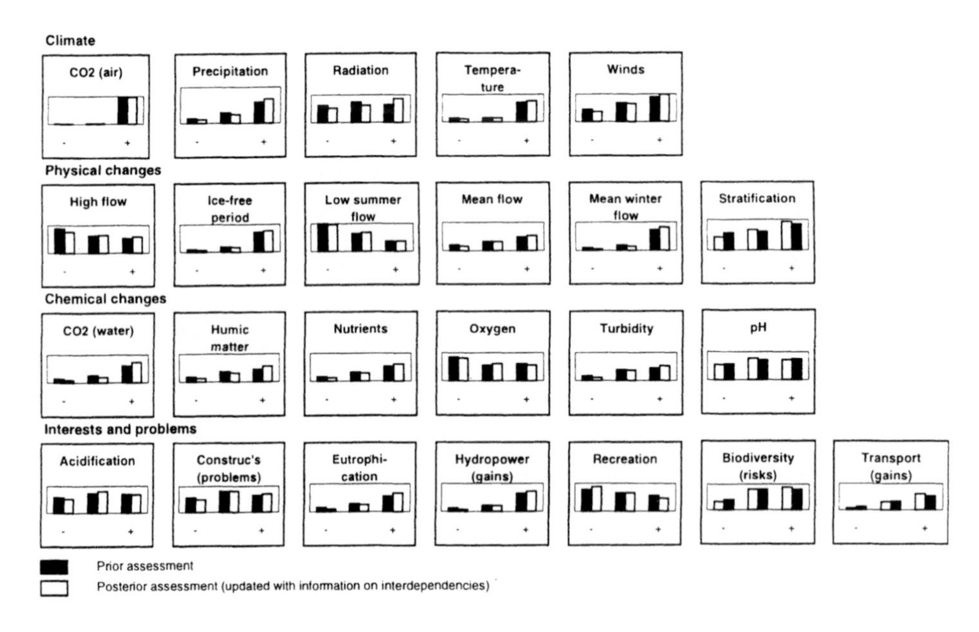

Fig. 10.8. Averaged results of the models of all the experts: prior and posterior probability distributions of the state variables.

season, mean winter flow, and the scope for hydropower production. The highest uncertainties, with the flattest probability distributions, were seen in possible changes in high flows, oxygen concentrations in water, pH, problems for construction, and total recreational value. Positive changes – judged, that is, with respect to currently prevailing social criteria – can be expected in transportation opportunities (length of the navigable season) and hydropower production. The latter is due to the expected increase in precipitation and mean flow. The most negative future was anticipated to occur in eutrophication, recreational values, and the growing risks to biodiversity.

Flat distributions, as well as inconsistencies between prior and posterior distributions, distinguish the unknowns from the better known variables. It is striking to observe in Figure 10.8 just how uncertain the overall outcome of our analysis of private expert beliefs is, in profound contrast to the crisp, deterministic results of Figure 10.3 – derived from the simulation models, by the very same individuals.

Two types of sensitivity analysis were performed: one for the state variables; the other for the causal/link structure of the model – reminiscent somewhat of the analyses of Sandberg (1976), as reported by Rosenhead (1989), which were touched upon in Chapter 5. For the cause–effect relationships (links), a small perturbation was made to each, one by one, and the subsequent change in all the variables observed. In addition, all the links were perturbed simultaneously. Figure 10.9 shows the impacts of perturbing the logic of three links on each of seven problem/interest variables. Disrupting the link from temperature to precipitation influenced hydro-

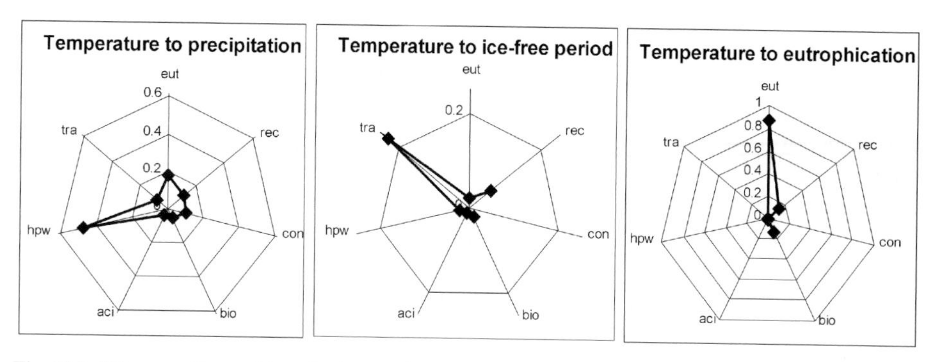

Fig. 10.9. "How are the problem (or interest) variables affected by uncertainties in some of the links?" Impacts are shown for three illustrative links, for the various variables of interest, where: eut = eutrophication; rec = recreation; con = construction; bio = biodiversity; aci = acidification; hpw = hydropower; and tra = water-borne transport. These results were generated using the averaged model; units employed are relative to the amount of perturbation.

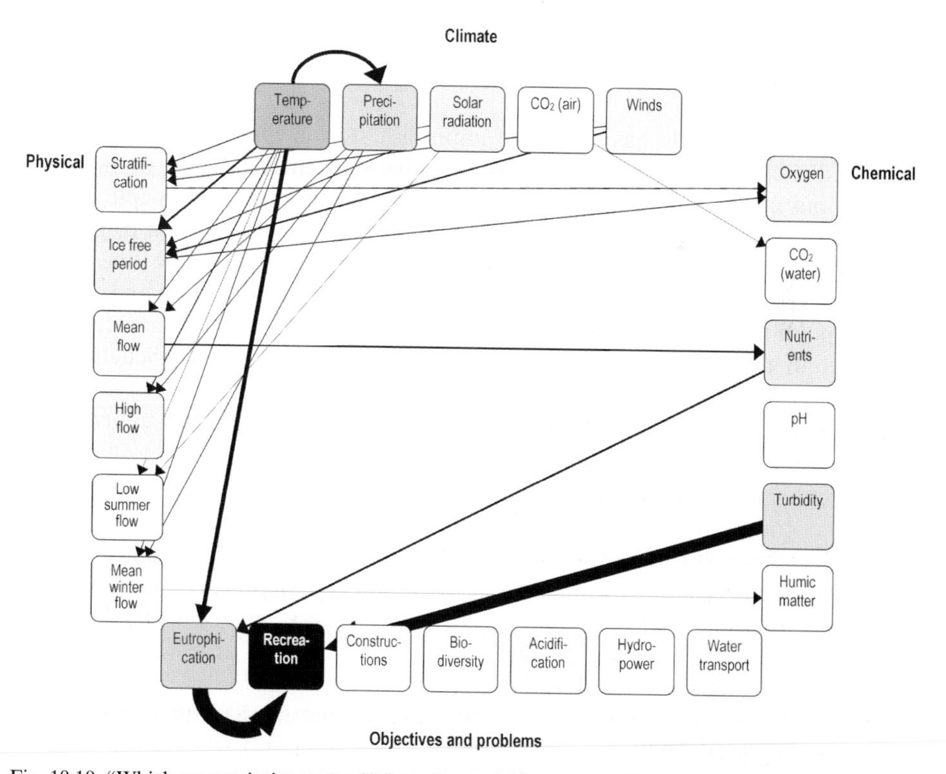

Fig. 10.10. "Which uncertainties matter?" Sample results from an analysis of sensitivity of the variable "recreation" to uncertainties of model structure and states. *Structure*: the heavier the weight of the arrow the more sensitive is the posterior distribution of recreation to uncertainty in that link. *States*: the darker the variable, the more sensitive is the posterior distribution of recreation to uncertainty in the given variable. These results were generated using the averaged model; similar results are available for other variables.

power production much more strongly than any other variable of interest, while the link from temperature to ice-free period influenced transport most strongly, and that from temperature to eutrophication had the most pronounced effect directly on eutrophication. When all the results from the structural (link) and state sensitivity analyses are combined, for example, for the interest variable of recreation, those uncertainties highlighted in Figure 10.10 (for both constituent state variables and links) are found to be crucial in controlling the uncertainty of the outcomes for recreation. Some causalities are more important than the others, such as the links from eutrophication and turbidity to recreation, as well as both the direct and various indirect links to eutrophication from temperature. In this case (Figure 10.10) only the links exceeding a defined level of sensitivity are shown, so that, for instance, no direct or indirect links from oxygen to eutrophication are visible. The impacts from oxygen to eutrophication were distributed through several links, none of which is especially sensitive; these are thus not among the important uncertainties in this respect.

10.4 BACK TO THE GLOBAL SCALE

The Finnish case study has been one important step in the development of our proposed methodology. It is a part of a global-scale investigation of the interconnections of water, food, poverty, climate, and urbanisation. In principle, it is not difficult to see how a method such as Belief Networks could be used to generate target futures, whose reachability might then be assessed along the lines of the following chapter. It is also obvious that the approach provides a very different counterpoint to the conventional computational models normally encountered, for example, in the International Geosphere–Biosphere Programme (Schellnhuber, 1999).

At the global scale, the analysis of the present chapter is designed to investigate the following issues in particular:

- The *most important* variables and variable clusters;
- The *relative importance* of different variables *among geographical regions*;
- The key *knowns*, *unknowns*, *threats*, *risks*, and *opportunities*;
- *Inconsistencies* in present comprehension of the issues, and in policy recommendations;
- Possible, interdisciplinary *cycles and feedbacks* that are difficult to find without an integrated analysis;
- Further *development of the methodology*.

As we have said in introducing this chapter, a "few dozen variables" must be settled upon to constitute the state space of the model. After a comprehensive literature review, such as materials from the UN summits of the 1990s, for example, the

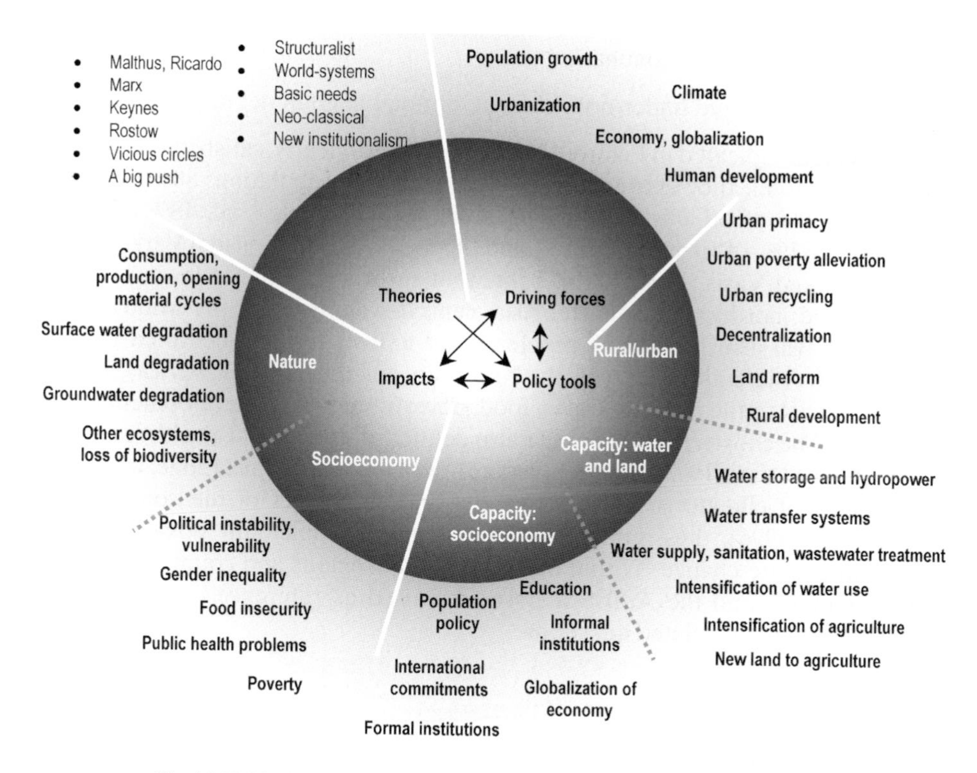

Fig. 10.11. Variables and variable clusters selected for the global-level study.

variables shown in Figure 10.11 appeared to constitute the best set for covering the problem domain with a reasonable resolution. They involve eleven theories of development, which generally form the backbone of policy strategies, although are rarely acknowledged as such in practice. Eighteen different policy tools are included, subdivided into those influencing the rural–urban parity, capacity building within water and land management, and socioeconomic capacity-fostering tools. Driving forces for the model have already been listed above and are taken to influence both society and its environment. These impacts, as well as the driving forces themselves, are the targets for control through the policy tools. In summary, the model contains 45 variables, with 1,980 links.

Although the scope of our analysis is primarily global, the case study is focused on a set of critical regions, chosen according to the following six criteria:

- water resources vulnerability and scarcity;
- densely populated (in a global context);
- population growth is rapid;
- urbanisation and mega-city growth is extensive;

- low/middle income dominance; and
- widely distributed under-nourishment, or notable grain net importers.

The five regions indicated in Figure 10.12 completely, or nearly, satisfy these criteria. Of the world total, they account for 59% of the total population, 80% of urban population, 60% of urbanisation, 7% of GNP, 34% of arable land, 48% of cereal production, 57% of irrigated land, and 45% of fertilizer use (Varis 1998c; Vakkilainen and Varis, 1998, 1999). Somewhat surprisingly, the following regions were excluded: north Africa, central Asia, central America, and the Middle East. These latter will also face water problems, yet unlike the five identified regions (China, south Asia, south-east Asia, the Nile basin, and the Sahel/W. Africa), appear not to be crucial to the issues of water, food, poverty, and urban development at the *global* scale. The chosen five regions are also defined as highly critical with respect to climate change (Varis 1998d). They cover fairly well the intertropical zone, where the predictability of climate is the most troublesome in respect of the conventional GCMs (IPCC 1996), thus rendering equally "unpredictable" any possible consequences of climate change. Within our study, other regions and countries, such as Mexico, Finland, and the Aral Sea Basin, are used as reference cases for some specific development-related questions.

Region	*1*	*2*	*3*	*4*	*5*	*6*
N Europe	—	—	—	—	—	
L and W Europe	+	+	—	—	—	
E Europe			—	—		
Ex-USSR/European	—		—	—		+
N America		+			—	—
C America	+		+	+		
S America	—	+	+			
N Africa, Egypt excl.	+	—	+			+
Nile Basin	+	+	+	+	+	+
W Africa	+	+	+	+	+	+
C Africa	—		+	+	+	+
S Africa	+		+	+		
N Asia	—	—				
C Asia, Kazakhstan	+	—	+			+
China	+	+	+	+	+	+
Middle East	+	+	+			+
S Asia	+	+	+	+	+	+
SE Asia	+	+	+	+		+
Japan	+		—	—	—	
Australia and NZ		—		—	—	—
Oceania	—	—		—	—	

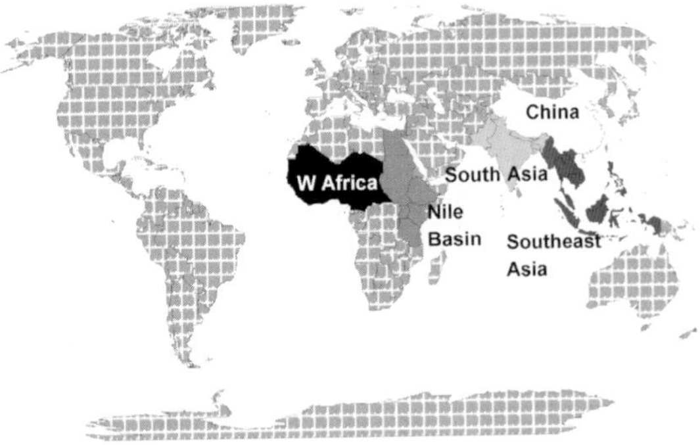

Fig. 10.12. Macro-regions of the world with respect to six critical factors: a – sign signifies "does not fulfil"; no mark signals "some risk"; and a + sign denotes "fulfils the criterion".

10.4.1 Interdisciplinary Analysis on a Global Scale: A Melange of Vicious Circles

When analyzing better-defined systems on a less comprehensive scale, we have been used to the notion of cause and effect operating unambiguously in one direction. With increasing complexity of the system, with issues addressed over ever longer time horizons, fewer relationships can be seen in such a simple way. Causalities are not very clear, among the circles and feedback loops which abound. The water sector is involved in countless ways in such phenomena (Varis, 1999). For example, infrastructure development in a city may boost immigration (Figure 10.13(a)), which then outpaces this development, unless some balancing actions in rural areas can be offered. Elsewhere, Figure 10.13(b) shows how the tendency to consider water solely as an economic good feeds negatively back to those having no money (the poor), that which cannot make use of money (nature and the environment), or those remaining outside formal society (the informal sector). Moreover, the view of water as an economic good may weaken still further already weak governments, in the sense that this may make water part of a formal economy over which such governments can exercise less and less control. There are many undisputed benefits in considering water as an economic good, but some important side-effects call for perspectives from other scales and dimensions, beyond mere financial rationality.

A number of "vicious" circles, which feed off each other, may therefore hinder social and economic development in many ways. It is not enough to break such circles

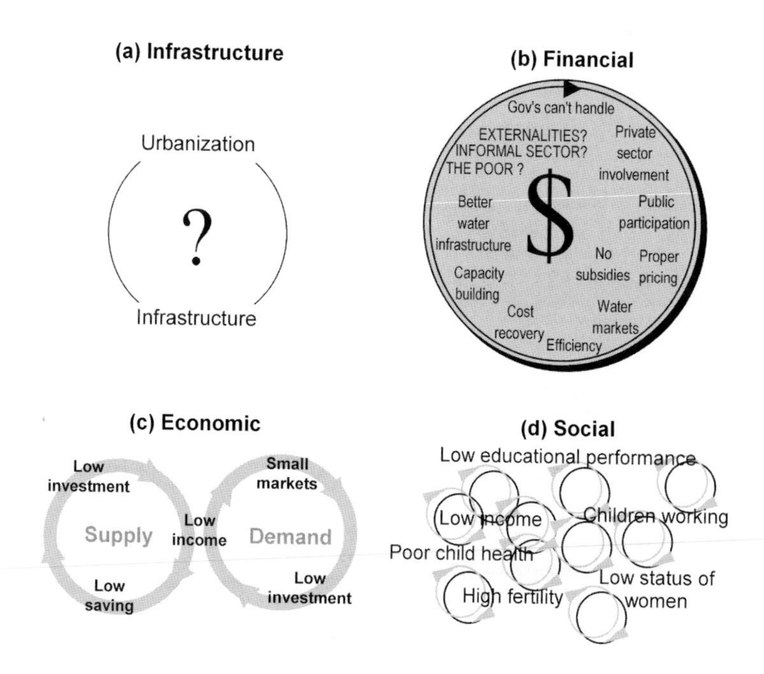

Fig. 10.13. Examples of vicious circles that hinder development.

one by one; several must be disrupted at the same time, jointly, by concerted action. Feedback loops and circular connections are readily appreciated in modern economic theory, at least to some extent, as presented in Figure 10.13(c) (Nafziger, 1997): essentially the rich get richer and the poor remain poor, although fortunately this is not always the case. There are ways to break such circles, e.g., by attracting foreign investments, or increasing savings rate, but they can seldom be realized within one sector or using the tools of just a single discipline. Vicious economic circles are linked with vicious circles in the whole society. Population growth is closely linked with poor child health (household water quality is crucial), low income (ability to access water services), and so on, as sketched in Figure 10.13(d). There exist multiple handles with which to grapple with the vicious, multi-generational circle of fertility, gender, poverty and education arising from so many complex, interconnected issues. But none looks all that effective on its own; what is important is their collective manipulation. Policy guidelines have thus come increasingly to include this notion of collectively breaking these vicious circles. There are many signs of the growing appreciation of cross-sectoral, cross-scale issues, in addition to those given in Figure 10.13. For example, Figure 10.14(a) shows the philosophy of breaking the vicious circles of poverty and affluence-based environmental degradation in order thus to work towards sustainability (Jalal, 1993), while Figure 10.14(b) summarizes the UNDP (1995) philosophy on human development as a tool of turning some of the vicious circles into virtuous ones.

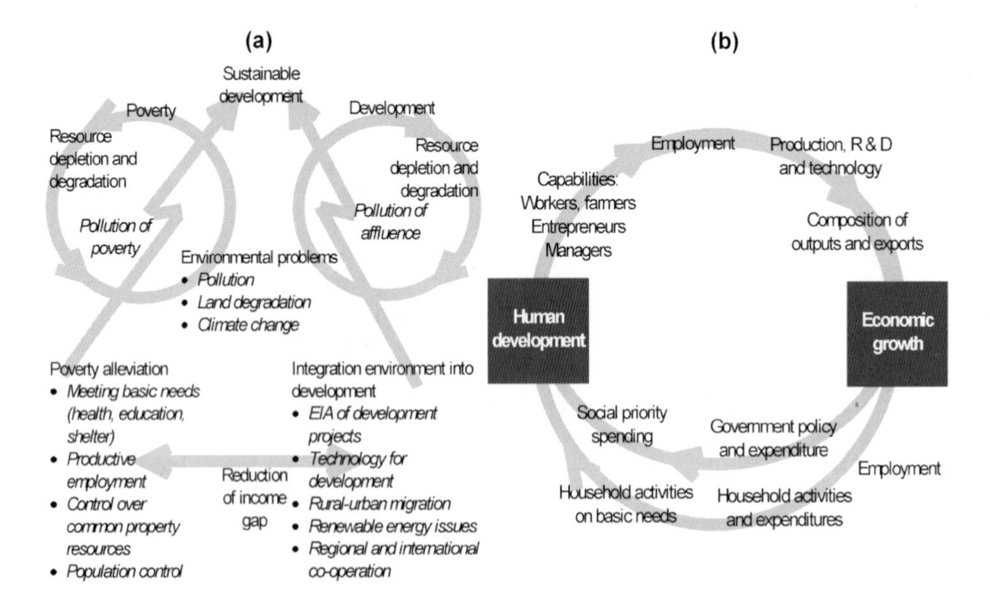

Fig. 10.14. Examples of how vicious circles can be turned into virtuous circles: in the left panel, according to Jalal (1993); in the right panel, according to UNDP (1996).

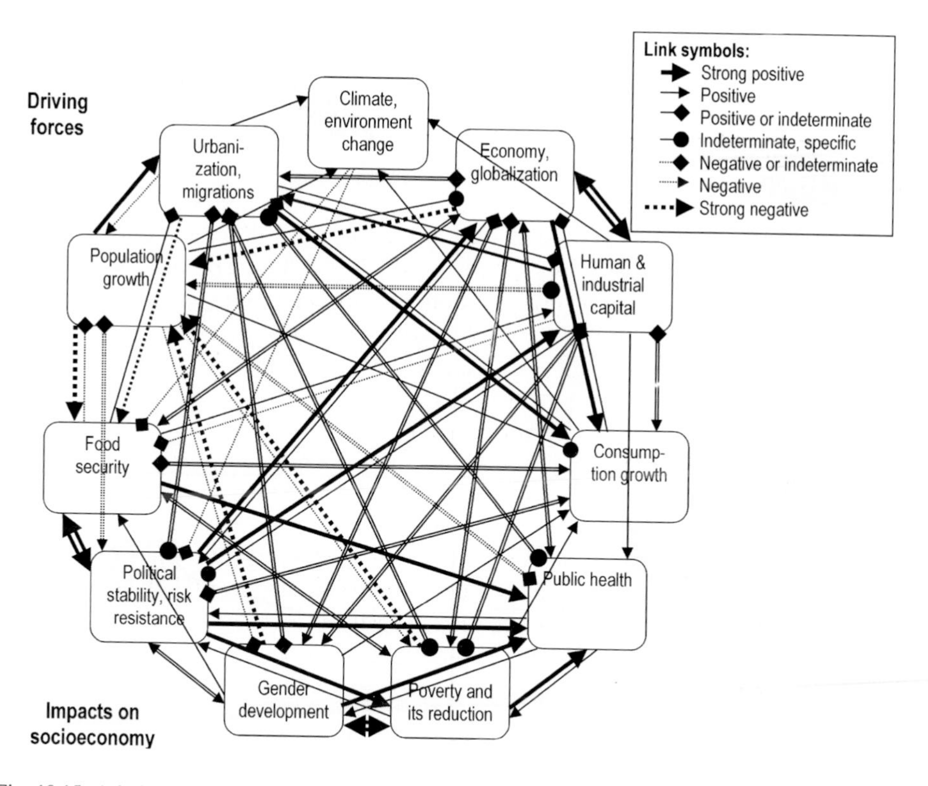

Fig. 10.15. A belief network model for analysis of the socio-economic impacts of global changes. The model is not specific to any region of the world; thus, for several links not even their directions of influence are indicated.

Vicious and virtuous circles have been acknowledged as crucial elements of the development process. However, they are mightily difficult to model using either statistical models or differential equation models. The nub of the argument of this chapter, therefore, is this: such interdisciplinary interconnections are too often ignored in forecasts, projections, even in policy making; and this, we assert, is partly due to the methodological shortcomings of forecasting environmental change.

With this much in mind, Figure 10.15 shows a belief network – part of a complex model embracing all the variables of Figure 10.11 – which allows the analysis of many of the important (and vicious) social and economic circles. Illustrative results for south Asia (including Bangladesh, India, Nepal, and Pakistan) are shown in Figures 10.16 and 10.17. Given the feature of belief networks that permits bi-directional

---

Overleaf: Fig. 10.16. Cause–effect sensitivity analysis among and between "driving forces" and "impacts on society and the economy". The impacts of a change in the prior distribution of each state variable on the posterior of the other states is shown. The sensitivities of the variables to changes in each link are also shown.

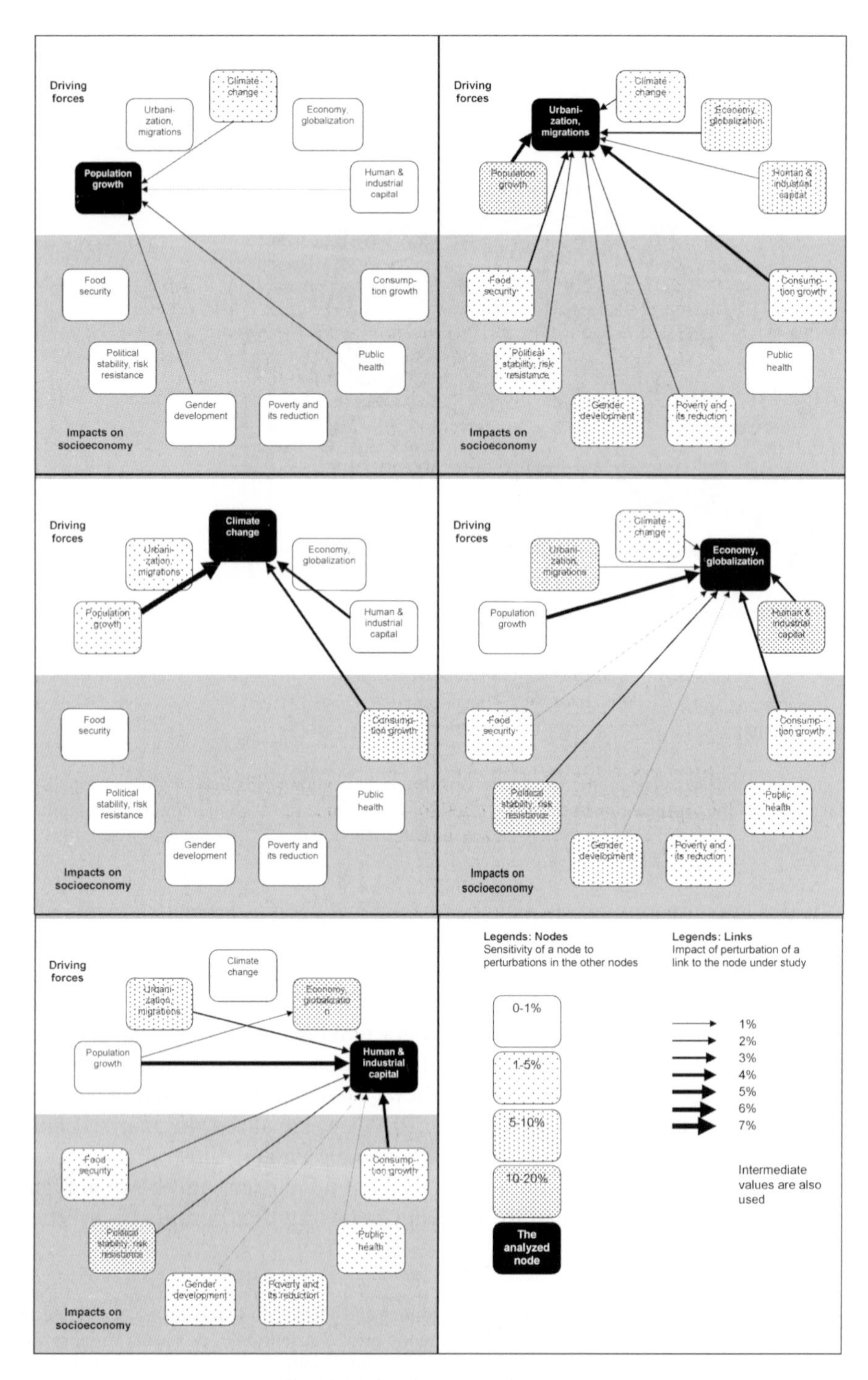

Fig. 10.16. Caption on previous page.

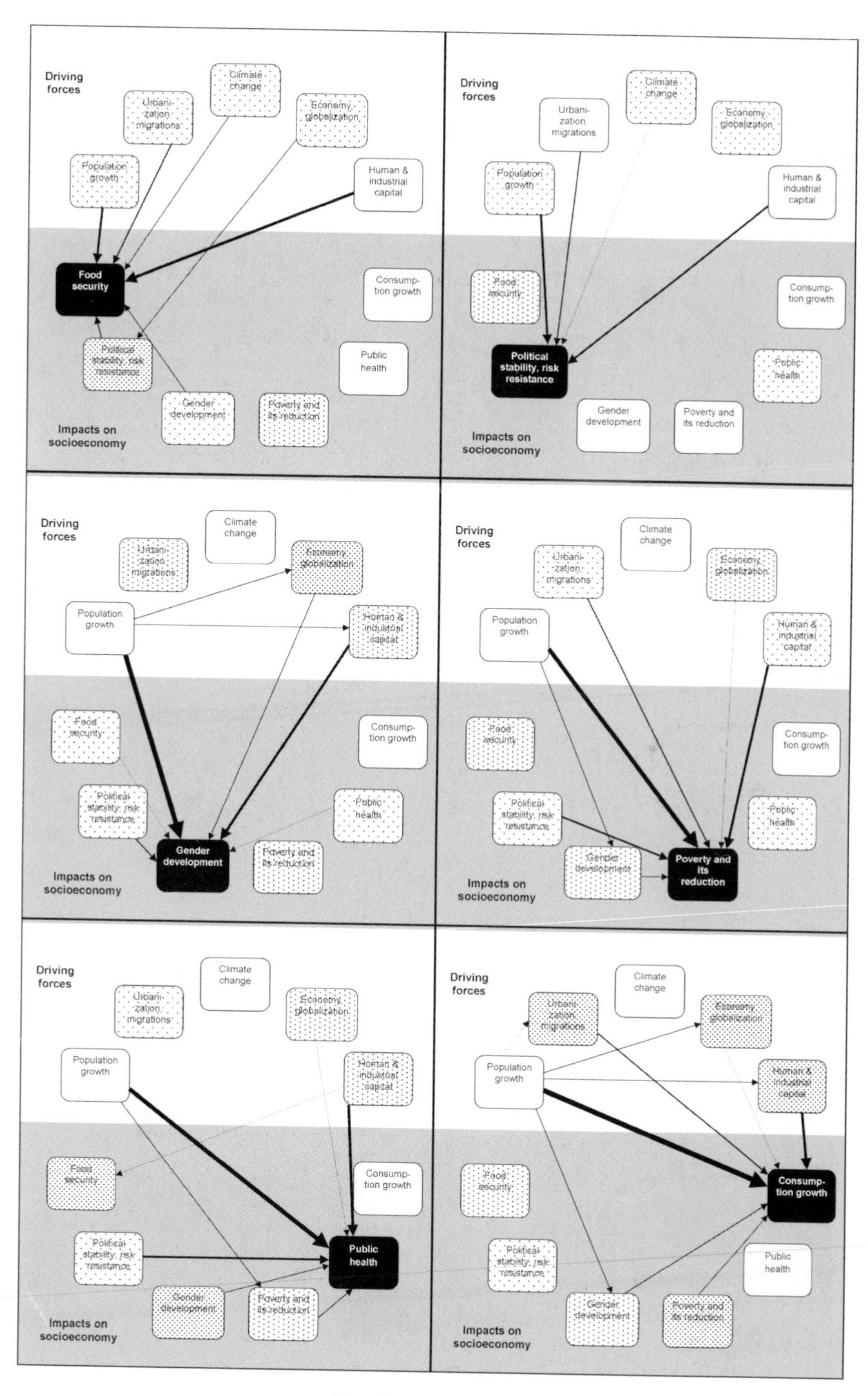

Fig. 10.16 (continued).

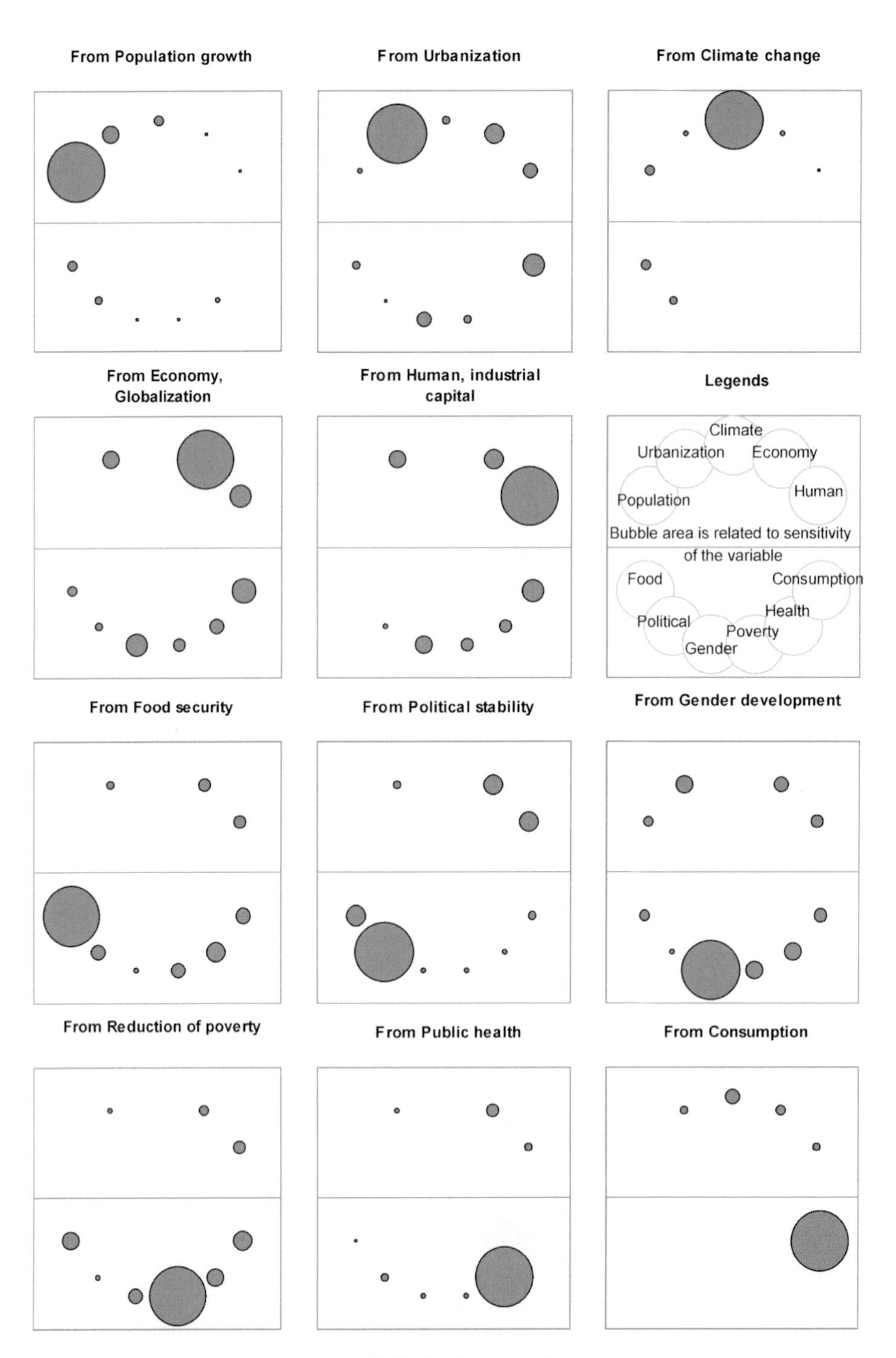

Fig. 10.17. Caption opposite.

Opposite: Fig. 10.17. Effect–cause sensitivity analysis among and between "driving forces" and "impacts on the economy". The impacts of a change in the prior distribution of each state variable on the posterior of the other states is shown.

---

propagation of uncertainty, it is easy to conduct sensitivity analyses in both directions: from effects to causes, and from causes to effects. While it is typical for a model to be constructed in the direction of (presumed) cause to effect, our results can be used in the opposite direction. Shachter and Heckerman (1987) have formulated this idea in the following words:

> First, many real-world problems involve causal interactions... Second, the direction of causality is most often opposite to the direction of usage... Third, experts are more comfortable when their beliefs are elicited in the causal direction...

When performed in the direction of cause–effect the results of the sensitivity analysis (Figure 10.16) reveal, for instance, that a change in population growth influences urbanisation most strongly. Almost all other variables are affected, food security and climate change somewhat more so than the others. The impacts flowing from urbanisation appear more pronounced, with respect to almost all other variables, than those of population growth. In the direction of effect-cause (Figure 10.17) the implications of our results, including those of a sensitivity analysis of the cause–effect relationships, are indeed more practical in character. In many cases, the links between those variables that are sensitive to one another are highlighted, but not always. For instance, public health is very sensitive to changes in food security, yet the corresponding causal link is not sensitive in the model, revealing thus that this link does not belong to the important structural uncertainties of the model.

10.5 EPILOGUE AND EXTENSIONS

The mainstream literature tends to give a fairly positive picture of the accuracy and certainty of forecasts and scenarios of regional and global environmental change projected into the future. But when one looks at the data and information available, the way they have been processed, and how the various forecasts mesh with one another (or not, as the case may be), it becomes clear that the levels of uncertainty are remarkable. The analyses tend to give an overconfident view of their results.

In the case history of forecasting impacts of climate change on inland waters in southern Finland, our analysis with Belief Networks revealed what has been observed elsewhere (van der Sluijs et al, 1998): scientists are in private far less certain than their scientific papers would suggest. In their formal studies, aided by conventional computational models, uncertainties are not closely scrutinised, and the structure underlying the system's behaviour is assumed to remain unchanged over the entire forecasting horizon. In our analysis, the scientific experts disagreed on the future development of many important factors, and were notably uncertain about how the values of variables would evolve, and uncertain about the nature of the interconnections between the variables – results in complete discord with the results the same individuals had published in scientific reports.

The same kinds of failings can be seen in the many attempts throughout the 1990s (following the 1992 Earth Summit in Rio de Janeiro) at grappling with the toughest of issues of environment, society, and climate change at a global scale. Our most pressing problems, it has to be said, cannot be adequately addressed using conventional approaches to model-building. We are suffering from some methodological weaknesses, but have here advocated Belief Networks as a means of making progress in the right direction. Given the data poverty and policy proximity of the issues touched upon here, the personal perceptions of the individuals who have studied these topics over many years is an important source of knowledge. Yet so little of this experience is included in mathematical and statistical models, or in scientific reports. There should be the means of facilitating studies capable of merging, summarising, and bringing the personal experiences of the scientific experts to a lay audience, including the policy makers. The feared dislocations and development pathways, chances, threats, and opportunities that will follow from the changing world cannot be detected from data alone, not by using structurally fixed simulation models alone, not with expert judgment alone, but through a balanced synthesis of all these constituents – the richness of variety of approach advocated at the very outset of the monograph, in Chapter 1.

REFERENCES

Amis, P., 1997. Indian urban poverty: where are the levers of its effective alleviation? *IDS Bulletin*, **28**(2), 94–105.

Bobrow, D.G., 1993. Artificial Intelligence in perspective: a retrospective on fifty volumes of the Artificial Intelligence Journal. *Artificial Intelligence,* **59,** 5–20.

Gordon, J. and Shortliffe, E.H., 1985. A method of managing evidential reasoning in a hierarchical hypothesis space. *Artificial Intelligence*, **26**, 323–357.

Horwitz, E.J., Breese, J.S. and Henrion, M., 1988. Decision theory in expert systems and artificial intelligence. *Int. J. Approx. Reasoning*, **2**, 247–302.

Howard, R.A., 1968. The foundations of decision analysis. *IEEE Trans. Syst. Sci. Cybernetics*, **4,** 211–219.

Howson, C. and Urbach, P., 1991. Bayesian reasoning in science. *Nature*, **350**, 371–374.

IPCC, 1996. *Climate Change 1995: The Science of Climate Change.* Cambridge University Press, Cambridge.

Jalal, K.F., 1993. *Sustainable Development, Environment, and Poverty Nexus*. Asian Development Bank, Manila.

Kämäri, J. (ed.), 1997. Special issue on climatic change impact assessment for Finland. *Boreal Environ. Res.,* **2**.

Korn, G.A. and Korn, T.M., 1968. *Mathematical Handbook for Scientists and Engineers.* McGraw-Hill, New York.

Kuikka, S. and Varis, O., 1997. Uncertainties of climatic change impacts in Finnish watersheds: a Bayesian network analysis of expert knowledge. *Boreal Environ. Res.,* **2**, 109–128.

Lutz, W. and Prinz, C., 1994. New world population scenarios. *IIASA Options*, Autumn '94, 4–10.

Morgan, M.G. and Henrion, M., 1990. *Uncertainty, A Guide to Dealing with Uncertainty in Quantitative Risk and Policy Analysis.* Cambridge University Press, Cambridge, MA.

Morgan, M.G. and Keith, D.W., 1995. Subjective judgments by climate experts. *Environ. Sci. Technol.,* **29**, 468–476.

Nafziger, E.W., 1997. *The Economics of Developing Countries.* 3rd Edn. Prentice-Hall, Upper Saddle River, NJ.

North, D.C., 1997. *The New Institutional Economics and its Contribution to Improving our Understanding of the Transition Problem.* 1st Annual Public Lecture, 7 March, United Nations University/World Institute for Development Economics Research, Helsinki.

Oliver, R.M. and Smith, J.Q. (eds.), 1990. *Influence Diagrams, Belief Nets and Decision Analysis*. Wiley, Chichester.

Pearl, J., 1986. On evidential reasoning in a hierarchy of hypotheses. *Artificial Intelligence*, **28**, 9–15.

Pearl, J., 1988. *Probabilistic Reasoning in Intelligent Systems: Networks of Plausible Inference*. Morgan-Kaufmann, San Mateo, CA.

Raiffa, H., 1968. *Decision Analysis*. Addison-Wesley, Reading, MA.

Räisänen, J., 1994. A comparison of the results of seven GCM experiments in Northern Europe. *Geophysica*, **30**, 3–30.

Rosenhead, J., 1989. Robustness analysis: keeping your options open. In: *Rational Analysis for a Problematic World* (J. Rosenhead, ed.). Wiley, Chichester, pp 193–218.

Rowe, A.J. and Watkins, P.R., 1992. Beyond expert systems—reasoning, judgment, and wisdom. *Expert Systems With Applications*, **4**, 1–10.

Sandberg, A., 1976. *The Limits to Democratic Planning.* Liberforlag, Stockholm.

Schellnhuber, H.J., 1999. 'Earth system' analysis and the second Copernican revolution. *Nature*, **402** (Supplement, 2 December) C19–C23.

Schneider, S.H., 1990. The global warming debate: Science or politics? *Environ. Sci. Technol.,* **24**, 432–435.

Shachter, R.D., 1986. Evaluation of influence diagrams. *Operations Res.,* **34,** 871–882.

Shachter, R.D., 1988. Probabilistic inference and influence diagrams. *Operations Res.,* **36,** 589–604.

Shachter, R.D. and Heckerman, D.E., 1987. Thinking backwards for knowledge acquisition. *AI Magazine,* Fall 1987, 55–61.

Shackley, S. and Wynne, B., 1995. Integrating knowledges for climate change. Pyramids, nets and uncertainties. *Global Environ. Change*, **5**, 113–126.

Shafer, G., 1990. Decision making. In: *Readings in Uncertain Reasoning.* (G. Shafer and J. Pearl, eds.). Morgan-Kaufmann, San Mateo, CA, pp. 7–13.

Shenoy, P.P. and Shafer, G., 1986. Propagating belief functions with local computations. *IEEE Expert,* Fall 1986, 43–52.

Shlyakter, A., Valverde, L.J. and Wilson, R., 1995. Integrated risk analysis of global climate change. *Chemosphere*, **30**, 1585–1618.

Smil, V., 1992. China's environment in the 1980s: some critical changes. *Ambio*, **21,** 431–436.

Stockholm Environment Institute, 1997. *Comprehensive Assessment of the Freshwater Resources of the World – Water Futures: Assessment of Long-Range Patterns and Problems.* Stockholm Environment Institute, Stockholm.

Szolovits, P. and Pauker, S.G., 1993. Categorical and probabilistic reasoning in medicine revisited. *Artificial Intelligence,* **59,** 167–180.

UNDP (United Nations Development Programme), 1996. *Human Development Report.* Oxford University Press, New York, USA.

Vakkilainen, P. and Varis, O., 1998. Water in the hungry, poor, and urbanizing world – a focus on uncertainty and interdisciplinarity. In: *Water: A Looming Crisis* (H. Zebidi, ed.). UNESCO, Paris, pp. 323–328.

Vakkilainen, P. and Varis, O., 1999. Will water be enough, will food be enough? *Technical Documents in Hydrology*, 24. UNESCO/IHP, Paris.

Van der Sluijs, J., van Eijndhoven, J., Shackley, S. and Wynne, B., 1998. Anchoring devices in science for policy: the case of consensus around climate sensitivity. *Social Studies in Science*, **28**(2), 291–323.

Varis, O., 1995. A belief network approach to optimization and parameter estimation in resource and environmental management models. IIASA WP-95-11, International Institute for Applied Systems Analysis, Laxenburg, Austria.

Varis, O., 1999. Water resources development: vicious and virtuous circles. *Ambio*, **28**, 599–603.

Varis, O., 1998a. A belief network approach to optimization and parameter estimation: application to resource and environmental management. *Artificial Intelligence* **101**, 135–163.

Varis, O., 1998b. *F.C. BeNe (Fully Connected Belief Networks): A Spreadsheet Toolkit for Problems with Several, Highly Uncertain and Interrelated Variables. Version 1.0 User's Guide.* Helsinki University Press, Helsinki.

Varis, O., 1998c. Water and Third World cities: the expanding puzzle. *Research in Progress* RIP-16, United Nations University/World Institute for Development Economics Research, Helsinki.

Varis, O., 1998d. What if the trade winds and monsoons change? In: *Proceedings of the 2nd Conference on Climate and Water*, 17–20 August, 1998, Helsinki, Finland (R. Lemmelä and N. Helenius, eds.). Edita, Helsinki.

Varis, O. and Kuikka, S., 1997a. BeNe-EIA: A Bayesian approach to expert judgment elicitation with case studies on climatic change impacts on surface waters. *Climatic Change,* **37**, 539–563.

Varis, O. and Kuikka, S., 1997b. Joint use of multiple environmental assessment models by a Bayesian meta-model: the Baltic salmon case. *Ecol. Modelling,* **102**, 341–351.

Varis, O. and Kuikka, S., 1999. Learning Bayesian decision analysis by doing: lessons from environmental and natural resources management. *Ecol. Modelling*, **119**, 177–195.

Varis, O., Kuikka, S. and Taskinen, A., 1994. Modeling for water quality decisions: uncertainty and subjectivity in information, in objectives, and in model structure. *Ecol. Modelling*, **74**, 91–101.

Varis, O. and Somlyódy, L., 1996. Potential impacts of climatic change on lake and reservoir water quality. In: *Water Resources Management in the Face of Climatic/Hydrologic Uncertainties* (Z. Kaczmarek, K. Strzepek, L. Somlyódy and V. Pryazhinskaya, eds.). IIASA, Laxenburg/Kluwer Academic Publishers, Dordrecht, pp. 46–69.

Zeitz III, F.H. and Maybeck, P.S., 1993. An alternate algorithm for discrete-time filtering. *IEEE Trans. Aerospace Elect. Systems*, **29,** 1123–1135.

APPENDIX: Computational Solution of Belief Networks

This appendix presents the basic numerical principles of belief networks (BNs) (Varis, 1998a). The basic approach is Bayesian; the information about the state of a variable is described as a probability distribution. The relations between variables are described as conditional probabilities. If two variables are in such a relation, the information from one node, presented as its prior probability distribution, is used to update the prior of the other node, yielding its posterior probability distribution. The updating can be done in two directions, i.e., from variable A to variable B and *vice versa*.

An extension in which the BN is used as one layer of a model (probabilistic layer), together with a layer of deterministic equations (state layer) is also described. This two-layered approach is called Generalised Belief Networks (GBNs). It allows inclusion of the structural uncertainty into deterministic models, and the use of meta-models, which include several uncertain models in the same, Bayesian framework. A way to use such two-layered networks in parameter estimation and optimisation in Gaussian systems is included thereafter.

The BN approach is based essentially on the work of Judea Pearl (1986, 1988), with some extensions. The GBN approach is the result of more recent development work (Varis, 1998a).

Propagation of Uncertainty in Belief Networks

The BN (constituting the probabilistic layer of a GBN) consists of nodes connected with links. Those properties of nodes, links, and networks that are relevant here are presented.

Nodes:

Each node i in a network contains

- A vector of (discrete) outcomes $\mathbf{y}_i$ that can be defined as inputs, or they may depend on the outcome values of other nodes.
- An evidence vector $\mathbf{e}_i$, with probabilities $e_1,...,e_k$ assigned to k outcomes. Here, the number of outcomes is three. The evidence vector transmits external information (data, targets, etc.) to the model.
- A posterior probability distribution $\mathbf{Bel}_i$.

The prior probabilities assigned to the outcomes are updated with information linked from other parts of the network, yielding the posterior probability distribution.

Links:

A probabilistic link (uncertainty link) transfers information from one node to another. It is defined as the link matrix $\mathbf{M}_{i|j}$ between two variables i and j, denoting the conditional probability of i given j. In the simplest case of a unidirectional chain, the link matrix equals a Markov chain state transition matrix.

Often, it is practical to give the strength of each link using a single parameter instead of inserting values for each matrix element separately. The following approach, as proved by Varis (1995), preserves all the moments of the distribution of j (expected value, skewness, kurtosis) except variance, which is increased correspondingly to the amount of uncertainty related to the respective relation. The link strength parameter is denoted as $\eta_{j|i}$, $i \neq j$. $\eta_{j|i} \in [-1, 1]$. A symmetric, $k \times k$ link matrix $\mathbf{M}_{j|i}$ is constructed as a function of $\eta_{j|i}$. η is now used as an input. For $\eta \geq 0$, the diagonal and off-diagonal elements of $\mathbf{M}$ are obtained by

$$m_{q,r} = \frac{1}{k} + \eta\left(1 - \frac{1}{k}\right), \qquad q = r = 1, \ldots, k \tag{10.A1a}$$

$$m_{q,r} = \frac{1}{k-1}\left[1 - \left[\frac{1}{k} + \eta\left(1 - \frac{1}{k}\right)\right]\right], \quad q \neq r \tag{10.A1b}$$

For $\eta < 0$,

$$m_{q,r} = \frac{1}{k} - \eta\left(1 - \frac{1}{k}\right), \qquad q = k - r \tag{10.A1c}$$

$$m_{q,r} = \frac{1}{k-1}\left[1 - \left[\frac{1}{k} - \eta\left(1 - \frac{1}{k}\right)\right]\right], \quad q \neq k - r \tag{10.A1d}$$

Network Propagation

The propagation of conditional probabilities is performed numerically. Since the updating can be done in two ways, two messages (Bayesian likelihoods) are computed, and the updated belief is obtained as the convolution product of these messages and the prior belief. The nodes are linked with link matrices that can be direction-specific. This approach does not update messages in cases where the propagation direction is changed. The calculation to the two propagation directions is performed symmetrically, but following the terminology of Pearl (1988), directions up and down are used here.

When propagating messages downwards in a BN, all messages coming to a node j, from an another node i, are denoted by $\mathbf{p}_{j|i}$ and messages leaving node i are denoted by $\boldsymbol{\pi}_i$. For any node j, preconditioned by any node i ($i < j$):

$$\mathbf{p}_{j|i} = \mathbf{M}_{j|i}\,\boldsymbol{\pi}_i \tag{10.A2}$$

The likelihood vectors $\mathbf{p}_{j|i}$ and $\boldsymbol{\pi}_i$ consist of the following elements:

$$\pi_i = \begin{bmatrix} \pi_i^1 \\ \pi_i^2 \\ \pi_i^3 \end{bmatrix} \quad \text{and} \quad \mathbf{p}_{j|i} = \begin{bmatrix} p_{j|i}^1 \\ p_{j|i}^2 \\ p_{j|i}^3 \end{bmatrix} \tag{10.A3}$$

For elements r, the π_i^r message is the scaled vector product (joint distribution) of the message $\pi^r_{i|1...i-1}$ and the evidence $\mathbf{e}_i^r$.

$$\pi_i^r\ \pi_{i|1...i}^r = \alpha e_i^r\, \pi_{i|1...i-1}^r \tag{10.A4}$$

where α is a scaling constant, scaling the sum of the k vector elements of $\boldsymbol{\pi}_i$ to unity. The incoming message $\boldsymbol{\pi}_{i|1...i-1}$ is the joint distribution of all the messages, $\mathbf{p}_{i|1}$ to $\mathbf{p}_{i|i-1}$, from the node's $i-1$ predecessors:

$$\pi_{i|1...i-1}^r = \prod_{k=1}^{i-1} p_{i|k}^r \tag{10.A5}$$

Starting from the first node, $\mathbf{p}_{1|0} = \mathbf{1}$ and $\boldsymbol{\pi}_1 = \mathbf{e}_1$, $\mathbf{p}_{2|0,1} = \mathbf{M}_{2|1}\boldsymbol{\pi}_i$, and so on.

Bottom-up propagation goes similarly. Only the direction is reversed. All messages coming to node i from node j are denoted by $\mathbf{l}_{i|j}$ and messages leaving j are denoted by $\boldsymbol{\lambda}_j$.

For each node j, the posterior belief distributions $\mathbf{Bel}_j$ can now be calculated on the basis of the prior distribution $\mathbf{e}_j$, updating it with the information from the sub-network above and below the node, i.e., vectors $\boldsymbol{\pi}_{j|1...j-1}$ and $\boldsymbol{\lambda}_{j|j+1...n}$, respectively:

$$\mathbf{Bel}_{j|1...j-1} = \gamma\, \boldsymbol{\pi}_{j|1...j-1} \cdot \mathbf{e}_j \cdot \boldsymbol{\lambda}_{j|j+1...n} \tag{10.A6}$$

where γ is a scaling constant.

GBN: A Two-layered Network

The applicability of BNs can be broadened remarkably by including deterministic relations $\mathbf{y}_j = \mathrm{f}(\mathbf{y}_i)$ between the outcomes of two conditioned nodes. Such relations can be algebraic equations, analytically solved differential equations, or for instance Boolean (monotonous) logic of a stakeholder, e.g., IF "more sewage" THEN "less fish".

Propagation in state layer (deterministic equations)

If the deterministic state equations are nonlinear, as very often in practice, the analytical propagation of uncertainty is usually too laborious. There are several approximate approaches to propagation of uncertainty in deterministic equations (Korn and Korn, 1968; Morgan and Henrion, 1990). A Taylor series expansion is among the most popular. The more accuracy is required, the more terms can be included. Consider here the first-order approximation, which in many cases is sufficiently accurate.

There are two specific cases in which rather practical equations for expected value and uncertainty of y can be derived: the weighted sums of components and products of powers of the components. In the case of weighted sums

$$y = \sum_{i=1}^{n} a_i x_i \tag{10.A7}$$

the mean and the variance can be obtained by

$$E[y] = \sum_{i=1}^{n} a_i E[x_i] \tag{10.A8}$$

$$\mathrm{var}[y] = \sum_{i=1}^{n} a_i^2 \,\mathrm{var}[x_i] + 2\sum_{i=1}^{n} \sum_{j=i+1}^{n} a_i a_j \,\mathrm{cov}[x_i, x_j] \tag{10.A9}$$

Accordingly, for product and power equations

$$y = \prod_{i=1}^{n} x_i^a \tag{10.A10}$$

the mean and the variance are

$$E[y] \approx \prod_{i=1}^{n} E[x_i]^{a_i} \tag{10.A11}$$

$$\mathrm{var}[y] \approx \sum_{i=1}^{n} \mathrm{var}[x_i] \left(\frac{a_i}{x_i^0} y^0 \right)^2 \tag{10.A12}$$

The variance equation can be processed in a more convenient form by using the coefficient of variation (cv)

$$\mathrm{cv}(z)^2 \equiv \frac{\mathrm{var}[z]}{E^2[z]}$$

hence,

$$\mathrm{cv}^2[y] \approx \sum_{i=1}^{n} a_i^2 \mathrm{cv}^2[x_i] \tag{10.A13}$$

Above, it was assumed that the model is structurally correct. In the present approach this does not need to be the case. An uncertainty estimate can be given of the model structure, expressed as the link strength η. The link strength can be augmented to the state layer in the following approximate manner:

$$\mathrm{cv}'[y] \approx \frac{\mathrm{cv}[y]}{\sqrt{\eta}} \tag{10.A14}$$

where cv′[y] is the coefficient of variation of the model prediction when the model structural uncertainty is included. In cv[y] it is excluded.

From state layer to probabilistic layer

The normally distributed model prediction y_i at point i (Figure 10A) is approximated by a discrete distribution with n equally likely intervals. Hence, in a GBN with no measured information, all distributions are uniformly distributed. This does not mean that there is no information, as the Bayesian concept of non-informative priors would suggest, but instead implies that the net contains no information contradicting the information propagated by the state layer of the GBN. If any external information (measured information, target level, etc.) differing from the model prediction is included, then non-uniform distributions reveal it in the net. A further, practical rationale for using uniform distributions in the probabilistic layer is that a vector product of two discrete uniform distributions is a uniform distribution. This feature is important when propagating information in the probabilistic layer.

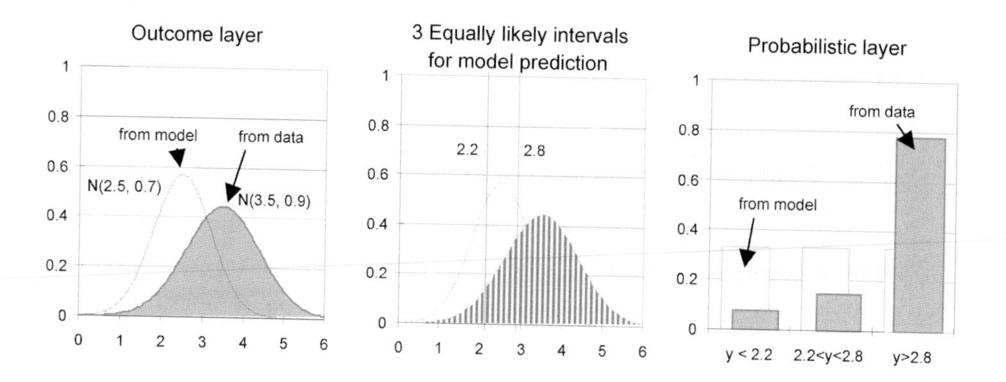

Fig. 10A. Discrete approximation of an observation.

Since the state layer uses continuous distributions and the probabilistic layer is in discrete form, discrete approximations of the continuous random variables are needed when taking them as priors to the probabilistic layer. The following approximation is used. First, define y_1 and y_2 such that

$$P(y \leq y_1) = 1/3$$
$$P(y_1 < y \leq y_2) = 1/3 \qquad (10.A15)$$
$$P(y > y_2) = 1/3$$

These values can be obtained by, e.g., using standard normal deviates:

$$y_1 = \mu_y - 0.4307\ \sigma_y$$
$$\qquad (10.A16)$$
$$y_2 = \mu_y + 0.4307\ \sigma_y$$

In other words, the model prediction is approximated with a discrete distribution with three equally likely intervals. These values can then be used to find the discrete approximation **e** for e^*. This will now be made using the intervals obtained above.

$$P(e_1) = P(e \leq y_1)$$
$$P(e_2) = P(y_1 < e \leq y_2) \qquad (10.A17)$$
$$P(e_3) = P(e > y_2)$$

These values are used as the evidence vector in the probabilistic layer.

From probabilistic layer to state layer

In the approach, there are two different paths of information from the probabilistic layer to the state layer:

- The link strength parameter η is involved in the propagation of uncertainty.
- The deviations between model prior distributions and posteriors $\mathbf{Bel}_i$ give important diagnostic information about the model. In parameter estimation or other adjustment of the model to fulfil given targets, the posteriors are iterated to make them uniform distributions.

The following quadratic/linear iteration scheme is based on comparison of the probabilities of the different outcome values of a control variable. They are iterated to be equal to one another.

$$\mu_i^* = \mu_i + a \cdot \mathrm{cv}_i \cdot (Bel_y^1 = Bel_y^3)[Bel_y^1 - Bel_y^3] \tag{10.A18a}$$

$$\eta_i^* = \eta_i + b \cdot (Bel_y^2 - 1/k) \tag{10.A18b}$$

where a and b are convergence parameters, Bel_y^r is the posterior probability of outcome r, k is the number of outcomes, μ_i is the mean of the prior distribution of node i (a control variable), η_i is the estimated link strength, and * refers to an updated iteration value.

Environmental Foresight and Models: A Manifesto
M.B. Beck (editor)

CHAPTER 11

Random Search and the Reachability of Target Futures

M.B. Beck, J. Chen and O.O. Osidele

11.1 INTRODUCTION

Without question, our concern is with contemplating the future. Yet forecasting just what that future might be, albeit couched in as much uncertainty as may be necessary for us to secure reasonable credibility, is *not* the focus of our enquiry. Consider this. In a conventional policy analysis we assemble the historical, empirical evidence of the system's observed behaviour, develop the model (with inputs and policy controls ($\boldsymbol{u}$), states ($\boldsymbol{x}$), parameters ($\boldsymbol{\alpha}$), and outputs ($\boldsymbol{y}$)), attempt to reconcile the model with the evidence, draw up future policy options/controls, predict the consequences of these possible actions, and identify therefrom which has the preferred response. Alternatively, we might specify some desired response, $\boldsymbol{y}_d$, and hence derive the corresponding policy action required to meet this goal. In both cases, the object of the analysis is ultimately the content of $\boldsymbol{u}$, the action to be taken. It is presumed that either structural change does not occur, or the structure of the model is fixed, or at least this structure can be given invariance by making the model so comprehensive as to render insignificant the possibility of (apparent) structural change arising inadvertently from our ignorance.

What we are about in this chapter is slightly different. Our point of departure is to elucidate and record speculations and fears about the future, as imagined by the lay person. We view the model as an assembly of candidate, constituent hypotheses woven into a web of interactions: a wide net, as it were, of relatively small mesh-size, to be cast over an appropriately large area of plausible science, with the intent of capturing principally those constituents on which the reachability of the imagined futures might turn. The primary object of the analysis is the content of $\{\boldsymbol{\alpha}^K(t^+)\}$, the subset of those model parameters identified as being the keys to discriminating whether or not the target behaviour is generated in the future (denoted t^+). In particular, we are curious about how the composition of this vector may differ from that of those parameters $\{\boldsymbol{\alpha}^K(t^-)\}$ found to be dominant in discriminating between behaviour matching, or not, the observed record of the past (t^-). What is to be scrutinised, or better understood, are as much changes with time from insignificance to significance as vice versa. Being oriented towards enquiry into the nature of behaviour in the distant future we must obviously be quite unsure of what is to be entrapped in the mesh of our analysis. Qualitative knowledge of an insecure character is not discarded; there is no over-reaching of ourselves in order to create the illusion of solidity and security of the knowledge encoded in what may be a relatively high-order model.

Beyond this, there is an asymmetry about the conventional forms of policy analysis. Attention is drawn towards reaching the optimum, towards what has to be done in order to hit the target, without concern for what it may take to miss the target. This one-sidedness is absent from what we shall propose. Identifying the key and redundant parameters is fundamentally dependent upon assessing the conditions necessary not only to succeed but also to fail in reaching the target behaviour. We are interested in what falls on either side of the divide between an acceptable/unacceptable match of the observed past or imagined future. Juxtaposition of the two informs the outcome of the analysis.

These ideas are not new, having first surfaced over a decade ago (Beck, 1987), inspired by the seminal contribution of Hornberger, Spear, and Young, which had been published several years earlier (Hornberger and Spear, 1980; Spear and Hornberger, 1980; Young et al., 1978). They have remained largely unchanged, except with regard to who exactly would be the authors of the scenarios for the feared and/or desired futures. In 1987 we appear to have had in mind "the analyst". Now it should be the "stakeholders", and not merely for reasons of the changing cultural milieu of environmental management since the early 1990s, as we shall see. In this chapter, we introduce the principles for an analysis of reachable futures, summarise the computational method, and present some illustrative results from a first, prototypical case study.

11.2 REACHABLE FUTURES

In 1995 the Science Advisory Board of the US Environmental Protection Agency produced a report entitled *Beyond the Horizon: Using Foresight to Protect the Environmental Future* (Science Advisory Board, 1995). In particular, the report observed that

> ... the [Environmental Protection] Agency has an obligation to search for the "weak signals" that portend future risk to human health and to ecosystems ...

and that any such search should be

> ... eclectic in its use of information sources ...

It recommended that

> ... EPA should establish an early-warning system to identify potential future environmental risks ...

and went on to discuss possible systems of enquiry for meeting this objective.

What, then, might be such a "system of enquiry"? There is, without belittling it, the obvious response: an area is to be identified in which scientific data are sparse and/or in conflict and the scientist conducting the enquiry is to submit – to the process of scientific peer review – an opinion on the interpretation of the data as portending some threat to the environment[1]. In other words, the extant historical record gathered within the paradigm of scientific enquiry is to be examined and interpreted by a practising scientist whose opinion will be judged by other practising scientists.

Worthy and necessary though this is, it is not the only thing that could be done. Cast in like terms, we introduce in this chapter a system of enquiry in which the public's concerns for the fate of the environment are assembled and their reachability then assessed through the device of a model, drawn unmistakably from the paradigm of science. The model may have to be recognised as a derivative map of the science base: a derivative of the ever-expanding primary knowledge base of atmospheric chemistry, aquatic biology, ecology, and so forth; a formally organised distillate of the raw material of relatively recently discovered facts. Yet some of these "facts", in fact, may be rather qualitative, if not speculative, in character, especially in the fields of biology and ecology (Shrader-Frechette, 1995). This second line of enquiry into the possible presence and character of problems just beyond the horizon, put cryptically, draws upon a combination of {scientific models & stakeholder imagination}. It differs in both elements from that of {scientific empirical observations & scientific opinion}, offering thus a wider search for the possibility of surprises and being undoubtedly eclectic, if unconventional, in the sources of information into which it taps. It has about it a whiff of the public directing upon which issues the torchlight of scientific enquiry is to be shone.

There will be speculation in both approximations of the "truth", in both the {scientific models} and the {stakeholder imagination}. The one is to be pitted against the other in order to identify, as we have already said, the content of $\{\boldsymbol{\alpha}^K(t^+)\}$. That the developer of the scientific model should be strictly distinct from the

---

[1] This appeared subsequently in a request for proposals for research on *FUTURES: Detecting the Early Signals* (1999 Science to Achieve Results (STAR) Program, National Center for Environmental Research and Quality Assurance, Office of Research and Development, US Environmental Protection Agency).

stakeholder with a creative imagination may therefore be vital to the success of such an analysis. For what one is attempting is the analog of reconciling past observed experience (data) with theory, in which one would like these two types of approximations of the truth to be drawn from maximally independent sources of knowledge/experience. So it should also be in reconciling fears about the future with the scientific model and not merely, as we can now see, as a reflection of the rightful rise to power of lay persons within the body of stakeholders (Darier et al., 1999; Korfmacher, 2001). After all, there would be little point in reconciling the scientific model with a future imagined by the constructor of the model, since that person would arguably be able to build into the thought experiments of the feared futures many of those things he/she understands to be critical to the scientific model. Of course, such a person would not be able to imagine a future entirely consistent with the model; we build mathematical models precisely because we cannot reason through the complexities of the system's apparent behaviour in our heads.

The challenges, in algorithmic terms, are therefore as follows. Given the current science base – uncertain, speculative, and flawed though it may be – are the feared/desired futures of the stakeholders technically reachable? Are these fears, in particular, groundless or not? If they are not groundless, on which of the current scientific unknowns might their reachability most critically depend? How indeed, going beyond the scope of this chapter (as subsequently touched upon in Chapter 15), might we design a model to maximise the earliest discovery of our ignorance of the problems just beyond the horizon? The intent of the Science Advisory Board's system of enquiry is to discern what the problems – the feared futures – might be. We shall instead take the imagined futures as given – in some way – and proceed to discriminating the plausible from the highly implausible among them. In order to do so we shall reach outside the conventional realm of scientific enquiry, yet nevertheless seek to be systematic and rigorous in our procedure.

In the wider setting, the challenges are to assess how community views (on preserving a given piece of the environment) may change over time as a function of iterative interaction with the science base, within an overall framework we might label "adaptive community learning" (for want of better words; Norton and Steinemann, 2001; Beck et al., 2001). We know what adaptive management is. In essence, policy therein fulfils two functions: to probe the behaviour of the environmental system in a manner designed to reduce uncertainty about that behaviour, i.e., to enhance learning about the nature of the *physical* system; and to bring about some form of desired behaviour in that system. Adaptive community learning ought both to subsume the principles of adaptive management (so defined) and include actions, or a process of decision-making, whereby the community of stakeholders experiences learning about *itself*, its relationship with the valued piece of the environment, i.e., the community-environment relationship, and the functioning of the physical environment. In this the community of stakeholders is interpreted in a much broader sense than merely stakeholders as policy persons/managers. Indeed, the scientifically lay stakeholder is pivotal.

11.3 A REGIONALISED SENSITIVITY ANALYSIS

We are strongly accustomed to the idea of the system's behaviour being specified in terms of a time-series of observations of the model's state (or output) variables. This is indispensable to matching past observations. But it is also a rather restrictive outlook, as admirably demonstrated in the work of Hornberger, Spear and Young in the late 1970s (Young et al., 1978; Hornberger and Spear, 1980; Spear and Hornberger, 1980). As their exploration of eutrophication in an estuarial inlet shows, it is possible to determine what might be the critical components of a model in discriminating between a match and a mis-match of *qualitatively* observed behaviour, and under gross uncertainty. This, of course, is not entirely what is needed for our present purposes, which is instead an exploration of the potential origins of imagined *future* patterns of behaviour. But in order to present their method, the reasons for its development, and the manner in which these prompt an approach to investigating the reachability of target futures, it is instructive to begin by reconstructing the contemporary knowledge base, hopes, and aspirations of those times, now some two decades ago.

11.3.1 Origins of the Method

In its simplest form the problem of eutrophication may be defined as follows. It is the artificial acceleration of the natural ageing of bodies of water as a result of an elevated rate of accumulation therein of nutrients (principally carbon-, nitrogen-, and phosphorus-bearing substances). This much was already understood by the late 1960s, as we now know from the case histories of Chapters 2 and 7. What was intended for the beneficial enhancement of primary biological production on the land surface was instead being diverted into enhanced primary production in the aquatic environment. This was manifest in the high-frequency, high-amplitude perturbations of sudden bursts of growth and rapid collapse in the biomass of phytoplankton – microscopic organisms at the base of the food chain in an aquatic ecosystem. Given a ready supply of the principal nutrients, it was well known that under an appropriate combination of environmental factors (solar irradiance, temperature, concentrations of other trace nutrients, and so on) populations of the various species of phytoplankton would grow rapidly, with a succession of species becoming dominant over the annual cycle. The phytoplankton would be preyed upon by zooplankton, which in turn were themselves prey to the fish. Dead, decaying, and faecal matter from all of this activity was equally well known to be re-mineralised by a host of bacterial species – operating either in the water column or in the bottom sediments of the body of water. Such re-mineralisation had an influence over the concentration of oxygen dissolved in the water and this, in its turn, exerted an influence over the degradation pathways operative in the overall process of microbially mediated re-mineralisation. Thus the cycling of elemental material fluxes was closed; thus the disturbances of excessive phytoplankton growth would be propagated around these cycles.

Such was the raw material of the primary science base of the time. It was rapidly being refined and encoded in mathematical form[2], typically in the standard terms of the following state variable dynamics,

$$\mathrm{d}\boldsymbol{x}(t)/\mathrm{d}t = \boldsymbol{f}\{\boldsymbol{x},\boldsymbol{u},\boldsymbol{\alpha};t\} + \boldsymbol{\xi}(t) \tag{11.1a}$$

with the observed outputs being defined as follows,

$$\boldsymbol{y}(t) = \boldsymbol{h}\{\boldsymbol{x},\boldsymbol{\alpha};t\} + \boldsymbol{\eta}(t) \tag{11.1b}$$

in which $\boldsymbol{f}$ and $\boldsymbol{h}$ are vectors of nonlinear functions, $\boldsymbol{u}$, $\boldsymbol{x}$, and $\boldsymbol{y}$ are the input, state, and output vectors, respectively, $\boldsymbol{\alpha}$ is a vector of model parameters, $\boldsymbol{\xi}$ and $\boldsymbol{\eta}$ are notional representations respectively of those attributes of behaviour and output observation that are not to be included in the model in specific form, and t is continuous time. We presumed then (in the 1970s) that as grew the order of the internal description of the system's behaviour, i.e., the order of $[\boldsymbol{x},\boldsymbol{\alpha}]$ in equation 11.1(a), so would the order of access to its external description, i.e., the order of $[\boldsymbol{u},\boldsymbol{y}]$ characterising observed behaviour. We expected the estimation of parameters to become "automated", subordinated to the formal algorithms of constrained optimisation instead of practised "by hand"; the observed features of the system's behaviour were expected, just as much, to become increasingly detached from our personal, qualitative, and arguably subjectively tainted, powers of observation (of sight, sound, and smell); the time series of data on $[\boldsymbol{u},\boldsymbol{y}]$ were expected to become increasingly available and, eventually, sampled at an appropriately fast frequency for a sufficiently long period; hence, as a consequence of all this, we would be able to make better forecasts of behaviour in the future.

Experience, however, was failing to bear out these expectations, and rather consistently so. To their credit, Hornberger, Spear, and Young were the first to recognise the situation as such and deal with it effectively. Their approach has come to be known as a Regionalised Sensitivity Analysis (RSA); it has been widely applied, yet little modified over the intervening years. There are traces of it here in Chapters 8, 12, 13, and 16; and we shall now expand upon the outline given of it earlier in Chapter 5.

11.3.2 Essence of the Analysis

Two presumptions underpin the classical approach to model parameter estimation, that is, to reconciling theory with *past* observed behaviour. First, there will be some "dots" through which the "curve" of the model's output can be made to pass closely (Figure 11.1(a)). Second, there will be a uniquely defined combination of values for the model's parameters ($\boldsymbol{\alpha}$) attaching to this optimal match of estimated and

[2] Culminating, to a degree, in the model CLEANER (Park et al., 1975).

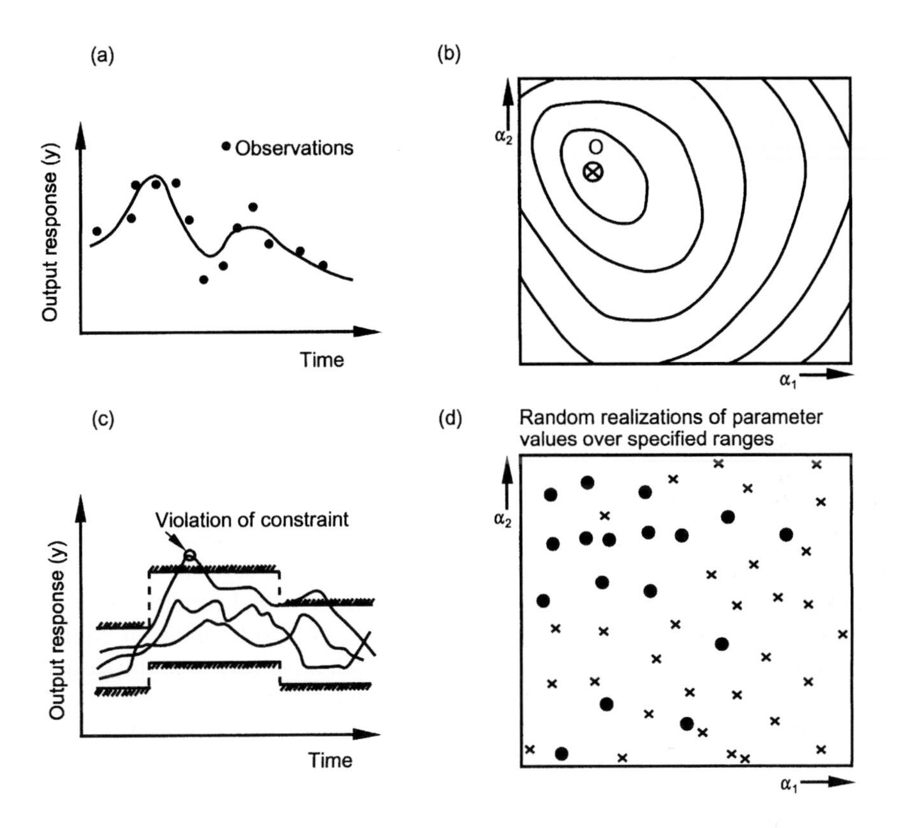

Fig. 11.1. A comparison of ((a) and (b)) the classical approach to model parameter estimation and ((c) and (d)) regionalised sensitivity analysis: (a) fitting the model response to the data; (b) contours of the fitting-function surface in the parameter space; (c) specification of the constraints on acceptable model response; and (d) analysis of model parameter values (dots indicate values giving rise to acceptable behaviour, and crosses indicate values giving rise to unacceptable behaviours).

observed behaviour (the point O in Figure 11.1(b)). Neither plays any role whatsoever in the RSA. The problem statement remains as inelegant and uncertain as it truly is, but *not* without the prospect of making meaningful progress, thus:

- The trajectories of the time-series observations $\{\boldsymbol{y}(t_0), \boldsymbol{y}(t_1), \ldots \boldsymbol{y}(t_N)\}$, against which the performance of the model would normally be evaluated, are replaced by a definition of (past) behaviour (B) in terms of less detailed, more qualitative, possibly subjectively derived, constraints, such as those shown in Figure 11.1(c).

- The conventional error-loss (objective) function for locating a uniquely best estimate ($\hat{\boldsymbol{\alpha}}$) of the parameter vector is replaced by a criterion for either accepting or rejecting a candidate parameterisation $\boldsymbol{\alpha}^j$ as giving rise to the defined behaviour (B) introduced in the foregoing (Figure 11.1(d)).

For example, in the original study of eutrophication in the Peel Inlet of Western Australia (Hornberger and Spear, 1980; Spear and Hornberger, 1980) one item of the behaviour definition (B) was chosen to constrain the estimated yearly peak biomass of the nuisance alga *Cladophora* ($\hat{y}_{max}$) to be greater than 1.5 times and less than 10.0 times its initial biomass at April 1 (defined as t_0), i.e.,

$$1.5\ \hat{y}(t_0) \leq \hat{y}_{max}(t) \leq 10.0\ \hat{y}(t_0);\ \text{for } t_0 \leq t \leq t_N \tag{11.2}$$

where $\hat{y}$ symbolises, in general, a model-generated estimate of the observed output. In addition, the ranges of permissible values from which the candidate model parameter vectors are to be drawn were specified as rectangular distributions with upper and lower bounds, i.e.,

$$\boldsymbol{\alpha}^l \leq \boldsymbol{\alpha}^j \leq \boldsymbol{\alpha}^u \tag{11.3}$$

The two types of inequalities (11.2) and (11.3) reflect the uncertainty of the empirical evidence and the uncertainty of the prior hypotheses, respectively. In the original study the pattern of the input disturbances $\boldsymbol{u}(t)$ and the initial conditions $\hat{\boldsymbol{y}}(t_0)$ were assumed to be known and not subject to uncertainty.[3] Both could, of course, be treated as parts of the parameterisation of the model, obviously so in the case of the latter, and possibly via the use of time-series models in the case of the former. This same principle could in fact be further extended to parameterisations of the processes $\boldsymbol{\xi}$ and $\boldsymbol{\eta}$ in equation 11.1, although this remains an as yet unexplored option.

The RSA comprises two steps, the first being a combination of straightforward Monte Carlo simulation with a binary classification of the candidate model parameterisations. A sample vector $\boldsymbol{\alpha}^j$ is drawn at random from its parent distribution and substituted into the model to obtain a sample realisation of the trajectory $\hat{\boldsymbol{y}}^j(t)$, which is then assessed for its satisfaction, or otherwise, of the set of constraints defined in the form of inequality (11.2). Repeated sampling of $\boldsymbol{\alpha}^j$, for a sufficiently large number of times, allows the derivation of an ensemble of parameter vectors giving rise to the behaviour (B) and a complementary ensemble associated with not-the-behaviour ($\bar{B}$). For this analysis, therefore, there is no meaningful interpretation of a degree of closeness to a uniquely best set of parameter estimates. Each sample vector $\boldsymbol{\alpha}^j$ giving rise to the behaviour is equally as good or as probable as any other successful candidate parameterisation of the model.[4]

Two sets of candidate parameterisations can therefore be distinguished: those m samples $\{\boldsymbol{\alpha}(B)\}$ that give the behaviour and those n samples that do not, i.e., $\{\boldsymbol{\alpha}(\bar{B})\}$.

---

[3] To all intents and purposes the states ($\boldsymbol{x}$) and model-generated estimates of the outputs ($\hat{\boldsymbol{y}}$) are here identical.

[4] A number of subsequent variations on this theme can be found in Keesman and van Straten (1990), Beven and Binley (1992), and Klepper and Hendrix (1994) (as well as in Chapter 12).

In the second step of the RSA the objective is to identify which among the hypotheses parameterised by $\boldsymbol{\alpha}$ are the significant determinants of observed past behaviour. Significance is here indicated by the degree to which the central tendencies of the marginal and joint distributions of the (*a posteriori*) ensembles of the "behaviour-giving" parameter values $\{\boldsymbol{\alpha}(B)\}$ and their complement $\{\boldsymbol{\alpha}(\bar{B})\}$ are distinctly separated. For each constituent parameter α_i the maximum separation, distance d_{max}, of the respective cumulative distributions of $\{\alpha_i(B)\}$ and $\{\alpha_i(\bar{B})\}$ may be determined and the Kolmogorov–Smirnov statistic, $d_{m,n}$, then used to discriminate between significant and insignificant such separations for a chosen level of confidence (Kendall and Stuart, 1961; Spear and Hornberger, 1980). Relatively large separation implies that assigning a particular value to the given parameter is key to discriminating whether the model does, or does not, generate the specified behaviour. Relatively small separation of the two distributions implies that evaluation of the associated parameter is redundant in so discriminating the performances of the model. For these latter it matters not, in effect, what value is given to the parameter; the giving or not giving of the specified behaviour is more or less equally probable whatever value of the parameter is realised. Thus, for instance, the distinct clustering of parameter combinations that give the behaviour, towards high values of α_2 and low values of α_1 in Figure 11.1(d), suggests that both parameters are key in the sense now defined. Indeed, one could also conclude that both of the constituent hypotheses associated with α_1 and α_2 are likely to be fruitful speculations in understanding the observed system behaviour. More generally, the parameter space of the model can be cleaved into those parameters found by this procedure to be key, $\{\boldsymbol{\alpha}^K\}$, and those found to be redundant, $\{\boldsymbol{\alpha}^R\}$.

11.3.3 Target Futures: Beliefs, Imagination, and Experience as Substitutes for Quantitative Observation

What we see is that the output of the model is required, as it were, to pass through a corridor of constraints with hurdles to be overcome (Figure 11.1(c)), and it either succeeds or fails for any given candidate parameterisation. In very general terms, passage through these constraints can be thought of as a design task to be matched (or fulfilled), or as a target to be struck, by the model. Much more importantly, indeed quintessentially, specification of the task may be drawn from the *beliefs, imagination, and experience* of an individual; from the expert field biologist, for example, who has walked around the estuary, or sailed across it, and witnessed the timing of the unsightly, rotting *Cladophora* biomass in each of the past five years or so. At the core of what we are proposing is this, then: the freedom to draw upon the qualitative features of an individual's imagination of *future* possibilities – and experience of the past – in order to specify target behaviours to be reached by the scientific model.

Hornberger and colleagues sought only to understand the past and, as far as we know, did not tap into the experience of any lay stakeholders in order to do so. In other cases, in particular that of the potential decline in quality of Lake Lanier in Georgia (just to the north of metropolitan Atlanta), we find that recreational anglers may have information vital to identifying the dominant mechanisms in the structure of the lake's foodweb. The anglers need shad fish as bait for catching bass. They know from experience – and the casual observations of their many fellow anglers – where to go at what time of the year to first catch their shad; and where the shad graze, so there must be the zooplankton and, by strong inference, the phytoplankton. Empirical experience such as this is not acquired by any conventionally scientific means, but it is information nevertheless material to circumscribing the target behaviour to be matched by the scientific model, above all, in respect of its biological components, for which there are so very rarely any scientific observations whatsoever.

In our search to discern the plausibility of potential future environmental risks, we shall tap into the imagination of the stakeholder. In this, in contrast to the assembly of classical empirical observations, access to the past is *not* utterly and fundamentally different from access to the future. We cannot in the present gather scientific observations of future behaviour. But we may well cherish something about the present and past and fear it may be lost to us in the future. And we may contemplate what might happen in the future just as we reflect on the experience of the past, expressing thoughts on either in much the same manner. Will there be outbreaks of *Cryptosporidium* in Lake Lanier; is it possible that the southward migrating zebra mussel could establish itself there, and so on?

These questions arising in our thoughts, or trailed before us by an articulate group of stakeholders – as admirably illustrated in the case of global climate change (Leggett, 1996) – are the target futures against which the scientific model is to be pitted. The model, for its part, provides a map of the problem-specific science base – its knowns, partially knowns, and unknowns. Each parameter resembles a tag, as it were, of the various items of raw material assembled into the model's structure from the several, perhaps many, conventional, primary disciplines. Again, in the case of Lake Lanier, we are beset by any number of domains, each subsuming clusters of parameters in the more detailed science map, where understanding and the availability of conventional empirical observations are lacking. A preliminary short-list of these domains would include: the coupled particulate-solute chemistry of, inter alia, Fe, Mn, Ca, and P; the bottom sediments and their interaction with exchange fluxes between the surface water and groundwater systems; the role of the microbial loop in the foodweb and, in particular, the positions of pathogenic micro-organisms in that web; stratification of the lake's water and the behaviour of one of the principal tributaries as a submerged jet; and the role of atmospheric deposition of nutrients (notably phosphorus) in hypolimnion primary production. What might it therefore be about the model, we could enquire, i.e., which domains and constituent parameters might it be, that will cause a "substantial decline in the striped bass population", a legitimate concern of Lake Lanier's community of anglers? Which features in the map of the science base, we would want to ask, i.e., which members in the assembled composite of all the constituent, parameterised hypotheses, are key and which redundant to the reachability of this target future?

It is a fact of life that we lack the resources for addressing all of the domains of potential importance with equal vigour. The stakeholders need to know whether their hopes and fears for the future may reasonably come to pass. The scientists need to know to which critical turning points in the map of the plausible knowledge base the limited resources of further scientific enquiry should be allocated. If the maximum specific growth-rate of mixotrophic flagellates appears to be key to the survival of a viable population of striped bass, i.e., this parameter is a member of the set $\{\boldsymbol{\alpha}^K(t^+)\}$, there would be little point in instead investigating in detail the seemingly redundant domains of stratification, submerged jets, atmospheric deposition, and so on (as contained in $\{\boldsymbol{\alpha}^R(t^+)\}$).

These, therefore, are the sorts of goals we would have in undertaking an analysis of reachable futures.

11.4 A CASE STUDY: AQUATIC FOODWEB IN A PIEDMONT IMPOUNDMENT

Lake Lanier lies in the fork of two major transportation corridors going northwards from the rapidly developing area of metropolitan Atlanta in the south-eastern state of Georgia in the USA. Situated upstream of Atlanta in the upper reaches of the Chattahoochee watershed, the pressures of development in the lake's immediate catchment are intense. In the region, the water resources of the Chattahoochee River, at the core of which are those of Lanier, are the focus of an inter-state dispute, between Georgia, Alabama, and Florida. Put simply, water has become a scarce commodity in the region's economy. In the midst of all this, not surprisingly, there is growing concern for preserving the well-being of Lake Lanier. In spite of such close attention, however, empirical experience of the behaviour of the lake's foodweb is far from the complete, quantitative coverage to which conventional scientific enquiry has become accustomed. Much better favoured in this respect is Lake Oglethorpe, a nearby, smaller impoundment. It has been the subject of extensive study since 1978 (see, for example, Porter, 1996; Porter et al., 1996) and affords us our prototypical case study. It is, likewise, the nursery of our analysis of futures specifically reachable for Lake Lanier; the platform from which extrapolation back to Lanier will eventually have to be made, no matter how grand this extrapolation might seem. For this is the very stuff of which so many contemporary environmental problems are made. They are not begun from the ideal of laboratory science; nor do they have the prospect of ever approaching that state.

11.4.1 Structure of the Model

Figure 11.2 shows the basic structure of one of the earlier versions of a model for Oglethorpe's aquatic ecology (Osidele and Beck, 2001a). At the macroscopic (food-chain) scale two hypotheses underpin this structure: (i) the dynamics of the system are controlled by both "bottom-up" and "top down" processes, i.e., by nutrient enrichment and primary production from below and by the influence of

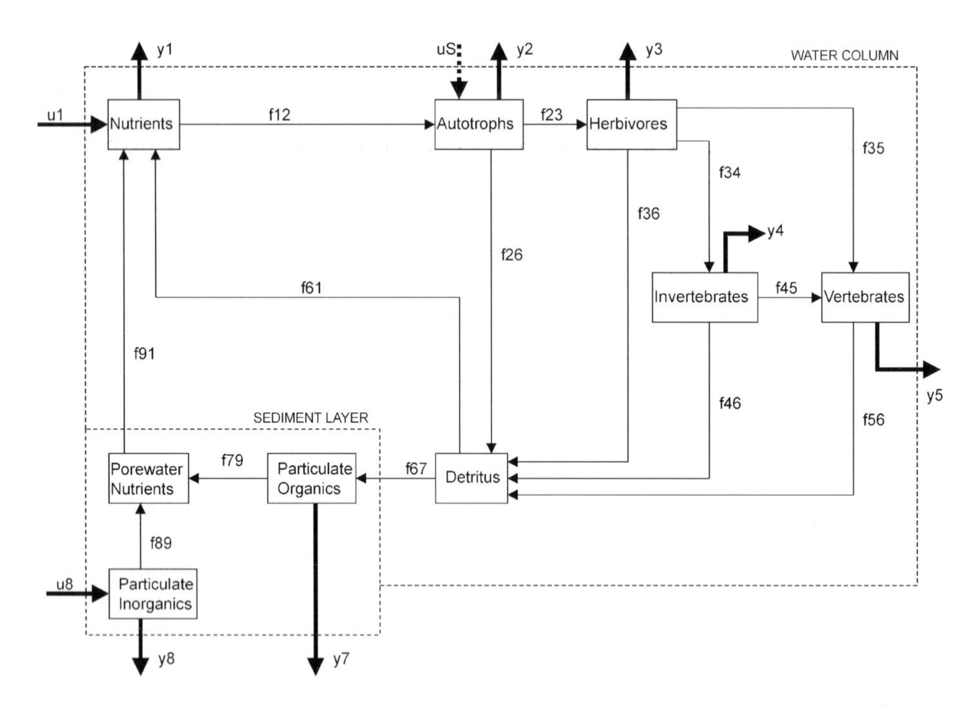

Fig. 11.2. Ecosystem components and interactions for Lake Oglethorpe, Georgia.

predators from above (Carpenter et al., 1985); and (ii) the availability of nutrients is influenced by the behaviour of sedimented material, which can act as both a repository and a source of nutrients, depending upon the physical and chemical status of the overlying water column.

For the purposes of argument, Figure 11.3 expresses the essence of the structure $[\boldsymbol{f},\boldsymbol{h},\boldsymbol{x},\boldsymbol{\alpha}]$ of Figure 11.2 in the branch-node network representation that has become (since Chapter 4) something of a convention in this monograph. In this, structure

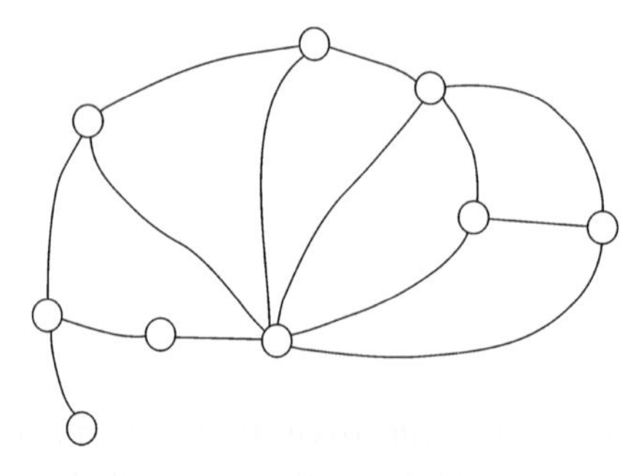

Fig. 11.3. Essence of the structure of Figure 11.2 expressed in the stylised metaphor of the branch-node network representation.

$[f,h,x,\alpha]$ has been extracted from its inter-connections with the system's inputs and outputs $[u,y]$. And in this particular case, its branches can but merely symbolise the parameterisation (α) of the model. Many of the interactions among the state variables are nonlinear, so that each of the branches may subsume more than just the symbolic convenience of a single parameter. A Monod expression for the kinetics of nutrient uptake by the autotrophs, for example, will contain three constituent parameters: a yield coefficient, the maximum specific growth-rate constant, and a half-saturation constant. Figure 11.3 is a device for conveying a conceptual argument and we must accordingly acknowledge the technical limitations from which it suffers.

In all, the model's structure contains 35 parameters, the uncertainties of each of which are to be bounded in accordance with equation 11.3. There are nine state variables (x) in the model, six describing the water column, the remainder describing the behaviour of the sediments. These are indicated by the nodes of Figure 11.3. Soluble reactive phosphorus is assumed to be the key nutrient limiting primary production in the form of the phytoplankton (the autotrophs), as is conventional; the herbivores represent the zooplankton community; in Oglethorpe, the invertebrates are predominantly *Chaoborus*; the vertebrates – the top predators – comprise all size and age classes of fish; and a state variable representing detritus accounts for the debris arising from excretion and (non-predatory) mortality of all the biological components in the model. In the sediment this organic detritus is denoted as particulate organics, the deposited river sediments as the particulate inorganics, and the phosphorus dissolved in the interstitial water of the sediments is referred to as the porewater nutrients. In the eventual cycle of things, river-borne particulate matter – predominantly clay minerals onto which phosphorus compounds readily adsorb – can be a significant source of nutrients for the phytoplankton.

11.4.2 Matching the Past and Reaching Future Behaviour

In general, our empirical experience of the lake's behaviour in the past may be expressed as

$$\bar{y}^l \leq \hat{y}(t^-) \leq \bar{y}^u \qquad (11.4a)$$

$$\underline{y}^l \leq \hat{y}(t^-) \leq \underline{y}^u \qquad (11.4b)$$

while our hopes and fears for its behaviour in the future may be expressed as

$$y_d^l \leq \hat{y}(t^+) \leq y_d^u \qquad (11.5a)$$

$$y_f^l \leq \hat{y}(t^+) \leq y_f^u \qquad (11.5b)$$

Here superscripts l and u denote respectively lower and upper bounds on the various domains of behaviour, i.e., those with acceptable ($\bar{y}$) and unacceptable ($\underline{y}$) similarity with the recorded experience of past behaviour (in equations 11.4(a) and 11.4(b) respectively), and those desired (y_d) and feared (y_f) in the future (in equations 11.5(a) and 11.5(b), respectively). Equations 11.4(a) and 11.4(b) might typically be complementary and collectively exhaustive, i.e., $\hat{y}(t^-)$ in equation 11.4(b) would be anything but that which satisfies equation 11.4(a), as here (although this does not always have to be so; Chapter 16). The same does not necessarily hold for considerations of the future. For example, desired behaviour would usually make reference to some domain quite distinct from feared behaviour, without excluding radically different behaviour, which may be (surprisingly) distinct from both.

From the rich store of empirical evidence – of a conventional, quantitative character (Porter, 1996; Porter et al., 1996) – definitions of past behaviour could, in principle, be developed for almost all of the state variables shown in Figure 11.3. In this prototypical test, however, constraints have been imposed on merely the past behaviour of the nutrient and autotroph state variables. The target behaviour of the future has equally simply been specified in terms of regulatory standards expected of the concentrations of the same two states, tailored to the growing season in Piedmont lakes of the south-eastern US (Raschke, 1993).[5]

On which constituent branches of the structure of Figure 11.3 hinges, then, the power to discriminate between whether past behaviour is matched or not? Are these same parameters key to the reachability, or otherwise, of the target, desired behaviour? Our responses are given in Figure 11.4, where we have used a Latin Hypercube sampling procedure as originally implemented in Chen (1993). These results refer to the matching (reaching) of past (future) annual cycles of behaviour in the lake. They also assume that the annual patterns of inputs ($\boldsymbol{u}$), which may include actions associated with policy, are identical for both the past and the future. As it happens, 176 out of a total of 3000 candidate parameterisations are found to match the recorded experience of the past; 1087 out of 3000 succeed in making the future pattern of behaviour reachable. Within the structure of the model, in overall terms, dominance can be seen to shift from the nutrient-autotroph-herbivore sub-web, $\{\boldsymbol{\alpha}^K(t^-)\}$ (Figure 11.4(a)), to the sediment–nutrient–autotroph sub-web, $\{\boldsymbol{\alpha}^K(t^+)\}$ (Figure 11.4(b)). Over time, between past and future, the significance of interactions with the herbivores has declined, while that of the interactions with the sediment has risen, from insignificance to significance.

We must now develop some prognoses from these raw results, in part for the community of scientists, and in part for the other stakeholders. From the perspective of the latter – in this hypothetical case, solely the regulatory body – reaching the desired future is indeed feasible, under current policies, but without developments in the watershed materially altering the current pattern of interplay between the watershed and the lake (as interpreted through the formal mathematical device of the inputs, $\boldsymbol{u}$, which here do not change between past and future). Such qualified

[5] Surveys of stakeholder concerns for the future, of both Oglethorpe and Lanier, have since been undertaken, along with the use of other instruments (such as a participatory "Foresight" workshop) for eliciting and generating the target patterns of behaviour.

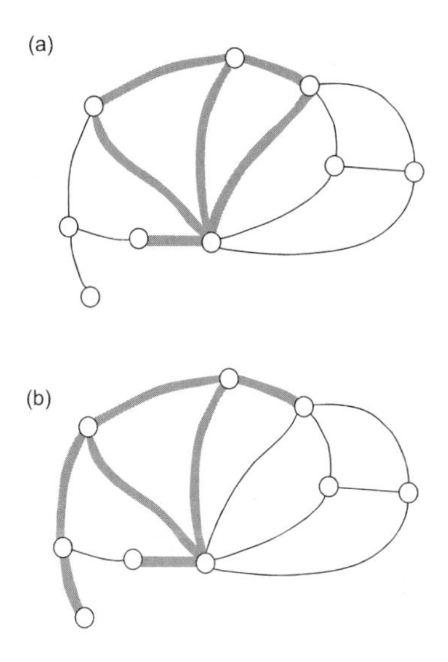

Fig. 11.4. A form of apparent structural change: (a) key parameters ($\{\boldsymbol{\alpha}^K(t^-)\}$; emboldened branches) associated with past behaviour attach primarily to the nutrient–autotroph–herbivore sub-web, (compare with Figure 11.2); (b) key parameters ($\{\boldsymbol{\alpha}^K(t^+)\}$; emboldened branches) associated with future behaviour attach primarily to the sediment–nutrient–autotroph sub-web (again compare with Figure 11.2).

reachability has no guarantee, of course. If we were to purchase more science, our results would provide guidance on how any available funds should best be spent, simply by juxtaposing the content of $\{\boldsymbol{\alpha}^K(t^+)\}$ with that of $\{\boldsymbol{\alpha}^R(t^+)\}$. From this we may conclude it is more important – under the current target behaviour definition – to focus attention on improving understanding of the circulation of phosphorus around the sediment than its propagation up the food chain beyond the autotrophs. In fact, we can already proceed to formulating a testable hypothesis within this domain of enquiry: that the empirically determined nature of phosphorus circulation within the sediment should be demonstrably consistent with the attributes of the set of candidate parameterisations $\{\boldsymbol{\alpha}^K(B);t^+\}$, i.e., the sample of particular realisations of the critical portions of the science base required for the desired behaviour (B) to be attained in the future. Evidence to the contrary would be rather thought-provoking. It could imply that the currently observed properties of the sediment–phosphorus behaviour are inconsistent with the desired target future being reachable, i.e., it corroborates not-the-behaviour ($\bar{B}$) being generated in the future. In this case the null hypothesis, so to speak, would be rejected *not* on the basis of being inconsistent with *current* conditions, but on the grounds of being inconsistent with specified behaviour in the *future*.

The prospect of having to reject the null hypothesis could just as easily imply that the structure of the model is flawed; and this would be unsurprising. In the analysis of reachable futures the model is not a static object, as it might be – for significant periods of time – in other contexts (of conventional, formal prediction). It is here a vehicle of exploration, designed to evolve in tandem with an ever-evolving problem landscape.

11.4.3 Taking Stock

What we have, then, is computational proof-of-concept and little else. Our results present us with just a sketch of how to cultivate environmental foresight. In this case, the sketch has been drawn from within the restricted palette of hypotheses encoded in a rather small {scientific model}; and the {stakeholder imagination}, into which we have tapped, would most probably attach to a member of Thompson's hierarchist social solidarity (Thompson, 1997), since it reflects largely that of a scientist-*cum*-policy-maker. It is doubtless less colourful than what might have been expected, had we had access to a scaremonger belonging, for example, to the egalitarian social solidarity (as we have already seen in Chapter 5).

We also have a number of open-ended issues to which some preliminary consideration must now be given. First, all manner of persons will number among the stakeholders, without consensus in their hopes and fears for the future. Each social solidarity – to simplify the vast variety of possibilities, for example – could legitimately be the authors of a target future. Would the coming to pass of each of this variety of futures turn thus on identical, critical unknowns in the science base; or would new science have to be pursued, in principle, on all manner of fronts? Consensus in describing the target behaviour is not essential; perhaps it is even undesirable. But in identifying how to allocate the scarce resources of further scientific enquiry, achieving both focus and consensus should be highly desirable.

Second, it seems likely that the science of the model and the hopes and fears of the stakeholders would tend to converge towards self-consistency over time. The whole purpose of the analysis of reachable futures is to explore whether the concerns of the community, in particular, for the fate of the environment in the longer term, are baseless or not. Iterative interaction with the science base is integral to this process of "adaptive community learning"; one of several attempts at moving beyond the mere skeleton of this concept can be found in Osidele et al. (2000). The virtue of strict separation – between the reasoning of the authors of the imagined futures, and that of the authors of the science incorporated into the model – might thus be gradually eroded. Would then the analysis of reachable futures be rendered impotent either by convergence, as we have now described it, or by this common (scientific) authorship of the two domains of knowledge and imagination, as noted earlier? Any response to such a question is clearly premature, but there are at least some grounds for believing impotence might not be the ultimate fate our analysis. Convergence, no matter how necessary for taking immediate action (in the short term), is unlikely ever to be accompanied by the complete absence of dissonance in a healthy, democratic community of stakeholders. Moreover, no matter how secure the knowledge bases at the core of the science encoded in the model, our experience of science is quintessentially that of its continual nurturing of speculation at the fringes. There will always be an egalitarian social solidarity (hopefully) capable of extrapolating from this fringe to enriched images of barely plausible futures, especially in the longer term.

Third, would the analysis be relegated to sterility through the following form of circular argument? Stakeholders, we may find, are greatly concerned about the presence and persistence of pathogenic bacteria in the future. To support investigation of these concerns a highly speculative sector of the science base – that of the place and function of pathogens in an aquatic foodweb – must be patched into the model. The RSA is then implemented, only to reveal parameterisation of the role of pathogenic bacteria as critical to the reachability of their persistence in the future. It will clearly be important to be wary of composing the model in a manner from which little can subsequently be learned.

Fourth, we have dealt only with identifying those changes of significance in the model's parameters ($\boldsymbol{\alpha}$) required for transferring the state of the system between the past and the future, not with determining the content of the policy actions and other interventions ($\boldsymbol{u}$) required to effect the same transfer. We are not disinterested in the latter, but exaggerating the alternative focus has been essential to achieving clarity of the arguments presented in this particular chapter (and those of Chapters 5 and 18). To enquire into the content of $\boldsymbol{u}$ is the familiar goal of conventional policy analysis. Adaptive management deals with choosing $\boldsymbol{u}$ not only for the purpose of controlling the system's behaviour, but also for reducing the uncertainty in $\boldsymbol{\alpha}$. Given an appropriate parameterisation of $\boldsymbol{u}$, our analysis might allow us to enquire into which items of external change, in concert with which bits of the science base, make the target futures reachable, or not (as outlined in Osidele and Beck, 2001b). In other words, which policy actions in conjunction with which changes in our understanding might make the feared/desired futures reachable? Which actions of policy are most (or least) likely to succeed with (or without) the acquisition of better understanding? Would a particular policy be risky, or plain ineffective, without a monitoring or research program designed to cope with a critical uncertainty?

Fifth, the method of RSA has remained essentially unchanged in the more than two decades since its first publication. Yet a crucial element of the analysis, i.e., discriminating key from redundant model parameters, is severely restricted by its heavy and usually exclusive dependence on the statistical assessment of univariate, marginal distributions. Widespread experience with models shows that the values of individual parameters may themselves not be as vital to replicating the defined behaviour of the system as are the values of certain clusters of two or more parameters. That this is so will be almost axiomatic in the discussion of the following chapter. Within these clusters the values of the constituent parameters may vary widely, but in a manner strongly correlated with each other – so that the cluster maintains a consistent, relatively invariant value. In order to identify parameters that may be critical to the reachability of the target future behaviours, yet be embedded in such clusters and therefore not detectable using univariate analyses, a Tree Structured Density Estimation (TSDE) algorithm may be implemented (Osidele and Beck, 2001b). This is one realisation of the required multivariate form of analysis (Spear et al., 1994).

Last, in preparing our mock prognoses no reference has been made to the quality of the model in performing its two tasks, of matching the past and reaching some future pattern of behaviour. We do not argue that such reference is essential, but cannot resist noting in passing the availability of a potentially useful measure of model quality (q), whose development is the subject of companion studies using exactly the same procedure of RSA as herein (Beck et al., 1997; Chen and Beck, 1998; Beck and Chen, 2000). q is generated from the maximum separations (d_{max}) of the behaviour-giving and not-behaviour-giving distributions for each of the ensemble of model parameters (and would be readily computable from the Oglethorpe results already discussed).

11.5 CONCLUSIONS

We have introduced a form of analysis designed to search for shifts in those elements of the model's structure deemed critical – in a specifically defined manner – as we turn our attention away from the experience of the past towards the imagined future pattern of the system's behaviour. For this, we have argued it is important for the structure of the model to be of a relatively high order, encompassing a large portion of the hypothetical knowledge bases (both secure and speculative) germane to the environmental problem at hand. Any so identified shifts of significance, over time into the future, will inevitably be referenced to what has been included within the structure of the model from the outset. Significance here does not imply that the values of the key parameters must be changing with time, merely that choosing a particular value for a given parameter may, for example, be quite immaterial to the model's replication of past experience, but vital to the generation of an imagined future pattern of behaviour. One could say that the patterns of behaviour, as they change with time, become sensitive in turn to different clusters of the constituent hypotheses incorporated in the (invariant) structure of the model. This – now revealed in detail – is apparent structural change of a character somewhat different from that which has hitherto been the focus of so much discussion, especially in Chapter 4. There, as subsequently in Chapter 15, structural change is gauged instead through temporal variations in the reconstructed estimates of parameters within (typically) a low-order model reflecting a compressed, macroscopic map of the underlying science base. In principle, the strong inference to be drawn is that (apparent) structural change in that case arises from what has not been adequately included within the model. Further clarification of the distinction thus emerging between the two kinds of structural change will in due course be resumed in Chapter 16.

These are early days for this approach to the analysis of reachable futures, as must be obvious from the questions raised above as we have taken stock of the tentative progress made across the prototype case study of Lake Oglethorpe. Subsequent experience with the approach is revealing important issues surrounding the means of eliciting the imagined (target) futures, with which we have not dealt in this monograph (other than in the preceding chapter, in Chapter 5, and by reference to the work of Leggett (1996)). It is clear, however, that an "industry" in generating such "environmental foresight" – in the words of the Science Advisory Board of the US Environmental Protection Agency (Science Advisory Board, 1995) – might well be emerging, witness the work of Schneider et al. (1998), for instance. It could be said that we ourselves are likewise engaged in developing "imaginable surprises", as they define what we herein would call the target futures. Certainly, in our notion of "examining the record of the past in a manner guided expressly by prior contemplation of the reachability of certain feared patterns of future behaviour" (Chapter 1) there are echoes of what Schneider and colleagues refer to as "backcasting scenarios from posited future states and/or reconstructing past scenarios in alternative ways to identify events or processes that might happen" (Schneider et al., 1998).

REFERENCES

Beck, M.B., 1987. Water quality modeling: a review of the analysis of uncertainty. *Water Resour. Res.*, **23**(8), 1393–1442.

Beck, M.B. and Chen, J., 2000. Assuring the quality of models designed for predictive tasks. In: *Sensitivity Analysis* (A. Saltelli, K. Chan and E.M. Scott, eds.). Wiley, Chichester, pp. 402–420.

Beck, M.B., Ravetz, J.R., Mulkey, L.A. and Barnwell, T.O., 1997. On the problem of model validation for predictive exposure assessments. *Stochastic Hydrol. Hydraul.*, **11**, 229–254.

Beck, M.B., Parker, A.K., Rasmussen, T.C., Patten, B.C., Porter, K.G., Norton, B.G. and Steinemann, A., 2001. Community values and the long-term ecological integrity of rapidly urbanizing watersheds. Annual Report 2000, Grant # R825758, US Environmental Protection Agency, Washington DC.

Beven, K.J. and Binley, A.M., 1992. The future of distributed models: model calibration and predictive uncertainty. *Hydrol. Proc.*, **6**, 279–298.

Carpenter, S.R., Kitchell, J.F. and Hodgson, J.R., 1985. Cascading trophic interactions and lake productivity. *Bioscience*, **35**, 634–639.

Chen, J., 1993. Modelling and control of the activated sludge process: towards a systematic framework. PhD Thesis, Imperial College of Science, Technology, and Medicine, London.

Chen, J. and Beck, M.B., 1998. Quality assurance of multi-media model for predictive screening tasks. Report EPA/600/R-98/106, US Environmental Protection Agency, Washington, DC.

Darier, E., Gough, C., De Marchi, B., Funtowicz, S., Grove-White, R., Kitchener, D., Guimarães-Pereira, Â., Shackley, S. and Wynne, B., 1999. Between democracy and expertise? Citizens' participation and environmental integrated assessment in Venice (Italy) and St. Helens (UK). *J. Environ. Policy Planning*, **1**, 103–120.

Hornberger, G.M. and Spear, R.C., 1980. Eutrophication in Peel Inlet, I: Problem-defining behaviour and a mathematical model for the phosphorus scenario. *Water Res.*, **14**, 29–42.

Keesman, K.J. and van Straten, G., 1990. Set-membership approach to identification and prediction of lake eutrophication. *Water Resour. Res.*, **26**, 2643–2652.

Kendall, M.G. and Stuart, A., 1961. *The Advanced Theory of Statistics*. Griffin, London.

Klepper, O. and Hendrix, E.M.T., 1994. A method for robust calibration of ecological models under different types of uncertainty. *Ecol. Modelling*, **74**, 161–182.

Korfmacher, K.F., 2001. The politics of participation in watershed modeling. *Environ. Manage.*, **27**(2), 161–176.

Leggett, J.K., 1996. The threats of climate change: a dozen reasons for concern. In: *Climate Change and the Financial Sector: The Emerging Threat, the Solar Solution* (J.K. Leggett, ed.). Gerling Akademie Verlag, Munich, pp. 27–57.

Norton, B.G. and Steinemann, A., 2001. Environmental values and adaptive management. *Environmental Values*, in press.

Osidele, O.O. and Beck, M.B., 2001a. Identification of model structure for aquatic ecosystems using regionalized sensitivity analysis. *Water Sci. Technol.*, **43**(7), 271–278.

Osidele, O.O. and Beck, M.B., 2001b. Analysis of uncertainty in model predictions for Lake Lanier, Georgia. In: *Proceedings AWRA Annual Spring Specialty Conference* (J.J. Warwick, ed.). TPS-01-1, American Water Resources Association, Middleburg, Virginia, pp. 133–137.

Osidele, O.O., Beck, M.B. and Fath, B.D., 2000. A case study in integrating stakeholder concerns with the water sciences. In: *Proceedings 8th National Symposium*, British Hydrological Society, pp. 1.53–1.59.

Park, R.A., Scavia, D. and Clesceri, N.L., 1975. CLEANER: The Lake George Model. In: *Ecological Modeling in a Resource Management Framework* (C.S. Russell, ed.). Resources for the Future, Washington, DC, pp. 49–81.

Porter, K.G., 1996. Integrating the microbial loop and the classic food chain into a realistic planktonic food web. In: *Food Webs: Integration of Patterns and Dynamics* (G.A. Polis and K. Winemiller, eds.). Chapman and Hall, New York, pp. 51–59.

Porter, K.G., Saunders, P.A., Haberyan, K.A., Macubbin, A.E., Jacobsen, T.R. and Hodson, R.E., 1996. Annual cycle of autotrophic and heterotrophic production in a small, monomictic Piedmont lake (Lake Oglethorpe): Analog for the effects of climatic warming on dimictic lakes. *Limnol. Oceanogr.*, **41**(5), 1041–1051.

Raschke, R.L., 1993. *Guidelines for Assessing and Predicting Eutrophication Status of Small South-eastern Piedmont Impoundments.* US Environmental Protection Agency, Athens, Georgia.

Schneider, S.H., Turner, B.L. and Garriga, H.M., 1998. Imaginable surprise in global change science. *J. Risk Res.*, **1**(2), pp 165–185.

Science Advisory Board, 1995. *Beyond the Horizon: Using Foresight to Protect the Environmental Future.* Report EPA-SAB-EC-95-007, Science Advisory Board, US Environmental Protection Agency, Washington, DC.

Shrader-Frechette, K., 1995. Hard ecology, soft ecology, and ecosystem integrity. In: *Perspectives on Ecological Integrity* (L. Westra and J. Lemons, eds.). Kluwer, Dordrecht, pp. 125–145.

Spear, R.C. and Hornberger, G.M., 1980. Eutrophication in Peel Inlet, II: Identification of critical uncertainties via generalised sensitivity analysis. *Water Res.*, **14**, 43–49.

Spear, R.C., Grieb, T.M. and Shang, N., 1994. Parameter uncertainty and interaction in complex environmental models. *Water Resour. Res.*, **30**(11), 3159–3169.

Thompson, M., 1997. Cultural theory and integrated assessment. *Environ. Modeling Assess.*, **2**, 139–150.

Young, P.C., Hornberger, G.M. and Spear, R.C., 1978. Modelling badly defined systems – some further thoughts. In: *Proceedings SIMSIG Conference*, Australian National University, Canberra, pp. 24–32.

Environmental Foresight and Models: A Manifesto
M.B. Beck (editor)

CHAPTER 12

Uncertainty and the Detection of Structural Change in Models of Environmental Systems

K.J. Beven

12.1 UNCERTAINTY AND CHANGE IN MODELLING ENVIRONMENTAL SYSTEMS

This monograph is fundamentally concerned with the detection and forecasting of future change in environmental systems: systems that are highly complex, with poorly controlled boundary conditions and generally subject to considerable uncertainty in our knowledge of the component processes and representation of that knowledge in predictive models. The question then arises as to whether, given such uncertainties, predictions of the impact of change will have any value. As we have said, at the close of Chapter 3, how can the subtleties of structural change be detected in the face of gross uncertainty? In this chapter, these questions are addressed directly within the context of a Monte Carlo based methodology for the assessment of uncertainty in the predictions of environmental models. This methodology explicitly recognises that due to limitations of both knowledge and observational data, it may be impossible to construct a unique model of such systems and such that it must be accepted that there will be many competing models that give acceptable predictions of the limited (and uncertain) observational data available for model testing. However, it will be shown that using the proposed methodology, even given this non-uniqueness in the potential representations of a system, it is still possible to assess the impacts of change within a risk assessment framework.

12.2 AN EXAMPLE

Consider the following example. A soil–vegetation–atmosphere-transfer (SVAT) model, similar to those currently used in global climate modelling, is being calibrated to some data collected at a point in a relatively homogeneous landscape. Within the model structure there are a number of parameters to be calibrated. These have physically significant names (interception capacity, minimum surface resistance, porosity etc.) but may be very difficult to measure or estimate *a priori*. There are internal states in the model (average soil moisture profiles, effective surface temperature, etc.) that also cannot be measured easily, and the input data to the model (net radiation, vapour pressure deficit, etc.) are also uncertain, especially close to dawn and dusk. Thus the modeller tries to fit the estimated fluxes of latent and sensible heat (there are no measurements available for subsurface drainage fluxes), where those estimates are derived from other measurements of temperature, humidity and wind speed. The modeller uses an optimisation routine to get the best fit of his/her model to the data (for example, Figure 12.1) but having some free computer time he/she also decides to carry out a sensitivity analysis using a Monte Carlo method.

To do so he/she chooses reasonable ranges for all the parameters to be varied and then selects random values of the parameters across those ranges. Having little *a priori* information about the most likely distribution of parameters within the range or of possible joint covariance of the parameters, he/she sets up the sampling to select independent uniform samples across the parameter ranges. For each run of the model a goodness of fit index is calculated. In common with many other such studies a simple coefficient of determination (r^2) index is used. The results are shown in Figure 12.2, where each point on the figures represents one run of the model.

The results are somewhat surprising. For any particular parameter values it is possible to obtain both good fits and bad fits of the model virtually all the way across the parameter range. Good fits are contingent on the parameter *set*, not on single

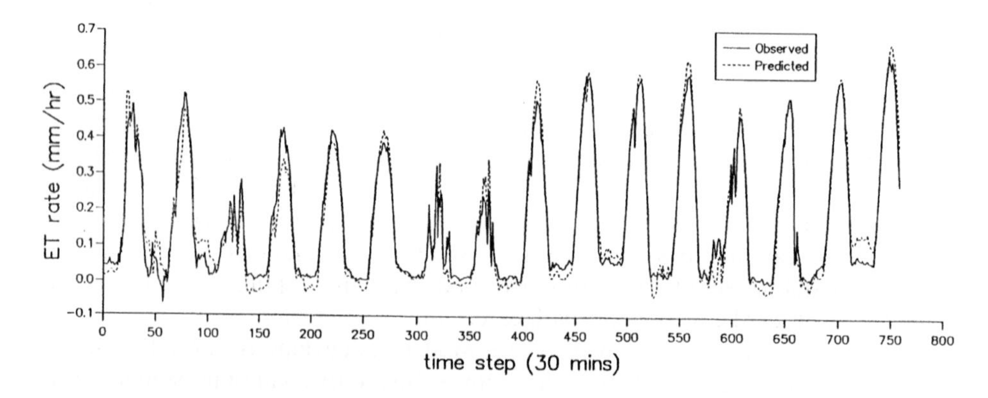

Fig. 12.1. Observed and predicted latent heat fluxes from the IFC3 (August 1987) period of the First ISCLCP Field Experiment (FIFE), Kansas, for the "best fit" model.

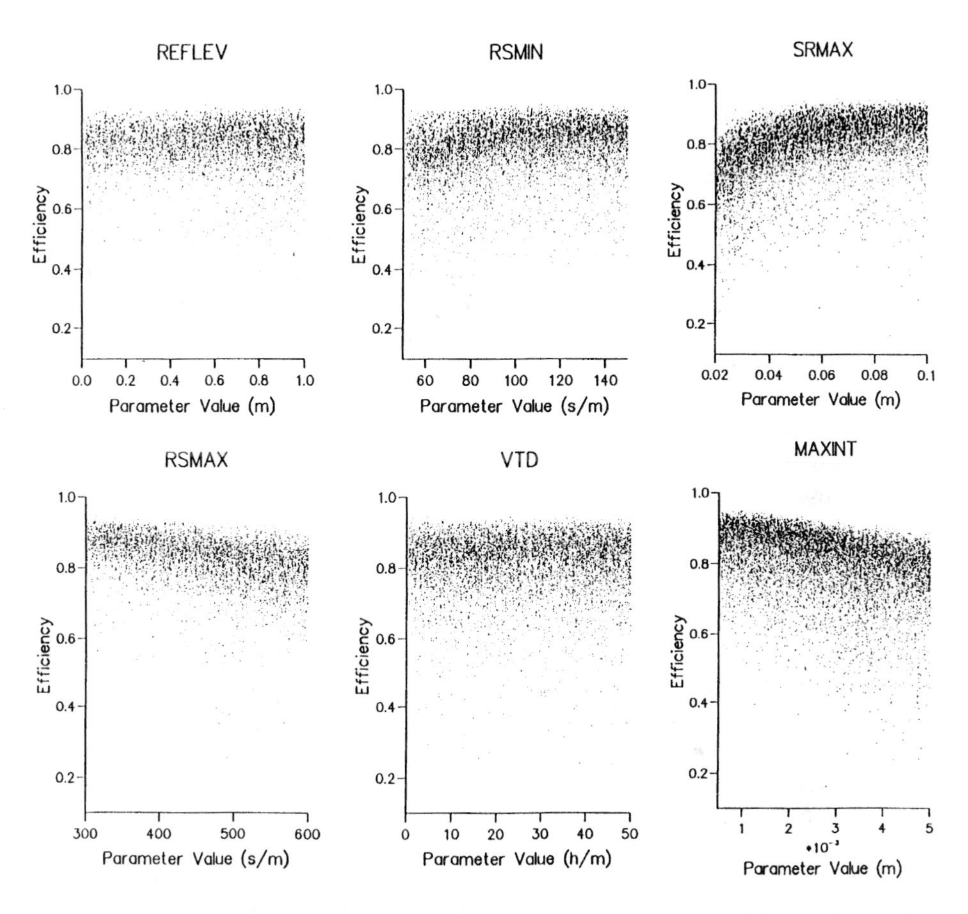

Fig. 12.2. Results of 10,000 Monte Carlo simulations for the TOPUP SVAT model applied to the IFC3 period of FIFE, summarised as model efficiency (r^2) values. Each point on the graph represents a simulation with a different set of randomly chosen parameter values. Independent uniform sampling across the ranges shown are used for all parameters.

parameter values. The interactions between the parameters are significant. One model run has the optimal parameter set, but there are many other parameter sets, that may be in very different parts of the parameter space, that give almost equally good fits. There is clearly potential for change in the optimal parameter values if a different calibration data set were used. It would appear that this is not a very well defined model when fitted in this way and that any "optimal" values of the parameters should be interpreted with care.

At first sight it might seem that this type of behaviour is particular to this particular system, with its not very well defined boundary conditions, system representation and "observed" variables. This is not the case. Such behaviour seems to be ubiquitous in models of environmental systems. Other examples are given for rainfall-runoff modelling by Duan et al. (1992), Beven (1993), Romanowicz et al. (1994), Freer et al. (1996), and Franks et al. (1998); in a SVAT model by Beven and Quinn (1994), Franks and Beven (1997), Lamb et al. (1998), Franks et al. (1998) and Cameron et al. (1999); in flood routing by Romanowicz et al. (1996, 1998) and Aronica et al. (1998); in geochemical modelling by Zak et al. (1997), and Zak and Ben (1999); in dispersion modelling by Hankin et al. (1998); and for a complex contaminant transport, fate and exposure model by Spear et al. (1994). Beven (1993) has used the term *equifinality* to describe the possibility of having multiple acceptable models of a system and in a number of papers has explored its implications for hydrological and geomorphological modelling (Beven, 1995, 1996). This term is meant to reflect a different concept to the commonly used *non-identifiability* (e.g. Beck, 1987) so that there is no implication that a unique model exists, if it could only be found. It is becoming increasingly clear that environmental modellers must develop techniques to deal with equifinality.

We should not really be surprised at this. Most environmental models are examples of what Morton (1993) calls *mediating models*. They mediate between an underlying theory, which may be partly qualitative or perceptual in nature, and quantitative prediction. As such, they have the general characteristics discussed in Morton's analysis: they have assumptions that are false and *that are known to be false* (for discussions in relation to the hydrological example of Darcy's law see Schrader-Frechette, 1989; Hofmann and Hofmann, 1992; Oreskes et al., 1994; and Beven, 1996); they tend to be purpose specific with parameters and auxiliary conditions that are different for different purposes; they reflect physical intuition but may contain some more or less arbitrary elements; they have real explanatory power but may never (nor should they be expected to) develop into full theoretical structures. They also have a history, in that successful model structures are copied and refined by later models. However, the predictions and parameters of such models should not be taken as valid outside of the context of the particular model structure being used.

12.3 EQUIFINALITY AND UNCERTAINTY IN ENVIRONMENTAL MODELS

It has long been recognised that there may be significant uncertainty associated with the predictions of environmental models (see, for example, Beck, 1987) but techniques for estimating that uncertainty have, for the most part, followed classical concepts of statistical theory, i.e., that there is some optimal set of parameters and that those parameters have some standard error or covariance matrix of estimation. The implied Gaussian uncertainty distribution on the parameters can, for linear models, be shown to lead to a Gaussian uncertainty distribution on the model predictions (see for example, Young, 1984). For nonlinear models, various approximate methods have been used to estimate the predictive uncertainty but within this same conceptual framework (see for hydrological examples Garen and Burges, 1981; Rogers et al., 1985; Kuczera, 1988; Beck and Halfon, 1991; Melching, 1992, 1995).

The equifinality amongst parameter sets described above requires a different set of concepts for the analysis of uncertainty. In fact the equifinality concept can be extended to allow for the fact that multiple model structures (and then parameter sets within those model structures) may give equally good or acceptable simulations of a particular system. Normally, of course, we do not evaluate multiple model structures in a simulation study but our reasons for choosing a particular model structure are sometimes rather ad hoc: "I used it with reasonable success in a previous study"; "it comes with a very impressive user interface and graphical outputs"; "it is readily (or cheaply) available on the local computer or over the Internet"; "it makes use of the GIS data I already have on the computer database"; "it was the model that I developed and so can be easily modified if necessary". Clearly all scientists will want to choose an *appropriate* model but it may be very difficult to decide *a priori* between different model structures and the last of these reasons normally takes precedence over other considerations.

So how do we learn to live with equifinality if it is, in fact, axiomatic of environmental models? It does introduce a number of problems, particularly when there is a requirement to give some physical significance to the values of parameters after they have been calibrated to some data set. In a multiple parameter model, the values derived in this way must be dependent on the values of the other parameters and should not therefore be taken as having physical significance outside of that context. They are values *conditional* on the model structure and calibration data set used.

This has some important implications. If the model structure is changed, the effective value of a parameter with the same name may change. Also, if values of parameters are taken from different sources in the literature, there is no guarantee that the resulting set of parameters will give good results. The published parameter values may have the same name but the different values will generally have been fitted within different contexts.

Another important conclusion is that equifinality implies uncertainty in prediction. Different models, all of which may be considered acceptable during calibration may produce very different results when used in prediction. This may not simply be statistical uncertainty around some modal behaviour in the classical sense but in models that, at least for some input sequences, are sensitive to their parameter combinations, may even result in different modes of behaviour in prediction. This is a well known feature of systems studied in nonlinear dynamics and, in general, environmental systems are complex nonlinear systems. Such changes in mode of behaviour are very difficult to predict beforehand. Retaining model/parameter set combinations as acceptable (or "behavioural" in the terminology of the Regionalised Sensitivity Analysis (RSA) of Hornberger and Spear (1981)) implies that *at that point in time* those models cannot be rejected as potential simulators of the system. As more data are collected it may well be possible to refine the model selection process and reduce the range of parameter sets that are retained. Until that time, however, it appears that it may be necessary to consider the predictions from many different, but behavioural, model/parameter set combinations.

This then raises the relationship between equifinality and computer power. The results of Figure 12.1 are based on 10000 runs of a relatively simple lumped storage model. This does not take too long to run on a modern workstation but can produce large files of results if long time series and multiple output variables are involved. More complex models would require much more computer time to run and many Gigabytes of storage for the results. It is not currently possible to explore equifinality using Global Circulation Models of the atmosphere or oceans, or the distributed "physically-based" models of hydrology, but equifinality in such models almost certainly exists. The true range of uncertainty in predictions of global warming is almost certainly far wider than the censored range of values reported by the Intergovernmental Panel on Climate Change (Houghton et al., 1995), as argued in Chapter 10, even if only variants on one of the models considered were used and fitted to current climate conditions – and as van der Sluijs et al. (1998) have noted, "solidarity" among the forecasts has fulfilled a sociological role.

Equifinality has only become apparent relatively recently because the computer power has become available to explore parameter response surfaces in some detail using either discrete or Monte Carlo methods. Discrete methods have some advantages in that uniform coverage of the specified parameter space is guaranteed. However even if only 10 increments are used on each parameter axis, varying two parameters would require 100 runs and varying five parameters would require 100,000 runs. Many models have many more parameters than five. Monte Carlo searches will at least sample the parameter space up to the maximum number of runs that can be supported by the available computing power, albeit that that sample may be sparse. Methods for increasing the efficiency of Monte Carlo sampling of behavioural simulations are also being explored (see, for example, Spear et al., 1994).

12.4 EQUIFINALITY AND CHANGE IN ENVIRONMENTAL SYSTEMS

If it is indeed necessary to reject the idea that a unique model of a system is attainable then there are also clear implications for the detection and forecasting of the impacts of change. If parameter values are only of significance within the context of a parameter set for a particular model structure it becomes much more difficult to know how to estimate changes in parameter values that might be necessary to try and simulate the likely impacts of changes to the system under study.

Further, if the nature of the changes being considered are gradual, then it will be necessary to introduce a model component to reflect the impacts of change on parameter values. Such changes might be natural, such as seasonal changes in vegetation growth and senescence, or they may be induced by man. Adding new model components, however, will inevitably add more parameters, and any such change in structure should also be expected to impact on any other parameter values. It will therefore be very difficult to estimate changes to parameter sets to reflect the impacts of known or expected changes to the system, especially if there is no strong or simple covariance structure in the parameters that can be used to guide such estimation (as often appears to be the case in behavioural parameter sets).

Detection of change will also require a different approach if there is no unique model of the system. One approach to the detection of change is time variable parameter estimation, such as that described in Chapter 15, which treats the parameters as stochastic variables, typically with multivariate Gaussian covariance structure. Where there are known points of change in the system, this approach can impose intervention points in the analysis (Young and Ng, 1989). Underlying this view is an assumption that there is some modal behaviour in the parameter space, around which the parameters may vary stochastically (an assumption redolent of the analysis later in Chapter 16). The implications of equifinality are that a wider view of the variation in behavioural parameter sets may be necessary (that may lead to wider uncertainties in prediction). It will be shown, however, that detection of change may also be possible while accepting the predictive uncertainty associated with multiple behavioural models.

12.5 THE GLUE METHODOLOGY: RATIONALE

One technique that allows for equifinality in model structures and parameter sets has been proposed by Beven and Binley (1992). The Generalised Likelihood Uncertainty Estimation (GLUE) methodology is a development of the RSA technique of Hornberger and Spear (1981) and has much in common with both other Monte Carlo analyses of the sensitivity of models to parameter variations (e.g., O'Neill et al., 1982; Rose et al., 1991; Dilks et al., 1992; Patwardhan and Small, 1992) and some set theoretic approaches to model uncertainty analysis (e.g., van Straten and Keesman, 1990). The GLUE methodology has been developed from an acceptance of multiple behavioural models (where "model" will from now on be used to indicate the combination of a particular set of parameter values within some particular model structure). If there is no unique optimal model then it follows that it will only be possible to rank the model structures/parameter sets considered as simulators of the system on some relative likelihood scale. One model of those considered will, of course, give the best fit to some period of calibration data, but it is also almost certain that if a second period of similar data is considered then the rankings of these possible models will change and that the best model found for the first period will not be the best for the second. If the system is experiencing change, due either to exogenous or internal effects, then the relative likelihoods of different models in reproducing observations of the system should also change.

Consequently, if at least a sample of the acceptable models are retained, a range of possible behaviours will be available in making predictions with these models. In the GLUE approach these are weighted according to their calculated likelihoods from the calibration period(s), and the weights used to formulate a cumulative distribution of predictions from which uncertainty quantiles can be calculated. It is worth noting that the term likelihood is being used here in a wider context than the likelihood functions of classical statistics which make specific assumptions about the nature of the errors associated with the model simulations. Romanowicz et al. (1994)

show that the classical likelihood functions can be used within the GLUE framework, although these functions, designed to find the optimum or maximum likelihood parameter set, give much greater weight to the better simulations and can result in response surfaces that are very peaked (but still not necessarily with a single "optimum").

It is important to stress that the parameters are never considered independently but only ever as sets of values. The likelihood measure $L(\boldsymbol{\theta}_i|\boldsymbol{Y})$ for the ith model is associated with a particular set of parameters $\boldsymbol{\theta}_i$ conditioned on the observed data variables $\boldsymbol{Y}$. It is possible to evaluate the sensitivity of individual parameters, either by looking at the distributions of "behavioural" and "non-behavioural" models defined by the associated likelihood weights, as in the RSA procedures of Hornberger and Spear (1981) (see also the hydrological example of Beven and Binley, 1992); or by evaluating the marginal distribution of likelihood for each parameter by integrating across the parameter space (see Romanowicz et al., 1994). However, it is worth stressing again that it is the contribution of each parameter value within an individual set of parameters that is important in the acceptability of a model.

The GLUE methodology also recognises that as more data or different types of data are made available it may be necessary to update the likelihood weights associated with different models. This is achieved quite easily using Bayes' equation which allows a prior distribution of model likelihoods to be modified by likelihood weights arising from the simulation of a new data set to produce an updated or posterior likelihood distribution (see below). Since each model is associated with a particular value of the chosen likelihood measure, the modification of the prior likelihood is achieved by applying Bayes' equation to each retained model independently. It has been found in applying this procedure that this tends to have the effect of reducing the number of retained models with significant likelihood values, i.e., as would be hoped, at least for a system that is not subject to significant change, more data will usually reduce the feasible region of the parameter space.

The recognition of model equifinality and the evaluation of models within the framework of predictive uncertainty allows for a different approach to the relationship between model conceptualisations, parameters and data. If, as in the GLUE procedure, the criterion of performance enters directly into the assessment of model uncertainty, there is the possibility of inverting the normal process of model calibration and validation and, rather than regarding model evaluation as a process of confirmation, pose the problem as one of falsification. Data are then used to reject "non-behavioural" models.

This type of methodology appears to sit somewhat uneasily between a number of philosophical frameworks for a scientific approach to modelling and theory generation. Most environmental modellers will agree that model or theory confirmation is a matter of degree of empirical adequacy (van Fraasen, 1980; Oreskes et al., 1994). It appears necessary to accept, however, that there are many different descriptions that are "adequate" in some sense, where adequacy itself may be limited or conditional, requiring further tuning or modification of ancillary conditions (such as parameter values) as more information becomes available. The idea of ranking the available models in terms of some likelihood measure would appear to result in a purely relativist attitude to the problem of modelling environmental systems, in keeping with the views of Feyerabend (1975) on the development of scientific thought (see Beven (1987) for a discussion in relation to hydrology) where there is no requirement for a necessary or strong correspondence between theory or reality. This is unacceptable, or at least unpalatable, to many scientists.

Alternatively, the rejection of "non-behavioural" models is also in the tradition of both logical empiricist and critical rationalist philosophical stances. Neither have traditionally, however, allowed for the possibility of multiple descriptions that remain "acceptable" even after a process of rejection based on goodness-of-fit, hypothesis testing or the rational appraisal of the underlying theory. The difficulties associated with rejection or falsification have been well aired in the philosophical literature (see, for example, Bhaskar, 1989) but it would appear to provide a useful way to proceed faced with the inherent difficulties of environmental modelling. The difficulties, as outlined above, suggest that a purely critical rationalist approach to environmental modelling is untenable, since it would lead to the rejection of all current models. Even a logical empiricist approach will require some fairly relaxed criteria of acceptability, particularly in respect of predictions of internal states of the system, to allow that some models can be retained as "acceptable".

If, however, acceptability is to be assessed in terms of some quantitative measure of performance or "likelihood" as in the Bayesian GLUE procedure described briefly above, it is clear that acceptability is necessarily a relative measure, with models having more or less support from the evidence (for a full exposition of the use of Bayesian methods in scientific explanation see Howson and Urbach, 1989). It is common to find that, in plots such as Figure 12.1, there is no clear demarcation between good simulations and poor simulations. The choice of a point of acceptability is therefore a subjective one and acceptability, however assessed, will be a relative measure.

There is an argument that if it is not possible to distinguish between models or theories on the basis of their predictions then *they are simply poor models.* It should then be the aim of the environmental scientist to improve the available techniques of observation and hypothesis testing that would allow competing models or hypotheses or parameter sets to be accepted or rejected (see, for example, Haines-Young and Petch, 1983). Such a view is not incompatible with the GLUE approach in which rejection is a fundamental process in constraining predictive uncertainty and improving model structure. The set of acceptable models could, in practice, be examined to determine if there are testable critical hypotheses that, with the collection of some additional data, would allow further models to be rejected (Beven, 2001).

12.5.1 Requirements of GLUE

The GLUE procedure requires first that the sampling ranges be specified for each parameter to be considered. This is not normally too great a problem since initially the ranges can be set to be as wide as considered feasible by physical argument or experience. It would normally be hoped that models at the edges of the parameter ranges will be rejected as non-behavioural after the first period of calibration (although our experience suggests that this is not always found to be the case for hydrological models, as shown in Figure 12.2 above).

Secondly, a methodology for sampling the parameter space is required. In most of the applications of GLUE carried out to date this has been done by Monte Carlo simulation, using uniform random sampling across the specified parameter range. The use of uniform sampling makes the procedure simple to implement and can be retained throughout the updating of likelihoods since the density distribution of the likelihoods is defined by the likelihood weight associated with each model. Effectively each set of chosen parameter values is always evaluated separately *as a set*, and its performance implicitly reflects any interactions and insensitivities of the parameters. This thereby avoids the necessity to assume a parameter covariance structure for the parameter values (although where such a structure can be justified, it can be used to make the Monte Carlo sampling more efficient).

Thirdly, the procedure requires a formal definition of the likelihood measure to be used and the criteria for acceptance or rejection of the models. This is a subjective choice, as with any choice of objective function. There may also be more than one objective function calculated from different types of data and it will then be necessary to specify how these will be combined. Bayes' equation provides a consistent framework for using these likelihood measures in both combining likelihoods calculated from different types of observed data and in updating the likelihoods associated with each model as more data become available.

12.5.2 Results of Using Different Likelihood Measures

The results of the GLUE methodology will depend on the choice of likelihood measures used. Examples of possible likelihood measures are discussed in Beven and Binley (1992), Romanowicz et al. (1994) and Freer et al. (1996) and Beven (2000). There are two related aspects to this choice: the choice of a particular measure of behaviour and also the choice of criteria for rejecting simulations as non-behavioural on the basis of that measure. The definition of the ranges of parameter values to be considered already involves an explicit *a priori* likelihood evaluation of zero for parameter values outside those ranges (while the choice of a particular model involves an *a priori* likelihood evaluation of zero for other possible models).

Freer et al. (1996) compare the results of using different likelihood measures in the application of a rainfall-runoff model to predict the discharges of the small Ringelbach catchment. These included the sum of squared errors, in the form:

$$L(\boldsymbol{\theta}_i|Y) = (1 - \sigma_i^2/\sigma_{\text{obs}}^2)^N; \text{ for } \sigma_i^2 < \sigma_{\text{obs}}^2 \qquad (12.1)$$

where $L(\boldsymbol{\theta}_i|Y)$ is the likelihood measure for the ith model conditioned on the observations, σ_i^2 is the associated error variance for the ith model, σ_{obs}^2 is the observed variance for the period under consideration and N is a shaping parameter.

For $N = 1$, equation 12.1 is equivalent to a coefficient of determination or the Nash and Sutcliffe (1970) efficiency criterion. Higher N values have the effect of accentuating the weight given to the better simulations. An extreme case of this is the likelihood function used by Romanowicz et al. (1994), based on a first-order auto-correlated Gaussian model for the prediction errors, which involves raising the error variance to the power of one half of the number of simulated time steps. Freer et al. (1996) also used the function:

$$L(\boldsymbol{\theta}_i | \boldsymbol{Y}) = \exp\{-N\sigma_i^2 / \sigma_{obs}^2\}; \text{ for } \sigma_i^2 < \sigma_{obs}^2 \quad (12.2)$$

which has the feature that in applying Bayes' equation for updating of the likelihood weights, the error variance for each period of data is given equal weight.

In applications of the methodology to date there has been no explicit account taken of measurement errors in either the input data or the observations against which the model is being calibrated. To do so would require some model of the measurement errors, however simple. It may be difficult to construct an appropriate model. In hydrology, for example, a basic input to any model is rainfall. Rain-gauges give a very poor sample of the rainfall falling on a catchment area. Radar measurements give a better spatial coverage but suffer from variable calibration characteristics. Thus estimates of catchment inputs provided to hydrological models will be in error to some unknown degree. Some events will be well estimated, some events may be grossly in error, but it may be very difficult to know which events are which except in extreme cases (see for example the "validation" event used by Hornberger et al., 1985). Even on small catchments, spatial variation in rainfalls can lead to simulation difficulties (see Faurès et al., 1995), although it may be more important to get the catchment average input volume correct than to know the spatial distribution (Obled et al., 1994). In what follows here, any effects of such errors will be reflected *implicitly* in the calculated likelihood values.

12.5.3 Calculation of Likelihood Distributions and Uncertainty Bounds

The next stage involves a decision about the criterion for model rejection. In most of the studies to date, a simple threshold on the likelihood measure has been used, though with very high values of the shaping parameter N in equation 12.1 or 12.2 this may not be necessary since poor simulations will, in any case, end up with very low weights. The uncertainty bounds associated with the retained simulations will depend on the choice of the likelihood measure (including the value of N) and rejection criterion.

Following rejection, the likelihood weights associated with the retained models can be rescaled to give a cumulative sum of 1.0. The rescaled weights can then be applied to their respective model discharges at each time step to give a cumulative distribution of discharges at that time step, from which the chosen discharge quantiles can be calculated to represent the model uncertainty. It is worth noting that if the rejection criterion is too strict, the resulting uncertainty bounds may not enclose a significant part of the observations.

A somewhat surprising result from the Freer et al. (1996) study was that the 90% confidence limits determined by the GLUE methodology for different likelihood measures were surprisingly similar. Increasing the value of N does tend to narrow the confidence limits, but this leads to a greater possibility of the observations lying outside of the confidence limits. There were also periods when the observations lay consistently outside of the estimated confidence limits for all the likelihood measures tried as a result of limitations in both model structure and input data (this was particularly evident following timing errors in the prediction of snowmelt in some years). It is not particularly helpful to suggest that all the models should be rejected due to these failures, but may be helpful in suggesting which model components or input series should be critically examined. Such periods do indicate that any assumptions about the nature of the modelling errors in this type of model must allow for nonstationarity in the bias, variance and other features of the error structure.

12.5.4 Updating of Uncertainty Bounds

Updating of the likelihood distribution as more data become available may be achieved by the application of Bayes' equation in the form

$$L_p(Y|\boldsymbol{\theta}_i) = L(\boldsymbol{\theta}_i|Y)L_0(\boldsymbol{\theta}_i)/C \tag{12.3}$$

where $L_0(\boldsymbol{\theta}_i)$ is a prior likelihood for the parameter set $\boldsymbol{\theta}$; $L(\boldsymbol{\theta}_i|Y)$ is the likelihood measure calculated with the set of observed variables Y; $L_p(Y|\boldsymbol{\theta}_i)$ is the posterior likelihood for the simulation of Y given $\boldsymbol{\theta}$ and C is a scaling constant calculated such that the cumulative of $L_p(Y|\boldsymbol{\theta}_i)$ equals unity. The procedure outlined in the previous section can be considered as an application of equation 12.3 with a prior distribution of equal likelihoods for every model and a zero likelihood measure for those models rejected as non-behavioural. Clearly it is not necessary that a uniform initial prior distribution be used (this is not necessarily even the best non-informative prior for certain functions; see, for example, Lee (1992)); prior beliefs about appropriate parameter values could also be incorporated.

However, given the results from the first period of simulation, the resulting posterior distribution can be used as the prior distribution for the next period. If, in fact, further observations from the next period are available, the resulting likelihood values can be used with equation 12.3 to update the posterior likelihood distribution for any further predictions. An important difference between the likelihood measures defined by equations 12.1 and 12.2 becomes apparent here. Because of the multiplicative nature of equation 12.3, the use of equation 12.1 as a likelihood measure means that after a number of updatings, the error variance associated with earlier periods carries less and less weight in the calculation of the posterior likelihoods. The use of equation 12.2, however, allows the error variances from additional periods to contribute linearly within the exponential. Thus only with equation 12.2 will the same likelihood distributions be obtained from using a period of data as a single entity compared with splitting it down into smaller periods and applying Bayes' equation after evaluating the likelihoods after each period. It is worth adding, however, that this feature of equation 12.1 might be desirable when a catchment (and by inference the model parameters) is undergoing gradual change. We will return to this in discussing prediction of the effects of change in Section 12.6 below.

12.5.5 On the Sensitivity to Individual Parameters

In the application of the GLUE methodology we have stressed that the behaviour of the model is assessed in terms of the performance of the *set* of parameters. It is the *combination* of parameter values that produces behavioural or non-behavioural simulations within the chosen model structure. The interaction between parameter values results in the broad regions of acceptable simulations when individual parameters are considered, as in Figures 12.1 and 12.2 (see also Duan et al., 1992; Beven, 1993). Such plots reveal little, however, about the sensitivity of the model predictions to the individual parameters, except where some strong change in the likelihood measure is observed in some range of a particular parameter.

More is revealed about sensitivity by an extension of the RSA of Hornberger and Spear (1981). They constructed distributions for each parameter conditioned on a classification of the Monte Carlo simulations into two classes, behavioural and non-behavioural. The criterion for differentiating between the two classes was a subjectively chosen value of a goodness of fit measure, as we have used here. Similarity between the cumulative distributions for the two classes suggests insensitivity to that parameter; strong differences between the distributions reveals a sensitive parameter.

A variant on this form of analysis is shown in Figure 12.3 which shows the model runs of Figure 12.2 presented in a different way, as cumulative distributions of each parameter for a number of likelihood bands created by taking the 1000 simulations with the highest likelihoods (set 10), then the 1000 with the next highest (set 9) and so on down. Strong differences between the distributions suggest sensitivity for that parameter (e.g. SRMAX, RSMIN); uniform distributions across the range of a parameter (a straight line on the plot) suggest insensitivity (e.g. REFLEV, VTD).

12.6 UNCERTAINTY AND PREDICTING THE EFFECTS OF CHANGE

Given predictive uncertainty of the type evaluated using such Monte Carlo experiments, is it possible to say anything about the nature or impact of environmental change? Clearly yes for the case of prediction into some future scenario of a modelled stationary system for which the parameter values have been conditioned on some historical calibration data. The results will simply be a distribution of likelihood weighted predictions of the variables of interest. Multiple scenarios could also be

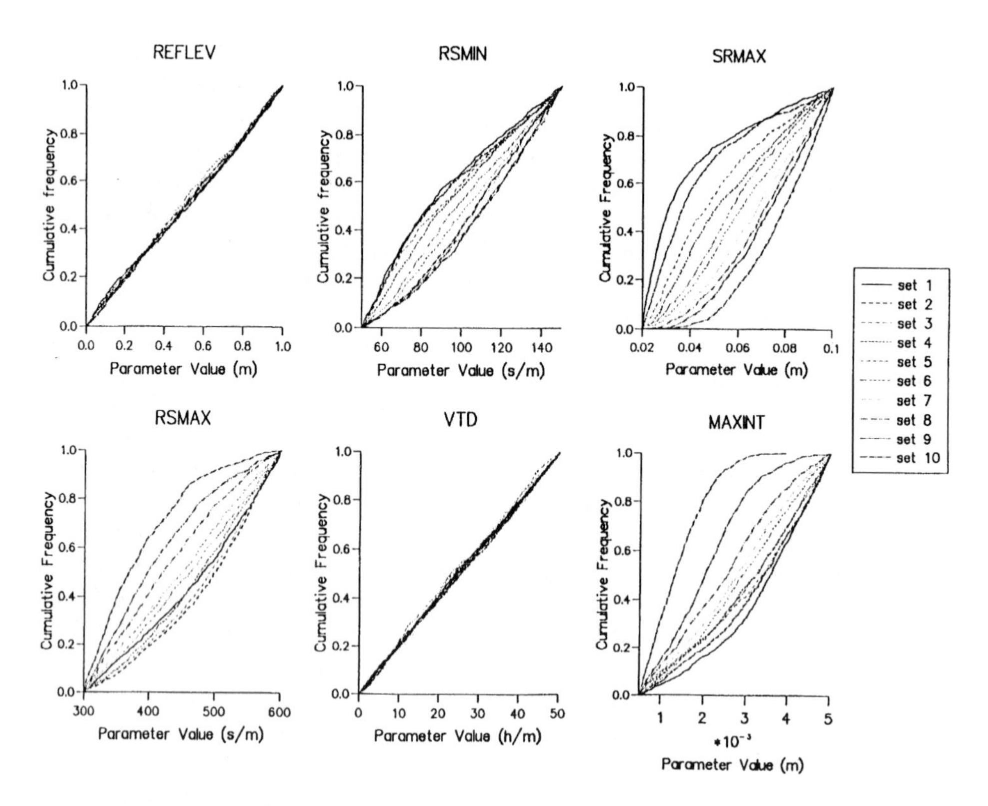

Fig. 12.3. Cumulative distributions of parameter values for 10 goodness-of-fit sets for the comparison of observed and predicted latent heat fluxes for the IFC3 period of FIFE; set 10 represents top 10% simulations, set 1 worst 10% simulations.

considered, perhaps associated themselves with some measure of likelihood determined either qualitatively or from some other model.

A more difficult prediction problem arises when some change in the system is expected that might lead to change in the appropriate parameter values or change in the appropriate model structure. This is the forecasting problem that is at the heart of the issues addressed in this monograph. Both parameter and structural changes may *a priori* be difficult to predict (note in particular that it is acceptable parameter *sets* that are required for the changed conditions, not necessarily just changes in single parameters). Indeed, it is probable that only a range of possibilities could be specified for such changes.

Consider first the case of parameter changes within a single model structure, where one or more parameters are expected to change as a result of either a natural event (such as fire or drought) or a change in management strategy. The model has been conditioned on historical data and the resulting "behavioural" sample parameter distributions have been determined. Incorporation of new parameter ranges is only simple if it is assumed that the parameters are essentially independent of each other or the covariance structure of the parameters is known and can be modelled. In the study of Binley et al. (1991), for example, in examining the prediction of the hydrological effects of a change in land use from grass to forest, independence of the parameters was assumed. In the general case it is not yet clear whether the parameter interactions that yield behavioural simulations (as demonstrated, for example, in Figure 12.2) can themselves be modelled (certainly a simple Gaussian covariance structure does not appear to be adequate, see, for example, Beven and Binley, 1992; Spear et al., 1994), nor whether that model could be assumed to be consistent for the changed parameter distributions.

One way of investigating these effects is through the use of examples where change is known to have occurred. In a modelling study of the hydrological effects of fire on the 1.46 km^2 Rimbaud subcatchment in the Maures massif in southern France, the hydrological model TOPMODEL (Beven et al., 1995; Beven, 1997) has been calibrated for a number of periods of record spanning the fire of August 1990 which burned some 85% of the catchment area (see Fisher and Beven, 1994). For some periods of record, particularly in predicting the response of storms occurring during the long summer drought and rewetting periods, the assumptions of TOPMODEL are not valid and generally non-behavioural simulations are obtained. For the wetter winter period it produces acceptable simulations, at least with some combinations of parameter values (Figure 12.4).

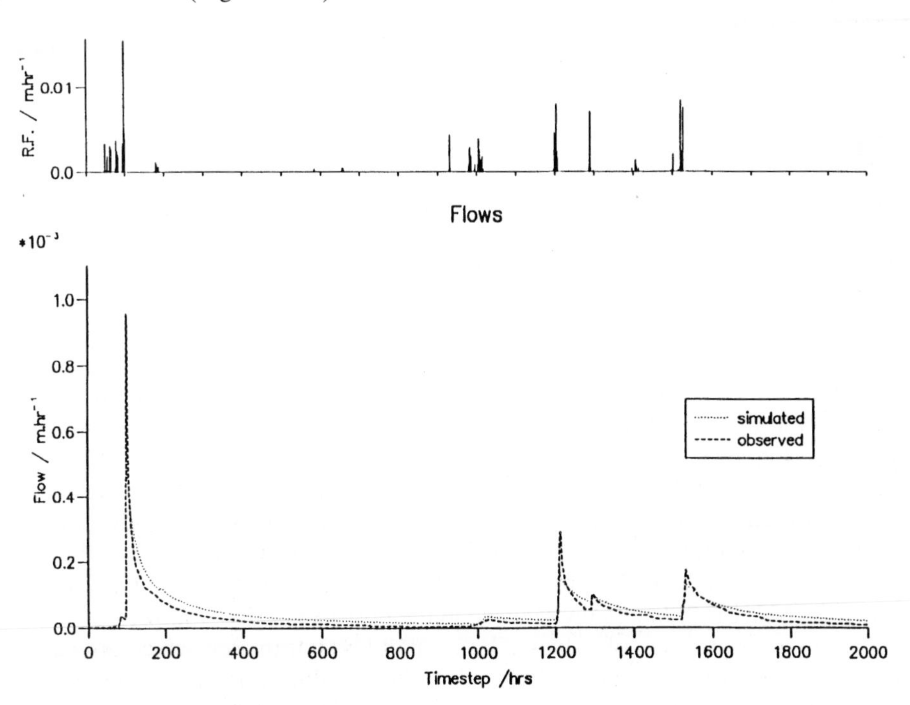

Fig. 12.4. Application of TOPMODEL to the Rimbaud catchment (1.46 km^2) in the Maures Massif, S. France, showing observed and "best fit" model predictions for the period 20.2.89 to 15.5.89.

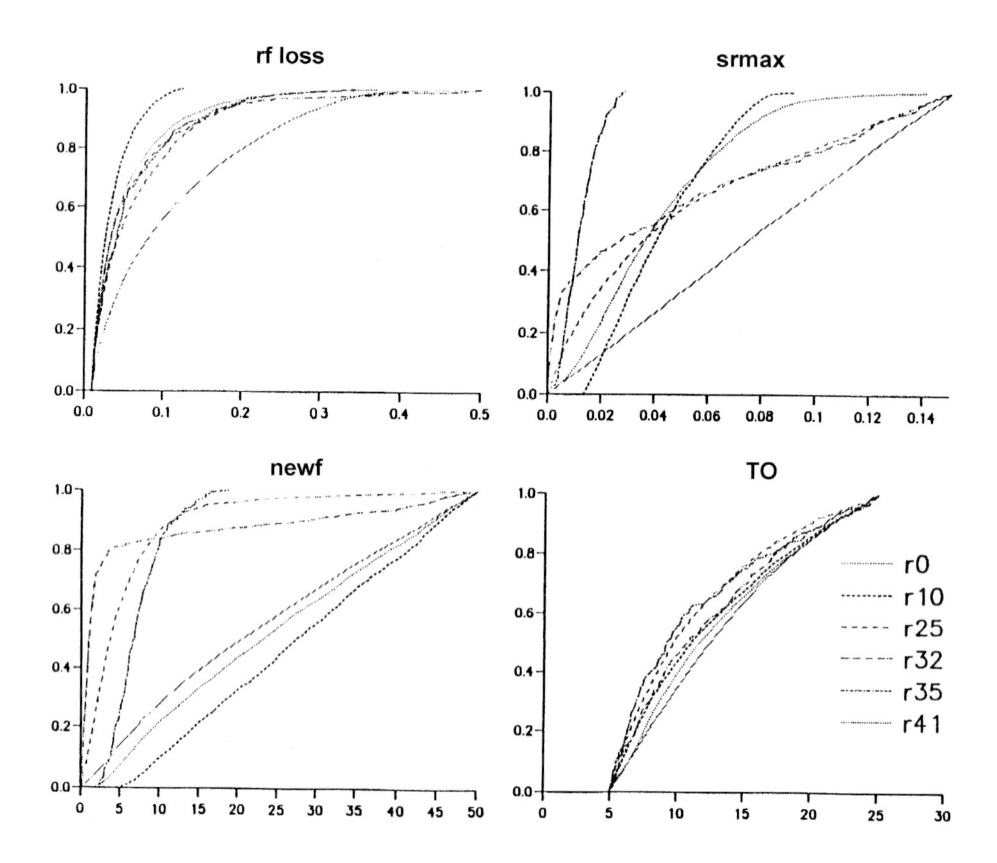

Fig. 12.5. Cumulative likelihood distributions for TOPMODEL parameters conditioned on each of six periods before and after the fire of August 1990. The periods are as follows: before the fire: *r*0 (1.1.88–15.6.88); *r*10 (20.2.89–15.5.89); after the fire: *r*25 (19.11.90–21.5.91): *r*32 (22.9.91–28.12.91); *r*35 (14.1.92–23.7.92); *r*41 (28.9.92–5.11.92).

Figure 12.5 shows the cumulative likelihood distributions for a number of the parameters in the model. Each line represents the distribution for the behavioural simulations for each period conditioned only on the data for that period. One of the parameters, T_O, the effective transmissivity of the soil, shows very little sensitivity to change between the periods. The NEWF parameter which controls the dynamics of the runoff contributing area, shows significantly smaller values post-fire and then a gradual recovery towards the pre-fire distribution. Similarly, the root zone storage capacity SRMAX is dramatically reduced in the post-fire period, followed by a later recovery to a wider range of values. The rainfall loss parameter, which affects the interception losses in the model, also tends to lower values post-fire but in the final period shows a distribution covering a wide range of values mostly higher than in the pre-fire period. This may be due to the greater interception loss efficiency of the re-growth cover but we have no other evidence at this stage to support this suggestion. These distributions can be tested for significance of the changes between periods using, for example, the nonparametric Kolmogorov–Smirnov *d* statistic.

It is, however, again necessary to take care in interpreting these changes for individual parameters, since it is the set of parameters with all the interactions implicit in the model dynamics, and in conjunction with any errors in the input data series, that determines whether a model is considered behavioural or not. It is hoped, of course, that if the model structure is a reasonable representation of the real processes, parameter values associated with different component processes should reflect change in those processes. However, Figure 12.2 suggests that these are not simple interactions and that it may be very difficult to describe the nature of change in parameter sets in any parsimonious way. But without such a description, it will be very difficult to use the knowledge gained from models of systems subject to historical change in attempts to predict the effects of future changes for similar sites.

A second example is given by the application of the SVAT model used in the prediction of evapotranspiration rates at the FIFE site in Kansas shown in Figures 12.1 and 12.2 to another period for which estimates of actual evapotranspiration rates were available. Figures 12.1 and 12.2 are both for the third Intensive Field Campaign (IFC3) in August 1987. Figure 12.6 compares parameter distributions for conditioning on IFC3 and IFC4 in October 1987 at the end of a long dry period. At this time the grass cover was showing signs of senescence and it is apparent from the parameter distributions that there has been a change in the behaviour of the vegetation cover. In particular, the results indicate that the best fit models are biased towards the lower end of the range of the maximum surface resistance RSMAX for the IFC3 period and towards the upper end of the range for IFC4. A similar disagreement was observed for the SRMAX available water capacity parameter. There is thus a suggestion here that a single set of parameter values cannot be used within this model structure to model the seasonal change in the effect of vegetation on evapotranspiration (but this has been widely done in the past in SVAT models used in GCMs). These results are reported in more detail in Franks et al. (1997).

Thus there are good indications in this case that the model structure should be changed to include some functional representation of that seasonal change. This will require one or more additional parameters (and consequent additional interactions with the existing parameters) to be added to an already parameter-rich model. This will therefore require that additional information be included in the calibration process, either measurements of the fluxes as used in the conditioning shown in Figures 12.2 and 12.6, or some index of the changing behaviour that can be used as an input to the model, such as the (approximate) leaf area index developed from remote sensing (Myneni and Williams, 1994; Kustas and Humes, 1996).

The lessons to be learned from these examples is that it may well be possible to detect or infer parameter and model structural change even within a framework that explicitly recognises model uncertainty *when data are available for conditioning*. The additional problem of predicting the effects of changes in parameters or structures *a priori* is not so clear cut, because of the very complex interactions that lead to the

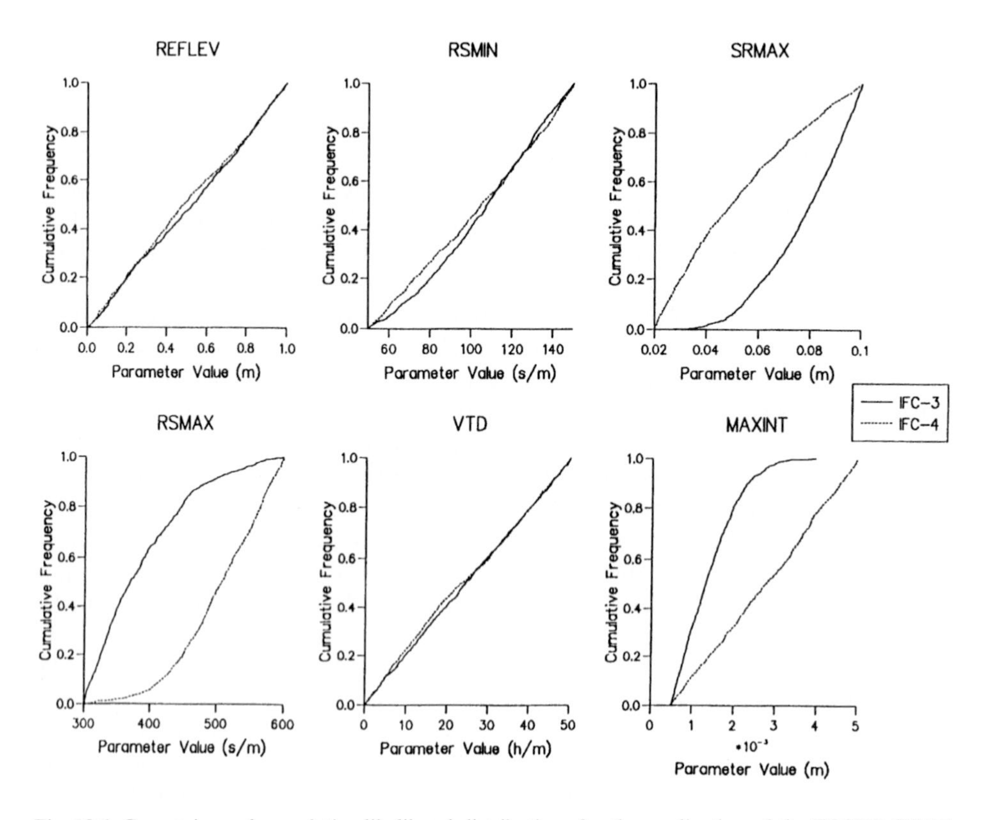

Fig. 12.6. Comparison of cumulative likelihood distributions for the application of the TOPUP SVAT model to the IFC3 (August 1987) and IFC4 (October 1987) periods of FIFE.

behavioural simulations. Clearly it can be done (in a related manner in Chapter 16). In principle here, however, by specifying new parameter distributions as "best estimate" prior distributions, even without taking any account of covariation between the parameters in conditioning on historical data, a range of likelihood weighted simulations can be made available for consideration in terms of desired/feared reachable futures.

However, one implication of the difficulty of knowing which sets of parameters will lead to behavioural simulations in the work described above is that prior estimation of parameter values will not necessarily be very reliable, even where parameter values might be available from other studies. Thus, if significant change is expected and improved predictions of the impacts of change are required in the context of differentiating between desired and feared futures, as in the previous chapter, then it is suggested that at least a minimal monitoring program should be put in place for critical variables so that some parameter sets can be rejected and the model predictions revised as new data are made available.

12.7 CONCLUSIONS

Given the seeming contradiction at the close of Chapter 3 – in acknowledging the gross uncertainty surrounding our analyses while yet seeking to detect the subtleties of structural change – our first conclusion must be that indeed such structural change *is* detectable. More broadly, however, this chapter has examined the question of uncertainty in environmental modelling in relation to the apparent equifinality of model structures and parameter sets in reproducing the observed behaviour of environmental systems. It would appear that equifinality should be considered as axiomatic of environmental models. Acceptance of this proposition requires a different approach to model calibration that more explicitly takes account of the conditional nature of values of individual parameters and uncertainty in the resulting predictions. The Generalised Likelihood Uncertainty Estimation (GLUE) methodology has been described in some detail as one approach towards achieving this end. Alternative approaches are being explored elsewhere (see for example, van Straten and Keesman, 1991; Klepper et al., 1991; Rose et al., 1991; Spear et al., 1994).

One implication of this work is that model rejection may be a more constructive strategy than model calibration. If many different models (and parameter sets within a particular model structure) can provide acceptable fits to the data, then it should be an aim to define critical observations or experiments that would allow some of those models to be rejected or to define model structures with tests for rejection in mind within the context of the required aims of the predictions for a particular application. This does not imply that there is some clear demarcation criterion for model rejection. Rather it recognises the opposite: that at one level *all* environmental models should be rejected as acceptable representations of reality, so that to allow any degree of predictability, model retention has to depend on some crude, subjective and relative measures of acceptability, such as the likelihood measures outlined above. New observations, particularly critical tests for a particular model, should lead to refinement of those criteria. And if such data lead to the rejection of all the possibilities being considered then the science should gain: either through the development of improved model structures that will meet the new test or by a greater understanding of the definition of acceptability and rejection relative to predictability for a particular purpose.

The applications reviewed above have shown that it is possible to consider the potential for change within an uncertainty framework. One important feature of the approach is that it is possible to study whether parametric change in the system is detectable within the constraints of model uncertainty. Significant parametric change suggests the possibility of structural change in the model, since it implies inadequacies of the model that can only be accommodated through assigning significantly different values to its parameters over different periods of time – when the model is reconciled with field observations. Thus, with a view towards exploring the propagation of such change into the future, one might seek to parameterise the form of the parametric change, in some way, within the existing model structure, which has been suggested in Chapter 5 (and will be further pursued in Chapter 14).

Predicting change in environmental systems before the event is a different type of problem. In that case it has to be assumed that the model structure does not change and will properly reflect the different modes of response of the system, even if speculative changes in the parameter sets and input data are tried, possibly outside the range of conditions to which the current behavioural models have been conditioned. An instructive example of such a study is that by Wolock and Hornberger (1991), who investigated the possible effects of future trends in climatic variables on runoff production and stream discharges. They carried out Monte Carlo experiments with stochastic input sequences but only a single ("best fit") set of parameter values in the hydrological model. Clearly the use of multiple parameter sets, as suggested here, would tend to increase the range of predictions produced under the future climatic scenarios leading to greater uncertainty.

One important implication of exploring the possibility of equifinality in environmental modelling is the focus on the importance of data for both conditioning, rejecting and critically testing models. It has already been noted that any predictions of future behaviour should be associated with a monitoring program so that those predictions can be refined (or rejected) as time progresses. The use of data in model rejection might also suggest ways of improving data collection procedures because it will certainly be the case that some data will be of far greater value in model rejection than others. Data that allow the internal dynamics of the model to be tested at an appropriate scale might be of particular value but very often the measurement techniques do not exist to do so because of a mismatch in scale between measurements and model predictions (see Beven, 1995, for a discussion in respect of hydrology).

There is still much to be learned about the best procedures for environmental modelling. The development of more and more complex models that incorporate more and more detail about processes, but which introduce more and more parameters that must be calibrated, does not appear to be the future (though the counter-argument will be promoted in Chapter 17). Unless values of those parameters can truly be determined independently of the model, the type of equifinality demonstrated here will be intrinsic to such models. Equifinality in modelled responses complicates matters, not least because it must throw doubt on parameter values for a wide variety of models that have been determined by calibration and reported in the literature as characteristic of a given vegetation type or soil or catchment. However, recognition of equifinality can also be seen as a very positive step in the attention on data that results. The future of environmental modelling will have to place more emphasis on the value of data, carefully collected for specific purposes, and on parametrically simple robust models, carefully designed for specific purposes.

ACKNOWLEDGEMENTS

The data presented in the figures comes from the work of Stewart Franks and James Fisher supported by the NERC TIGER programme and the EU DM2E Project respectively. Input data for the models were provided by Jay Famiglietti, Merilyn Gentry, Jacques Lavabre and Flavie Cernesson.

REFERENCES

Aronica, G., Hankin, B.G. and Beven, K.J., 1998. Uncertainty and equifinality in calibrating distributed roughness coefficients in a flood propagation model with limited data. *Adv. Water Resour.*, **22**(4), 349–365.

Beck, M.B., 1987. Water quality modelling: a review of the analysis of uncertainty. *Water Resour. Res.*, **23**, 1393–1442.

Beck, M.B. and Halfon, E., 1991. Uncertainty, identifiability and the propagation of prediction errors: a case study of lake Ontario. *J. Forecast.*, **10**, 135–161.

Beven, K.J., 1987. Towards a new paradigm in hydrology. In: *Water for the Future: Hydrology in Perspective*. IASH Publn. No. 164, 393–403.

Beven, K.J., 1989a. Interflow. In: *Unsaturated Flow in Hydrologic Modelling* (H.J. Morel-Seytoux, ed.). Proc. NATO ARW, Arles, Reidel, Dordrecht, pp. 191–219.

Beven, K.J., 1989b. Changing ideas in hydrology: the case of physically-based models. *J. Hydrol.*, **105**, 157–172.

Beven, K.J., 1993. Prophecy, reality and uncertainty in distributed hydrological modelling. *Adv. Water Resour.,* **16**, 41–51.

Beven, K.J., 1995. Linking parameters across scales: subgrid parameterisations and scale dependent hydrological models. *Hydrol. Process.*, **9**, 507–525.

Beven, K.J., 1996. Equifinality and uncertainty in geomorphological modelling. In: *The Scientific Nature of Geomorphology* (B.L. Rhoades and C.E. Thorn, eds.). Wiley, Chichester, pp. 289–313.

Beven, K.J., 1996. A discussion of distributed modelling, Chapter 13A. In: *Distributed Hydrological Modelling* (J.-C. Refsgaard and M.B. Abbott, eds.). Kluwer, Dordrecht, pp. 255–278.

Beven, K.J. (ed.), 1997. *Distributed Modelling in Hydrology: Applications of TOPMODEL*. Wiley, Chichester.

Beven, K.J., 2000. *Rainfall-runoff Modelling: the Primer*, Wiley, Chichester.

Beven, K.J., 2001. How far can we go in distributed hydrological modelling? *Hydrol. Earth System Sci.*, **5**, 1–12.

Beven, K.J. and Kirkby, M.J., 1979. A physically based variable contributing area model of basin hydrology. *Hydrol. Sci. Bull.*, **24**(1), 43–69.

Beven, K.J. and Binley, A.M., 1992. The future of distributed models: model calibration and uncertainty prediction. *Hydrol. Process.*, **6**, 279–298.

Beven, K.J. and Quinn, P.F., 1994. Similarity and scale effects in the water balance of heterogeneous areas. *Proc. AGMET Conference on The Balance of Water - Present and Future*, Dublin.

Beven, K.J., Lamb, R., Quinn, P.F., Romanowicz, R. and Freer, J., 1995. TOPMODEL, In: *Computer Models of Watershed Hydrology* (V.P. Singh, ed.). Water Resource Publications, Highlands Ranch, CO, pp. 627–668.

Beven, K.J., Freer, J., Hankin, B. and Schulz, K., 2001. The use of generalized likelihood measures for uncertainty estimation in high order models of environmental systems. In: *Nonlinear and Nonstationary Signal Processing* (W.J. Fitzgerald, R.L. Smith, A.T. Walden and P.C. Young, eds.). Cambridge University Press, Cambridge.

Binley, A.M., Beven, K.J., Calver, A. and Watts, L., 1991. Changing responses in hydrology: assessing the uncertainty in physically-based predictions. *Wat. Resour. Res.*, **27**, 1253–1262.

Binley, A.M. and Beven, K.J., 1991. Physically-based modelling of catchment hydrology: a likelihood approach to reducing predictive uncertainty. In: *Computer Modelling in the Environmental Sciences* (D.G. Farmer and M.J. Rycroft, eds.). IMA Conference Series, Clarendon Press, Oxford, pp. 75–88.

Bhaskar, R., 1989. *Reclaiming Reality*, Verso, London.

Cameron, D., Beven, K.J., Tawn, J., Blazkova, S. and Naden, P., 1999. Flood frequency estimation by continuous simulation for a gauged upland catchment (with uncertainty). *J Hydrology*, **219**, 169–187.

Dilks, D.W., Canale, R.P. and Maier, P.G., 1992. Development of Bayesian Monte Carlo techniques for water quality model uncertainty. *Ecol. Modell.*, **62**, 149–162.

Duan, Q., Sorooshian, S. and Gupta, V., 1992. Effective and efficient global optimisation for conceptual rainfall-runoff models. *Water Resour. Res.*, **28**, 1015–1031.

Faurès, J.-M., Goodrich, D.C., Woolhiser, D.A. and Sorooshian, S., 1995. Impact of small-scale spatial rainfall variability on runoff modelling. *J. Hydrol.,* **173**, 309–326.

Feyerabend, P.K., 1975. *Against Method: Outline of an Anarchistic Theory of Knowledge*. New Left Books, London.

Fisher, J. and Beven, K.J., 1994. Application of TOPMODEL and Bayesian uncertainty estimation framework to Rimbaud catchment to investigate the consequences of desertification, Final Report, EU Project, Ecosystem Studies: Modelling the Effects of Desertification in Mediterranean Regions, Lancaster University.

Franks, S.W., Beven, K.J., Quinn, P.F. and Wright, I.R., 1997. On the sensitivity of soil–vegetation atmosphere transfer (SVAT) schemes: equifinality and the problem of robust calibration. *Agric. For. Met.,* **86**, 63–75.

Franks, S.W. and Beven, K.J., 1997. Bayesian estimation of uncertainty in land surface–atmosphere flux predictions. *J. Geophys. Res.*, 102 (D20), 23991–23999.

Franks, S.W., Gineste, Ph., Beven, K.J. and Merot, Ph., 1998. On constraining the predictions of a distributed model: the incorporation of fuzzy estimates of saturated areas into the calibration process. *Water Resour. Res.*, **34**, 787–797.

Freer, J., Beven, K.J. and Ambroise, B., 1996. Bayesian estimation of uncertainty in runoff production and the value of data: an application of the GLUE approach. *Wat. Resour. Res.*, **32**(7), 2161–2173.

Garen, D.C. and Burges, S.J., 1981. Approximate error bounds for simulated hydrographs. *J. Hydraul. Div. ASCE*, **107**(HY11), 1519– 1534.

Haines-Young, R.H. and Petch, J.R., 1983. Multiple working hypotheses: equifinality and the study of landforms. *Trans. Instn. Br. Geogr.*, **8**, 458–466.

Hankin, B. and Beven, K.J., 1998. Modelling dispersion in complex open channel flows: 1. Equifinality of model structure. *Stochastic Hydrol. Hydraul.*, **12**(6), 377–396.

Hofmann, J.R. and Hofmann, P.A., 1992. Darcy's law and structural explanation in hydrology. In: *PSA 92* (D. Hull, M. Forbes and K. Okruhlik, eds.). Philosophy of Science Association, East Lancing, Mich., 1, pp. 23–35.

Hornberger, G.M., Beven, K.J., Cosby, B.J. and Sappington, D.E., 1985. Shenandoah watershed study: calibration of a topography-based, variable contributing area hydrological model to a small forested catchment. *Water Resour. Res.*, **21**, 1841–1850.

Houghton, J.T., Meira Filho, L.G., Callander, B.A., Harris, N., Kattenberg, A. and Maskell, K., 1995. *Climate Change 1995, The Science of Climate Change*. Cambridge University Press, Cambridge.

Howson, C. and Urbach, P., 1989. *Scientific Reasoning: The Bayesian Approach.* Open Court Press, La Salle, Illinois.

Klepper, O., Scholten, H. and van der Kamer, J.P.G., 1991. Prediction uncertainty in an ecological model of the Oosterschelde Estuary. *J. Forecast.*, **10**, 191–209.

Kuczera, G., 1988. On the validity of first order prediction limits for conceptual hydrologic models. *J. Hydrol.*, **103**, 229–247.

Kustas, W.P. and Humes, K.S., 1996. Sensible heat flux from remotely-sensed data at different resolutions. In: *Scaling up in Hydrology Using Remote Sensing* (J.B. Stewart, E.T. Engman, R.A. Feddes and Y. Kerr, eds.). Wiley, Chichester, pp. 127–145.

Lamb, R., Beven, K.J. and Myrabø, S., 1998. Use of spatially distributed water table observations to constrain uncertainty in a rainfall-runoff model. *Adv. Water Resour.*, **22**(4), 305–317.

Lee, P.M., 1989. *Bayesian Statistics: an Introduction*. Edward Arnold, London.

Melching, C.S., 1992. An improved first-order reliability approach for assessing uncertainties in hydrologic modelling. *J. Hydrol.*, **132**, 157–177.

Melching, C.S., 1995. Reliability estimation. In: *Computer Models of Watershed Hydrology* (V.P. Singh, ed.). Water Resource Publications, Highlands Ranch, CO, pp. 69–118.

Morton, A., 1993. Mathematical models: questions of trustworthiness. *Br. J. Phil. Sci.*, **44**, 659–674.

Myneni, R.B. and Williams, D.L., 1994. On the relationship between FAPAR and NDVI. *Remote Sens. Environ.*, **49**, 200–211.

Nash, J.E. and Sutcliffe, J.V., 1970. River flow forecasting through conceptual models, 1. A discussion of principles. *J. Hydrol.*, **10**, 282–90.

Obled, Ch., Wendling, J. and Beven, K.J., 1994. The role of spatially variable rainfalls in modelling catchment response: an evaluation using observed data. *J. Hydrol.*, **159**, 305–333.

O'Neill, R.V., Gardner, R.H. and Carney, J.H., 1982. Parameter constraints in a stream ecosystem model: incorporation of *a priori* information in Monte Carlo error analysis. *Ecol. Modell.*, **16**, 51–65.

Oreskes, N., Schrader-Frechette, K. and Belitz, K., 1994. Verification, validation and confirmation of numerical models in the earth sciences. *Science*, **263**, 641–464.

Patwardhan, A. and Small, M.J., 1992. Bayesian methods for model uncertainty analysis with application to future sea level rise. *Risk Analysis*, **12**, 513–523.

Rogers, C.C.M., Beven, K.J., Morris, E.M. and Anderson, M.G., 1985. Sensitivity analysis, calibration and predictive uncertainty of the Institute of Hydrology Distributed Model. *J. Hydrol.*, **81**, 179–191.

Romanowicz, R. and Beven, K.J., 1998. Dynamic real-time prediction of flood inundation probabilities. *Hydrol. Sci. J.*, **43**(2), 181–196.

Romanowicz, R., Beven, K.J. and Tawn, J., 1994. Evaluation of predictive uncertainty in non-linear hydrological models using a Bayesian approach, In: *Statistics for the Environment. II. Water Related Issues* (V. Barnett and K.F. Turkman, eds.). Wiley, Chichester, pp. 297–317.

Romanowicz, R., Beven, K.J. and Tawn, J., 1996. Bayesian calibration of flood inundation models. In: *Floodplain Processes* (D.E. Walling and P.D. Bates, eds.). Wiley, Chichester, pp. 333–360.

Rose, K.A., Smith, E.P., Gardner, R.H., Brenkert, A.L. and Bartell, S.M., 1991. Parameter sensitivities, Monte Carlo filtering and model forecasting under uncertainty. *J. Forecast.*, 10, 117–133.

Schrader-Frechette, K.S., 1989. Idealised laws, antirealism and applied science: a case in hydrogeology. *Synthese*, **81**, 329–352, 1989.

Spear, R.C. and G.M. Hornberger, 1980. Eutrophication in Peel Inlet. II. Identification of critical uncertainties via Generalised Sensitivity Analysis. *Water Res.*, **14**, 43–49.

Spear, R.C., Grieb, T.M. and Shang, N., 1994. Parameter uncertainty and interaction in complex environmental models. *Water Resour. Res.*, **30**(11), 3159–3170.

Van Fraasen, B.C., 1980. *The Scientific Image*. Clarendon, Oxford, 235 pp.

Van der Sluijs, J., van Eijndhoven, J., Shackley, S., and Wynne, B., 1998. Anchoring devices in science for policy: the case of consensus around climate sensitivity. *Social Stud. Sci.*, **28**(2), 291–323.

Wolock, D.M. and Hornberger, G.M., 1991. Hydrological effects of change in levels of atmospheric carbon dioxide. *J. Forecast.*, **10**, 105–116.

Young, P.C., 1994. *Recursive Estimation and Time Series Analysis*. Springer-Verlag, Berlin.

Young, P.C. and Ng, C.N., 1989. Variance intervention. *J Forecasting*, **8**, 399–416.

Zak, S.K. and Beven, K.J., 1999. Equifinality, sensitivity and uncertainty in the estimation of critical loads. *Sci. Total Environ.*, **236**, 191–214.

Zak, S.K., Beven, K.J. and Reynolds, B., 1997. Uncertainty in the estimation of critical loads: a practical methodology. *Soil, Water, Air Pollut.*, **98**, 297–316.

Environmental Foresight and Models: A Manifesto
M.B. Beck (editor)

CHAPTER 13

Simplicity Out of Complexity[1]

P.C. Young and S. Parkinson

13.1 INTRODUCTION

The environment is a complex assemblage of interacting physical, chemical and biological systems, many of which are inherently nonlinear dynamic processes, with considerable uncertainty about both their nature and their interconnections. It is surprising, therefore, that stochastic dynamic models are the exception rather than the rule in environmental science research. One reason for this anomaly lies in the very successful history of physical science over the last century: modelling in deterministic terms has permeated scientific endeavour over this period and has led to a pattern of scientific investigation which is heavily reductionist in nature. Such deterministic reductionism appears to be guided by a belief that physical systems can be described very well, if not exactly, by deterministic mathematical equations based on well known scientific laws, provided only that sufficient detail can be included to describe *all* the physical processes that are *perceived* to be important by the scientists involved. This leads inexorably to large, nonlinear models reflecting the scientist's perception of the environment as an exceedingly complex dynamic system.

To some scientists who believe in this deterministic approach, the explicit introduction of uncertainty appears almost as an admission of defeat, despite the fact that they recognise the many difficulties (and indeed uncertainties) associated with the modelling of such complex phenomena. They argue that, since the environmental processes are nominally so complex, they cannot be described adequately by simple

[1] This chapter is based in part on Young et al. (1996).

"systems" models which, to them, appear to be largely "black box" in character. Perhaps because there have been some very bad examples of environmental systems analysis in the past, some noted scientists even consider that simpler stochastic models are, in some sense, unscientific and associated with Weinberg's "republic of transcience" (Philip, 1975) – as will be readily apparent from Chapter 2. This is despite the fact that the search for simplicity has characterised much scientific endeavour since William of Occam first enunciated its virtues in his maxim of parsimony at the beginning of the fourteenth century.

Of course, there is no doubt that deterministic reductionism is a paradigm that has served us well in the past: the startling advances in science that have occurred over the past century are testimony to its effectiveness. But we must remember that the paradigm was built on experimental physics and the ability to design and conduct carefully planned experiments. Indeed, the first author once asked a very eminent physicist why his fraternity were not all that interested in using more sophisticated statistics when analysing experimental results. "The answer is very simple", he replied, "we have little need for statistics because our experiments are so well planned that the noise on the experimental data is almost non-existent" – a little tongue in cheek, perhaps, but there is a fundamental truth in his remarks. If we are able to plan experiments very well, then the need for statistics and stochastic models is diminished (although not removed), so that the efforts of statisticians in recovering small signals or parameterised patterns of behaviour (models) from noisy data is not so important.

Certainly, the requirement for statistical rigour would become less significant in the case of *simple* dynamic models *if* the planning and execution of the *dynamic* experiments[2] could be sufficient to excite and separately identify *all* of the important dynamic modes of behaviour in the system, with little noise or uncertainty on the experimental data. But these situations are quite rare: tight experimentation is usually unrealistic in practice, even in the confines of a laboratory; and it is clearly much more difficult, if not impossible, within the natural environment (in spite of the attempts at such in the accounts given in Chapters 2 and 8 of research into surface-water acidification). Moreover, deterministic reductionist models are normally anything but simple in dynamic terms, so that there is a strong possibility of redundant, poorly identifiable parameters in the model that will not be exposed properly unless stochastic methods are employed.

The reliance of physical scientists on extremely well planned and executed experiments has, we believe, had a rather insidious effect when such physicists turn to modelling the natural world. Like engineers who actually build their systems and are often (although not always) confident about the nature of the mathematical equations that can be used to describe them, *some* scientists working on the dynamic modelling of complex natural phenomena *appear* to believe that they can confidently

---

[2] Note we stress "dynamic" here: static systems (if they are practically relevant) can pose less problems in experimental design.

formulate the equations of the system and expect to validate them (or at least fail to falsify them) by comparison with only monitored data. Note that we emphasise the word "appear" here because the discussion of the results obtained from such large, deterministic models of the environment sometimes tends to reflect an acknowledgement of uncertainty which belies their deterministic nature. The recent exorbitant interest in the modelling of "chaotic" processes, for example, could reflect an implicit acceptance of uncertainty, combined with an inherent (and, in the circumstances, understandable) urge to account for such uncertainty in the favoured deterministic terms. Indeed, it is often argued that uncertainty is simply due to lack of knowledge and can be reduced to acceptable levels, or even explained, if the model is expanded to include more detail and a "better" characterisation of the constituent physical phenomena. We know, from Chapter 2, of the intention of the International Geosphere–Biosphere Programme to succeed in "reducing the uncertainties" precisely in this manner. Given the power of modern super computers, this has led inexorably to larger and larger deterministic models.

Perhaps the most topical example of such an approach is the global modelling exercises being conducted by many scientists all over the world in the study of global climatic change. Here, the giant General Circulation Models (GCMs) and their 'poorer relations' (in development and operational cost terms), the much less complex Global Carbon Cycle (GCC) models, are totally deterministic. The question of uncertainty is investigated mainly by "scenario" or "sensitivity" studies. Here, the effects of uncertain future behaviour in some of the input variables (e.g. fossil fuel related inputs) are investigated by comparing the effects of various future deterministic patterns of variation assumed by the analyst. This is all well and good, but hardly an acceptance of wider stochasticity or an acceptable scientific exercise in statistical terms (see Shackley et al., 1998).

There are, however, some signs that attitudes may be changing. We believe there is a growing realisation that, despite their superficially rigorous scientific appearance, simulation models of the environment based on deterministic concepts are more extensions of our mental models and perceptions of the real world than necessarily *accurate* representations of the world itself. The problems of validating such complex speculative models are such that the uncertainty, or even ambiguity, in their predictions can sometimes make them more "prophecies" (Beven, 1993) or "educated guesses" than scientific predictions in which we can have great faith. Despite these limitations, however, such models are undoubtedly very useful as a step in the scientific process and cannot be dismissed, even by the most ardent statistician. After all, prior to the collection of data, they are all that we have in modelling terms and they undeniably reflect the state-of-the-art scientific understanding of the system dynamics. Rather, we must try to understand them better; identify their strengths and weaknesses; and look for improvements in methodology which will allow us to develop and use these modern "Delphic Oracles" in a sensible, probabilistic fashion which reflects satisfactorily their inherent uncertainty and even acknowledges the possibility of ambiguity.

In this chapter, we outline a general approach to the modelling of stochastic dynamic systems that attacks some of the problems discussed above. It is an approach that reflects our previous experience with the modelling of such systems and emphasises the importance of stochastic methods and statistical analysis in the derivation of the models. The various methodological issues and the philosophical approach to modelling that they underpin are illustrated by an important practical example. Reflecting the current importance of global climate research, this study is concerned with the global carbon cycle, starting with a fairly large, deterministic model that has been used in the deliberations of the Intergovernmental Panel on Climate Change (IPCC) and, for very obvious reasons, has not been the subject of any planned experiments (although the observed disturbances to the global carbon-14 (^{14}C) balance caused by nuclear explosions in the atmosphere have been used to some extent in their derivation).

This example rejects the concept of *deterministic* reductionism as the major approach to environmental modelling, but recognises the value of simulation models and reductionist thinking in the overall modelling process. More importantly, however, we stress the importance of explicitly acknowledging the basic uncertainty that is essential to any characterisation of physical, chemical and biological processes in the natural environment, and argue for the greater utilisation of data-based statistical methods in environmental modelling. We feel strongly that this *Data-Based Mechanistic* (DBM) modelling philosophy (Young and Lees, 1992; Young, 1998 and the prior references therein) is essential if mathematical models are to prove useful in environmental decision making.

13.2 A GENERAL DATA-BASED MECHANISTIC (DBM) APPROACH TO MODELLING COMPLEX ENVIRONMENTAL SYSTEMS

If we are to use large environmental simulation models sensibly, then it is necessary to consider them afresh in the context of the uncertainty that not only surrounds much of our knowledge of the environment but also affects the observational data obtained from monitoring exercises. The first author and his colleagues have attempted to draw attention to the limitations of large deterministic models of the environment for many years (Young, 1978, 1983, 1992a,b, 1993a, 1998, 1999b; Young and Minchin, 1991; Young and Lees, 1992; Young and Beven, 1994; Shackley et al., 1998; Beck, 1983) and, in this chapter, we continue with this task by considering how a rational combination of *stochastically defined* simulation models and data-based statistical methods can provide valuable insight into the nature of poorly defined environmental systems.

This general approach – summarised in outline in Chapter 5 – is built on the assumption that the dynamic modelling of environmental systems should involve two basic model types: speculative and normally quite complex simulation models which represent the current, state-of-the-art, scientific understanding of the environmental system; and DBM models obtained initially from the analysis of observational time-series but only considered credible if they can be interpreted in physically meaningful terms. The term 'data-based mechanistic modelling' was first used in Young and Lees (1992) but the basic concepts of this approach to modelling dynamic systems have developed over many years. It was first applied seriously within a hydrological context in the early 1970s, with application to modelling water quality in rivers (Beck and Young, 1975).

The objective statistical derivation of these much simpler DBM models contrasts with the rather subjective formulation of the complex simulation models. However, the two, apparently quite different, types of model are brought together in a rather novel phase of the analysis where the DBM methodology is used to simultaneously linearise and reduce the order of the complex simulation model, so exposing its 'dominant modes' of dynamic behaviour. In the two preceding chapters, we have already seen evidence of the profitable use of stochastically defined simulation models; and we shall see further examples of the successes of DBM models in the following two chapters.

The five major phases in the proposed DBM modelling strategy are as follows:

1. The important first step in any modelling exercise is to define the objectives and to consider the type of model that is most appropriate to meeting these objectives. Since the concept of DBM modelling requires adequate data if it is to be completely successful, this stage also includes considerations of scale and the likely data availability at this scale, particularly as they relate to the defined modelling objectives.

2. In the initial phases of modelling, it is likely that observational data will be scarce, so that any major modelling effort will have to be centred on simulation modelling, normally based on largely deterministic concepts, such as dynamic mass and energy conservation. In the proposed approach, which is basically Bayesian in concept, these deterministic simulation equations are converted to a stochastic form by assuming that the associated parameters and inputs are inherently uncertain and can only be characterised in some suitable stochastic form, such as a probability distribution function (pdf) for the parameters and a time-series model for the inputs. The subsequent stochastic analysis uses Monte Carlo Simulation (MCS) to explore the propagation of uncertainty in the resulting stochastic model, and sensitivity analysis of the MCS results to identify the most important parameters which lead to a specified model behaviour (as is the case also in Chapter 11).

3. The initial exploration of the simulation model in stochastic terms is aimed at revealing the relative importance of different parts of the model in explaining the dominant behavioural mechanisms. This understanding of the model is further enhanced by employing a novel method of combined statistical linearisation and model order reduction. This is applied to time-series data obtained from planned experimentation not on the system itself but on the *simulation model* which, in effect, becomes a surrogate for the real

system. This rather unusual analysis is exploited in order to develop low-order, dominant mode approximations of the simulation model; approximations that are often able to explain its dynamic response characteristics to a remarkably accurate degree (e.g. coefficients of determination R_T^2, based on the model response error, of greater that 0.999: i.e. greater than 99.9% of the output response explained by the model output[3]). Conveniently, the statistical methods used for such linearisation and order reduction exercises are the same as those used for the DBM modelling from real time-series data that follows as the next stage in the modelling process.

4. The DBM methods were developed primarily for modelling environmental (and other) systems from normal observational time-series data obtained from monitoring exercises (or planned experimentation, if this is possible) carried out on the real system. In this stage of the proposed modelling approach, therefore, they are used to enhance the more speculative simulation modelling studies once experimental data become available. In this manner, the DBM models represent those *dominant modes* of the system's behaviour that are clearly identifiable from the observational time-series data and, unlike the simulation models, the efficacy of the DBM models is heavily dependent on the quality of these data. This is, of course, both their strength and their weakness in practical terms. At this stage, the importance of utilising powerful methods of statistical estimation becomes paramount, since the models may be nonstationary, nonlinear or a combination of both. Currently, this involves the exploitation of both *Time Variable Parameter* (TVP: see Young, 1999a) and *State Dependent Parameter* (SDP: see Young, 2000) methods.

5. The final stage of model synthesis should always be an attempt at model validation (see e.g Young, 2001). The word 'attempt' is important since validation is a complex process and even its definition is controversial. Some academics (e.g. Konikow and Brederhoeft, 1992, within a ground-water context; Oreskes et al., 1994, in relation to the whole of the earth sciences) question even the possibility of validating models. However, statistical evaluation of the model by confirming that statistical diagnostics are satisfactory (e.g. no significant autocorrelation in the residuals or cross correlation between the residuals and input variables; no evidence of un-modelled nonlinearity, etc.) is always possible and can engender greater confidence in the efficacy of the model. Also, one specific, quantitative aspect of validation is widely accepted; namely 'conditional predictive validation', in which the predictive potential of the model is evaluated on data other than that used in the identification and estimation stages of the

[3] Note that this is a much more discerning measure of the model's adequacy than the conventional coefficient of determination R^2 based on the one-step-ahead prediction errors.

analysis. When validated in this narrow sense, it can be assumed that the model represents the best theory of behaviour currently available that has not yet been falsified in a Popperian sense.

These are the five major stages in the process of DBM model synthesis but they are not the end of the modelling process. If the model is to be applied in practice (and for what other reason should it be constructed?) then, as additional data are received, they should be used to evaluate further the model's ability to meet its objectives. Then, if possible, both the model parameters and structure can be modified if they are inadequate in any way. This process, sometimes referred to as 'data assimilation', can be achieved in a variety of ways. Since most data assimilation methods attempt to mimic the Kalman Filter (KF)[4], however, it is likely to involve recursive updating of the model parameter and state estimates in some manner, as well as the use of the model in a predictive (forecasting) sense. This process of data assimilation is made simpler in the DBM case because the favoured estimation methods used in DBM modelling are all inherently recursive in form and so can be used directly for on-line data assimilation (Young, 1984).

Recursive estimation is, of course, the embodiment of Bayesian estimation and the KF has long been interpreted in these terms. In the case of linear stochastic systems with Gaussian disturbances (or linear systems having only input nonlinearities: see Young and Tomlin, 2000), the KF provides the ideal Bayesian data assimilation and on-line forecasting algorithm. For nonlinear systems, however, other approaches become necessary, some of which can be related directly to the KF: e.g. the Extended Kalman Filter (e.g Beck, 1979) or the SDP version of the KF (Young, 2000). Increasingly, given the power of modern computers, Monte Carlo simulation is exploited for on-line Bayesian updating of probability distribution functions and evaluation of predictive uncertainty, as mentioned later in the Conclusions.

Note, finally, that the DBM approach emphasises the importance of observational data obtained from experiments or monitoring exercises on the real system and the need, wherever possible, to carefully plan such "dynamic experiments" (see, e.g., Goodwin and Payne, 1977) so that the dominant modes of system behaviour are clearly identifiable from these data. In this latter connection, the concept of scale in the measurements is very important for both data-based and simulation modelling. It is clearly not sensible to assume that a "parameter" measured at the micro scale (where "micro" is, of course, relative to the size of the complete system) is the same as that which might be applicable at the aggregate, macro level. DBM modelling, whether it is applied directly to real time-series data or used for developing simple, dominant mode representations of a simulation model, yields mathematical relationships that are directly related *to the scale of the time-series measurements used in their derivation*. This needs to be acknowledged carefully when drawing deductions from the modelling results and interpreting the model in physically meaningful terms.

---

[4] Since, in the case of linear stochastic systems with Gaussian disturbances, the KF provides the ideal data assimilation algorithm.

13.3 METHODOLOGICAL BASIS OF THE DBM APPROACH

Most of the computational procedures of our DBM methodology have been described elsewhere and, therefore, they will only be discussed briefly at this stage (ample reference to their more detailed usage will be provided elsewhere).

13.3.1 The Stochastic Simulation Model and Regionalised Sensitivity Analysis (RSA)[5]

In general, the high-order, non-linear simulation model can be represented by a set of state equations in continuous or discrete-time. We will initially consider the former, since the conservation equations upon which many conceptual environmental models are based are often (although not always) more naturally formulated in differential equation terms[6]. These state equations can be represented by the following set of lumped-parameter, vector equations (essentially those defining the discussion throughout many of the other chapters),

$$\frac{d\boldsymbol{x}(t)}{dt} = \boldsymbol{f}(\boldsymbol{x}(t), \boldsymbol{u}(t), \boldsymbol{\zeta}(t), \boldsymbol{\vartheta}) \tag{13.1a}$$

$$\boldsymbol{y}(t) = \boldsymbol{g}(\boldsymbol{x}(t), \boldsymbol{u}(t), \boldsymbol{\xi}(t), \boldsymbol{\vartheta}) \tag{13.1b}$$

In equation 13.1(a), $\boldsymbol{x}(t) = [x_1, x_2, ..., x_n]^T$ is, in general, an n dimensional, *non-minimal state vector* (see e.g. Young, 1993a; Young et al., 1987) that characterises the environmental system; $\boldsymbol{u}(t) = [u_1, u_2, ..., u_m]$ is an m dimensional vector of input or exogenous variables that perturb the system; $\zeta(t) = [\zeta_1, \zeta_2, ..., \zeta_q]^T$ is a q dimensional vector of unmeasurable stochastic inputs; and $\boldsymbol{\vartheta} = [\vartheta_1, \vartheta_2, ... \vartheta_r]^T$ is an r dimensional vector of parameters that characterise the system within the non-linear structure defined by $\boldsymbol{f}(\cdot)$. In general, not all of the state variables in $\boldsymbol{x}(t)$ will be either observed or important to the problem at hand and so it is assumed that there is a $p < n$ dimensional vector $\boldsymbol{y}(t) = [y_1, y_2, ..., y_p]^T$ of output variables defined by equation 13.1(b), where $\xi(t) = [\xi_1, \xi_2, ..., \xi_p]^T$ is a vector of observational or measurement noise, again defined in stochastic terms.

For the present purposes, the elements of the parameter vector $\boldsymbol{\vartheta}$ are assumed to be stochastic variables defined by a specified pdf. For instance, the constituent parameters may be derived from a process of statistical estimation based on the usual Gaussian assumptions, so that $\boldsymbol{\vartheta}$ would then be represented stochastically by its first two statistical moments: a vector of mean values $\hat{\boldsymbol{\vartheta}}$ (its estimate) and an associated

---

[5] The procedure has been referred to as Generalised Sensitivity Analysis, or as a Regional Sensitivity Analysis; we adhere to the label of Regionalised Sensitivity Analysis for consistency across the monograph as a whole.

[6] We do not explicitly discuss distributed-parameter, partial differential equations here, although they are often used in environmental modelling, but a similar (although more complex) approach could be developed in this case.

covariance matrix $\boldsymbol{P}_\vartheta$. More likely in the case of large environmental simulation models, ϑ would be represented by some mean value $\bar{\vartheta}$, with the associated uncertainty specified by common measures such as standard deviation, variance or 95% confidence intervals about this mean value. Typically, in this latter case, the statistical measures of uncertainty are obtained either by reference to the scientific literature on the subject or by some form of experiment. Such a simplistic specification of uncertainty is much more likely in practice, since the scientist normally has little prior information on the statistical relationships between the model parameters. But clearly, this specification is limited by the implicit assumption that the parameters are statistically independent and do not arise from a joint distribution. The arguments of the preceding chapter have already indicated this may lead to unrealistic conclusions and, therefore, it needs to be taken into account when evaluating the results of any subsequent uncertainty analysis.

Of course, environmental simulation models are most often formulated in deterministic terms, as in the classical approach to modelling that we have alluded to previously, where ϑ is assumed to be known exactly and the stochastic inputs $\zeta(t)$ and $\xi(t)$ are absent. Here, however, we will consider the richer stochastic setting of equation 13.1(a). Within this formulation, there are two sources of uncertainty: first, the stochastic inputs $\zeta(t)$, which can be a time series specified in some statistical form over a particular simulation period; and second, the uncertain parameters ϑ. Although, in itself, the problem of analytically solving equation 13.1(a) in the purely deterministic case can be quite difficult, if not impossible, software such as MATLAB/ SIMULINK™ (Mathworks, 1994) with its iconographic facilities, have made the numerical solution in this deterministic situation fairly straightforward. On the other hand, the complexity of the associated Fokker–Plank equations that need to be solved in the stochastic case virtually rules out analytical solution and resort must be made to alternative numerical methods, such as stochastic Monte Carlo Simulation (MCS).

Until comparatively recently, MCS was difficult to justify because it absorbed a large amount of computing power and occupied unacceptably large computing time. Unless the model is very large, as in the case of the GCMs, this is no longer the case, however, since the rapid improvements in computing power have even made it possible to carry out MCS studies for most moderately sized simulation models on desktop or portable computers. Larger models, such as the 26th-order, non-linear global carbon cycle model of our case study, can still present problems for desktop machines or workstations but the analysis can be carried out quite straightforwardly on parallel processing computers, which are ideally suited to the generation of the multiple realisations required by MCS. Here, the number of such realisations n required to yield a specified accuracy d in the results is defined using the Kolmogorov–Smirnov statistic, i.e., $n = (K_\forall/d)^2$, where $K_\forall$ is defined for a confidence level (e.g., Spear, 1970).

To reiterate elements of earlier chapters, in the basic form of MCS, each random realisation is generated by first randomly sampling the parameters from their assumed parent pdfs and then computing a model simulation with the model defined by these randomly chosen constant parameters and perturbed by any deterministic or stochastic disturbance inputs. In the latter case, the inputs must also be randomly generated for each realisation but now as time-series spanning the simulation time interval. Clearly this operation requires a complete stochastic model for the disturbance: for example, it might be as complicated as a vector AutoRegressive (AR) or AutoRegressive, Moving Average (ARMA) process; or as simple as a constant mean value, white noise process.

More specifically, in line with the discussion of Chapter 11, the MCS can also form the basis for RSA (Young, 1983; Spear, 1993; Parkinson and Young, 1998), where it provides a means of assessing the relative importance of the parameters in explaining specified modes of behaviour. RSA was first developed by the author's Group in the Centre for Resource and Environmental Studies at the Australian National University, in connection with a study of the Peel Inlet in Western Australia (Hornberger and Spear, 1981; Young et al., 1978). A major objective of this study was to consider which environmental and biological mechanisms were primarily responsible for the accumulation of a macro-algal weed on the beaches of the Inlet.

In RSA, the range of output responses from MCS are categorised into two groups: one designated as *acceptable* behaviour (B); and the other as *unacceptable* behaviour ($\bar{B}$). Here "acceptable" is defined to suit the nature of the problem at hand. For instance, in the case of the Peel Inlet, it defined a response where the macro-algal population predominated in comparison with the competing phytoplankton population. In the case of climatic change, on the other hand, (B) could represent those model responses that are close, in some statistical sense, to the observations over the historical record and exceed certain specified levels of change at some time in the model predicted future behaviour (so acting as an amalgam of both the past and future, in contrast to the deliberate separation of past from future in Chapter 11). The sets of model parameter values that lead to acceptable behaviour in this sense are then statistically compared with those that do not. Finally, and most importantly, those parameters whose sample cumulative distribution functions are found to be significantly different between the B and $\bar{B}$ groups are identified as being the most important in determining the specified behaviour. Usually the number of "important" parameters in this sense is small when compared with the total number of parameters in the model.

Beyond this, MCS can also provide a method for stochastic model optimisation: i.e. for adjusting the prior model parameter pdfs to *a posteriori* distributions which produce model responses that match the observational data in some specified statistical sense. For instance, in the case study of the global carbon cycle model, the mean values of the pdfs associated with the most significant parameters obtained from the RSA are optimised by a simple search routine to maximise the number of acceptable behaviour (B) realisations, where now "acceptable" is defined as a model response which lies within two standard deviations of the observations over the whole historic record. More sophisticated approaches to such optimisation – or "shaping" of the posterior distributions of acceptable parameter values – have already been presented in detail in the preceding chapter, as well as elsewhere, in Keesman and van Straten (1990) and Beven and Binley (1992). Still more complex numerical procedures based on Markov-Chain Monte Carlo (MCMC: see e.g. Gamerman, 1997) methods are now available but they are not easy to use. As we shall see, however, even the simple procedure used in this chapter provides valuable insight into the nature of the complex model and its most important parameters.

13.3.2 Linearisation and Order Reduction of the Non-linear Simulation Models

The linearisation of non-linear models has been popular for many years as a means of better understanding the nature and stability of such models when perturbed from some equilibrium condition defined, for example, by a suitable steady-state solution of equation 13.1(a) in the deterministic situation (i.e., when $d\mathbf{x}(t)/dt = \mathbf{0}$ for constant $\mathbf{u}(t)$ and $\zeta = 0$). Conventionally, large, non-linear simulation models are linearised by a process involving Taylor series expansion about these specified equilibrium solutions. With the advent of powerful computers, such linearisation procedures are now automated in software packages, such as MATLAB/SIMULINK. When applied to complex, high-order models, however, they sometimes fail and, in any case, yield large sets of linear differential or difference equations that are almost as difficult to handle as their non-linear progenitors. There are various methods available to reduce the order of such linearised models, while retaining their dominant modes of behaviour (e.g. Kailath, 1980; Moore, 1981). Here, however, we will outline a data-based approach in which special statistical methods of identification and estimation are exploited to derive the reduced-order, linear model approximation in a single operation based on experimental data obtained by perturbing the *simulation model* about its equilibrium solution using special test signals.

In contrast to the more conventional two-step procedure, this statistical *Dominant Mode Analysis* (DMA) not only yields a reduced-order linear model which explains well the observed dynamic behaviour of the complex simulation model, but it also provides a good indication of the "small perturbational" limits of the model; i.e., the level of input perturbation that can be tolerated before the non-linear behaviour of the parent model negates the utility of its linearised approximation. From our experience with DMA, however, this domain of applicability is often much larger than one might expect from prior considerations. Often the linear model can closely mimic the behaviour of its complex progenitor for quite large perturbations (although this depends upon the nature of the non-linearities and, of course, cannot be guaranteed in general). Moreover, a series of such linear models with smoothly changing parameters can often characterise the non-linear model over its whole operational envelope.

As in conventional linearisation, it is most convenient if this combined model-order reduction and linearisation analysis is restricted to the deterministic situation where the parameters are defined by their *a priori* (or optimised *a posteriori*) mean values. The equilibrium solutions used in the analysis are usually defined for a range of constant input values over the operating envelope of the system and they can be obtained by either using an automatic numerical technique, such as the TRIM procedure in MATLAB, or by running the simulation model to a steady-state solution with the inputs maintained at their specified constant values. Of course, modifications to this approach are required if the system is characterised by a "dynamic" steady state (e.g., a limit cycle) or by unstable modes, but this is not the case in the examples discussed later.

For each equilibrium condition defined in the above manner, each of the input variables to the simulation model is perturbed about its constant value by a signal that is *sufficiently exciting* (see, e.g., Young, 1984) to generate a set of input-output time-series data that are adequate to identify and estimate *unambiguously* the reduced-order, linear model. Although this model can be identified in any form, there are advantages in considering the multivariable Transfer Function Matrix (TFM) description relating the m inputs and p outputs of the original higher-order model. Such a TFM can be defined in terms of the three major operators: the continuous-time differential (or Laplace) operator, s; the discrete-time equivalent of this, the discrete-differential or delta operator, δ; and, finally, the backward shift operator, z^{-1}. In this chapter, we restrict attention mainly to the latter backward shift operator. Here, in the deterministic case when both disturbance vectors ($\xi(t)$ and $\zeta(t)$) are absent, the vector of output variables $\boldsymbol{y}(k)$, sampled at some sampling interval Δt that allows for adequate characterisation of the dominant dynamics of the model, is related to the vector of similarly sampled input variables $\boldsymbol{u}(k)$ by the following linearised, discrete-time, TFM description:

$$
\boldsymbol{y}(k) = \begin{bmatrix} \frac{B_{11}(z^{-1})}{A_{11}(z^{-1})} & \frac{B_{12}(z^{-1})}{A_{12}(z^{-1})} & \vdots & \frac{B_{1m}(z^{-1})}{A_{1m}(z^{-1})} \\ \frac{B_{21}(z^{-1})}{A_{21}(z^{-1})} & \frac{B_{22}(z^{-1})}{A_{22}(z^{-1})} & \vdots & \frac{B_{2m}(z^{-1})}{A_{2m}(z^{-1})} \\ \cdots & \cdots & \ddots & \cdots \\ \frac{B_{p1}(z^{-1})}{A_{p1}(z^{-1})} & \frac{B_{p2}(z^{-1})}{A_{p2}(z^{-1})} & \vdots & \frac{B_{pm}(z^{-1})}{A_{pm}(z^{-1})} \end{bmatrix} \boldsymbol{u}(k) \tag{13.2}
$$

where the individual transfer functions $B_{ij}(z^{-1})/A_{ij}(z^{-1})$; $i = 1,2,\ldots,p$; $j = 1,2,\ldots,m$, are ratios of polynomials in z^{-i}, where $z^{-i}y(k) = y(k-i)$, whose orders are defined by the statistical identification procedure discussed below.

In order to ensure sufficient excitation, the normally recommended test signals used for running such experiments are Pseudo-Random Binary (or multi-level) Sequences (PRBS) with a step length chosen appropriately in relation to the dominant frequency response characteristics of the simulation model. However, other inputs may be preferable in certain circumstances, as we shall see in our case study of the global carbon cycle model. In the multivariable case, the linearised model of equation 13.2 with m inputs and p outputs can be obtained in three ways: either using the full multivariable description directly; by decomposing the multivariable system into mp Single-Input, Single-Output (SISO) sub-systems, each of which is identified and estimated in turn; or by decomposing it into p Multiple-Input, Single-Output (MISO) sub-systems. Here, however, we utilise the multiple SISO approach, since it is simpler and entirely appropriate for the example discussed later.

Although other parameter estimation algorithms could be utilised, we prefer the Simplified Refined Instrumental Variable (SRIV) approach (as outlined in Appendix 1 and described fully in Young, 1984, 1985) for model order identification and parameter estimation, since it not only yields consistent estimates of TFM parameters, but also exhibits close to optimal (minimum variance) performance in the model-order reduction context, where the model residuals are very small indeed. In addition, it has proven very successful in many practical applications over the past few years. The SRIV algorithm is also available in a recursive form that allows for the possibility of variation in the model over the observation interval parameters (TVP estimation), or the presence of non-linearity in the model (SDP estimation).

For identification and estimation purposes, the ijth SISO transfer function in equation 13.2 relating the ith input to the jth output is represented in the form,

$$y_j(k) = \frac{B_{ij}(z^{-1})}{A_{ij}(z^{-1})} u_i(k) + \varepsilon(k) \tag{13.3}$$

where the TF polynomials are defined as,

$$\begin{aligned} A_{ij}(z^{-1}) &= 1 + a_{1ij} z^{-1} + a_{2ij} z^{-2} + \cdots + a_{nij} z^{-nij} \\ B_{ij}(z^{-1}) &= b_{0ij} + b_{1ij} z^{-1} + b_{2ij} z^{-2} + \cdots + b_{mij} z^{-mij} \end{aligned} \tag{13.4}$$

with the polynomial orders nij and mij identified by minimising the YIC or AIC criteria (both of which can prove useful in this regard; see Appendix 1); and any pure time delay of δ samples affecting the relationship is accommodated by setting the δ leading coefficients of $B_{ij}(z^{-1})$ to zero. Note that in this model-order reduction and linearisation context, the noise term $\varepsilon(k)$ represents the residual lack of fit or *response error* between the reduced-order linear model output $\hat{y}_j(k)$ and the full non-linear response $y_j(k)$, where,

$$\hat{y}_j(k) = \frac{B_{ij}(z^{-1})}{A_{ij}(z^{-1})} u_i(k) \tag{13.5}$$

and it is not associated with the measurement noise $\xi(k)$ in equation 13.1(b). Since the simulation model is deterministic, however, this noise will normally be very small indeed, as it represents only the residual effects of the higher-order, non-dominant modes. Consequently, the coefficient of determination associated with this response error (see Appendix 1) will be very close to unity if the model order reduction analysis is successful (as it has been in all examples we have considered to date).

13.3.3 DBM Modelling from Real Data

Conveniently, the same SRIV approach to model identification and estimation used for combined linearisation and order reduction of the simulation models can also form the basis for DBM modelling from real time-series data (see Appendix 1). Here the initial model takes a similar form to equation 13.3, but with $y_j(k)$ and $u_i(k)$ now denoting the ith measured input and jth measured output of the *real* system and the response error term $\varepsilon(k)$ term replaced by discrete-time measurement noise $\xi(k)$, i.e.,

$$y_j(k) = \frac{B_{ij}(z^{-1})}{A_{ij}(z^{-1})} u_i(k) + \xi(k) \tag{13.6}$$

The instrumental variable aspect of the SRIV method ensures that it is robust to noise or uncertainty on the data; while the adaptive prefiltering, which is required for optimality, not only helps to reduce noise effects still further but also ensures that the model represents the data within a frequency band which is most relevant to the definition of the dominant modes (Young, 1984).

In situations where the system exhibits non-stationary or non-linear dynamic behaviour, the DBM approach exploits either the recursive-iterative form of the SRIV algorithm, to estimate time variation in the parameters (TVP estimation), or the related SDP estimation to identify the nature of the non-linearity prior to final parametric non-linear model estimation (see Appendix 1 and Young, 2000). The fully recursive form of the SRIV and related algorithms also allows for the development of on-line adaptive methods of model parameter updating which can be used in adaptive environmental management and data assimilation: e.g in adaptive flood warning (Lees et al, 1994).

Note that, unlike the situation in model linearisation and order reduction, the input signal $u_i(k)$ is not necessarily under the control of the scientist, except where planned experiments are possible, as in environmental applications such as tracer experimentation (e.g. Young, 1999b). Consequently, it may be impossible to ensure that the input is sufficiently exciting and this may limit the identifiability of the component TF model of equation 13.6. In circumstances where the model is identifiable, however, it can provide two primary functions in modelling terms. First, in line with the DBM philosophy, it provides the basis for mechanistic model inference: that is to say, defining appropriate dynamic mechanisms that are consistent with the TFM form and can be interpreted in physically meaningful terms. Second, it can be used to evaluate further the merits of the simulation model. For example, if the estimated DBM model is similar in a statistical sense to the reduced-order, dominant mode approximation of equation 13.4, and can be compared mechanistically with the dominant mechanisms of the simulation model, then this enhances the credibility of the simulation model (and *vice versa*).

13.4 CASE STUDY: THE ENTING–LASSEY GLOBAL CARBON CYCLE MODEL

Over the last 100–130 years the global average of the annual-mean surface temperature of the Earth is believed to have risen by approximately 0.5°C (IPCC, 1990). Whilst this level is within the boundaries of natural climatic variation, there is some unconfirmed evidence to support the theory that the rise is, at least in part, due to the activities of humans. Such activities, e.g., fossil fuel burning and deforestation, release large amounts of *greenhouse gases* into the atmosphere. These gases are believed to affect the natural radiative balance by trapping heat in the lower atmosphere: leading to *global warming*. If this is confirmed, then the resulting temperature increases may have serious consequences: for example, in terms of sea-level rises which could force large-scale human migration from coastal areas; or changes in precipitation and soil moisture content which could have dramatic effects on agriculture.

The main human-produced, or *anthropogenic*, greenhouse gases are carbon dioxide, methane, nitrous oxide and the halocarbons (which include CFCs), whose atmospheric levels have been rising consistently since about 1765.[7] Since the emissions of these gases seem set to continue for the foreseeable future and the potential effects could be so dramatic, there is a need to understand how the greenhouse gases may affect the climate both now and in the future. At present, the main method used for such research is the construction of computer-based mathematical models. Although such models can be of a very simple, empirical type, there seems – as we have said – to be a preference amongst the scientific community studying climate change for more complex and normally deterministic, dynamic simulation models. The largest of these are the gigantic General Circulation Models (GCMs), whose mass and energy conservation equations, in the form of distributed parameter, partial differential equations, are so complex that they need to be solved by some form of numerical approximation in a super computer (e.g. Oglesby and Saltzman, 1990). Rather less complicated, but still of reasonably high dynamic order by most standards, are the Global Carbon Cycle (GCC) models. Here, the movement of carbon in the global environment is described by a set of dynamic mass and/or energy conservation relationships in the form of lumped-parameter, ordinary differential equations.

As pointed out above, such simulation models are, almost exclusively, deterministic. Indeed, there appears to be a view that probabilistic models are "second best" (Huggett, 1993). In this section, we consider a typical deterministic, non-linear GCC model developed by Enting and Lassey (1993; hereinafter referred to as EL) on the basis of an original box-diffusion model suggested by Oeschger et al. (1975). A simplified block diagram of the model in shown in Figure 13.1 and a brief

[7] 1765 is considered to be the start of the "industrial period" (IPCC, 1990). This is the period over which human activities are believed to have emitted sufficient quantities of trace gases to have had a global climatic effect. It is an arbitrary date, but seems consistent with changes in the observed levels of these gases.

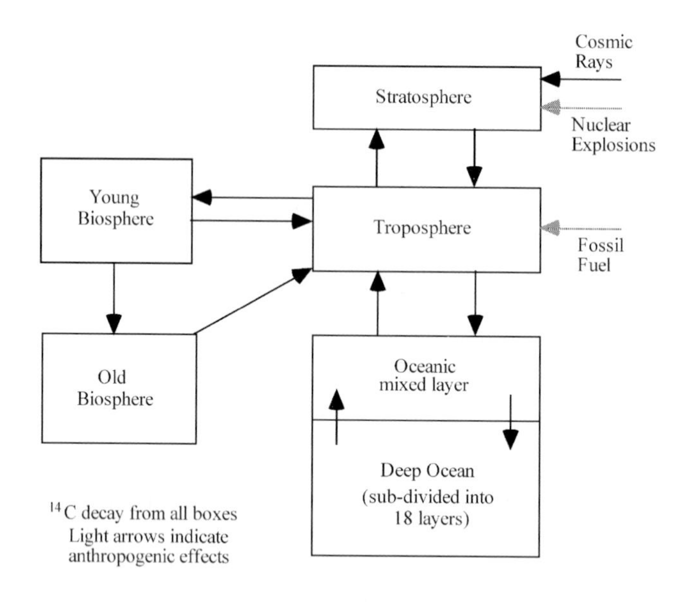

Fig. 13.1. Block diagram of the Enting–Lassey (EL) Global Carbon Cycle Model.

description of its major aspects is provided in Appendix 2. The model is typical of the non-GCM models used by the Intergovernmental Panel on Climate Change (IPCC); the body set up to assess current scientific thinking on climate change and to advise on internationally coordinated policy in this area.

13.4.1 Initial Stochastic Model Formulation and Simulation

Following the modelling approach discussed in previous sections, the first task is to convert the EL model into stochastic form by assigning pdf's to all the thirty-two model parameters, as well as stochastic descriptions for four uncertain inputs. These have been obtained by reference to the latest literature on the subject and from information supplied by scientists working on global climate change. In the detailed research (Parkinson, 1995; Parkinson and Young, 1998), thirty-two different versions of the EL model were investigated. Here, however, attention is restricted to just two of these: E19, which is modified to include two extra vertical ocean fluxes to model detrital rain-out, and an extra atmosphere–biosphere flux to account for enhanced plant growth at high CO_2 levels (CO_2 fertilisation effect); and E29, which includes the three extra fluxes of E19, as well as an extra atmosphere–biosphere flux due to effects such as deforestation (collectively known as "land-use changes").[8] The response of the former was found to be closest to the observed variation in atmospheric ^{13}C

[8] These are discussed further in Appendix 2.

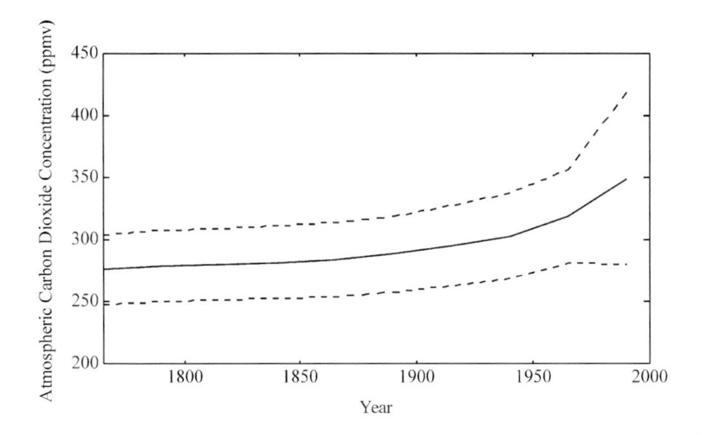

Fig. 13.2. The EL model version, E29: variation of stochastic uncertainty in the atmospheric CO_2 concentration p_a. 95% confidence limits shown as dashed lines.

levels; whilst that of the latter was found to be closest to the observed variation in atmospheric CO_2 levels (Parkinson and Young (1998) discuss this in more detail).

Having converted the EL model into stochastic form, simple MCS immediately indicates how the assumed uncertainty is propagated in the model. Figure 13.2 is typical of the results obtained from such analysis: it shows the evolution of the stochastic uncertainty in the atmospheric CO_2 concentration p_a of model version E29 over a period of 225 years from 1765 to 1990, using a time step of 0.5 years in the numerical solution of the model equations. As can be seen, p_a increases from 275 ± 30 to 351 ± 43 ppmv.[9] The results are based on 160 realisations, as defined by the Kolmogorov–Smirnov statistic, which ensures that comparison between the outputs of two model versions can detect differences of greater than 5% of the change in the output over the simulation period (Parkinson, 1995). The other major model outputs display similar stochastic behaviour: the ensemble mean of atmospheric ^{13}C depletion, $\delta_{a(2)}$, reduces gradually from approximately –6.5 to –7.7‰, with a relatively constant uncertainty of about ±0.5‰; while the ensemble mean and 95% confidence limit of the atmospheric ^{14}C depletion, $\Delta_{a(3)}$, are both approximately constant at 120‰ and ±850‰, respectively, until 1950 and then the mean increases to 650‰ at 1965, before decreasing to 400‰ by 1987. In all cases, the output distributions are approximately Gaussian in form, showing that the model is behaving linearly and suggesting that the nonlinear modes of behaviour are not being excited to any large extent during the simulation period. These results can be compared with the measured values of the variables given in Table 13.1, where the uncertainties quoted are 95% confidence intervals (±2 standard deviations).

---

[9] These uncertainties, in common with all those quoted in this section, are 95% confidence limits (± 2 standard deviations).

Table 13.1

Observational uncertainties for comparison with Enting–Lassey MCS results

Output	Observed Uncertainty	Time
Atmospheric CO_2 concentration/ppmv	353±2	1990.5
Atmospheric ^{13}C Depletion/‰	−7.24±0.1	1978.0
Atmospheric ^{14}C Depletion/‰	175±40	1987.5

In Table 13.1, the values for atmospheric CO_2 concentration come from IPCC (1990) and are based on world-wide measurements of the gas; whilst the values for the atmospheric ^{13}C depletion come from Enting and Pearman (1983). The values for the atmospheric ^{14}C depletion are taken from Enting and Lassey (1993). The most obvious thing to be noted when these observations are compared with the MCS results is the much smaller level of uncertainty in the observed values. For p_a and $\Delta_{a(3)}$, the MCS uncertainty is about twenty times larger; whilst for $\delta_{a(2)}$, the factor is about five. Even allowing for the fact that we have had to assume independent pdfs for the parameters (see the discussion on this in Parkinson and Young, 1998), these results suggest either that the specified uncertainties in the parameter values (which, it will be recalled are obtained from the current literature on the subject and presumably reflect reasonable scientific opinion) are too large; or that there are limitations in the model.

The latter possibility is strengthened if similar MCS results for other model versions are considered. In particular, we define a *model acceptance criterion* based on the observational values. This states that a model version is acceptable if the 95% confidence intervals of *all* of its outputs overlap the corresponding limits of the observations, and rejects any that do not. Using this criterion, it is found that, despite the means of the three outputs of many of the thirty-two versions being distant from the observed (many standard deviations of the observations away) only five can be rejected as unacceptable because their 95% confidence intervals are so wide. Further, when a simple comparison between the means of the MCS results of the twenty-seven "accepted" versions and the observations is carried out, there is conflict between those versions which are closest for one output and those which are closest for another. For example, those versions which include both land-use changes and CO_2 fertilisation are closest for total carbon (with E29 being the "best" of these); whilst, for ^{13}C, those versions which include CO_2 fertilisation but *not* land-use changes are closest (with E19 being the "best" of these).

Clearly, future research is needed to either drastically reduce the uncertainty in the parameter values used in MCS or to attempt to obtain better estimates of the model parameters from some sort of statistical inference based on observational data. The latter is unlikely to succeed because the paucity of observational data, combined with the complexity of the model and the large number of parameters, conspire to make such analysis very difficult. Indeed, given the lack of sufficient excitation in the observed input signals to the model (see later discussion), it is almost certain that the parameters are not identifiable from the observed data. In this situation, model optimisation is not feasible unless many of the 'less important' parameters are constrained to *a priori* assumed values. This pragmatic approach can provide useful results: for example, Enting and Lassey (1993) use it in a careful and reasonable manner. However, even their careful analysis has some questionable aspects, as pointed out in Parkinson and Young (1998). Certainly, constrained optimisation of this type is a procedure that lends itself to the imposition of prejudicial prior assumptions about the values of the 'well known' parameters and so is fraught with danger. Certainly, it cannot be recommended on objective statistical grounds.

It is possible that some reduction in the *a priori* uncertainty in the parameters could be achieved by trying to obtain more accurate field measurements; or it could be that the discovery of dependence between groups of parameters would reduce the combined effect of their uncertainty on the model and the subsequent MCS results (as discussed in Beck (1987) and explored more fully in Beck and Halfon (1991)). However, as Parkinson and Young (1998) demonstrate, the assumption of independent pdfs for the parameters can actually lead to a *lower* variance ensemble of MCS model responses than that obtained if the covariances are taken into account. As a result, a parametric independence assumption does not necessarily mean higher levels of predictive uncertainty and, indeed, it could result in estimates of the uncertainty levels that are too conservative (see also Beck and Halfon, 1991).

The major problem with the EL model, however, remains its size and complexity in comparison with the poverty of the observational data base: the two are clearly in conflict from a statistical point of view. If no radical improvement in the *a priori* parametric specification is possible, then one must question whether the present Enting-Lassey model is entirely appropriate for this kind of stochastic analysis. If not, its use in predicting carbon cycle behaviour over long periods into the future also becomes questionable. One alternative is to consider how the *a priori* values of the most important model parameters can be adjusted by stochastic model optimisation (see earlier), although this is not really solving the fundamental identifiability problem. Another is to obtain a more efficiently parameterised model by direct analysis of the observed data. We shall examine both options below. Before this, however, it is necessary to see if we can ascertain which of the simulation model parameters are "most important" using an RSA.

13.4.2 Regionalised Sensitivity Analysis (RSA)

Having obtained a general idea of the EL model's large sensitivity to parametric uncertainty, we can attempt to evaluate which parameters are most important in defining the dynamic behaviour of the model. Here, an acceptable behaviour, B, is defined with reference to observations, in a similar way to the model acceptance criteria described above. Initially, the observed values and their standard errors were used to provide bands of acceptable behaviour. However, because of the large discrepancy between the uncertainty in the MCS results and the observations, the use of these bands produces only a small number of realisations exhibiting acceptable behaviour. RSA is, of course, very difficult under such conditions and so either the widths of the behavioural bands need to be enlarged or the uncertainty in the parameters reduced.

Using the former approach (described in Parkinson, 1995), the RSA results suggest that the pre-industrial atmospheric level of CO_2, $(p_a)_{pi}$, and the ^{14}C production rate from cosmic rays, ω, are both significant in E19 and E29. In addition, the CO_2 compensation rate of C-3 plants, c_c, is significant in E19; whilst, in E29, the other significant parameters are the pre-industrial atmospheric ^{13}C depletion, $(\delta_{a(2)})_{pi}$, the air–sea exchange time, τ_{am}, the pre-industrial size of the old biosphere, $(N_{o(1)})_{pi}$, and the net primary production, P_N. Unfortunately, there does not seem to be any common link between all the significant parameters in either model version, so it must be concluded that there is no one section of the model which is particularly responsible for the uncertainty seen in all the variables at the output of the model.

The lack of a completely clear message from the RSA results is probably due, once again, to the large uncertainties on the *a priori* assumed parameter values and it can only be corrected by better definition of these uncertainties. Bearing in mind the literature on RSA, these results are not too surprising, as similar problems have been encountered in other studies of large models. It is possible that some improvements could be made to the RSA approach: Spear (1993), for example, suggests a method for checking that all the significant parameters have been found using this technique. He argues that there is a possibility of the effects of two or more parameters cancelling each other out, and describes a technique for examining parameter covariance matrices to discover if this is the case (see also Osidele and Beck, 2001).

However, the RSA approach has served a useful purpose and established those parameters, from the very many that characterise the EL model, that are important in allowing the model to match the observed data in stochastic terms. Moreover, comparison between the results of this stochastic RSA and those obtained using a simple deterministic sensitivity analysis (Parkinson, 1995) show that RSA can detect significant parameter interactions which the more conventional approach cannot. Even the use of much more sophisticated deterministic methods encounters difficulties in trying to find such interactions in highly complex models.

13.4.3 Stochastic Model Optimisation

Having used RSA to establish the parameters in the EL model that appear most important in generating acceptable behaviour, the next step in this stochastic investigation of the model is to see how the pdfs of these important parameters need to be adjusted to maximise the number of realisations exhibiting such acceptable behaviour. In other words, allowing the mean and standard deviations of the model output probability distributions to match those of the observed data, so that the model has the highest probability of this behaviour. We seek, therefore, to discover the "best" model in this stochastic sense, simply by adjusting the means of the pdfs of

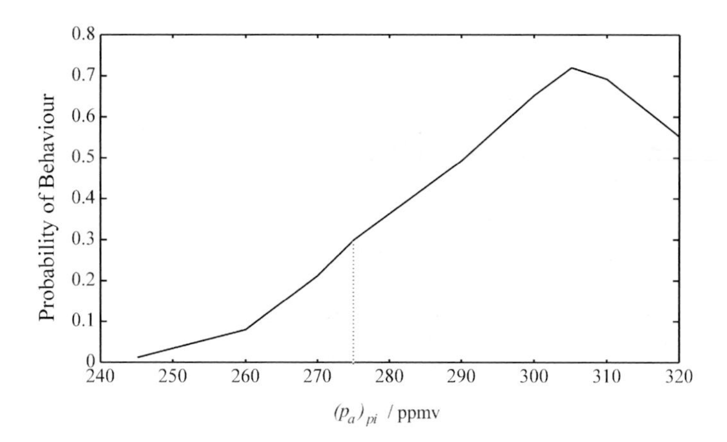

Fig. 13.3. The EL model version, E19: variation of the probability of acceptable behaviour with the mean of the pdf of the pre-industrial atmospheric CO_2 concentration $(p_a)_{pi}$.

the significant parameters from their *a priori* values until a peak of this probability is found in the parameter space of each model under analysis. Clearly this is an alternative to more conventional model parameter estimation, which seems more appropriate in this complex model situation.

From the preceding analysis of sensitivity, three parameters are found to be significant in model version E19: $(p_a)_{pi}$, ω and c_c. Figure 13.3, which is typical of the results obtained in this analysis, shows the variation of the probability of acceptable behaviour with the mean of the pdf for the pre-industrial atmospheric CO_2 concentration, $(p_a)_{pi}$. Note that there is no *a priori* physical reason why the three important parameters should be dependent on one another, so none has been assumed. Note also that, as in the case of $(p_a)_{pi}$ shown in Figure 13.3 (see the dashed line), the standard parameter means (ω = 6.0 kgy^{-1}, c_c = 80 ppmv) for the other parameters do not give a particularly high probability of acceptable behaviour. Indeed, considering all three parameters, a large improvement is gained by changing their mean values: namely, increasing $(p_a)_{pi}$ to 305 ppmv; reducing ω to 5.0 kgy^{-1}; and reducing c_c to 40 ppmv. And these adjustments are relatively large: two standard deviations or more in the cases of $(p_a)_{pi}$ and c_c.

In model E29, six parameters are found to be significant: $(p_a)_{pi}$, $(\delta_{a(2)})_{pi}$, τ_{am}, ω, $(N_{o(1)})_{pi}$ and P_N. Parameter interaction was suspected between $(p_a)_{pi}$ and $(\delta_{a(2)})_{pi}$, the pre-industrial levels of CO_2 and ^{13}C in the atmosphere; and also between $(N_{o(1)})_{pi}$, the pre-industrial size of the old biosphere and P_N, net primary production. The former interaction was confirmed, whilst the latter was denied by the application of ANOVA testing (Parkinson, 1995). Hence, $(p_a)_{pi}$ and $(\delta_{a(2)})_{pi}$ are plotted against each other as a surface plot in Figure 13.4 (the other four parameters are considered singly, but the results are not shown). It is found that a high probability of acceptable behaviour is obtained by using the default means of parameters $(p_a)_{pi}$, ω and P_N; whilst the means of τ_{am}, $(N_{o(1)})_{pi}$ and $(\delta_{a(2)})_{pi}$ all need to be adjusted quite considerably to give good

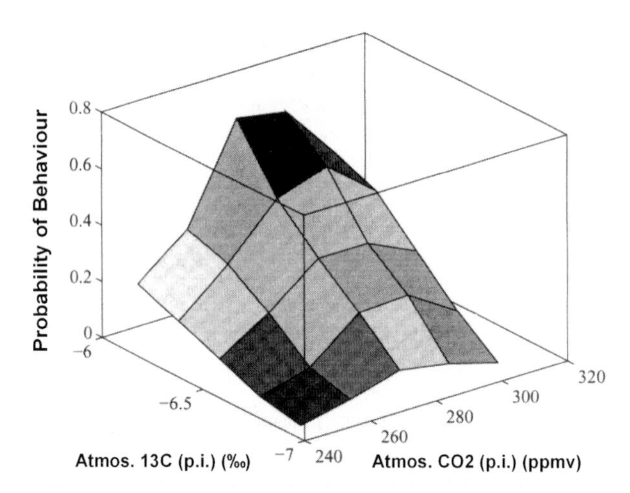

Fig. 13.4. The EL model version, E29: variation of the probabilities of acceptable behaviour with the pdf means for the pre-industrial atmospheric CO_2 concentration, $(p_a)_{pi}$; and the pre-industrial atmospheric ^{13}C depletion $(\delta_{a(2)})_{pi}$.

results (again, by two standard deviations or more). Parameter mean values that produce the "best" model are $(p_a)_{pi}$ = 275 ppmv, $(\delta_{a(2)})_{pi}$ = −6.1‰, τ_{am} = 7.0 y, ω = 3.0–4.0 kgy^{-1}, $(N_{o(1)})_{pi}$ = 2600 GtC, P_N = 180 $GtCy^{-1}$. However, in this case, the default values, ω = 6.0 kgy^{-1} and P_N = 100 $GtCy^{-1}$, yield only a small reduction in the probability of acceptable behaviour.

To summarise, the probability of acceptable behaviour in both model versions E19 and E29 is approximately doubled by optimisation of the means of the pdfs for some or all of the RSA-defined significant parameters. Moreover, to achieve this high level of probability, relatively large changes, of two standard deviations or more, need to be made to certain of these parameters. This helps to emphasise again the shortcomings of both model versions: E19 in its modelling of total carbon and ^{14}C; and E29 in its modelling of ^{13}C and ^{14}C.

13.4.4 Uncertainty in Model Predicted Future Levels of Atmospheric Carbon Dioxide

Attention can now be directed to the problem of projecting possible atmospheric CO_2 increases across the present century, as part of the investigation into possible anthropogenic climate change. In particular, version E29 of the EL model is used because of its status as the "best" model for total carbon (Parkinson, 1995). It must be emphasised that this version is used with its original, or *a priori* pdfs, rather than the optimised *a posteriori* values discussed in the preceding section. This is because the model incorporating these adjusted values has not been validated against independent data.

Figure 13.5 shows the stochastic evolution of the atmospheric CO_2 concentration, p_a, produced by E29 when forced by fossil fuel and land-use change inputs according

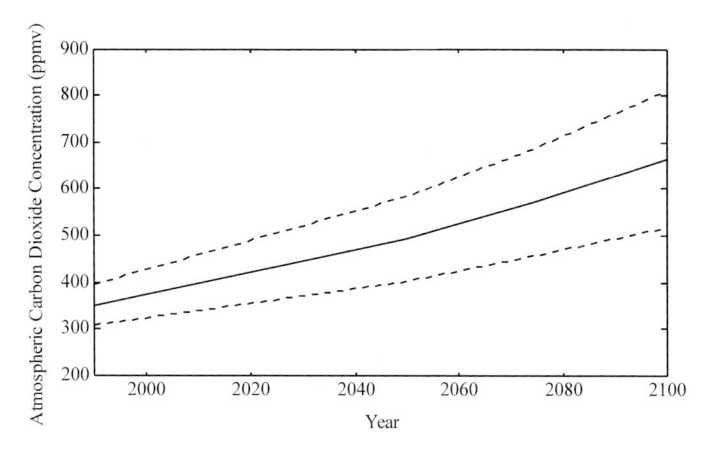

Fig. 13.5. The EL model version, E29: evolution of stochastic uncertainty of atmospheric carbon dioxide concentration forced by IPCC Scenario IS92a. Dashed lines are the uncertainty bands (95% confidence limits) obtained with the *a priori* uncertainties.

to IPCC emissions scenario IS92a (IPCC, 1992). The dashed line shows the uncertainty (95% confidence limits) in the evolution. As expected, this grows steadily throughout the simulation period from 351±43 ppmv in 1990 to 664±144 ppmv in 2100. If it is assumed that the parametric and input uncertainties in the EL model have been overestimated: in this instance, by an arbitrary factor of two, then these confidence limits are reduced to about half those shown in the figure[10]. Table 13.2 compares these two levels of uncertainty with those from three deterministic modelling exercises carried out by other research workers. It is clear that the uncertainty produced here is much larger than that based on deterministic methods, despite the reduction in the assumed parametric and input uncertainties to compensate for the fact that the pdfs of the parameters are being assumed independent (see previous discussion). The level of output uncertainty generated with the original parametric and input uncertainties is between five and ten times larger; and even when these uncertainties are assumed to be half of the originally assumed values, the amount of output uncertainty is still more than twice that derived by deterministic studies. Obviously, the deterministic studies are not picking up as many sources of uncertainty as the MCS and this needs serious consideration in any evaluation of the deterministic model results being considered by the IPCC.

[10] IPCC scenarios calculate anthropogenic gas emissions from 1990–2100 based on assumptions about future political and socio-economic factors. IS92a is a "non-extreme" scenario. It is assumed, in this study, that there is an uncertainty of ±25% in converting a political scenario into yearly fossil fuel emissions and of ±50% in converting into land use change flux. However, the simulation was repeated with zero uncertainty on these two inputs between 1990 and 2100, and no significant reduction in the evolution uncertainty was found.

Table 13.2

Comparison of uncertainty in future CO_2 levels due to IPCC Scenario IS92a from various modelling exercises

Year	Uncertainty in Atmospheric CO_2		Source
	/ppmv	Range	
2050	494 to 510	16	Enting (1994) pers. comm
	520 to 550	30	Wigley and Raper (1992)
	491±86	172	Figure 13.5 dashed line
	487±44	88	Figure 13.5 dotted line (uncertainties halved)
2100	667 to 719	52	Enting (1994) pers. comm
	740 to 800	60	Wigley and Raper (1992)
	615 to 683	68	Enting and Lassey (1993)
	664±144	288	Figure 13.5 dashed line
	658±71	142	Figure 13.5 dotted line (uncertainties halved)

13.4.5 Dominant Mode Analysis (DMA): Model Linearisation and Order Reduction

Although five versions of the EL model have been subjected to DMA (Parkinson, 1995), only version E29 is considered in detail, again because it appears to be the "best" version in describing total carbon. Also, for simplicity, only the SISO relationship between fossil fuel input, S_t^F, and the atmospheric CO_2 level, p_a, is explored, due to its potential importance in the role of anthropogenically induced climate change. In the analysis, the simulation model is set initially at an equilibrium condition with p_a = 275 ppmv. S_t^F is then perturbed by a small perturbation signal and the output response, p_a, to this S_t^F perturbation is recorded. Note, however, that since the analysis is concerned *only* with the fossil fuel–atmospheric CO_2 relationship, the effect of land-use change is not considered explicitly. This needs to be taken into account if the output of the model is to be compared with observational data (as discussed later).

The choice of the input signal is important since, nominally, it should be *sufficiently exciting* to identify unambiguously the major modes of dynamic behaviour that characterise the simulation model. In this example, however, the choice of input signal is complicated by the dynamic characteristics of the EL model. In particular and not surprisingly given the physical nature of the simulated system, the experimental results confirm that the EL model is a "stiff" system of differential equations, in the sense that its response characteristics are dominated by an integrator (arising from the conservativity of the carbon balance system) and time constants with widely different values, one of which is very close to the unit circle in the complex z plane for the discrete-time TF model.

In this situation, the selection of a suitable PRBS perturbation signal is difficult since the fundamental step length needs to be very long to accommodate the longest time constants and the experimental time period becomes very large indeed. Moreover, it is also necessary to constrain the dominant eigenvalue associated with the integrator to unity in the complex z plane, in order to avoid the estimation problems associated with the nonstationarity of the response. And simple differencing of the data does not solve this problem because the integrator acts *in parallel* with the other compartmental responses. For this reason, it is simpler to apply a single unit (1 GtC) *impulsive* input and use the power of the SRIV algorithm to identify the linear TF model from these impulse response data. Although the impulse input does not *persistently* excite the model dynamics, we will see subsequently that it does provide sufficient information to identify the dominant modes of behaviour provided the simulation period is made sufficiently long for the system to return to equilibrium.

Although, for clarity, the unit impulse response of the nonlinear model is shown in Figure 13.6(a) only over the first 500 years, the identification and estimation analysis is based on a full 3000 year simulation. The use of the impulse response in this manner conveniently allows the effect of the parallel integrator to be removed by simple subtraction of the final steady-state level of the response (0.0772 ppmv) achieved after the 3000 years, so allowing for the remaining dominant eigenvalues to be better identified and estimated. The best linear discrete-time TF model approximation is identified by reference to the YIC, R_T^2 and AIC criteria obtained from repeated SRIV estimation based on a wide variety of TF model structures (see Appendix 1). The identification results for the best 20 models, based on the YIC, are shown in Table 13.3.

Table 13.3

SRIV identification results

Den	Num	Del	YIC	R_T^2	AIC
3	3	1	–20.523	0.9998	–16.005
3	4	1	–16.319	0.9999	–16.456
2	2	0	–17.2418	0.993055	–12.4637
4	3	1	–16.180	0.9999	–16.431
2	2	1	–16.165	0.9939	–12.608
5	3	1	–13.472	0.9999	–16.627
2	3	1	–13.026	0.9963	–16.627
3	2	1	–12.371	0.9958	–12.987
5	4	1	–11.693	0.9999	–16.939
4	2	1	–10.949	0.9969	–13.288
3	5	1	–9.7541	0.9999	–16.656
2	5	1	–9.5241	0.9983	–13.860
5	2	1	–10.737	0.9976	–13.525
2	5	1	–9.2827	0.9976	–13.552
1	1	1	–9.2096	0.7191	–8.7778
2	4	1	–8.8256	0.9975	–13.504
5	4	1	–8.2221	0.9999	–16.4223
1	2	1	–6.8317	0.7554	–8.9153
2	1	1	–5.3946	0.7334	–8.8294
1	3	1	–4.2442	0.7935	–9.0839
1	4	1	–3.6248	0.8325	–9.2927
1	5	1	–3.3845	0.8681	–9.531

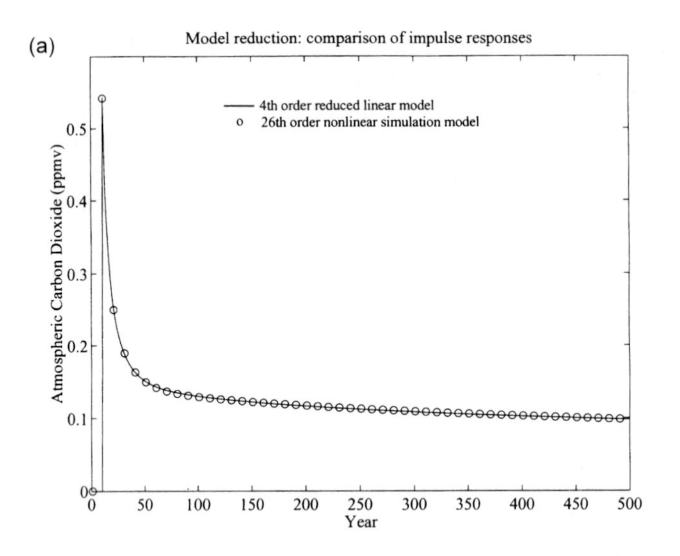

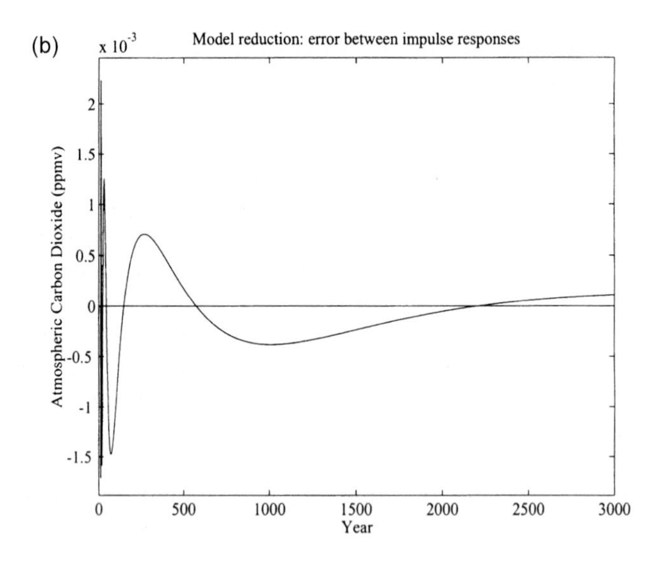

Fig. 13.6. Combined model linearisation and reduction: (a) comparison between the unit impulse responses of the fourth order, linear model and the 26th order nonlinear simulation model; (b) error between the unit impulse responses of the fourth order, linear model and the 26th order nonlinear simulation model (note scale change in comparison to plot in (a)).

We see that the [3,3,1] model has by far the lowest YIC (–21.66) and explains 99.98% of the impulse response data (i.e. $R_T^2 = 0.9998$). Other higher-order models, such as the [5,4,1] which minimises the AIC, explain the data a little better ($R_T^2 = 0.9999$) but have much less negative YIC values (–11.69). Moreover, when some of these higher-order models are investigated, they tend to have very similar (and sometimes unstable) roots in the numerator and denominator polynomials of the TF (i.e., they exhibit pole-zero cancellation). This is a clear indication of over-parameterisation and is almost certainly the main reason for the higher YIC values in these cases. Note how these results also illustrate the disadvantages of the AIC in this TF model identification context. Despite the fact that it is often used for this purpose (incorrectly, we believe), it is defined primarily for AR models. The final SRIV estimated [3,3,1] TF model takes the form,

$$\hat{x}(k)=\frac{\hat{b}_1 z^{-1}+\hat{b}_2 z^{-2}+\hat{b}_3 z^{-3}}{1+\hat{a}_1 z^{-1}+\hat{a}_2 z^{-2}+\hat{a}_3 z^{-3}}u(k) \tag{13.7}$$

where,

$$\hat{a}_1=-2.6987(0.0013);\ \hat{a}_2=2.4118(0.0019);\ \hat{a}_3=-0.7130(0.0012);$$
$$\hat{b}_1=0.4633(0.0003);\ \hat{b}_2=0.8528(0.0006);\ \hat{b}_3=0.3905(0.0004);$$

Here and later, the figures in parentheses are the approximate standard errors on the parameter estimates obtained from the SRIV estimation; $\hat{x}(k)$ denotes the model generated atmospheric CO_2 level p_a, and $u(k)$ is the fossil fuel input S_t^F, both at the kth sampling instant.

The complete reduced order model is the 4th order combination of (13.7) and the mass conservation integrator. When a multi-order TF model such as (13.7) is identified and estimated, the normal DBM approach is to consider how the model can be decomposed in a manner which has some meaningful physical interpretation. However, as in the case of all TF models of greater than first order, this model can be decomposed into a number of forms comprising different combinations of series, parallel and feedback interconnections of first-order processes. In systems theoretic terms, none of these decompositions are individually *identifiable*, in the sense that, without further *a priori* information, all of the decompositions are feasible descriptions of the input–output behaviour. Taking into account the additional integration effect, the simplest decomposition in computational terms is the following one, obtained by straightforward partial fraction expansion:

$$\hat{x}(k)=\frac{0.0772}{1-z^{-1}}u(k)+\frac{0.0600z^{-1}}{1-0.9978z^{-1}}u(k)+\frac{0.1728z^{-1}}{1-0.9438z^{-1}}u(k)+\frac{0.2305z^{-1}}{1-0.7571z^{-1}}u(k) \tag{13. 8}$$

This suggests that the reduced order model can be composed of four *parallel* transfer function pathways. The steady state gains G_i, residence times (time constants) T_i and parallel partition percentages P_i of the three first order blocks, for i = 1,2,3, are calculated as follows (see e.g. Young, 1984, 1992a,b), with all figures rounded to 4 decimal places or less:

$$G_1 = 0.0600/(1\text{–}0.9978) = 27.9;\ T_1 = -1/\log_e(0.9978) = 464 \text{ years};\ P_1 = 87.4\%$$

$$G_2 = 0.1728/(1\text{–}0.9438) = 3.1;\ T_1 = -1/\log_e(0.9409) = 17.3 \text{ years};\ P_1 = 9.6\%$$

$$G_3 = 0.2305/(1\text{–}0.7571) = 0.95;\ T_1 = -1/\log_e(0.7540) = 3.6 \text{ years};\ P_1 = 3.0\%$$

These residence time estimates make good physical sense, as discussed later.

Figure 13.6(a) compares the first 500 annual samples of the response of the model (13.8) to the 1 GtC impulse of fossil fuel input with the full nonlinear model impulse response; while Figure 13.6(b) shows the error between the responses over the complete 3000 samples used for identification and estimation. As can be seen, the residual error is very small (less than 0.002 ppmv) but highly structured. Clearly the reduced-order model has captured the dominant modal behaviour and this error represents the combined effects of all the higher-order, non-dominant dynamics, as well as residual nonlinear effects.

Figure 13.7 compares the response of the reduced-order linear model of equation 13.8 with the full nonlinear model response to the measured fossil fuel input S_t^F over the period 1840–1990 (from CDIAC, 1991). Both models respond almost identically at first and the error, which is always very small, grows only slightly as the amplitude increases in response to the increasing fossil fuel input. These results are to be expected: the reduced-order, linear model can only be expected to approximate the much higher-order, nonlinear model very well around the equilibrium level and

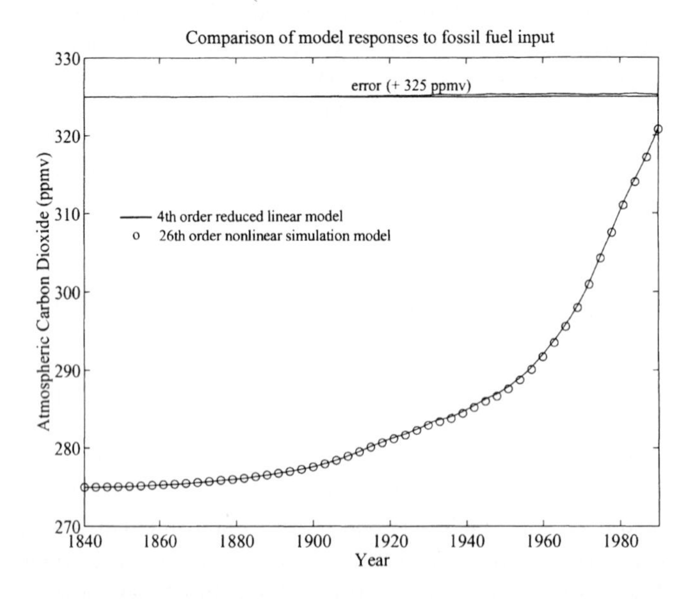

Fig. 13.7. Combined model linearisation and reduction: comparison between the responses of the fourth order, linear model and the 26th order nonlinear simulation model to fossil fuel input (error shown above (+325 ppmv).

some divergence is inevitable as the input perturbation leads to greater excursions from this equilibrium. Nevertheless, the differences are very small indeed given the level of perturbation and the simplicity of the reduced-order model.

This same model linearisation and order-reduction analysis has been applied to the other five versions of the EL model and [3,3,1] models are identified in all cases, with the estimated parameter values reflecting the differences between both the model versions and the selected equilibrium conditions (Parkinson, 1995). However, the estimated gains and time constants change smoothly over the operating envelope of the nonlinear models and indicate clearly the efficacy of the linearisation and model-order reduction. Perhaps the most surprising aspects of these results is the drastic reduction in model order and the similar parallel decomposition form obtained for all operating points and versions of the model. In other words, the SRIV analysis has very effectively administered Occam's razor to the complex global carbon balance model and exposed its very few, well behaved, dominant modes of dynamic behaviour.

Finally, it is necessary to end with a caveat: although the model order reduction experiments have yielded very interesting results and provided greater insight into the nature of the nonlinear EL model, the reduced-order model parameter values should be treated with some caution. The results show that the EL model is a stiff system of differential equations, in the sense that its response characteristics are dominated by three time constants with widely different values, as well as the pure integration element. Moreover, at the discrete-time sampling interval of one year, the eigenvalue associated with the longer time constant is very close to the unit circle in the complex z plane. On the other hand, direct SRIV identification and estimation of a continuous-time, differential equation model yields the following 3rd order model, which can be compared with the discrete-time model in (13.7):

$$\begin{aligned}\frac{d^3y(t)}{dt^3}+0.284\frac{d^2y(t)}{dt^2}+0.0126\frac{dy(t)}{dt}+0.0000255y(t)=0.255\frac{d^3u(t)}{dt^3}\\ +0.518\frac{d^2u(t)}{dt^2}+0.0668\frac{du(t)}{dt}+0.000815u(t)\end{aligned} \tag{13.8a}$$

and partial fraction expansion in this case produces comparable estimates of the residence times (4.4, 19.1 and 467 years).

In fact, these latter continuous-time estimates are superior in statistical terms to those of the discrete-time model estimates. This is because the continuous-time model is more appropriate to the continuous-time EL model formulation and the offending, near unity, root is no longer a problem. For instance, Figure 13.8 is a plot of the normalised histograms for the derived residence times associated with the model (13.8a). These were computed by MCS using 50000 realisations based on the SRIV estimated covariance matrix for the parameters in this model. It is clear that the residence times of all three pathways are very well defined.

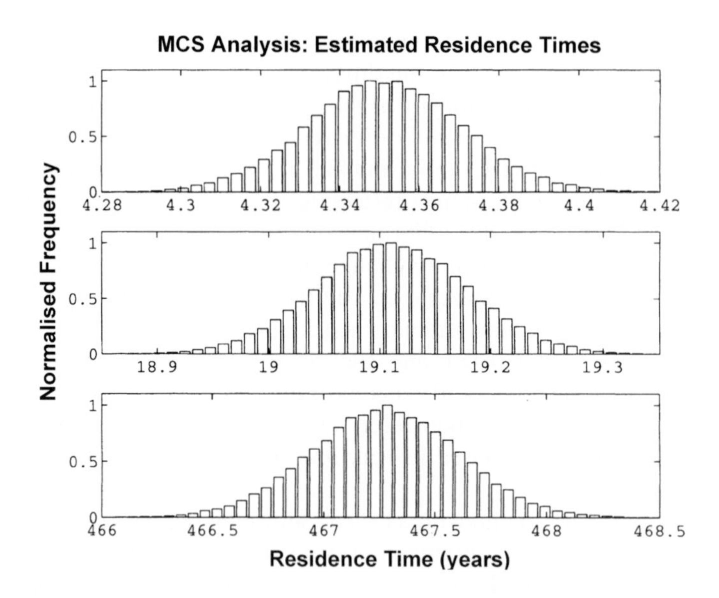

Fig. 13.8. Normalised histograms for the derived residence times associated with the model (13.8a). computed by MCS using 50000 realisations based on the SRIV estimated covariance matrix for the parameters in this model.

13.4.6 Simple Mechanistic Interpretation of the Reduced-Order Model Results

Within the DBM philosophy, it is very important to consider whether the reduced-order model provides a credible description in physical terms. In this regard, the parallel form of the model and the associated impulse response of Figure 13.6(a) both make reasonable physical sense. The former suggests that the atmospheric CO_2 variations resulting from fossil fuel inputs arise from three pathways with time constants of approximately 459, 16.4 and 3.5 years, all of which seem sensible given the different compartments represented in the high-order simulation model. The estimated shorter time constants are not dissimilar to the air–sea exchange time of 11 ± 4 years and the troposphere–stratosphere exchange time of 6 ± 2 years, respectively, as assumed in the simulation model. And the larger time constant is almost certainly associated with the assumed value of 900 ± 250 years for the deep-ocean turnover time. Naturally, the complex interconnected nature of the simulation model modifies the *effective* values of these time constants when the model is integrated and so the identified dominant-mode residence times seem very reasonable.

The integration element is obviously required in the reduced model to ensure the conservativity of the system, as assumed in the EL model formulation: in the case of the impulse response, for example, the final equilibrium level of 0.0772 ppmv is controlled by the gain of this integrator and corresponds to the final global atmospheric CO_2 concentration achieved after the 1 GtC of fossil fuel input has been distributed around the various other compartments in the global system. The transient to this final level is defined entirely by the three, first-order, parallel transfer functions which reflect the distribution of atmospheric CO_2 to the other compartments as time progresses.

Finally, the rise in the atmospheric concentration of CO_2 in the *first year* following this 1 GtC impulsive input of fossil fuel is computed from the reduced model of equation 13.8 as 0.47 ppmv. This is exactly equal to the computed value of 0.47 ppmv for the rise in the atmospheric concentration of CO_2 induced by an input of 1 GtC of carbon for 1 year if all the carbon stays in the atmosphere (Enting, 1991).

So, when all these comparisons are taken into account, we can conclude that the much simplified, reduced order linear model not only reproduces the high order, nonlinear model behaviour to a remarkable degree, it also makes reasonable physical sense, as required by DBM analysis. Of course, the parallel decomposition used in the above discussion is not the only one that makes reasonable physical sense in this example, although it is the simplest to compute. If we consider the block diagram of the EL model in Figure 13.1, for instance, it is clear that the reduced order models (13.7) or (13.8) are compatible with a simplified, linear version of this large model and this could be derived straightforwardly, if required. More importantly, however, the DMA results suggest that the EL model could have been formulated in this much simpler but still physically meaningful manner without any degradation in dominant modal dynamic terms.

13.4.7 Another Example of Model Reduction

The approach to model reduction outlined in the previous section has been used also in an evaluation of the ANU-BACE global carbon cycle model (Taylor, 1996). Here the reduced order linear model obtained from the impulse response of the full, high order, nonlinear model is 5th order and, once again, this low order model represents the dynamics of its much higher order parent very well over the whole of the historic period. In this case, however, some minor evidence of nonlinearity appears, in the form of a growing error between the reduced and full order model outputs, as the present day is approached. Nevertheless, the error is very small and growing smoothly, suggesting that the 5th order dynamics are adequate and a Time Variable Parameter (TVP) or piece-wise linear model (i.e. a series of constant parameter linear models which are each defined at different defined operating points) should be able to characterise the high order nonlinear model over much longer periods of time.

13.4.8 Comments

As regards the prediction of atmospheric CO_2, the EL model reduction results, taken together with those for the ANU-BACE model, must call into question the need for such complex representations of carbon balance at the global scale and suggests that a much simpler representation with fewer, or at least lower order, sub-systems (e.g. the EL model has 18 compartmental levels in the ocean sub-system alone) could have produced very similar results. Moreover, such a reduced order model would be more appropriate to the amount of observational climate data available which, in itself, makes the assumption of a high order model rather questionable on statistical grounds. Indeed, as pointed out previously, the EL model can only be fitted to the available data with constraints applied to many of the parameters, a common indicator of severe over-parameterisation.

It is clear that the nonlinearities in the EL model are not being activated to any large degree over the historical period considered here, otherwise its behaviour could not have been mimicked by a low order linear model. This has important implications regarding the validity of the model as a characterisation of the global carbon balance dynamics into the future and, therefore, its potential utility in predictive terms. If the nonlinearities are not required to explain the historical observations, then it is clear that they will not have been adequately identified and estimated on the basis of these observations. In other words, they constitute a largely unvalidated part of the model that represents no more than a current scientific conjecture about the nature of the nonlinear mechanisms operative at the global scale.

There is nothing wrong with conjecture, of course, provided it is made clear that the model is conjectural, and therefore limited, in this sense. Unfortunately, caveats of this type are not normally emphasised and the deterministic nature of global climate models militates against the quantification of uncertainties and their effect on predictions (as already observed in Chapter 10). Encouragingly, however, questions about both model uncertainty and simplification are now starting to attract attention in the climate change community. For example, in a comment on 'climate change research after Kyoto', Hasselmann (1997) concludes that, "*Researchers can help to build on ... models by simplifying their climate models for optimisation studies... This will not be easy, but several promising reduction techniques have been explored*". And he refers to two recent papers (Joos et al., 1996; Hasselmann et al., 1997) which describe related approaches to DMA, although using a different methodological approach.

13.4.9 Modelling From Real Data

DBM modelling based on real data is quite difficult in the case of the global carbon balance, since it requires consistent sets of time-series data that are not easy to obtain. To exemplify the approach, however, we will consider the analysis of the fossil fuel–atmospheric CO_2 data shown in Figure 13.9. It must be emphasised, however, that this analysis is mainly illustrative and it is should not be taken too seriously as an exercise in climate change research.

Note that these data are the *changes* in the variables derived from the original measured level data (fossil fuel input from CDIAC, 1991; atmospheric CO_2 level from Enting, 1991). In the case of the atmospheric CO_2, which varies in a fairly

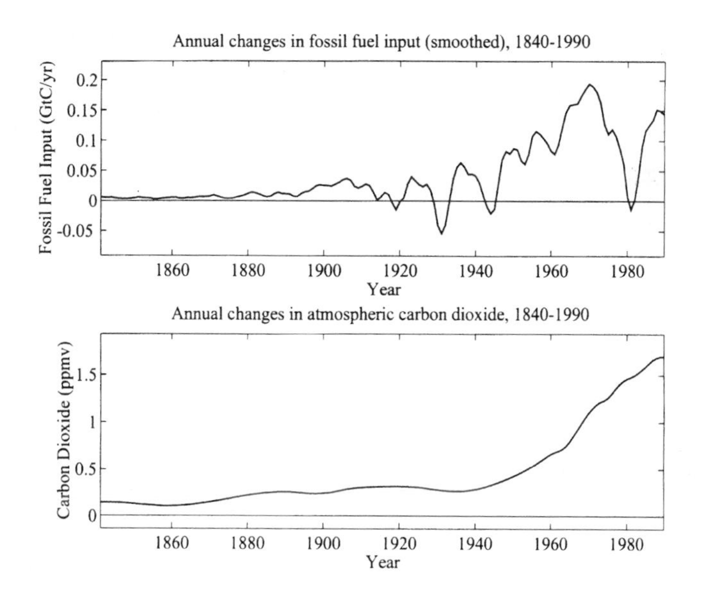

Fig. 13.9. DBM modelling: annual changes in fossil fuel input and atmospheric carbon dioxide output (1840–1990).

smooth fashion, these changes can be obtained by simple differencing. The fossil fuel data are, however, quite noisy and simple differencing amplifies the noise to an unacceptable degree. For this reason, both series were differentiated, using a fixed interval smoothing algorithm (see Young, 1984, 1993a, 1999a). This successfully prevents high frequency noise amplification on the fossil fuel data and, in the case of the atmospheric CO_2 data, yields a series that is indistinguishable from the simple differenced series, demonstrating that the processing is not removing any useful information from the data.

SRIV identification and estimation applied to the input–output data in Figure 13.9 yields the following second order [2,1,0] model,

$$\hat{x}(k) = \frac{\hat{b}_1 z^{-1}}{1 + \hat{a}_1 z^{-1} + \hat{a}_2 z^{-2}} u(k) \qquad (13.9)$$

where,

$$\hat{a}_1 = -1.8432(0.0127); \hat{a}_2 = 0.8461(0.0127); \hat{b}_1 = 0.0736(0.0044).$$

This is the best identified model in terms of the YIC and explains 98.98% of the output data (R_T^2). Figure 13.10 compares the output of the model of equation 13.9 applied to the original *level* data: here the R_T^2 = 0.9979 and the model captures closely the main characteristics of the atmospheric CO_2 increase over the observation

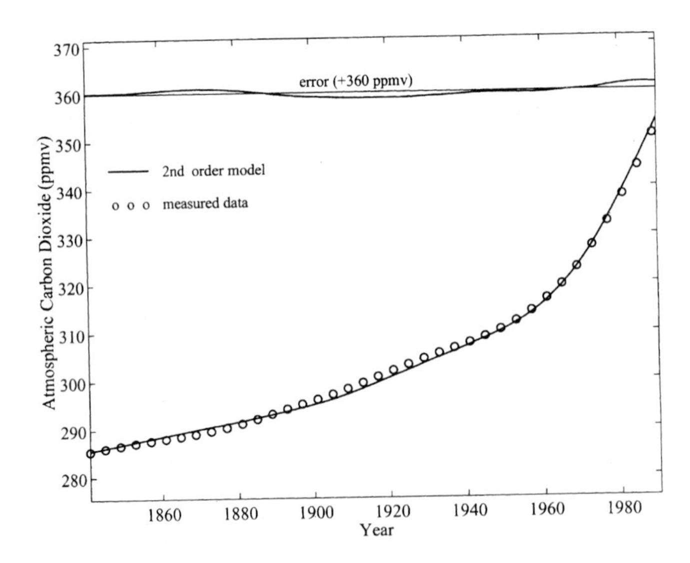

Fig. 13.10. DBM modelling: comparison of SRIV estimated model output with measured atmospheric carbon dioxide (1840–1990) (error +360 ppmv shown above).

period. Unlike the reduced order TF of equation 13.8 obtained from the perturbation experiment with the full nonlinear model, however, this model can be decomposed only into the following *serial* connection of two first order processes,

$$\hat{x}(k) = \frac{0.271z^{-1}}{1+0.9786z^{-1}} \cdot \frac{0.271z^{-1}}{1+0.8647z^{-1}} u(k) \tag{13.10}$$

where the steady-state gains and residence times are given by,

$$G_1 = 0.271/(1–0.9786) = 12.66;\ T_1 = –1/\log_e(0.9786) = 46.2 \text{ years;}$$

$$G_2 = 0.271/(1–0.8647) = 2.0;\ T_2 = –1/\log_e(0.8647) = 6.9 \text{ years.}$$

There is clearly considerable difference between this TF model and the linear reduced-order approximation of the full nonlinear simulation model. The parallel structure of the latter contrasts with the serial connection of the former; there is no apparent evidence of a large time constant in equation 13.10; and the integrator which dominates equation 13.8 is not present at all in 13.10, implying non-conservativity. Of course, these results are not really surprising: given the relatively short time series on which the model is based, we might expect that, even if they are present, the longer-term dynamic characteristics will not be clearly identifiable.

So which model should we believe? Many scientists might well dismiss equation 13.10 simply as a "black box" representation devoid of physical meaning. A minority might point out that, although the full nonlinear model has much clearer physical meaning, it is quite speculative in nature and, in contrast to 13.10, it has not been estimated rigorously in statistical terms from the observational data. We believe, however, that the correct attitude is to see both models as useful steps in our attempt to understand a complex physical system. Neither is to be totally accepted or rejected, since they both represent models of a very uncertain physical system.

To emphasise the uncertain nature of the model of equation 13.10, let us consider it further in stochastic terms, based on the statistical properties of the parameter estimates and concentrating on its use in a predictive context. Perhaps rather surprisingly, its prediction of immediate future mean variations in atmospheric CO_2, under a similar future fossil fuel input scenario, is not that different from that obtained from the nonlinear model and shown in Figure 13.6. Figures 13.11 and 13.12 take this analysis a little further by considering the model's prediction of future atmospheric CO_2 variations based on a very hypothetical and optimistic scenario of future fossil fuel utilisation. Although such analysis is highly speculative and not meant to be taken too seriously, it does help to raise certain important points about the deterministic and stochastic aspects of the TF model of equation 13.10.

Figure 13.11 shows the basic deterministic simulation and prediction: here, the model input $u(k)$ is based on the historic fossil fuel series between 1840 and 1990 but, after 1990, the input is varied as shown in the lower graph of Figure 13.11, so that the

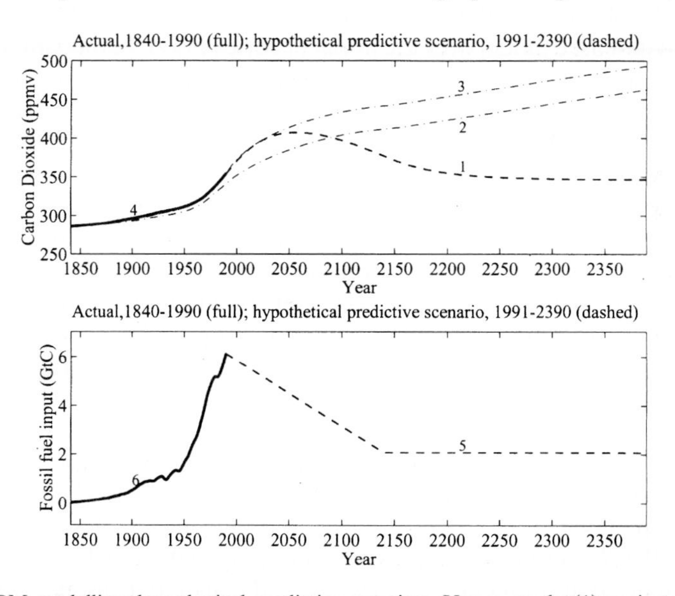

Fig. 13.11. DBM modelling: hypothetical predictive exercises. Upper graph: (1) projected changes in atmospheric carbon dioxide for hypothetical, optimistic future changes in fossil fuel input; (2) unconstrained reduced order model projection for same input scenario (dash-dot); (3) reduced order model projection for same input scenario but constrained to the measured data for period 1840–1990 (dash-dot). Note: in (2) and (3) the linear reduced order model response is virtually the same as the response of the full nonlinear model; (4) atmospheric CO_2 data (1840–1990) shown as heavy line. Lower graph: (5) optimistic future scenario for fossil fuel (1990–2390) shown dashed; (6) fossil fuel input (1840–1990) shown as heavy line.

predicted CO_2 output eventually returns to an equilibrium value similar to that applying in 1990. Put simply, we have developed a (politically and practically impossible) control strategy for fossil fuel usage which, on the basis of the deterministic part of the model of equation 13.10, would stabilise atmospheric CO_2 concentrations to present levels by about the year 2300, following an unavoidable transient rise to over 400 ppmv in the intervening period (as an echo of the projections depicted earlier in Figure 2.3 of Chapter 2).

This is very speculative, of course, but the results are interesting, particularly if they are compared with those obtained under a similar input scenario with the linear, reduced-order model of equation 13.8, as shown by dash–dot lines in Figure 13.11. Here, curve 2 is the basic simulation result from 1840, whereas in curve 3, the output is constrained to be equal to the measured atmospheric CO_2 over the initial historic period 1840–1990. Note that almost identical curves to these are obtained from the full nonlinear model simulation, once again showing the efficacy of the model linearisation and reduction analysis. Two curves are shown here because the basic unconstrained result (curve 2) is biased downwards because of two facts. First, the "land-use change" effects are not included in the simulation and, without these, the CO_2 fertilisation depresses the CO_2 concentration below the observations. Second, the reduced-order model (and its full nonlinear parent) have not been estimated in relation to the observational data and so do not, in any case, fit the observations as well as the data-based model over the initial historic period. In this situation, curve 3 provides a better idea of the simulation model prediction from the conditions observed in 1990.

Curves 2 and 3 in Figure 13.11 also expose the fundamental differences between the data-based model of equation 13.10 and the simulation models (both the full nonlinear model and its reduced-order approximation). In particular, it is clear that the presence of the additional, longer-term, dominant modes in the latter models leads to considerably different predictions in the period from about 2030 onwards. The second-order, data-based model reacts quite rapidly (curve 1) to the reduction in fossil fuel input (its maximum time constant is only 46.2 years and it has no integration element). In contrast, the simulation model outputs merely continue with decreasing growth rates (curve 3) until the predicted atmospheric CO_2 rises to a level of almost 500 ppmv by 2390. This helps to emphasise that, unlike the simulation models, the simple data-based transfer function (13.10) is not conservative. In other words, based on the limited historic data in Figure 13.9, carbon is being lost from the system in some manner and there is no significant evidence of any very long-term dynamics in the system. Of course, this does *not* mean that these dynamics are absent; rather it implies that if long-term behaviour and complete conservativity are to be included in the physically-based simulation models, then we must be very circumspect about any longer term predictions, particularly if they are computed on a deterministic basis, as in Figure 13.11.

As we have seen in previous sub-sections, such deterministic analysis can be quite misleading, since the propagation of the assumed uncertainty in the model parameters often results in large predictive uncertainties. In the present data-based

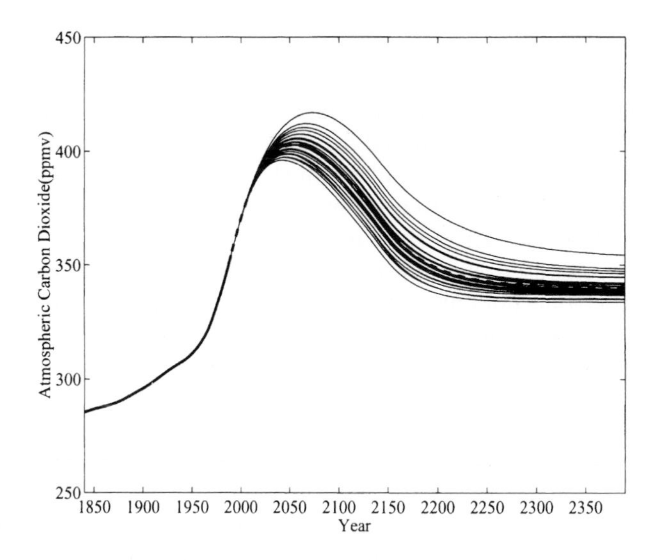

Fig. 13.12. DBM modelling: ensemble of the 30 Monte Carlo predictions based on SRIV estimated second order model (1990–2390). Data plotted as heavy line prior to 1990. Prediction based on nominal (mean) parameter estimates plotted as heavy, dashed line after 1990.

context, this becomes clear if we take note of the estimation statistics and consider the model predictions in stochastic terms, as illustrated in Figure 13.12. This shows the MCS results for the second-order model of equation 13.10 using the same scenario as Figure 13.11, but with the uncertainty in the model parameter estimates, as defined by the covariance matrix obtained from SRIV estimation, used to generate the random realisations.

Naturally, given the relative naïvety of this simple data-based analysis of the fossil fuel–CO_2 time series, it is not possible to draw any significant conclusions from the exercise. But it does raise some important questions about climate modelling and the nature of current GCC models. First, it is clear that the information content of the available data will only support a second order linear model, so that any more complex, nonlinear model based on these same data should be used with considerable circumspection. Secondly, and perhaps more controversially, it makes us question the assumption of conservativity, as implemented in the EL model.

No one believes that carbon is actually being lost from the system but one presumes it can be immobilised for extremely long periods of time such that, to all intents and purposes, it plays little part in the observed shorter-term dynamic behaviour of atmospheric CO_2 that is of interest to us here. On the other hand, the EL model suggests that no losses are occurring and so, in order to *stabilise* the atmospheric CO_2 level using this model, it is necessary to cease completely all fossil fuel-related emissions, otherwise the concentration will continue to rise. And in order to *reduce* the concentrations below current levels, we would have to actually remove carbon from the system in some manner!

13.5 CONCLUSIONS

Given the major advances in computer hardware and software that have occurred over the past decade, computer-based dynamic models are becoming progressively easier to construct. Environmental scientists have been particularly adept at employing these advances to successfully develop large, nonlinear, environmental simulation models which reflect their perception of the environment as a complex dynamic system composed of many, interacting sub-systems. In this paper, we have drawn attention to some limitations of the deterministic, reductionist philosophy that characterises this type of modelling. And we have argued for a more discerning, stochastic approach that exploits advanced statistical and systems methods to enhance our understanding of large environmental models.

This Data-Based Mechanistic (DBM) modelling strategy achieves these objectives in two major ways. First, by acknowledging explicitly the uncertainty that characterises our knowledge of the environment; and second, by attempting to extract some dynamic simplicity from the apparently complex, high-order dynamics of the large environmental simulation models. Of course, we are not claiming that the modelling methodology underlying this approach is fully developed. For instance, there is a need for further research on model reduction techniques, particularly for application to large multivariable (multi-input, multi-output) and nonlinear models. Also, we have used one particular approach to sensitivity analysis but there has been considerable research on this topic in recent years and other techniques (see e.g. Saltelli et al., 2000) may well yield superior results. Yet another topic that requires further attention is MCS analysis and DBM modelling applied to the distributed-parameter (partial differential equation) and other spatio-temporal models that are widely used in environmental science. By raising these important issues, however, we hope that this chapter will help to stimulate such research in the future. To this end, most of the analysis described here has been carried out using numerical methods that are available in our MATLAB *CAPTAIN* Toolbox (see http://www.es.lancs.ac.uk/cres/captain/), as well as other tools in MATLAB.

Our message may be controversial, perhaps even provocative to some. But it is not *intended* to provoke antagonism: on the contrary, we hope this chapter will promote constructive discussion on these matters. Certainly, there should be no doubt in the reader's mind that we believe strongly in the value of physically-based simulation models, as revealed by the case study we have chosen to discuss. Indeed, we see such models as an essential aspect of modern scientific endeavour. Above all, and perhaps more so than in any other chapter in this monograph, interpretation of the case study depends essentially on being able to approach the given problem from a variety of perspectives – with both low-order models (LOMs) and high-order models (HOMs). When seen in its broader context, this chapter marks something of a pivotal point in the approach now being set out in this part of the monograph. It is preceded by two chapters with a focus towards the use of HOMs, and it will be followed by two chapters whose focus is towards the use of LOMs.

This having been said, however, we cannot accept that deterministic simulation models are the only ones that are scientifically credible (Shackley et al., 1998); or even that they are entirely appropriate in the face of the uncertainty that characterises our understanding of environmental systems. Rather we favour a more catholic approach; one in which the objectives of the modelling exercise are carefully defined and models are selected objectively to satisfy these objectives. In general, a judiciously chosen *mixture* of several models will be called for, as illustrated by our case study of the global carbon cycle.

As pointed out previously, all DBM-type models can be converted to an on-line form. In this context, it sometimes makes sense to formalise the concept of using several models in what have been called Bayesian 'multi-process' models (originally suggested by Harrison and Stevens in 1971: see West and Harrison, 1989). Here, it is assumed that no single model can be uniquely defined for a real system for all time, perhaps because of nonstationarity, but also arising from problems such as identifiability and non-uniqueness. Consequently, multiple feasible models of the system are continually processed in parallel and mixed for inferences with respect to their estimated prior or posterior probabilities. This idea is appealing, particularly for someone with a Bayesian turn-of-mind, and so it has received continuing attention over the past 30 years.

The multi-process idea is mainly applicable to the large, over-parameterised simulation models that suffer from lack of uniqueness problems, such as the global carbon cycle models discussed in this chapter. In one sense, the multi-process model is ideal for those scientists and engineers who recognise the problems of deterministic reductionism but, for various reasons, want to retain the large models despite these problems. For instance, recognising over-parameterisation as a severe problem in hydrological modelling, and referring to such lack of uniqueness as "equifinality", Beven and his co-workers (see e.g. Beven et al., 2000 and the references therein) have recently reincarnated the multi-process idea. They utilise it in association with their Bayesian, Generalised Likelihood Uncertainty Estimation (GLUE) technique and apply this to hydrological catchment models. Of course, the same idea can be used with DBM models: for example, nonstationary models of this type will need to be adapted over time and a multi-process formulation provides one means of allowing for this. On the other hand, standard recursive estimation and on-line updating within a Kalman filter context, or the nonlinear equivalent of this, seems a more attractive proposition.

In this chapter, DBM ideas have been applied mainly to the analysis of the large EL global carbon cycle model. The modelling from real data discussed in Section 13.4.9 is introduced simply as an illustrative and rather 'tongue-in-cheek' example: it certainly cannot be considered as true DBM modelling. It must be emphasised, therefore, that the DBM modelling strategy has been applied successfully in many other practical contexts. These range from the BOD-DO modelling of Beck and Young (1975), through the *Aggregated Dead Zone* (ADZ) characterisation of solute transport in rivers (Beer and Young, 1983; Wallis et al., 1989), to the variety of examples outlined in Young (1998).

An excellent recent example of practical DBM modelling is a study of mass and energy transfer in agricultural buildings (Price et al., 1999). Here, the dominant modal dynamics are identified from the experimental data and the identifiable parameters in this second order differential equation model are estimated by the continuous-time SRIV algorithm. This model is then interpreted, in a physically meaningful manner, as a feedback process that can be compared directly with equations obtained from standard heat transfer theory. However, unlike those in the DBM model, the parameters of this conventional model are not identifiable from the data. In other words, the DBM modelling has provided an *identifiable* formulation of heat transfer theory that can replace the conventional, *non-identifiable* formulation in applications such as this.

To conclude, in this chapter we have tried to demonstrate the merits of DBM modelling and statistical methods in all aspects of environmental systems analysis. When experimental data are available, such an approach ensures that the identifiable modes of dynamic behaviour are defined and estimated objectively in the most parsimonious manner. However, when data are scarce, so that complex simulation modelling may be called for, these same techniques allow for the identification of the dominant dynamic modes of the *model*, so facilitating linearisation and order reduction. In both cases, simplicity is extracted from apparent complexity and the scientist is able to understand more clearly the dynamic nature of the system under study at the scale that is most relevant to the experimental data.

ACKNOWLEDGEMENT

An earlier version of this chapter has been published as:

Young, P.C., Parkinson, S. and Lees, M.J., 1996. Simplicity out of complexity: Occam's razor revisited. *J. Appl. Statistics*, **23**, 165–210.

Where such material is reproduced here, it is done so with permission. The original source can be found through the website of the *Journal of Applied Statistics* at http://www.tandf.co.uk.

REFERENCES

Akaike, H., 1974. A new look at statistical model identification. *IEEE Trans. Auto. Control*, **AC-19**, 716–723.

Beck, M.B., 1979. Model structure identification from experimental data. In: *Theoretical Systems Ecology* (E. Halfon, ed.). Academic Press, New York, pp. 259–289.

Beck, M.B., 1983. Uncertainty, system identification and the prediction of water quality. In: *Uncertainty and Forecasting of Water Quality* (M.B. Beck and G. van Straten, eds.). Springer-Verlag, Berlin, pp. 3–68.

Beck, M.B., 1987. Water quality modeling: a review of the analysis of uncertainty. *Water Resour. Res.*, **23**(8), 1393–1442.

Beck, M.B. and Halfon, E., 1991. Uncertainty, identifiability and the propagation of prediction errors: a case study of Lake Ontario. *J. Forecasting*, **10**(1&2), 135–161.

Beck, M.B. and Young, P.C., 1975. A dynamic model for BOD-DO relationships in a non-tidal stream. *Water Res.*, **9**, 769–776.

Beer, T. and Young, P.C., 1983. Longitudinal dispersion in natural streams. *Proc. Am. Soc. Civil Eng., Jnl. Env. Eng.*, **109**, 1049–1067.

Beven, K.J., 1993. Prophecy, reality and uncertainty in distributed hydrological modelling. *Adv. Water Resour.*, **16**, 41–51.

Beven, K.J., Freer, J., Hankin, B. and Schulz, K., 2000. The use of generalised likelihood measures for uncertainty estimation in high order models of environmental systems. In: *Nonlinear and Nonstationary Signal Processing* (W.J. Fitzgerald, A. Walden, R. Smith and P.C. Young, eds.). Cambridge University Press, Cambridge, pp. 115–151.

Beven, K.J. and Binley, A., 1992. The future of distributed models: model calibration and uncertainty prediction. *Hydrol. Proc.*, **6**, 279–298.

Box, G.E.P. and Jenkins, G.M., 1970. *Time Series Analysis Forecasting and Control*. Holden-Day, San Francisco.

CDIAC, 1991. *Trends 91: A Compendium of Global Change* (T.A. Boden, R.J. Serpanski and M.P. Farrell, eds.). Carbon Dioxide Information and Analysis Center, Oak Ridge, Tennessee.

Enting, I.G., 1991. Calculating Future Atmospheric CO_2 Concentrations. Division of Atmospheric Research, Technical Paper 22. CSIRO, Melbourne, Australia.

Enting, I.G., 1994. Personal Communication.

Enting, I.G. and Pearman, G.I., 1983. Refinements to a One-Dimensional Carbon Cycle Model, Division of Atmospheric Research, Technical Paper 3. CSIRO, Melbourne, Australia.

Enting, I.G. and Lassey, K.R., 1993. Projections of Future CO_2. Division of Atmospheric Research, Technical Paper 27. CSIRO, Melbourne, Australia.

Gamerman, D., 1997. *Markov Chain Monte Carlo: Stochastic Simulation for Bayesian Inference*. Chapman and Hall, London.

Goodwin, G.C. and Payne, R.L., 1977. *Dynamic System Identification: Experiment Design and Data Analysis*. Academic Press, New York.

Hasselmann, K., 1998. Climate–change research after Kyoto. *Nature*, **390**, 225–226.

Hasselmann, K., Hasselmann, S., Giering, R., Ocana, V. and Storch, H. v., 1997. Sensitivity study of optimal emission paths using a simplified structural integrated assessment (SIAM). *Climate Change*, **37**, 345–386.

Hornberger, G.M. and Spear, R.C., 1981. An approach to the preliminary analysis of environmental systems. *J. Environ. Manage.*, **12**, 7–18

Huggett, R.J., 1993. *Modelling the Human Impact on Nature: Systems Analysis of Environmental Problems*. Oxford University Press, Oxford.

IPCC, 1990. *Climate Change: The IPCC Scientific Assessment*, J.T. Houghton, G.J. Jenkins, and J.J. Ephraums (eds. for the Intergovernmental Panel on Climate Change), Cambridge University Press, Cambridge.

IPCC, 1992. *Climate Change 1992: The Supplementary Report to the IPCC Scientific Assessment*, J.T. Houghton, B.A. Callender and S.K. Varley (eds. for the Intergovernmental Panel on Climate Change), Cambridge University Press: Cambridge.

Joos, F., Bruno, M., Fink, R., Siegenthaler, U., Stocker, T.F., Le Quere, C. and Sarmiento, J.L., 1996. An efficient and accurate representation of complex oceanic and biospheric models of anthropogenic carbon uptake. *Tellus*, **48B**, 397–417.

Kailath, T., 1980. *Linear Systems*. Prentice Hall, New York.

Keesman, K.J. and van Straten, G. (1990) Set-membership approach to identification and prediction of lake eutrophication. *Water Resour. Res.*, **26**(11), 2643–2652.

Konikow, L.F. and Bredehoeft, J.D., 1992. Ground water models cannot be validated. *Adv. Water Resources*, **15**, 75–83.

Lees, M.J., Young, P.C., Ferguson, S., Beven, K.J. and Burns, J., 1994. An adaptive flood warning scheme for the River Nith at Dumfries. In: *Second International Conference on River Flood Hydraulics* (W.R. White and J. Watts, eds.). Wiley, Chichester, pp. 65–77.

Ljung, L. and T. Söderström, 1983. *Theory and Practice of Recursive Estimation*. MIT Press, Cambridge, Mass.

Mathworks, Inc., 1994. *MATLAB* Reference Guide. Mathworks, Natick, MA

Moore, B., 1981. Principle component analysis in linear systems: controllability, observability, and model reduction. *IEEE Trans. Automatic Contr.*, **26**, 17–32.

Oeschger, H., Siegenthaler, U., Schottere, U. and Gugelmann, A., 1975. A box diffusion model to study the carbon dioxide exchange in nature. *Tellus*, **27**, 168–192.

Oglesby, R.J. and Saltzman, B., 1990. Sensitivity to the equilibrium surface temperature of a GCM to systematic changes in atmospheric carbon dioxide. *Geophys. Res. Lett.*, **17**, 1089–1092.

Oreskes, N., Shrader-Frechette, K. and Belitz, K., 1994. Verification, validation, and confirmation of numerical models in the earth sciences. *Science*, **263**, 641–646.

Osidele, O.O. and Beck, M.B., 2001. Analysis of uncertainty in model predictions for Lake Lanier, Georgia. In: *Proceedngs AWRA Annual Spring Specialty Conference* (J.J. Warwick, ed.). TPS-01-1, American Water Resources Association, Middleburg, Virginia, pp. 133–137.

Parkinson, S., 1995. The Application of Stochastic Modelling Techniques to Global Climate Change, Ph.D Thesis, Lancaster University, U.K.

Parkinson, S. and Young, P.C., 1998. Uncertainty and sensitivity in global carbon cycle modelling. *Climate Res.*, **9**, 157–174.

Philip, J.R., 1975. Some remarks on science and catchment prediction. In: *Prediction in Catchment Hydrology* (T.G. Chapman and F.X. Dunin, eds.). Australian Academy of Science, Canberra, Australia.

Price, L.E., Young, P.C., Berckmans, D., Janssens, K. and Taylor, J., 1999. Data-based mechanistic modelling and control of mass and energy transfer in agricultural buildings. *Annual Rev. Contr.*, **23**, 71–82.

Saltelli A., Chan, K. and Scott, M., eds., 2000. *Sensitivity Analysis*. Wiley, Chichester.

Shackley, S., Young, P.C., Parkinson, S. and Wynne, B., 1998. Uncertainty, complexity and concepts of good science in climate change modelling: are GCMs the best tools? *Climatic Change*, **38**, 159–205.

Spear, R.C., 1970. The application of Kolmogorov–Renyi statistics to problems of parameter uncertainty in systems design. *Int. J. Contr.*, **11**(5), 771–778.

Spear, R.C., 1993. Regional sensitivity analysis in environmental systems. In: *Concise Encyclopaedia of Environmental Systems* (P.C. Young, ed.). Pergamon Press, Oxford, pp. 476–479.

Taylor, J.A., 1996. Fossil fuel emissions required to achieve stabilisation using ANU-BACE: a box-diffusion carbon cycle model. *Ecol. Modelling*, **86**, 195–199.

Wallis, S.G., Young, P.C. and Beven, K.J., 1989. Experimental investigation of the aggregated dead zone model for longitudinal solute transport in stream channels. *Proc. Inst. Civil Engrs, Part 2*, **87**, 1–22.

Wellstead, P.E., 1978. An instrumental product moment test for model order estimation.

Automatica, **14**, 89–91.
West, M. and Harrison, J., 1989. *Bayesian Forecasting and Dynamic Models*. Springer-Verlag, New York.
Wigley, T.M.L. and Raper, S.C.B., 1992. Implications for climate and sea level of revised IPCC emissions scenarios. *Nature*, **357**, 293–300.
Young, P.C., 1978. A general theory of modelling for badly defined dynamic systems. In: *Modeling, Identification and Control in Environmental Systems* (G.C. Vansteenkiste, ed.). North-Holland, Amsterdam, pp. 103–135.
Young. P.C., 1983. The validity and credibility of models for badly defined systems. In: *Uncertainty and Forecasting of Water Quality* (M.B. Beck and G. van Straten, eds.). Springer-Verlag, Berlin.
Young, P.C., 1984. *Recursive Estimation and Time-Series Analysis*. Springer-Verlag, Berlin.
Young, P.C., 1985. The instrumental variable method: a practical approach to identification and system parameter estimation. In: *Identification and System Parameter Estimation: Vol. 1 & 2* (H.A. Barker and P.C. Young, eds.). Pergamon Press, Oxford.
Young, P.C., 1989. Recursive estimation, forecasting and adaptive control. In: *Control and Dynamic Systems* (C.T. Leondes, ed.). Academic Press, San Diego, pp. 119–166.
Young, P.C., 1992a. Parallel processes in hydrology and water quality: a unified time series approach. *J. Inst. Water Environ. Manage.*, **6**, 598–612.
Young, P.C., 1992b. Parallel processes in hydrology and water quality: objective inference from hydrological data. Chapter 2 in: *Water Quality Modelling* (R.A. Falconer, ed.). Ashgate, Vermont, pp. 10–52.
Young, P.C., 1993a. Time variable and state dependent modelling of nonstationary and nonlinear time series. Chapter 26 in: *Developments in Time Series Analysis* (T. Subba Rao, ed.). Chapman and Hall, London, pp. 374–413.
Young, P.C., ed., 1993b. *Concise Encyclopaedia of Environmental Systems*. Pergamon Press, Oxford.
Young, P.C., 1998. Data-based mechanistic modelling of environmental, ecological, economic and engineering systems. *Environ. Modelling and Software*, **13**, 105–122.
Young, P.C., 1999a. Nonstationary time series analysis and forecasting. *Progr. Environ. Sci.*, **1**, 3–48.
Young, P.C., 1999b. Data-based mechanistic modelling, generalised sensitivity and dominant mode analysis. *Computer Phys. Commun.*, **117**, 113–129.
Young, P.C., 2000. Stochastic, dynamic modelling and signal processing: time variable and state dependent parameter estimation. In: *Nonlinear and Nonstationary Signal Processing* (W.J. Fitzgerald, A. Walden, R. Smith and P.C. Young, eds.). Cambridge University Press, Cambridge, pp. 74–114.
Young, P.C., 2001. Data-based mechanistic modelling and validation of rainfall-flow processes. In: *Model Validation: Perspectives in Hydrological Science* (M.G. Anderson and P.D. Bates, eds.). Wiley, Chichester, pp. 117–161.
Young, P.C. and Beven, K.J., 1994. Data-based mechanistic modelling and the rainfall-flow nonlinearity (special issue on Environmental Time Series Analysis). *Environmetrics*, **5**, 335–363
Young, P.C. and Lees, M., 1992. The active mixing volume: a new concept in modelling environmental systems. In: *Statistics and the Environment* (V. Barnett and R. Turkman, eds.). Wiley, Chichester, pp. 3–43.
Young, P.C. and Minchin, P.E.H., 1991. Environmetric time-series analysis: modelling natural systems from experimental time-series data. *Int. J. Biol. Macromolecules,* **13**, 190–201.

Young, P.C. and Tomlin, C., 2000. Data-based mechanistic modelling and adaptive flow forecasting. In: *Flood Forecasting: What Does Current Research Offer the Practitioner?* (M.J. Lees and P. Walsh, eds.). British Hydrological Society (BHS) Occasional Paper No. 12, Centre for Ecology and Hydrology on behalf of BHS, Wallingford, U.K., pp. 26–40.

Young, P.C., Chotai, A. and Tych, W., 1991. Identification, estimation and control of continuous-time systems described by delta operator models. In: *Identification of Continuous-Time Systems* (N.K. Sinha and G.P. Rao, eds.). Kluwer, Dordrecht, pp. 363–418.

Young, P.C., Jakeman, A.J. and McMurtrie, R., 1980. An instrumental variable method for model order identification. *Automatica,* **16**, 281–294.

Young, P.C., Parkinson, S. and Lees, M.J., 1996. Simplicity out of complexity: Occam's razor revisited. *J. Appl. Statistics*, **23**, 165–210.

Young, P.C., Parkinson, S. and McIlveen, J.F.R., 1993. Computer simulation. In: *Concise Encyclopedia of Environmental Systems* (P.C. Young, ed.). Pergamon, Oxford, pp. 113–118.

Young, P. C., Spear, R. C. and Hornberger, G. M. (1978) Modeling badly defined systems: some further thoughts, Proc. SIMSIG Conference, Canberra, Australia.

Young, P.C., Behzadi, M.A., Wang, C.L. and Chotai, A., 1987. Direct digital and adaptive control by input–output, state variable feedback. *Int. J. Control*, **46**, 1861–1881.

APPENDIX 1: SRIV Identification and Estimation

This Appendix describes briefly the SRIV method for Single–Input, Single–Output (SISO) systems and outlines the associated identification and optimisation procedures referred to in the main text. Note that very similar SRIV identification and estimation procedures are available for both continuous-time differential equation models, and for discrete-time delta operator models (see e.g. Young et al., 1991), as well as for multiple-input single output (MISO) and full multiple-input, multiple-output (MIMO) models.

A1.1 The SRIV Method

The Simplified Refined Instrumental Variable (SRIV) method of identification and estimation for constant parameter TF models of stochastic SISO systems (Young, 1985) is applied to either the model (13.3), for simulation model linearisation and order reduction; or to the model (13.6) for DBM modelling. In both cases, it is assumed that the SISO model takes the following general form, where the *ij* subscripts have been omitted for convenience,

$$y(k) = \frac{B(z^{-1})}{A(z^{-1})} u(k) + e(k) \qquad (13.A1)$$

The associated estimation model can be written in the alternative vector form,

$$y(k) = z(k)^T \boldsymbol{a} + \eta(k); \; k = 1,2,\ldots,N \qquad (13.A2)$$

where $\eta(k) = A(z^{-1})e(k)$; while $y(k)$, $z(k)$ $e(k)$ and $\boldsymbol{a}$ are defined accordingly. For instance, in the case of the model (13.3),

$$y(k) = y_j(k);\; z(k)^T = [-y_j(k-1) - y_j(k-2) \ldots -y_j(k-nij)\; u_i(k) \ldots u_i(k-mij)]^T;$$

$$e(k) = \varepsilon(k);\; \boldsymbol{a}(k) = [a_{1ij}\, a_{2ij} \ldots a_{nij}\, b_{0ij} \ldots b_{mij}]^T;$$

while for model (13.6) the definitions are the same except for the fact that $y_j(k)$ and $u_i(k)$ now represent the time series data obtained from experiments or monitoring exercises on the *real system* rather than the *simulation model*; and the noise $e(k)$ is the associated measurement noise $\xi(k)$ on these real data.

For a sample size N, the non-recursive (*en-bloc*) SRIV estimate $\hat{\boldsymbol{a}}$ of the parameter vector $\boldsymbol{a}$ is obtained by the solution of the following "IV normal equations",

$$\left[\sum_{k=1}^{k=N} \hat{\boldsymbol{x}}^*(k) \boldsymbol{z}^*(k)^T\right] \hat{\boldsymbol{a}} = \sum_{k=1}^{k=N} \hat{\boldsymbol{x}}^*(k) y^*(k) \tag{13.A3}$$

These are a modification of the associated and well known least squares normal equations for the same model. For each input–output pair, the data vectors in (13.A3) are defined as follows, where again the ij subscripts have been omitted for convenience,

$$\begin{aligned} z^*(k)^T &= [-y^*(k-1) - y^*(k-2) \cdots - y^*(k-n) u^*(k) \cdots u^*(k-m)]; \\ \hat{\boldsymbol{x}}^*(k)^T &= [\hat{x}^*(k-1) \hat{x}^*(k-2) \cdots \hat{x}^*(k-n) u^*(k) \cdots u^*(k-m)]. \end{aligned} \tag{13.A4}$$

Here, $\hat{x}(k)$ is the *instrumental variable*, defined as an estimate of the "noise free" system output and obtained from the following adaptive "auxiliary model",

$$\hat{x}(k) = \frac{\hat{B}(z^{-1})}{\hat{A}(z^{-1})} u(k) \tag{13.A5}$$

where the polynomials $\hat{A}(z^{-1})$ and $\hat{B}(z^{-1})$ are adaptive estimates of the TF model polynomials; and the star superscripts indicate that the associated variables are adaptively pre-filtered in the following manner,

$$y^*(k) = \frac{1}{\hat{A}(z^{-1})} y(k); \quad u^*(k) = \frac{1}{\hat{A}(z^{-1})} u(k); \quad \hat{x}^*(k) = \frac{1}{\hat{A}(z^{-1})} \hat{x}(k) \tag{13.A6}$$

The adaption of both the auxiliary model and prefilters is performed within a three-step iterative (or relaxation) procedure (Young, 1984, 1985 and the references therein). Subsequent to the final iteration, the following two matrices are generated, in addition to the estimated parameter vector $\hat{\boldsymbol{a}}(N)$:

(i) The inverse of the Instrumental Product Matrix (IPM),

$$\hat{P}(N) = \left[\sum_{k=1}^{k=N} \hat{\boldsymbol{x}}^*(k) \boldsymbol{z}^*(k)^T\right]^{-1}$$

(ii) The Covariance Matrix,

$$P^*(N) = \hat{\sigma}^2 \left[\sum_{k=1}^{k=N} \hat{x}^*(k)\hat{x}^*(k)^T \right]^{-1}$$

where $\hat{\sigma}^2$ is the variance of the model residuals $\hat{e}(k)$, i.e.,

$$\hat{e}(k) = y(k) - \hat{x}(k); \hat{\sigma}^2 = \frac{1}{N} \sum_{k=1}^{k=N} \hat{e}(k)^2 \tag{13.A7}$$

The matrix inverse in the definition of $P^*(N)$ is generated separately after estimation is complete, with the required $\hat{x}^*(k)$ and $u^*(k)$ variables obtained from the auxiliary model and prefiltering operations based on the final iteration estimate of $\hat{\boldsymbol{a}}(N)$. In the case of Gaussian white residuals, it can be shown that $P^*(N)$ is an estimate of the covariance matrix associated with the parameter estimate vector $\hat{\boldsymbol{a}}(N)$ obtained at the final iteration. Consequently the square root of its diagonal elements provide an estimate of the standard error on the elements of $\hat{\boldsymbol{a}}(N)$. Strictly, these results do not apply in other more general circumstances but MCS analysis indicates that $P^*(N)$ provides a useful conservative measure of the uncertainty on the SRIV parameter estimates. Finally, $\hat{P}(N)$ is utilised to generate the YIC criterion, as discussed below in A1.2. The recursive-iterative version of this SRIV estimation procedure follows straightforwardly from the above non-recursive equations (see e.g Young, 1985) and this can be used in situations where the model parameters may change over the observation interval. Here, the estimate $\hat{\boldsymbol{a}}(k)$ of the parameter vector $\boldsymbol{a}$ at the kth sample is updated as a function of its prior estimate $\hat{\boldsymbol{a}}(k-1)$ at the (k–1)th sample and the latest prediction error, so allowing for parametric change between samples. Also, such recursive algorithms can be converted into a Fixed Interval Smoothing (FIS) form, where the FIS estimate $\hat{\boldsymbol{a}}(k/N)$ at the kth sample is obtained in a two-pass recursive operation (forward filtering and backward smoothing), so that it is based on all N samples of the available data (see Young, 1999a). If necessary, such FIS algorithms can be used to obtain non-parametric estimates of nonlinearities during the identification stage of nonlinear DBM models (see Young, 2000 and the prior references therein). These can then be parameterised (e.g. in a state dependent parameter form) prior to their full parametric estimation within the identified structure of the nonlinear dynamic model.

Note finally that, if the noise $e(k)$ is not white but can be assumed to have rational spectral density and follow an AutoRegressive Moving Average (ARMA) process, then the related but more complex Refined Instrumental Variable (RIV) algorithm is optimal and can be used instead. In more general situations, the IV aspect of the analysis ensures that the simpler SRIV estimate $\hat{\boldsymbol{a}}(N)$ is always consistent, provided the IV assumptions are applicable (e.g. Young, 1984). Moreover, experience over many years suggests that the SRIV algorithm is very robust in practice and often yields good results when the more sophisticated RIV algorithm (and other related optimal algorithms, such as the ML approach of Box–Jenkins (1970) or the Prediction Error Minimisation (PEM) approach of Ljung and Söderström (1983)) fail to yield acceptable models because of their need to *simultaneously* estimate the noise process.

A1.2 Model Order Identification

Model order identification is based around the R_T^2, *YIC* and *AIC* criteria, which are defined as follows:

$$\text{(i) } R_T^2 = 1 - \frac{\hat{\sigma}^2}{\sigma_y^2};\quad \sigma_y^2 = \frac{1}{N}\sum_{k=1}^{k=N}[y(k)-\bar{y}]^2;\quad \bar{y} = \frac{1}{N}\sum_{k=1}^{k=N} y(k)$$

$$\text{(ii) } YIC = \log_e \frac{\hat{\sigma}^2}{\sigma_y^2} + \log_e\{NEVN\};\quad NEVN = \frac{1}{np}\sum_{i=1}^{i=np}\frac{\hat{\sigma}^2\cdot\hat{p}_{ii}}{\hat{a}_i^2}$$

$$\text{(iii) } AIC(np) = N\log_e \hat{\sigma}^2 + 2\cdot np$$

where $np = n + m + 1$ is the number of estimated parameters in the $\hat{\boldsymbol{a}}(N)$ vector; $\hat{p}_{ii}$ is the ith diagonal element of the $\hat{P}(N)$ matrix (so that $\hat{\sigma}^2\cdot\hat{p}_{ii}$ can be considered as an approximate estimate of the variance of the estimated uncertainty on the ith parameter estimate); and $\hat{a}_i^2$ is the square of the ith parameter in the $\hat{\boldsymbol{a}}(N)$ vector.

We see that the coefficient of determination R_T^2 is a statistical measure of how well the model explains the data. If the variance of the model residuals $\hat{\sigma}^2$ is low compared with the variance of the data σ_y^2, then R_T^2 tends towards unity; while if $\hat{\sigma}^2$ is of similar magnitude to σ_y^2 then it tends towards zero. Note, however, that R_T^2 is based on the variance of the model errors $\hat{e}(k)$ and it is *not* the more conventional coefficient of determination R^2 based on the variance of the one step ahead prediction errors. R_T^2 is a more discerning measure than R^2 for TF model identification: while it is often quite easy for a model to produce small one step ahead prediction errors, since the model prediction is based on past measured values of the output variable $y(k)$, it is far more difficult for it to yield small model response errors, where the model output is based only on the measured input variable $u(k)$ and does not refer to $y(k)$.

The YIC is a more complex, heuristic criterion. From the definition of R_T^2, we see that the first term is simply a relative measure of how well the model explains the data: the smaller the model residuals the more negative the term becomes. The second term, on the other hand, provides a measure of the conditioning of the IPM, which needs to be inverted when the IV normal equations (1.A2) are solved. If the model is over-parameterised, then it can be shown (e.g. Wellstead, 1978; Young et al., 1980) that the IPM will tend to singularity. And then, because of its ill-conditioning, the elements of the inverse $\hat{P}(N)$ will increase in value, often by several orders of magnitude. When this happens, the second term in the YIC tends to dominate the criterion function, indicating over-parameterisation. An alternative justification of the YIC can be obtained from statistical considerations (see e.g. Young, 1989). Although heuristic, the YIC has proven very useful in practical identification terms over the past ten years: it should not be used as a sole arbiter of model order, however, and improvements in its statistical definition are being researched.

The Akaike Information Criterion (AIC) is a well known identification criterion for purely stochastic AR processes (Akaike, 1974) and it is used here to identify the order of AR models for the noise process $\xi(k)$, based on the analysis of the model residuals $\hat{e}(k)$ (although it can sometimes assist in identifying the TF model order if used carefully). Here, the first term is a measure of how well the model explains the data; while the second term is simply a penalty on the number of parameters in the model. Thus, as in the YIC, the AIC seeks a compromise between the degree of model fit and the complexity of the model; i.e. it helps to ensure parsimony and simplicity in the overall TF model.

APPENDIX 2: The Enting–Lassey Global Carbon Cycle Model

This appendix presents an outline of the non-linear global carbon cycle model developed by Enting and Lassey (1993) on the basis of the earlier box diffusion model of Oeschger et al. (1975). The model has been developed by the present authors on two computer platforms: in the SIMULINK simulation program on standard SUN workstations and Macintosh/PC computers; and in 3L parallel PASCAL form on an IBM-PC with Transputer boards fitted. SIMULINK, which is an extension to the MATLAB™ numerical computation package, is particularly useful for simulation model development since it allows large dynamic models to be simulated using rather novel 'iconographic' programming techniques (see Young et al., 1993). However, the parallel computer is exceptionally well suited for MCS and provides the greatly reduced computation times required for such analysis.

Figure 13.1 in the main text shows a box diagram of the standard version of the EL model. The model attempts to explain the movement of total carbon, and its two minor isotopes, ^{13}C and ^{14}C, throughout the global environment on time scales of decades to centuries. The minor isotopes are included to take advantage of their role as tracers in the system.

The global environment, and therefore the EL model, can be considered as being composed of three main compartments: the atmosphere; the ocean; and the land-based, or terrestrial, biosphere. The ocean and the terrestrial biosphere exchange large amounts of carbon with the atmosphere, but little with each other. Before the start of the human industrial period (taken to be 1765 – IPCC, 1990) these exchanges are assumed to be in equilibrium. After this time, however, an input mainly due to fossil fuel combustion, enters the atmosphere and is dissipated throughout the system, although much remains in the atmosphere.

In order to try to model this effect accurately, as perceived by EL, the model is divided into 23 boxes as shown in Figure 13.1 (the deep ocean is composed of 18 boxes), each characterised by levels of total carbon, ^{13}C and ^{14}C. Three inputs are considered in the standard version of the model: two into the troposphere, total carbon and ^{13}C from fossil fuel emissions; and one into the stratosphere, ^{14}C, from nuclear weapons detonations. In addition, there is a natural source of ^{14}C into the stratosphere due to the action of cosmic rays on the edge of the atmosphere; and a natural sink for ^{14}C from all boxes due to radioactive decay.

The net exchange of total carbon between the atmosphere and the terrestrial biosphere is assumed in the model to be zero, i.e., at equilibrium, for all *natural* biospheric changes. Due to a fractionation process, however, which discriminates against the uptake of ^{13}C and ^{14}C, this is not the case for the heavier isotopes. Their exchange is mainly governed by the turnover time, τ_o, of the old biosphere.

The air–sea exchange is presumed to be proportional to Δp_{CO_2}, the difference between the CO_2 partial pressures of the atmosphere and the ocean surface layer. Thus

$$F_{am} = \kappa \Delta p_{CO_2} \quad (13.A8)$$

where F_{am} is the net flux between the atmosphere and the surface mixed layer and κ is the gas exchange coefficient. Calculation of the surface layer partial pressure is complicated by the fact that much of its carbon is suspended in ionic form and does not contribute to this partial pressure. This "buffering" action limits the rate at which the ocean can take up anthropogenic CO_2.

The deep ocean is modelled by 18 boxes, of equal depth. These represent a discrete approximation to the diffusion process, which is believed to dominate ocean circulation (Oeschger et al., 1975):

$$\frac{\partial c}{\partial t} = K \frac{\partial^2 c}{\partial z^2} \quad (13.A9)$$

where c is the concentration of a given substance, in this case, carbon; and K is the coefficient of diffusion in the direction given by the distance parameter z (in this case, vertically).

In order to try to model the global carbon cycle more accurately, Enting and Lassey have added five modifications to the standard form of the model:

1. downward flux of carbon in the ocean due to organic detrital movement;
2. downward flux of carbon in the ocean due to inorganic (carbonate) detrital movement;

3. using a non-linear relation between the buffer factor and the surface layer partial pressure, rather than a constant value;
4. adding a flux (based on observed data) from the old biosphere to the troposphere to account for land-use changes, e.g., deforestation;
5. adding a flux from the troposphere to the old biosphere to account for an uptake in carbon due to accelerated plant growth at higher CO_2 levels – the CO_2 fertilisation effect.[11]

The effect of these five modifications, both individually and in combination, on the atmospheric carbon levels is investigated in detail by Parkinson and Young (1998).

---

[11] The CO_2 fertilisation effect is a possible solution to the 'missing sink' problem in global carbon cycle modelling (IPCC, 1990, 1992).

Environmental Foresight and Models: A Manifesto
M.B. Beck (editor)

CHAPTER 14

Structural Effects of Landscape and Land Use on Streamflow Response

T.S. Kokkonen and A.J. Jakeman

14.1 INTRODUCTION

At present there are no credible models to predict the effect on hydrological response of land-use change in gauged catchments or of climate change in ungauged catchments. It seems to be increasingly accepted within the hydrological community that instead of developing complex process representations of hydrological response, more effort should be directed towards measurements of hydrological phenomena and landscape attributes that drive such phenomena (Goodrich and Woolhiser, 1991; Vertessy et al., 1993; and Chapter 12). Such efforts are required to further understanding of the mechanisms and controls of hydrological processes, and to provide estimates of water fluxes within the landscape that are frequently required to assess impacts of different management practices.

Improving understanding is an issue having more of a scientific nature, of course, and it can be pursued at a relatively small scale in experimental sites specifically established for research purposes. But even in extensively instrumented research sites hydrological systems present appreciable difficulties for modellers; in the wakes of Chapters 8 and 12, this will come as no surprise. These difficulties arise from the sheer complexity of internal system behaviour involving dynamic and multi-dimensional interactions that are physical, and possibly chemical and biological in character. In addition, such interactions vary spatially due to heterogeneity of the media where they occur and – of crucial significance for this monograph – often also temporally, reflecting changing dominant controls of the system (Jakeman et al., 1994), as we have already seen in Chapter 5. Compounding this complexity is our incapacity to measure internal system states as comprehensively and accurately as we would like and – again, as we have now seen from the arguments of the preceding chapter – to control the inputs in order to excite the system in a deliberate manner for the purposes of enhanced learning.

Providing estimates of water fluxes in the landscape is a more practical matter. Key scientific advances required for such problems include the capacity to predict hydrological fluxes in ungauged catchments and in response to land-use changes. In these cases, where the scale involved is usually much larger, the need for a sufficient amount of observations becomes even more critical. Remotely sensed data can provide some assistance here. It is foreseeable that acquisition of spatially distributed data will become easier and more affordable with constantly improving sensing techniques. However, remotely sensed data tend to represent only indirect measurements of the phenomena of interest, and thus sufficiently long records of on-site measurements of hydrological fluxes will continue to be needed. In short, currently, and in the near future, hydrologists are often faced with questions that need to be answered with only restricted amounts of data available.

When only limited data are at hand it is of paramount importance to deal with uncertainties inherent in the modelling results. To do this in an objective, statistically rigorous manner we argue in this chapter that the model applied to simulate the hydrological system should preferably be of a relatively simple structure. Statistical techniques may then be applied to determine whether the model has captured the dynamics of the system sufficiently well. In many disciplines, such as industrial control and economics, it is the norm for models to be statistically justifiable. In hydrology, largely due to the problems mentioned above, this kind of approach has gained only restricted advocacy.

A goal of hydrological modelling should be to capture, or to quantify, as parsimoniously as possible, the hydrological response of a catchment. Hydrological response is understood here as the manner in which a catchment reacts, in terms of streamflow volume and dynamics, to a specified climatic forcing. Thus it is a climate-independent or standardised property, and is a function of landscape and land-use characteristics. However, substantial climate dependency exists in the actual relationships between the factors driving the system (precipitation, evapotranspiration) and the output of the system (streamflow). This is due to two factors. Firstly, water storage within the catchment induces memory in the system, so that soil moisture conditions antecedent to a precipitation event have a significant effect on the resulting streamflow magnitudes. Secondly, there is a long-term effect of climate on catchment biota structure and function, the state of which will affect storage and dynamic response characteristics of the catchment (Eagleson, 1978). Such a long-term effect is not considered here.

Removing the climate signal from observed hydrological data enhances the likelihood of revealing relationships between the hydrological response of a catchment and its terrain and land-use properties. Herein, we adopt this strategy to separate out the different factors influencing the hydrological behaviour of a catchment. As a result we are able to explore in two case studies how catchment-scale runoff processes relate to (a) different terrain properties between catchments belonging to the same region and (b) land-use changes occurring through time in the same catchment.

Our approach differs from the more standard procedures in hydrological modelling, where all the anticipated controls are included in complex models *a priori*, and parameters subsequently estimated. Here, the emphasis is on identification of separate controls on hydrological response: climate, landscape, and land use. The knowledge gained at the identification step can then lead to more model complexity, if significant controls worth quantifying are detected. This is consistent with the "data and theory to model" approach in Jakeman et al. (1994). It is an alternative approach to modelling environmental systems – to begin with simple assumptions, and build up the level of model detail by testing additions and refinements to the structure and prior assumptions, by recourse to evaluation of their consistency with system observations and theory. In a sense, looking back over Chapters 11, 12, and 13, one can see a migration from the use of somewhat higher-order models (HOMs) to relatively low-order models (LOMs) in this chapter – and we shall indeed return to the role of HOMs in Chapter 17. Here, we demonstrate how to use our approach (with LOMs) in order to address problems encountered in catchment hydrology, but the lessons clearly have wider applicability, as illustrated for instance by Dietrich et al. (1989), Young and Lees (1992), and Young (1993). The interplay between the two approaches, of the "large" and the "small", has already been at the focus of the preceding chapter; and we shall echo elements of that discussion, of finding the dominant modes of a system's behaviour, in what we have to say in this chapter. Two case studies will be used. The first shows how parameterisation of the rainfall–runoff relationship can be made a function of the catchment's landscape attributes. The second considers how construction of farm dams has altered the structure of a catchment's hydrological behaviour. Irrespective of the means of incorporating additional process mechanisms, the objective is to create a model which could be applied to conditions where no observed data are available, that is, to be capable of extrapolating process behaviours into the future, or predicting responses for areas where no sufficient amount of measurements exist (the remarks of Chapter 8 notwithstanding).

14.2 THE MODEL AND SOME PRELIMINARIES

The first of our two case studies is taken from the Coweeta Hydrologic Laboratory located in the Nantahala Mountain Range of North Carolina, in the south-eastern United States. The 2185 ha laboratory consists of two adjacent basins, of which Coweeta Basin (1626 ha) has been the primary site for hydrological experimentation. Since the establishment of Coweeta in the early 1930s, 32 weirs have been installed to monitor stream flows. Currently 17 catchments, with areas ranging from 9 to 61 ha, are gauged for streamflow (Post et al., 1998). Average annual precipitation, ranging from 1870 mm at low elevations to 2500 mm at higher elevations, is relatively evenly spread, having a slight maximum in March, and minimum in October. Less than 2%

of the total precipitation falls as snow (Post et al., 1998). Deciduous oak species dominate canopy vegetation, with an abundant evergreen undergrowth consisting of rhododendron and mountain laurel. A comprehensive description of the Coweeta Hydrologic Laboratory is given in Swank and Crossley (1988).

The Yass catchment (388 km^2) in the Murrumbidgee Basin, New South Wales, Australia, forms our second case study. The predominant land use is grazing, but there are also some areas of cropping. Average annual rainfall (633 mm) has a slight summer maximum and a winter minimum. Most of the runoff (mean annual 220 mm) occurs in the winter and autumn months from June to November.

14.2.1 Model

The IHACRES rainfall-runoff model is used as the basis for our analyses. Its development derives from the idea of selecting the model to encapsulate the response characteristics of a catchment with a complexity matching the information content of the available data. Additional model complications are only incorporated if they both result in a more accurate reproduction of the observations and do not cause the parameter covariation to become intolerably large (Jakeman et al., 1994).

The model consists of two modules. The nonlinear rainfall loss module converts rainfall to rainfall excess or effective rainfall, which is defined to be the share of rainfall that eventually becomes streamflow. The linear module represents the transformation of rainfall excess to streamflow. The nonlinear modules used in the two case studies to be presented in this paper differ in structure. In the first case study, which discusses relationships between the hydrological response of a catchment and its terrain properties, the nonlinear module has the form as described in Jakeman and Hornberger (1993). The effective rainfall at time step k, u_k, is computed from

$$u_k = s_k r_k \tag{14.1}$$

where r_k is rainfall, and s_k is a catchment wetness index at time step k. The index s_k is calculated by a weighting of the rainfall time series, the weights decaying exponentially backward in time, namely

$$s_k = (1/c)r_k + (1 - \tau_w^{-1})s_{k-1} \tag{14.2}$$

The parameter τ_w is approximately the time constant, or inversely, the rate at which the catchment wetness declines in the absence of rainfall. The parameter $1/c$ represents the increase in storage index per unit rainfall in the absence of evapotranspiration. It is not really a free parameter, but is chosen so that the volume of effective rainfall is equal to the total streamflow volume over the calibration period. To account for seasonal fluctuations in evapotranspiration, the following simple function of temperature (T_k, at time step k) is used

$$\tau_w(T_k) = \tau_w \exp[(20 - T_k)f] \quad (14.3)$$

where f is a temperature modulation parameter on the rate of evapotranspiration.

In the second case study, where effects of land use on the catchment response are investigated, the model has a nonlinear module structure as described in Evans and Jakeman (1998). The basic difference is that this module is capable of delivering, in addition, an estimate of catchment evapotranspiration; it is basically a store that is recharged by precipitation P and depleted by evapotranspiration E and effective rainfall U. However, the state of the store is not expressed as a water level but as a catchment moisture deficit CMD. The catchment moisture store accounting scheme is given at time step k by

$$CMD_k = CMD_{k-1} - P_k + E_k + U_k \quad (14.4)$$

Evapotranspiration, E_k at time step k, is characterised here as a function of potential evaporation (or temperature T_k, as a surrogate for potential evapotranspiration, PET_k) and catchment moisture deficit. Effective rainfall is assumed to be dependent on catchment moisture deficit only. The parameterisations for evapotranspiration and effective rainfall are defined as

$$E_k = c_1 T_k \exp(-c_2 CMD_k) \quad (14.5)$$

$$U_k = -(c_3/c_4)CMD_k + c_3, \text{ for } CMD_k < c_4; \quad U_k = 0, \text{ for } CMD_k \geq c_4 \quad (14.6)$$

$U_k = c_3 - CMD_k$, for $CMD_k < 0$; where c_1, c_2, c_3, and c_4 are model parameters. Interested readers are referred to Kokkonen and Jakeman (2001) for further discussion of different IHACRES nonlinear modules.

The linear module of IHACRES allows for a flexible configuration of linear stores connected in parallel and/or series. Use of statistical identification procedures suggests that the most appropriate form for the Coweeta basin catchments is two stores in parallel (Jakeman and Hornberger, 1993). For the Yass catchment, due to its ephemeral nature, only one store can be identified. When the linear module has two parallel stores it is fully described by three parameters, which are:

τ_q (days), the time constant governing the rate of recession in the quicker of the two parallel stores;

τ_s (days), the time constant governing the rate of recession in the slower of the two stores;

v_q (dimensionless), the partitioning coefficient between the two stores, i.e., the proportion of quickflow to total flow.

Only one parameter, a time constant, is required to describe the linear module in the case of a single store model structure.

The equations underlying the linear module, and its link to a transfer function model, can be found in Jakeman et al. (1990) and Jakeman and Hornberger (1993).

14.2.2 Assessment Criteria

On this occasion the deviations of observed flow from modelled flow are assessed by using two statistics. These were efficiency (F), defined as

$$F = 1 - (\sigma_e^2 / \sigma_q^2) \tag{14.7}$$

where σ_e^2 is the variance of the residual errors $e_k = q_k - \hat{q}_k$ between the observed (q_k) and modelled ($\hat{q}_k$) streamflow and σ_q^2 is the variance of the observed discharge, and the relative water balance error ($RWBE$), given as

$$RWBE = 100[(\sum \hat{q}_k - \sum q_k) / \sum q_k]\% \tag{14.8}$$

14.2.3 Removing Climate Dependency

When investigating the nature of the controls exerted by landscape and land use on the hydrological response of a catchment, it is desirable to filter out first the influence of other effects, such as climate variability, since this may obscure the sought-for relationships. Gan and Burges (1990) reported that there may be considerable climate dependency in rainfall-runoff model parameters, indicating that the model has not been capable of removing climatic effects appropriately. In an attempt to find out how well our model performs in terms of eliminating climatic dependencies from further analysis, the model was calibrated to the same catchment using data from two climatically very different periods. The catchment selected for this exercise is Coweeeta 36, one of the experimental catchments at the Coweeta Hydrologic Laboratory. The first period of time is from 4 October, 1947, to 17 October, 1950, and is a wet period having an average rainfall of 7.2 mm/d, well above the average of the entire record (6.0 mm/d). It also includes the largest value for daily runoff in the entire record, 115 mm. The second period is relatively dry, this being from 20 September, 1984, to 1 November, 1987. Average rainfall is 4.6 mm/d, while the largest daily runoff in this period is half that of the first period, 57.5 mm.

The most obvious and simplest way of assessing whether the different climatic sequences in the two periods affect calibration results is to examine how well the parameters calibrated on one of the periods are capable of reproducing the runoff time series for the other. Statistics of fit for all four combinations are listed in Table 14.1. In terms of efficiencies model performance is affected little when the parameters calibrated on one period are transferred to the other. When dry-period parameters are used for predicting runoff for the wet period, an efficiency of 0.84 is achieved, which is only marginally inferior to the efficiency yielded by the calibration on the wet period (0.86). The same is true when dry-period runoff is predicted with wet-period parameters. In this case the efficiency drops to 0.76 when compared with the value of 0.78 obtained in calibration. The absolute value of the relative water balance error increases to 2.68% from 1.76% when wet-period parameters are applied to the dry period instead of using the parameters calibrated on the dry

Table 14.1

Fit statistics for calibrations and simulations on two climatologically different periods.

Prediction period	Parameters calibrated on the dry period		Parameters calibrated on the wet period	
	F	*RWBE* (%)	*F*	*RWBE* (%)
Dry	0.78	1.76	0.76	–2.68
Wet	0.84	7.6	0.86	–0.22

Table 14.2

Linear module parameter 95% confidence intervals for two climatologically different periods

	τ_q	τ_s	v_q
Dry	1.50–1.82	35.78–48.09	31.76–36.59
Wet	1.47–1.67	43.99–62.96	41.26–45.37

period. And when the wet-period runoff is estimated with the dry-period parameters, a relative water balance error of 7.60% is obtained, as opposed to the almost negligible value of –0.22% when calibration is performed on the wet period.

In conclusion, the model performance drops substantially only when the relative water balance error obtained by application of dry-period parameters to the wet period is considered and even in this case the relative water balance error would be considered tolerable in many cases.

Another way of investigating the climate independence of the model is to consider the parameter values yielded by the two climatically different calibration periods. The optimisation method for determining linear module parameters also delivers an estimate of the covariance between the parameters, which enables an objective comparison of the similarity between the two sets of linear module parameters. Table 14.2 lists 95% confidence intervals for the linear module parameters for both calibration periods. Neither of the time constants (τ_q and τ_s), at the given confidence level, is significantly different when calibrated on the two time periods, yet the partitioning coefficient (v_q) appears to be higher when calibrated on the wetter period. Figure 14.1 shows confidence intervals for the instantaneous unit hydrographs (IUHs) for both calibration periods. These impulse response functions depict the speed with which effective rainfall entering the linear routing module of the model emerges as streamflow. As such, they reflect the combined effects of all three model parameters in a single figure. A lack of identifiability is known to be present in the model's structure, in the sense that there may be both relatively high estimation error variances associated with some individual parameters and high covariances amongst the errors attaching to two or more of these parameters. The IUHs reflect succinctly all the consequences of these variances and covariances. They are a more complete assessment of the similarity of the linear module parameters, since the uncertainty of individual parameters is increased by inter-relations between them. The two IUHs are distinct from each other, although their confidence intervals overlap in part of the domain. The main difference is that the IUH for the wet period has a higher peak than the IUH calibrated on the dry period, indicating thus that in the wet period the transformation of rainfall excess into streamflow occurs more quickly. This result is consistent with the earlier finding that the parameter depicting the share of quickflow, v_q, is higher for the wet period. Perhaps the hydrological response is indeed somewhat "flashier" during wet periods and, since this is not accounted for in the model structure (the climate history only affects the volume of generated runoff, not the timing), climate dependency *is* accordingly detected in model parameters. The parameter shift, although significant, is relatively small, however. In fact, it is much smaller than that when catchments exhibiting clearly different hydrological behaviour are calibrated to a common time period. Figure 14.2 shows IUH confidence intervals for three Coweeta catchments with substantially different hydrological responses, with calibration of each being achieved for the period from 1 January 1955 to 30 October 1958.

In spite of the fact that we cannot convince ourselves of being able to remove completely the effects of climate on the hydrological response, we nevertheless assert that the climatic "signature" can be removed to an extent sufficient to facilitate detection of how other factors may control the hydrological behaviour of a

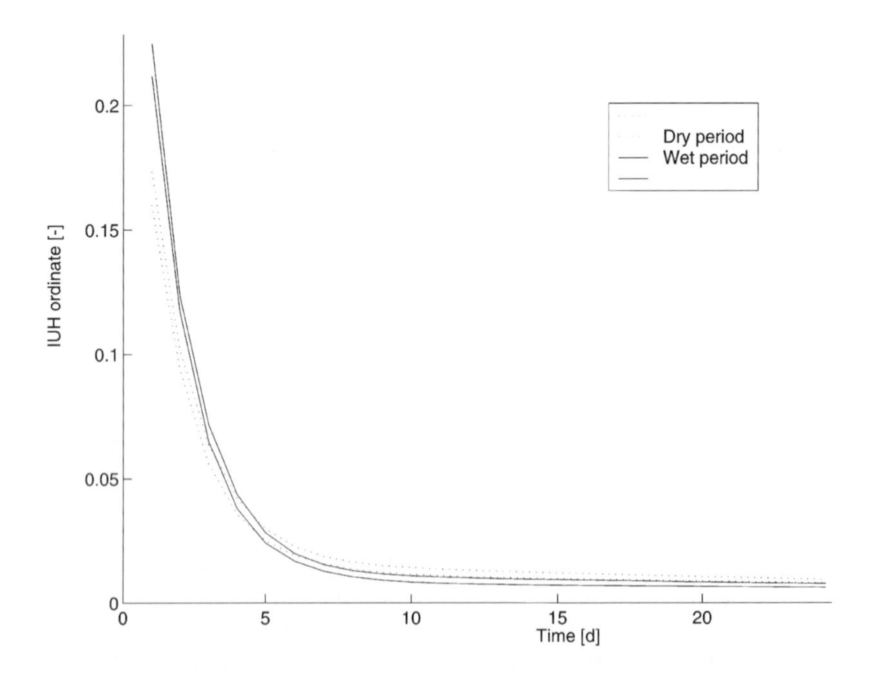

Fig. 14.1. Confidence intervals for the Instant Unit Hydrographs (IUHs) as calibrated for Coweeta catchment 36 over two periods of time (one dry, the other wet).

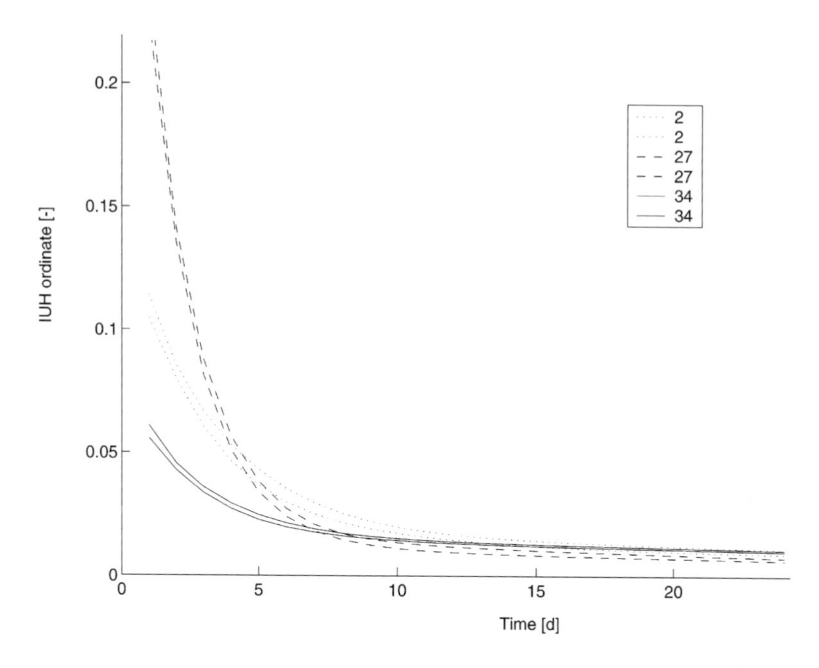

Fig. 14.2. IUH confidence intervals for Coweeta catchments 2, 27, and 34.

catchment. This assumption is grounded by the following: (a) the model performs reasonably well when parameters used for runoff prediction are adopted from a period with a substantially different climate sequence; and (b) the differences between parameters calibrated for two climatologically different periods are small.

14.3 CASE STUDY 1: COWEETA, USA

Regionalisation can be defined as the transfer of information from one catchment to another (Blöschl and Sivapalan, 1995). Such a transfer is called for when no flow records of sufficient length are available at the site of interest; it is often achieved by establishing a relationship between the hydrological response of a catchment and its physical attributes. In the present case the hydrological response can be taken as having been quantified through the parameters of IHACRES, while the physical attributes include the weir elevation and the mean slope of the catchment. These two attributes were screened from a much larger set of potential explanatory variables on the grounds of statistical significance and physical plausibility of their influence over the rainfall-runoff process. In essence, what we are seeking is a relationship suggested by the general form of equation 5.4 in Chapter 5, i.e., parameterisation of any identified parametric variation (in this case from one catchment to another, as opposed to changes with time).

IHACRES has been calibrated for all 13 catchments at Coweeta. Ideally, we should have been able to choose one common span of time for all the catchments under study, with complete sets of observations, but this would have restricted the number of catchments available to only eight. By choosing two windows of the record, 13 catchments, having data for either (or both) of the periods, can be included in the analysis. Period 1 is from 1 November 1937 to 31 August 1939; the second refers to 1 January 1955 to 30 October 1958. The model fits well the discharge of all the catchments, with efficiencies ranging from 0.84 to 0.92, and the absolute value of the relative water balance error always being less than 10%. All the fit statistics are shown in Table 14.5 (first column).

Table 14.3 lists correlations between the catchment attributes and model parameters. Those correlations which are significant (two-tailed test) at the 5% level are given in bold-face type. Elevation appears to be a major driver of hydrological response in the Coweeta catchments, correlating significantly with four out of six parameters, while slope is identified as having significant correlations with three model parameters. Two of these relationships are shown as scatter plots in Figures 14.3 and 14.4, respectively for the partitioning parameter v_q *versus* slope, and for the water balance parameter c *versus* weir elevation. In fact, we take this line of exploration further through regression analysis. Each catchment is treated as if it were ungauged and its data excluded from the regressions. Since there are thirteen catchments in total, thirteen different regionalisation equation sets can thus be formed. Only small differences, if any, however, may be expected in the screening of significant explanatory variables for the thirteen different data sets. Having analysed three of the thirteen data sets, always arriving at an identical set of explanatory variables, these explanatory variables were accepted to reflect the regression relationships across all the catchments. In the full account of the analysis (Kokkonen et al., 2002), three different regression methods were employed, of which two were capable of taking into account correlation between dependent variables (i.e., the rainfall-runoff model parameters in this case). No significant difference in performance was detected between any of the regression methods, however. Accordingly, only the results yielded by the simplest method are repeated here. In this method ordinary least squares (OLS) regression is applied to explain each parameter individually using those catchment attributes that maximise the significance of the overall regression as independent variables. The following list gives the identified relationships: τ_w from weir elevation; f constant; c from weir elevation; τ_q from weir elevation; τ_s from weir elevation; and v_q from slope. Recalling the cross correlations

Table 14.3

Correlations between catchment characteristics and IHACRES parameters. Those that are significant at the 5% confidence level are shown in bold.

	τ_w	f	c	τ_q	τ_s	v_q
ELEV	**–0.68**	–0.12	**–0.78**	–0.56	–0.68	0.37
SLOPE	**–0.56**	–0.42	–0.51	–0.12	**–0.62**	0.66

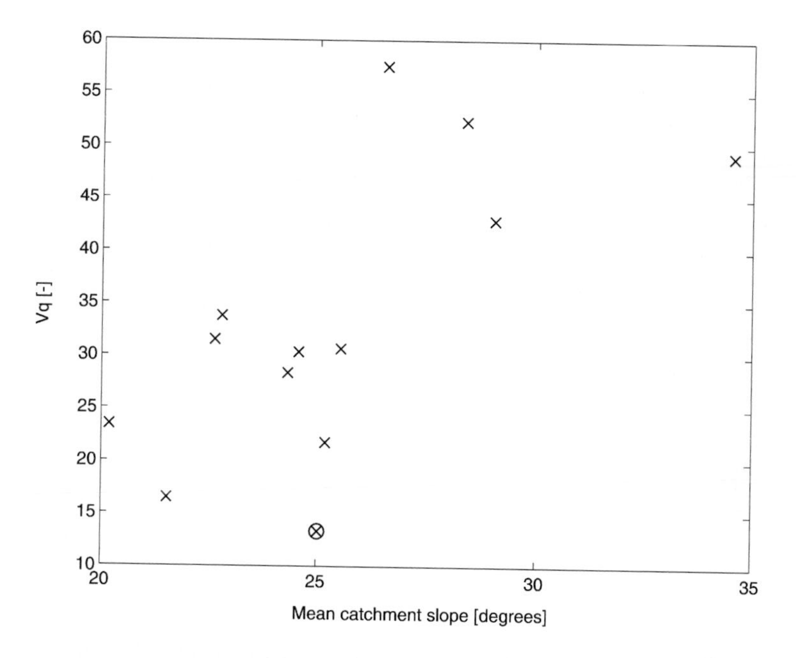

Fig. 14.3. Parameter v_q plotted against the mean catchment slope (catchment 34 is circled).

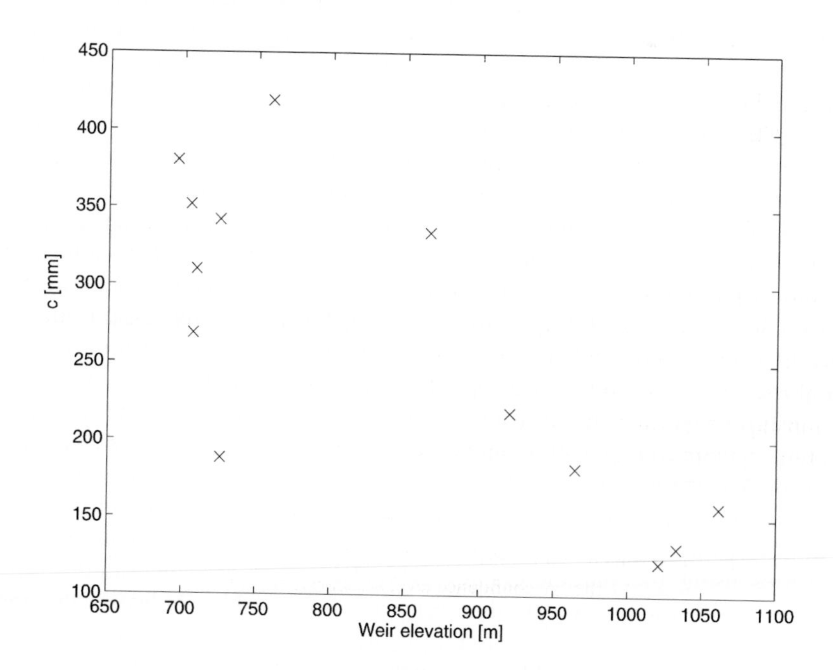

Fig. 14.4. Parameter c plotted against weir elevation.

Table 14.4

IHACRES parameter values as predicted from regression. Calibrated values shown in parentheses. No refers to a catchment number.

No.	τ_w [d]	f [–]	c [mm]	τ_q [d]	τ_s [d]	v_q [–]
1	12.0 (15.1)	12.8 (14.0)	331 (352)	3.1 (3.0)	49.3 (48.9)	30.7 (28.3)
2	12.7 (10.5)	13.1 (10.4)	336 (310)	3.1 (3.0)	50.4 (40.8)	33.7 (30.7)
6	12.4 (14.4)	12.6 (17.3)	332 (381)	3.2 (2.8)	49.8 (48.3)	25.8 (31.5)
13	11.5 (15.6)	12.6 (16.4)	321 (342)	3.0 (3.2)	48.7 (46.6)	25.9 (33.8)
14	12.9 (9.9)	13.0 (12.1)	344 (269)	3.3 (2.1)	49.8 (45.1)	19.6 (23.4)
17	11.1 (13.2)	13.0 (11.6)	291 (419)	2.6 (5.3)	47.0 (49.0)	33.9 (57.5)
18	13.2 (4.1)	13.2 (9.5)	343 (188)	3.1 (2.6)	47.4 (55.3)	33.6 (21.8)
27	4.4 (5.6)	12.8 (14.7)	134 (157)	1.8 (1.8)	38.0 (31.5)	38.6 (52.3)
28	6.9 (6.4)	12.9 (13.8)	195 (182)	2.1 (2.2)	40.0 (37.8)	31.2 (30.3)
32	7.7 (8.5)	12.9 (13.7)	217 (218)	2.4 (1.7)	40.6 (47.4)	25.3 (16.5)
34	8.5 (14.1)	12.9 (13.2)	239 (334)	2.5 (2.6)	42.4 (53.5)	33.8 (13.3)
36	6.1 (3.5)	13.1 (10.3)	173 (121)	1.9 (2.1)	36.4 (42.4)	41.8 (42.9)
37	5.8 (3.7)	13.1 (11.1)	163 (131)	1.9 (1.8)	39.5 (29.7)	64.0 (49.1)

between catchment attribute and model parameter from Table 14.3, these relationships can be explained in words as follows. Elevation correlates negatively with c, which means that higher elevated catchments with smaller c values have lower storage and a more rapid increase in catchment wetness index in response to a rainfall event, relative to the lower catchments. Hence more of the rainfall is converted into effective rainfall and these catchments have higher runoff coefficients. Higher elevated catchments also have a quicker response, which is indicated by the decrease in both time constants of the linear module, τ_q and τ_s, with elevation. The partitioning coefficient, v_q, increases with slope, so that the more steeply sloping catchments have a larger share of quickflow, and hence exhibit a flashier behaviour than catchments with shallower slopes. The temperature modulation parameter, f, was not successfully related to any of the catchment characteristics and appears, therefore, as a constant in the regression equations. Table 14.4 lists estimated IHACRES parameter values as produced by the corresponding regression relationship – from catchment attributes – along with the estimates previously identified from the original catchment (precipitation, streamflow) data. The regression coefficients themselves for all thirteen sets of equations are not shown, as the cross correlations between the catchment characteristics and model parameters (Table 14.3) essentially provide the same information.

An obvious means of validating these results is to predict the daily streamflow time series using the parameters for IHACRES that are generated from the regression relationships. Recall that separate relationships between the model parameters and attributes were derived for the thirteen different subsets of twelve Coweeta catchments. Here daily runoffs are reproduced for all catchments using the regressed parameters, hence considering each to be ungauged for streamflow. Since

Table 14.5

Performance statistics using calibrated and predicted IHACRES parameters.

Catchment No.	Calibrated		Predicted	
	F	*RWBE* (%)	*F*	*RWBE* (%)
1	0.90	–3.5	0.90	–5.6
2	0.88	5.6	0.85	–5.0
6	0.89	–5.2	0.76	17.8
13	0.91	–4.5	0.90	–3.5
14	0.89	1.4	0.83	–8.1
17	0.90	–4.1	0.84	17.9
18	0.84	9.5	0.84	4.7
27	0.89	–2.2	0.87	6.1
28	0.92	–1.3	0.91	2.2
32	0.91	0.0	0.86	–3.1
34	0.92	–0.2	0.52	–1.2
36	0.89	–1.4	0.88	–9.3
37	0.87	–2.1	0.85	0.6

these catchments are in fact gauged, the accuracy of the predictions may be assessed by comparison against the observed flows. Table 14.5 lists fit statistics (*F* and *RWBE*) for all catchments and for both calibrated and estimated parameter values. Overall the regionalised parameters produced daily runoff predictions not greatly inferior to those given by calibration. Figure 14.5 shows predicted streamflows, based on the regionalised parameters, and measured streamflows for one catchment (number 1), where good results were obtained. However, there are two catchments whose runoffs were poorly reconstructed when regionalised parameters were applied. Most noticeably, even though the overall water balance for catchment 34 is adequately reproduced, peak values are consistently over-predicted (Figure 14.6). This occurs because the predicted value for v_q is much greater than it should be according to calibration (Table 14.4). Figure 14.3, which plots v_q against slope, shows catchment 34 as a clear outlier, having a far smaller value for v_q than the slope–v_q relationship would suggest. Another catchment where regionalised parameters perform inadequately is catchment 6, which has the largest f parameter value of all the catchments. Since f was found not to be related to any catchment characteristics, there is no way a regionalisation procedure can account for this feature. The large f value could be due to an unknown control, but more likely it is merely an artifact of calibration. It is worth pointing out that calibration was performed in this case study on a completely objective basis according to a fit criterion where no expert reasoning was taken into account.

While elevation and slope have been found useful in explaining the hydrological response of the Coweeta catchments, some of the other characteristics, such as those describing the stream network structure of the catchments, were rejected in the screening analysis as not being plausible in differentiating among catchments in

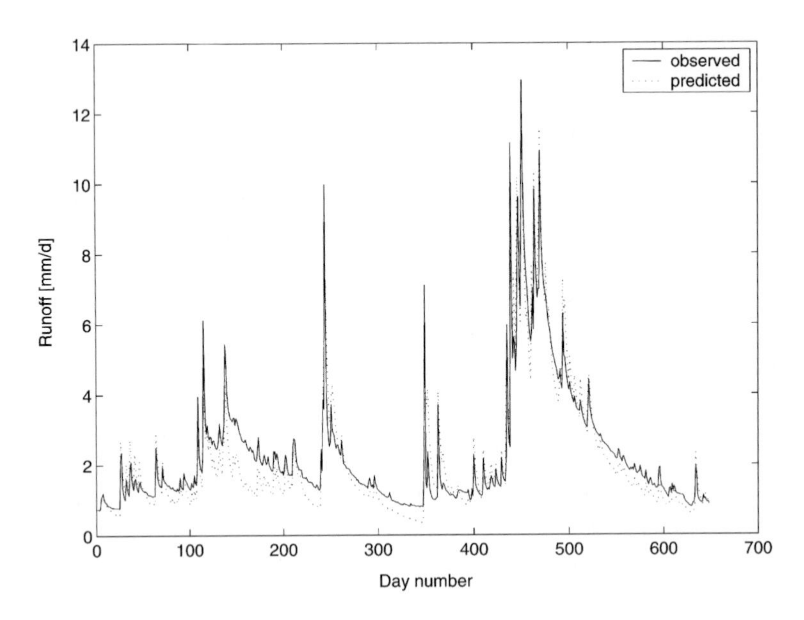

Fig. 14.5. Observed and predicted runoff for Coweeta catchment 1 for November 1, 1937, to August 31, 1939 (using regionalised parameters).

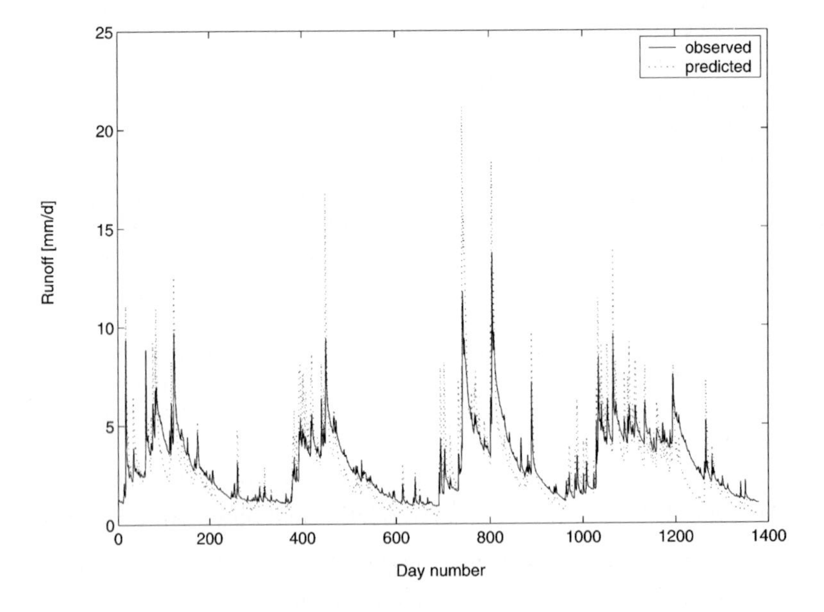

Fig. 14.6. Observed and predicted runoff for Coweeta catchment 34 for January 1, 1955, to October 30, 1958 (using regionalised parameters).

terms of their hydrological behaviour. This result reinforces the complex, system-dependent nature of factors influencing catchment response. In other words, factors which may logically seem important – at the outset – in determining the hydrological behaviour can be insignificant in the area of interest, for instance, because their effects have been masked by more powerful drivers of the hydrological response. This outcome casts doubt on the underpinning of complex models where so-called understanding is imposed *a priori* upon model structure and variables.

14.4 CASE STUDY 2: YASS RIVER

One might hypothesize that diverting water for irrigation purposes should result in an increase in evaporative losses within a catchment, and that the share of precipitation seen as streamflow should correspondingly diminish. Whether this is the case for the catchment of the Yass River, which has been subject to intensive farm dam development in the recent years, is the central issue now to be addressed (as discussed more fully in Schreider and Jakeman, 1999).

Our analysis is based on a concept of potential streamflow response (PSR), which is defined to be the ability of a catchment to generate streamflow in response to climatic excitation under given land use and vegetation conditions. As this is a climate-independent characteristic the effect of the climatic signal on streamflow must again be first removed. Suppose the rainfall-runoff model – here IHACRES – is calibrated for a part of the record, thus capturing the PSR at that time. Suppose further the model is then applied over the entire record using the climate time series as input. If the PSR of a catchment is reduced (after the calibration period) as a consequence of the increasing number of farm dams, the model should clearly overestimate streamflows when applied under the changed conditions of land use. Similarly, if prior to the calibration period the PSR has been greater due to fewer farm dams, the streamflow estimates yielded by the model when applied to that prior time should be smaller than the observed. The identification problem here is to investigate whether there is a significant trend in the model residuals, indicating a constant reduction in the PSR of a catchment, which could be attributed to the increased storage capacity of the farm dams. The approach adopted to address this problem can be outlined as follows:

- Calibrate IHACRES to a part of the existing record of daily streamflow and climatic data for the catchment of interest (calibration period was one year here);
- Using the calibrated parameters, simulate streamflow over the entire record;
- Compute daily residuals between modelled and observed streamflow;
- Calculate moving-average time series for the daily residuals (averaging, or smoothing, period being two years);
- Test whether there is a statistically significant trend in the time series constructed at the previous step.

Some further comments are warranted to clarify the procedure outlined above. Firstly, there is a trade-off in selecting the length of the calibration period. A long time period is preferred in order to minimise residual and parameter error variance in the calibration, yet a short period would be better in terms of ensuring stationarity with respect to the level of farm dam development. Secondly, the model residuals need to be smoothed to account for the periods of no rainfall. The Yass catchment streamflow is highly ephemeral, which leads to long periods where model residuals are zero, irrespective of whether the PSR of a catchment has changed. Because there is seasonality involved when a catchment experiences dry spells, the smoothing period needs to be long enough to extend over the entire seasonal cycle. Finally, fitting a trend to the model residuals revealed discrepancies between the linear trend and the model residuals that were found to be auto-correlated. Ordinary least squares regression estimates are inefficient when serial correlation is present and the standard error estimates are biased, so that judgements about the significance of the trend line may be seriously undermined. For these reasons, the estimated generalised least squares method (Judge et al., 1980) was employed to fit the trend line.

Figure 14.7 shows observed and modelled moving averages for the streamflow in the Yass River, along with the moving average of the model residuals and a trend line fitted to the residual time series. The model was calibrated to a one year time period starting on 1 January 1979. Visual inspection of the figure suggests that the model indeed increasingly overpredicts the flow, indicating that the PSR of the catchment is decreasing with time. This result is supported by statistical analysis, according to which the trend line is significant at the 5% risk level. On the basis of this trend line

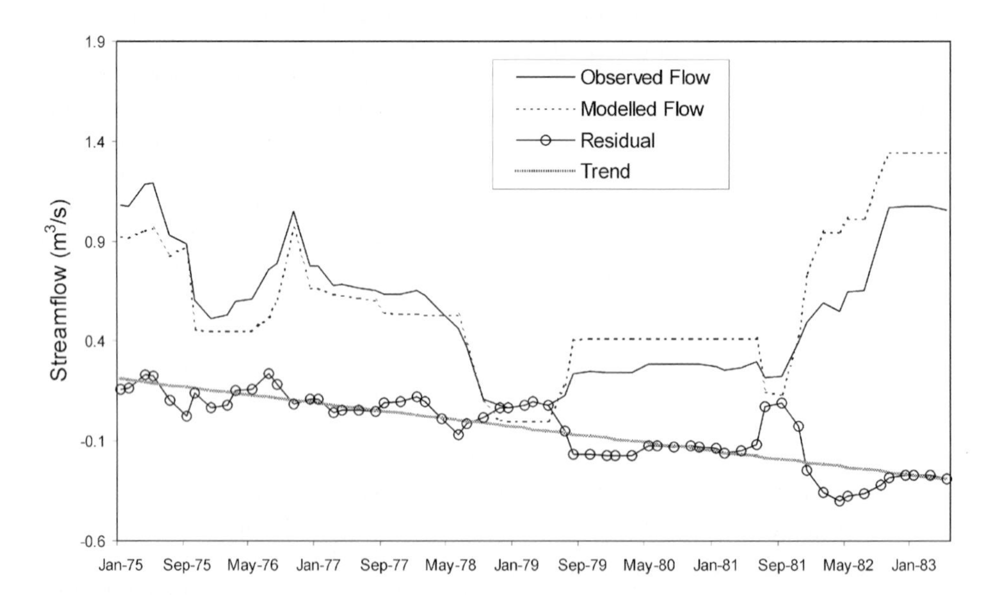

Fig. 14.7. Observed, modelled, and residual moving averages of streamflow in the Yass catchment. The trend of the residual time series is shown.

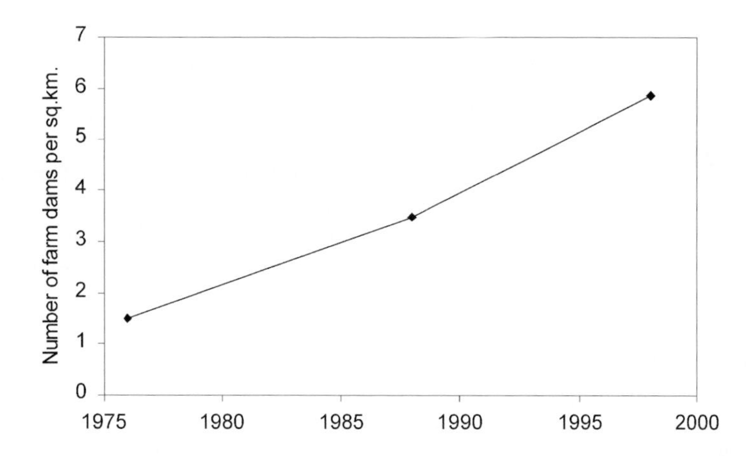

Fig. 14.8. Development of farm dam density in the Yass catchment.

the runoff volume is found to be decreasing by 9.5% every year. Figure 14.8 shows the increase in numbers of farm dams over the period of analysis, as estimated from aerial photography. Four representative cross-sections of the catchment were analysed, the average values of which are depicted in the figure. Other changes potentially affecting the water balance in the Yass River catchment include hobby farming associated increases in tree planting, and more use of perennial pasture which has higher water requirements than native pasture. However, it is assumed that the predominant factor in reduction of the PSR will have been the farm dams.

Our analysis has allowed us to gain an appreciation of the effects of farm dams on streamflow response. Given that there is a significant trend it ought to be possible to develop a more complex model accounting explicitly for these impacts. In fact, this can be achieved by extending the model structure through an additional module, the storage of water in farm dams "upstream", as it were, of the nearest stream-gauge (where it would be recorded as streamflow). In order to avoid identification problems, however, it will be necessary to keep the number of additional parameters requiring calibration in this module as low as possible. Further, given the lumped nature of the basic rainfall-runoff model, the impact of all farm dams must be represented as a single, equivalent (conceptual) farm dam. This is essentially an additional store within the catchment, which is recharged by direct precipitation and runoff from other parts of the catchment, and depleted by irrigation water consumption and evaporation. The total surface area of the dams controls two of the above mentioned processes – direct precipitation to the dams and evaporation. An estimate of changes in total surface area of dams through time can be extracted from aerial photography, and then inserted into the model. Spatial analysis incorporating digital elevation models and dam locations might be used to estimate the proportion of catchment area draining to the dams, which is required to assess the amount of water the dams receive as runoff from the surrounding area. Estimates of consumption of stored water could, in principle, be retrieved from farmers.

14.5 CONCLUSIONS

The various factors influencing the hydrological response of a catchment can be identified separately. The effects of a climate signal on the streamflow data can first be extracted, thus improving the chances of revealing how other factors, such as landscape properties and land use, control the hydrological response of a catchment. Our approach has been to start with a basic LOM, thus to identify parametric variation, or structural inadequacy, all the more crisply, and then to extend the model accordingly (Jakeman et al., 1994).

Although it may not be possible to eliminate the effect of climate completely, the rainfall-runoff model of our two case studies proved to be relatively insensitive to the climate of the calibration period. In the first of these, parameterisation of the relationship between landscape attributes and the *parameters* of a simple low-order model (LOM) could be demonstrated for a catchment in the Coweeta Basin, North Carolina. The weir elevation and the mean slope of a catchment were identified as causal variables and could be used to extrapolate hydrological response from one point in space (if not time) to another. Overall these predictions were only slightly inferior to those given by direct calibration of the same model to the streamflow data. Some of the landscape attributes one might beforehand have presumed would control hydrological behaviour of the catchment (such as those that describe the stream network structure), were not found to be important.

For the Yass catchment – our second case study – changes with time (as opposed to space) were more significant. But in this instance they were revealed in the residual errors of model fit instead of through the model's parameters, which in turn suggested a means of adapting and improving the model's structure. It seems likely the effects of the upward trend in farm dam construction, once accounted for and parameterised within the extended model structure, for example, through a time constant (τ_f, say) and a partitioning coefficient of some kind (v_f, say), could then be extrapolated into the future. We would be able to explore structural change in the hydrological behaviour of the catchment through this parametric variation, $\tau_f(t^+)$ and $v_f(t^+)$, into the future (t^+).

REFERENCES

Beck, M.B., Jakeman, A.J. and McAleer, M.J., 1993. Construction and evaluation of models of environmental systems. In: *Modelling Change in Environmental Systems* (A.J. Jakeman, M.B. Beck and M.J. McAleer, eds.). Wiley, Chichester, pp. 3–35.

Blöschl, G. and Sivapalan, M., 1995. Scale issues in hydrological modelling – a review. *Hydrol. Process.*, **9**(3–4), 251–290.

Dietrich, C.R., Jakeman, A.J. and Thomas, G.A., 1989. Solute transport in a stream-aquifer system, 1, Derivation of a dynamic model. *Water Resour. Res.*, **25**(10), 2171–2176.

Eagleson, P.S., 1978. Climate, soil, and vegetation, 6, Dynamics of the annual water balance. *Water Resour. Res.*, **14**, 749–764.

Evans, J.P. and Jakeman, A.J., 1998. Development of a simple, catchment-scale rainfall-evapotranspiration-runoff model. *Environ. Modelling & Software*, **13**.

Gan, T.Y. and Burges, S.J., 1990. An assessment of a conceptual rainfall-runoff model's ability to represent the dynamics of small hypothetical catchments, 2, hydrologic responses for normal and extreme rainfall. *Water Resour. Res.*, **26**, 1605–1619.

Goodrich, D.C. and Woolhiser, D.A., 1991. Catchment hydrology. U.S. Report on Hydrology to the International Union of Geodesy and Geophysics 1987–1990, *Rev. Geophys.*, Suppl., 202–209.

Jakeman, A.J. and Hornberger, G.M., 1993. How much complexity is warranted in a rainfall-runoff model? *Water Resour. Res.*, **29**, 2637–2649.

Jakeman, A.J., Littlewood, I.G. and Whitehead, P.G., 1990. Computation of the instantaneous unit hydrograph and identifiable component flows with application to two small upland catchments. *J. Hydrol.,* **117**, 275–300.

Jakeman, A.J., Post, D.A. and Beck, M.B., 1994. From data and theory to environmental model: The case of rainfall-runoff. *Environmetrics*, **5**, 297–314.

Judge, G.G., Griffiths, W.E. , Hill, R.C. and Lee, T.C., 1980. *The Theory and Practice of Econometrics*. John Wiley and Sons, USA.

Kokkonen, T.S. and Jakeman, A.J., 2001. A comparison of metric and conceptual approaches in rainfall-runoff modeling and its implications. *Water Resour. Res.*, **37**, 2345–2352.

Kokkonen, T.S., Jakeman, A.J. and Young, P.C., 2002. Predicting daily flows in ungauged catchments – model regionalization from catchment descriptors at Coweeta. *Hydrol. Processes*, submitted.

Post, D.A., 1996. Identification of relationships between catchment scale hydrologic response and landscape attributes. PhD Thesis, Australian National University, Canberra.

Post, D.A., Jones, J.A. and Grant, G.E., 1998. An improved methodology for predicting the daily hydrologic response of ungauged catchments. *Environ. Modelling and Software*, **13**, 395–403.

Schreider, S. Yu. and Jakeman, A.J., 1999. Impacts and implications of farm dams on catchment yield. Report to Murray-Darling Basin Commission, Australia.

Swank, W.T. and Crossley, D.A., Jr. (eds.), 1988. *Forest Hydrology and Ecology at Coweeta*. Springer-Verlag, New York.

Vertessy, R.A., Hatton, T.J., Oshaughnessy, P.J. and Jayasuriya, M.D.A., 1993. Predicting water yield from a mountain ash forest catchment using a terrain analysis based catchment model. *J. Hydrology*, **150**(2–4): 665–700.

Young, P.C., 1993. Time variable and state dependent modelling of nonstationary and nonlinear time series. In: *Developments in Time Series Analysis* (T. Subba-Rao, ed.). Wiley, Chichester, pp. 374–413.

Young, P.C. and Lees, M., 1992. The active mixing volume: a new concept in modelling environmental systems. In: *Statistics and the Environment* (V. Barnett and R. Turkman, eds.). Wiley, Chichester.

Environmental Foresight and Models: A Manifesto
M.B. Beck (editor)

CHAPTER 15

Elasto-plastic Deformation of Structure

M.B. Beck, J.D. Stigter and D. Lloyd Smith

15.1 INTRODUCTION

Many models of the behaviour of environmental systems can be defined according to the following (lumped-parameter) representation of the state variable dynamics,

$$d\boldsymbol{x}(t)/dt = \boldsymbol{f}\{\boldsymbol{x},\boldsymbol{u},\boldsymbol{\alpha};t\} + \boldsymbol{\xi}(t) \tag{15.1a}$$

with observed outputs being defined as follows,

$$\boldsymbol{y}(t) = \boldsymbol{h}\{\boldsymbol{x},\boldsymbol{\alpha};t\} + \boldsymbol{\eta}(t) \tag{15.1b}$$

in which $\boldsymbol{f}$ and $\boldsymbol{h}$ are vectors of nonlinear functions, $\boldsymbol{u}$, $\boldsymbol{x}$, and $\boldsymbol{y}$ are the input, state, and output vectors, respectively, $\boldsymbol{\alpha}$ is a vector of model parameters, $\boldsymbol{\xi}$ and $\boldsymbol{\eta}$ are notional representations respectively of those attributes of behaviour and output observation that are not to be included in the model in specific form, and t is continuous time. Should it be necessary, spatial variability of the system's state can be assumed to be accounted for by, for example, the use of several state variables of the same attribute of interest at the several defined locations.

We have argued – in particular, in Chapter 4 – that structural change is a function of that which has not been included adequately within such a model and we must now examine more carefully how the sources of this change can be detected. Put simply, structural error, inadequacy, or approximation may enter into equation 15.1 through $\boldsymbol{\alpha}$, $\boldsymbol{\xi}$, and $\boldsymbol{\eta}$, although these points of entry differ in their interpretation and significance. The principal distinction is between $\boldsymbol{\alpha}$, embedded within the choices for

$[\boldsymbol{x},\boldsymbol{\alpha},\boldsymbol{f},\boldsymbol{h}]$, which signify that which we presume (or wish) to know of the system's behaviour, relative to the purpose of the model, and $[\boldsymbol{\xi},\boldsymbol{\eta}]$, which acknowledge in some form that which falls outside the scope of what we believe we know. The difference between the two is as the difference between what we call respectively the {presumed known} and the {acknowledged unknown}. Much, of course, must be subsumed under the latter, that is, under the definitions of $\boldsymbol{\xi}$ and $\boldsymbol{\eta}$. We may have chosen to exclude from the model some of that which was known beforehand (but which was judged not to be significant); there may be features for which there are no clear hypotheses (and therefore no clear mathematical expressions), other than that these may in part be stochastic processes with presumably quantifiable statistical characteristics; there may be yet other features of conceivable relevance, but of which we are simply ignorant; and, as is most familiar, there may be factors affecting the processes of observation such that we are unable to have uncorrupted, perfect access to knowledge of the values of the inputs, states, or outputs.

15.1.1 Apparent Structural Change: A Lesser Form of Evolution

Typically thus, *structural error* may be thought of as a measure of the extent to which the expression of what is "known", i.e., $[\boldsymbol{x},\boldsymbol{\alpha},\boldsymbol{f},\boldsymbol{h}]$, diverges from the "truth". What we might refer to as our "ignorance" is conventionally lumped conceptually under the labels of $[\boldsymbol{\xi},\boldsymbol{\eta}]$. In the gap between the tractable known (the model) and the truth (reality) lies the difference, as Allen (1990) would argue, between the behaviour of mechanical and evolutionary systems:

> [I]f the world is viewed as some kind of 'machine' made up of component parts which influence each other through causal connections, then instead of simply asking how it 'works', evolutionary theory is concerned with how it got to be as it is.
>
> The Newtonian paradigm was not about this. It was about mechanical systems either just running, or just running down.
>
> The key issue is centred on the passage between detailed microscopic complexity of the real world, which clearly can evolve, and any aggregate macroscopic 'model' of this.
>
> The central question which arises is that in order even to think about reality, to invent words and concepts with which to discuss it, we are forced to reduce its complexity. We cannot think of the trillions of molecules, living cells, organisms, individuals and events that surround us, each in its own place and with its own history. We must first make a taxonomic classification, and we must also make a spatial aggregation.
>
> [I]f, in addition to our basic taxonomic and spatial aggregations, we assume that only average elements make up each category, and that only the most probable events actually occur, then our model reduces to a 'machine' which represents the system in terms of a set of differential equations governing its variables.
>
> But such a 'machine' is only capable of 'functioning', not of evolving. It cannot restructure itself or insert new cogs and wheels, while reality can!

What Allen imagines is the possibility of the structure of the web of interactions, of which we conceive in our models, dissolving, as it were, and then re-crystallizing into some other structure, with a different number of states and parameters and different inter-connections between the states. And what Allen asks is: can we discover the rules by which the system will re-structure itself?

We do not presume to answer such a question herein. Instead, we ask the questions: assuming the arrangement of the interactions among the state variables in the model is fixed and invariant, i.e., the content of $[\boldsymbol{f},\boldsymbol{h}]$ is invariant, can we detect a change in those inter-connections, i.e., a change in $\boldsymbol{\alpha}$, and from this could we fathom a different implied number and arrangement of the system's state variables and their interactions? We argue that this is likely to be a real and common problem, because our models are bound to be approximations of the unknowable truth. What may appear to have dominated behaviour in the past – from interpretation of the empirical record – may decline into insignificance in the future. Conversely, what may previously have been below the resolving power of the model, buried perhaps in the residual noise of the historical observations, may come to dominate behaviour in the future. We might call this *apparent* structural change, brought about because the resolving power of our models can never be as fine-grained as that of the truth.

Much of the chapter is devoted to the conceptual elaboration of this issue, an extensive explanation of why a novel form of recursive estimation algorithm will be needed for its investigation, and a presentation of some simple case-study results, which illustrate both the potential and the limitations of current solution algorithms. We shall draw back therefore from addressing the problem of attempting to predict, in effect, the discrete event of the birth of a new state equation in the model, which is implied in Allen's question. But we must now be more specific about the problems we *shall* address.

15.2 A MOTIVATING EXAMPLE

Metropolitan growth has the potential for imposing substantial stress on the aquatic environment through the discharge of a host of more or less easily degradable organic pollutants. We may denote these collectively as the "substrate". They can be transformed into simple end-products through microbial activity, typically into water, carbon dioxide, and methane. The consequence of such activity is manifest in a depressed concentration of oxygen dissolved in the water body. And this dissolved oxygen (DO) concentration, notably its departure from saturation, can be used to gauge the healthy state or otherwise of the aquatic environment. Historically, a deficit of DO would have been considered essentially the only form of "ill health". A level of DO above saturation may also be considered unhealthy, however, since this will in general be apparent only in the presence of excessive growths of phyto-plankton stimulated by the relative abundance of nutrient material in the waste discharges, principally N- and P-bearing substances.

In former times, the need was to relate the DO concentration in a stressed river (x_2) to the concentration of substrate therein (x_1). Hence one could determine the extent to which the presence of the latter should be reduced – through the construction of a wastewater treatment plant – in order to achieve an acceptable, or healthier, level of the former. Without yet descending any further into particular details, this relationship can be expressed as the following model for the dynamics of the two state variables (see, for example, Beck and Young, 1976):

Structure I (Resolving Power 0):

$$\dot{x}_1^0(t) = (u_1^0(t)/\alpha_1^0)u_2^0(t) - (u_1^0(t)/\alpha_1^0)x_1^0(t) - \alpha_2^0 x_1^0(t) + \xi_1^0(t) \tag{15.2a}$$

$$\begin{aligned} \dot{x}_2^0(t) = {} & (u_1^0(t)/\alpha_1^0)u_3^0(t) - (u_1^0(t)/\alpha_1^0)x_2^0(t) - \alpha_2^0 x_1^0(t) \\ & + \alpha_3^0[\alpha_4^0 - x_2^0(t)] + \xi_2^0(t) \end{aligned} \tag{15.2b}$$

Here superscript 0 signifies an arbitrary datum for the resolving power of the model (in this instance, a relatively low power) and the dot notation in $\dot{x}_i$ denotes differentiation with respect to time t. The first pair of terms on the RHS of both equations 15.2(a) and 15.2(b) deals with an account of the mass transport of material into and out of the defined element (stretch) of river. The third term, $\{-\alpha_2^0 x_1^0(t)\}$, which appears in both equations, represents the focal point of the interaction between the two state variables. It expresses the hypothesis that the removal of substrate and DO from the system occurs at the same rate and is simply proportional to the concentration of the substrate. In other words, the relationship between x_1 and x_2 is effectively one of mere chemistry. In practice, together with the fourth term, $\{+\alpha_3^0(\alpha_4^0 - x_2^0(t))\}$, this relationship has long been a part of the accepted canon of managing the quality of the aquatic environment.

Yet we have been primed by the preamble to this specification of the model to believe that the behaviour of *bacteria* and *phytoplankton* should somehow have a significant bearing on the behaviour of the substrate and DO. At some more refined resolving power, the parameters $\boldsymbol{\alpha}$ of the prior theory embedded in Structure I and the vector ξ – of labels acknowledging what must surely have been omitted in our ignorance – can be revealed as having more specific, more detailed forms. For example, if we were to magnify the image of our system, inserting three further state variables into it, that is, the concentrations of a bacterial biomass (x_3), a phytoplankton biomass (x_4), and detrital, degradable organic matter (x_5), we could derive the following five-state model:

Structure II (Resolving Power 1):

$$\begin{aligned} \dot{x}_1^1(t) = {} & (u_1^1(t)/\alpha_1^1)u_2^1(t) - (u_1^1(t)/\alpha_1^1)x_1^1(t) - \alpha_2^1 x_1^1(t)x_3^1(t)/[\alpha_5^1 + x_1^1(t)] \\ & + \alpha_6^1 x_5^1(t) + \xi_1^1(t) \end{aligned} \tag{15.3a}$$

$$\begin{aligned} \dot{x}_2^1(t) = {} & (u_1^1(t)/\alpha_1^1)u_3^1(t) - (u_1^1(t)/\alpha_1^1)x_2^1(t) - \alpha_7^1 x_1^1(t)x_3^1(t)/[\alpha_5^1 + x_1^1(t)] \\ & + \alpha_3^1[\alpha_4^1 - x_2^1(t)] + \alpha_8^1 x_4^1(t)u_7^1(t)/[\alpha_9^1 + u_7^1(t)] - \alpha_{10}^1 x_4^1(t) + \xi_2^1(t) \end{aligned} \tag{15.3b}$$

$$\dot{x}_3^1(t) = (u_1^1(t)/\alpha_1^1)u_4^1(t) - (u_1^1(t)/\alpha_1^1)x_3^1(t) + \alpha_{11}^1 x_1^1(t)x_3^1(t)/[\alpha_5^1 + x_1^1(t)] - \alpha_{12}^1 x_3^1(t) + \xi_3^1(t) \quad (15.3c)$$

$$\dot{x}_4^1(t) = (u_1^1(t)/\alpha_1^1)u_5^1(t) - (u_1^1(t)/\alpha_1^1)x_4^1(t) + \alpha_{13}^1 x_4^1(t)u_7^1(t)/[\alpha_9^1 + u_7^1(t)] - \alpha_{14}^1 x_4^1(t) + \xi_4^1(t) \quad (15.3d)$$

$$\dot{x}_5^1(t) = (u_1^1(t)/\alpha_1^1)u_6^1(t) - (u_1^1(t)/\alpha_1^1)x_5^1(t) + \alpha_{15}^1 x_4^1(t) - \alpha_{16}^1 x_5^1(t) + \xi_5^1(t) \quad (15.3e)$$

Relative to Structure I, it is important to note how the {presumed known} and {acknowledged unknown} of that earlier structure have been unfolded and unpacked into the refinements of Structure II, with its greater resolving power (denoted by superscript 1). Thus:

(i) The {presumed known} of $\{-\alpha_2^0 x_1^0(t)\}$ in equations 15.2(a) and 15.2(b) has been unfolded into the terms $\{-\alpha_2^1 x_1^1(t)x_3^1(t)/[\alpha_5^1 + x_1^1(t)]\}$ and $\{-\alpha_7^1 x_1^1(t)x_3^1(t)/[\alpha_5^1 + x_1^1(t)]\}$ of equations 15.3(a) and 15.3(b) *and* the now explicit dynamics of an additional state, x_3, i.e., equation 15.3(c).

(ii) The {acknowledged unknown} of $\xi_2^0(t)$ in equation 15.2(b) has been partly filled by the terms $\{+\alpha_8^1 x_4^1(t)u_7^1(t)/[\alpha_9^1 + u_7^1(t)]\}$ and $\{-\alpha_{10}^1 x_4^1(t)\}$ in equation 15.3(b) *and* the now explicit dynamics of x_4 (equation 15.3(d)).

(iii) The {acknowledged unknown} of $\xi_1^0(t)$ in equation 15.2(a) has been partly filled by the term $\{+\alpha_6^1 x_5^1(t)\}$ in equation 15.3(a) *and* the now explicit dynamics of x_5 (equation 15.3(e)), which in turn reaffirms the need for the state equation for x_4.

That there could be a yet further, similar unfolding of the {presumed known} and {acknowledged unknown} into a more refined Structure III is entirely possible, of course.

15.2.1 Problem Definition

Figuratively, we might once more approximate Structure I by the branch-node $((\boldsymbol{\alpha})-(\boldsymbol{x}))$ network of Figure 15.1(a).[1] On this occasion, however, the pictorial metaphor has been extended: to include – as dashed branches – the labels ξ for our {acknowledged unknown} in equation 15.2. The problems of essential interest in this chapter can now be defined as follows:

[1] If one is to be pedantic about such representations, a more complex structural map is required. In particular, such a map would have both directed branches – for the influence of one state over another – and branches emanating from and returning to the same state node (for parameters appearing in expressions where the dynamics of the given state are governed by the value of that same state). Constructing and visualising these maps computationally is a matter for further research.

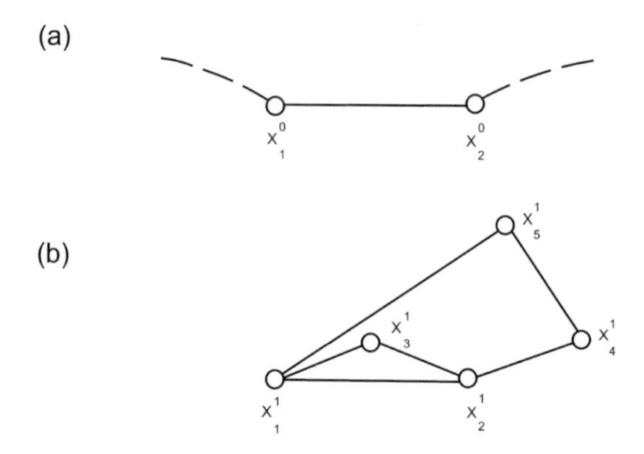

Fig. 15.1. Representation of model structures in the metaphor of the branch-node (($\boldsymbol{\alpha}$)–($\boldsymbol{x}$)) network: (a) **Structure I**, in which the dashed branches account for the putative influences of the {acknowledged unknown}; (b) Structure II.

(a) How could we discern, from interpretation of observed behaviour through the device of Structure I, any tendency for behaviour to appear to be evolving from this structure to that of, for example, Structure II (in Figure 15.1(b))? In other words, supposing the three additional state variables of Structure II (x_3^1, x_4^1, x_5^1) were effectively invariant over some initial period, such that the parameterisation of Structure I might justifiably be deemed (initially) invariant, how could we detect the onset of significant changes in these variables – from the test of Structure I alone?

(b) How, more specifically, could we discriminate – through the use of Structure I – between an apparent change of structure due to the more refined features subsumed under $\boldsymbol{\alpha}$ in the {presumed known} and an apparent change due to our {acknowledged unknown} under the rubric of ξ? Typically, x_3^1 might be significantly non-zero but changing slowly in the longer term, whereas x_4^1 and x_5^1 might be essentially zero for long periods, but then rise rapidly to substantial values (and just as quickly collapse back to zero). How could we detect, again from a test of Structure I alone, that the variable x_3^1 has mistakenly been subsumed under the definition of the invariant parameter α_2^0, *not* under the label ξ_1^0? How, conversely, could we discern the fact that the non-invariant behaviour of x_4^1 and x_5^1 attaches to something excluded entirely from Structure I, *not* to an explicit, but poor, approximation of the truth covered by the vector of parameters $\boldsymbol{\alpha}^0$? Figuratively, how might we apprehend the possibility that the (parametric) branch connecting x_1^0 and x_2^0 in Figure 15.1(a) implies the existence of x_3^1 in Figure 15.1(b), but that the dashed branches emanating from x_1^0 and x_2^0 in Figure 15.1(a) imply the presence of x_4^1 and x_5^1 in Figure 15.1(b)?

In general, and in contrast to the analysis of reachable futures set out in Chapter 11, the principle herein is to employ a relatively low-order structure for the model in processing the empirical observations, with the expectation that the "true" behaviour is, in fact, of a higher order (just as in the preceding chapter). However, we must also be alert to the possibility of parts of the structure declining into insignificance, suggesting thus evolution towards (in parts) a simpler structure, even for test models that are themselves not especially complex. For example, had we begun with Structure II, supposing x_3^1 to be both non-zero and substantially changeable with time, how could we detect – in the absence of direct observations of x_3^1 (as is customary in practice) – its decline to an invariant, negligibly small value (as can well be the case in practice; see the discussion of rehabilitating the Rhine River in Malle, 1994).

These, then, are the kinds of problems we wish to address. We must now introduce and elaborate a metaphor: in order to conceive of the means to solve these problems; to identify why the framework of recursive estimation seems so appropriate; yet to reveal why at the same time it has hitherto been somewhat mis-guided; and to arm ourselves with a conceptual framework with which to prise open the gates to a potentially more fertile field of algorithmic development. This will not be a brief discussion, however. Nor should it be, since the problems elucidated in the foregoing are central to this monograph as a whole.

15.3 A METAPHOR

The web, or mesh, of interactions among the state variables ($\boldsymbol{x}$) of the model, in Figure 15.1(b), can be looked upon as the frame of an aircraft or as an engineering structure, with its columns, beams, cantilevers, struts, trusses and so forth (Chapters 3, 4, and 5; Beck, 1987; Beck et al., 1990; or Beck and Halfon, 1991). The constituent hypotheses of which the model is composed, that is, the branches of the web of interactions in which the parameters $\boldsymbol{\alpha}$ appear as defining features, are as the component members (the individual columns and beams) of the engineering structure. Just as the air-frame or building has rigidity and the capacity to maintain a stable shape, so too has the structure of the model. To this extent the arrangement one to another of the nodes of the model's structure – the chosen logic of the interactions among its state variables – is fixed, invariant, and incapable of evolution.

Imagine now that the air-frame is indeed the aircraft in flight, subject to all the buffeting of the turbulence in the medium through which it is passing. The air-frame, of course, is designed to resist plastic deformation of its structure in the face of these disturbances. We most certainly do not wish to experience any localised bulging or compression of the fuselage, nor catastrophic failure in any component struts of the air-frame. The constituent members of the frame should nevertheless be elastic in their behaviour, undergoing deflection from time to time – with respect to some reference coordinate system – yet returning to an "equilibrium" position after passage of the load that induced the deflection. From a distance, all aluminium struts, i.e., all members of the structure, would appear to have identical material properties. But we know that at a very fine scale the crystal structure of the material will differ from part to part, with flaws in the desired regularity that may eventually lead to very different macroscopic properties among the constituent members of the air-frame assembly (Ashby, 1987). While these mechanical metaphors for the structure of a model cannot evolve – they cannot develop a new aileron out of thin air, so to speak – they *can* change, sometimes elastically, sometimes in a plastic manner, within the confines of the way in which their component members have been pinned together. And this is a very close, albeit inexact, analog of what has been introduced herein as apparent structural change in the abstract world of the mathematical model.

Our problem is now therefore this: how, within the fixed logical juxtapositions of the interactions among the model's state variables, can we detect strains in the members of those interactions as the model's structure is dynamically loaded by the divergence of its outputs from observed behaviour? This is our analog of the aircraft in flight. Further, since the model's structure will contain constituent hypotheses with differing intrinsic strengths and weaknesses, and since some of them will clearly be flawed, our expectation is that elasto-plastic deformation of the structure will be a function both of the loads imposed and the material properties of the structure's components. We can construct a mapping between the two problems: from that of detecting apparent structural change in a state–space model of a dynamical system, to that of analysing elasto-plastic deformation of an engineering structure subjected to a dynamic loading. For the former we do not have even a good problem-solving framework, let alone a method of solution. Yet for design against plastic deformation of an engineering structure there is a wealth of problem-solving procedures (Munro and Lloyd Smith, 1972; Ashby, 1987; Lloyd Smith, 1990). What can be learned, then, from the problem-solution couple of this metaphor; can any of its problem-solving procedures be mapped back into the field of detecting apparent structural change?

15.3.1 The Conceptual Framework of Recursive Prediction

Let us first step back to reflect on the nature of *adaptive control*, whence derives (historically) the fundamental notion of parametric change and the technical means to track it (as already discussed in Chapters 3 and 5).

Some of the elements of the input vector $\boldsymbol{u}$ in equation 15.1(a) may be subject to manipulation, by which they justify being labelled "controls", as opposed to "disturbances". In the engineering of control we are at liberty, within certain bounds, to choose the values for these control variables such that the state $\boldsymbol{x}(t)$, or output $\boldsymbol{y}(t)$, follows a desired trajectory $\boldsymbol{x}_d(t)$, or $\boldsymbol{y}_d(t)$, through time (whatever the fluctuating character of the disturbances). In adaptive control, the parameterisation of the model, $\boldsymbol{\alpha}$, is acknowledged as *not* being invariant with time; and the intent of the control scheme is to choose the controls (within $\boldsymbol{u}(t)$) such that the changes $\boldsymbol{\alpha}(t)$ can be more readily tracked while yet achieving convergence between $\boldsymbol{y}(t)$ and $\boldsymbol{y}_d(t)$. The premises are these: better control will follow from a better understanding of the relationships between inputs and outputs; this improved understanding will follow from prudent (but not excessive) probing of the nature of these relationships, using deliberate perturbations of the controls; and, at the next available opportunity for such probing, say at the instant t_k in time, the choices of any changes in the controls will be based on all the experience accumulated up to the previous assessment of success, at say t_{k-1}, in improving understanding and achieving the desired control.

Control itself, however, is *not* our problem herein. Our task, given the past observations $[\boldsymbol{u}(t), \boldsymbol{y}(t)]$, is in part to choose $\boldsymbol{\alpha}(t)$, such that when these parameter trajectories are substituted into equations 15.1, the model-derived estimate of what should be observed, i.e., $\hat{\boldsymbol{y}}(t) = \boldsymbol{h}\{\boldsymbol{x},\boldsymbol{\alpha};t\}$, is (retrospectively) maximally close to what was actually observed, that is, $\boldsymbol{y}(t)$. In this manner we may then trace the deformation induced in the model's structure as it is placed under the dynamic loading, as it were, of the divergence between $\boldsymbol{y}(t)$ and $\hat{\boldsymbol{y}}(t)$. Unlike the problem of control engineering, we are here not free to choose the values of the inputs and, at least for the record of the past, neither are we free to choose some desired output trajectory $\boldsymbol{y}_d(t)$; $\boldsymbol{y}(t)$ is given and immutable. Armed with a technique for estimating the trajectories of $\boldsymbol{\alpha}(t)$, however, we might be able to reveal the inadequacies of what we have presumed is known, within the choices of $[\boldsymbol{x},\boldsymbol{\alpha},\boldsymbol{f},\boldsymbol{h}]$ appearing explicitly in the model.

But this is only a part of our problem. For we must also develop the means to detect changes of *significance* – *not* of pure chance – in the expanse between this presumed known and the unknowable truth. $\boldsymbol{\xi}$ and $\boldsymbol{\eta}$ are fictions, being themselves unknowable by virtue of this self-same definition, as mere labels for the divergence between the model and reality. In order to make any further progress we must reformulate the representation of our model (equation 15.1), in fact, putting it in the naturally recursive format of adaptive control, or in what is called an innovations representation of the system's behaviour (around which we began to build our discussion in Chapter 5), as follows:

$$\mathrm{d}\boldsymbol{x}(t \mid t_{k-1})/\mathrm{d}t = \boldsymbol{f}\{\boldsymbol{x}(t \mid t_{k-1}), \boldsymbol{u}(t), \boldsymbol{\alpha}\} + \boldsymbol{K}\boldsymbol{\varepsilon}(t \mid t_{k-1}) \tag{15.4a}$$

$$\boldsymbol{y}(t_k) = \boldsymbol{h}\{\boldsymbol{x}(t_k \mid t_{k-1}), \boldsymbol{\alpha}\} + \boldsymbol{\varepsilon}(t_k \mid t_{k-1}) \tag{15.4b}$$

Here the argument $(t \mid t_{k-1})$ signals a predicted value of the associated quantity at some (future) time t utilising the model and all observed information, in particular, in respect of the observed output $\boldsymbol{y}$, up to and including that available at the most recent sampling instant, t_{k-1}. $\boldsymbol{\varepsilon}(t_k \mid t_{k-1})$ is the innovation, i.e., the mismatch between now the predicted and observed values of the output at the next sampling instant in discrete time, t_k, in equation 15.4(b); $\boldsymbol{\varepsilon}(t \mid t_{k-1})$ in equation 15.4(a) is the value of this quantity at times not coincident with the sampling instant. $\boldsymbol{K}$ is a weighting matrix and can be thought of as a device for distributing the impacts of the innovations among the constituent representations of the various state variable dynamics, i.e., the representations $f_i\{\cdot\}$ for each state x_i. Precisely how $\boldsymbol{K}$ is to be chosen, and the role it might play in detecting apparent structural change, are clearly matters of central interest.

Before turning to that issue, it is important to note that whereas $\boldsymbol{\xi}$ and $\boldsymbol{\eta}$ were not computable quantities, $\boldsymbol{\varepsilon}$ is, at least up to a point. Having fixed the structure of the model, including assigning values to $\boldsymbol{\alpha}$ and $\boldsymbol{K}$, and having made assumptions about the values of x and ε at some initial time t_0, acquisition of the observation $y(t_1)$ permits $\varepsilon(t_1|t_0)$ to be computed as the difference between this observation and its (one-step-ahead) prediction, $y(t_1|t_0) = h\{x(t_1|t_0),\alpha\}$, in which $x(t_1|t_0)$ has been obtained as the value of the state of the system at t_1 computed from the solution of equation 15.4(a) over the interval $t_0 \rightarrow t_1$. To obtain $x(t_1|t_0)$ a subtle but not insignificant implication is the requirement for access to knowledge of how $u(t)$ and $\varepsilon(t)$ evolve over the interval of integration. The common, pragmatic assumption is that they do not change, but remain constant at their respective values at the beginning of the interval, here $u(t_0)$ and $\varepsilon(t_0)$.[2] Given $\varepsilon(t_1|t_0)$, the process can be repeated upon acquisition of yet another observation $y(t_2)$ at time t_2, and so on.

To summarise, as our candidate model structure (equation 15.4) is piloted through the observed record, subsequent predicted values of the state variables are adapted in the light of preceding errors of prediction. These errors, of course, gauge not the "absolute" divergence of the model from the truth; yet the divergence, whatever its nature, is at least computable. In this formulation, the trajectories of the state variables (*not* the model's parameters) are adapted in the light of the model's capacity to reflect what has been observed (*not* the truth). According to our metaphor, the shape of the aircraft here remains entirely intact, without even elastic deformation, although its trajectory through the atmosphere may exhibit high-frequency judder in its precise position relative to an otherwise pre-determined path. As such, the innovations representation is but a modest advance on what would otherwise be an entirely non-adaptive form of model. Were our recursive system of prediction to be given no access to observations of the outputs $y(t_k)$, equation 15.4(b) would be redundant, and the values of the state variables would simply have to be generated from equation 15.4(a) in the familiar, non-adaptive, deterministic manner with no innovations. This, then, would be the "pre-determined path" to which we have just referred.

15.3.2 Tracing Structural Error in the {Presumed Known} and the {Acknowledged Unknown}

The divide between what we know relatively well – the {presumed known} – and what we do not know at all well – the {acknowledged unknown} – lies now between the first and second terms on the RHS of equation 15.4(a). It resides likewise between the first and second terms on the RHS of equation 15.4(b), in the sense that $h\{\cdot\}$ may not capture well the precise nature of how the observing process gains access to the state variables of the model (although this might on many occasions be considered a minor imperfection). As opposed to the more familiar form of model of

[2] Unless more sophisticated assumptions are to be made, it is clear that the performance of the innovations representation should in general improve as the interval between the sampling instants decreases.

equation 15.1, the innovations representation opens up the possibility of detecting structural error not only via the estimated behaviour of the conventional model parameters $\boldsymbol{\alpha}$ (as previously), but also through the computable values of the amalgam of $\boldsymbol{K}\boldsymbol{\varepsilon}(t\,|\,t_{k-1})$.

We might – but will not – assign values to the elements of $\boldsymbol{K}$ by assumption, prior to starting analysis of the performance of the model against the observed record. At the most elementary level, for instance, if n state variables were connected logically (via $\boldsymbol{h}\{\cdot\}$) to m observed outputs, thus contributing to the composition of an m-dimensional innovations vector $\boldsymbol{\varepsilon}$ in equation 15.4(b), it would make sense to assume that the non-zero elements of $\boldsymbol{K}$ (an $n \times m$ matrix) should "inversely" re-distribute the impacts of these innovations back to the dynamics of the n state variables in equation 15.4(a). Further, if we thought, *a priori*, that we knew well (poorly) the dynamic equations of some of the state variables, the associated elements of $\boldsymbol{K}$ might be assigned relatively small (large) values. Alternatively, if we were prepared to invoke several more sophisticated assumptions, for example, about the variance-covariance properties of the various sources of uncertainty entering into the model of equation 15.1, $\boldsymbol{K}$ could be computed indirectly from an accompanying representation of the manner in which the uncertainty attaching to the estimated values of the state variables is propagated through time. This would be tantamount to using the celebrated Kalman filter.[3] There might be some debate about whether employing the filter truly constitutes assigning values to $\boldsymbol{K}$ by prior assumption, especially in the case of an extended Kalman filter (EKF). Our experience has shown that the performance of the EKF is very probably impaired by the manner in which $\boldsymbol{K}$ is computed and that this impairment – again, very probably – has much to do with the fact that the computation can indeed be largely (if not entirely) a function of the prior assumptions (Beck, 1994). The present discussion is in fact a direct consequence of the search for an algorithm that is not so dominated in this manner.

Instead of making any such assumptions, we would prefer the consequences of juxtaposing the model with observed behaviour to be minimally prejudiced by what must be assumed to be known at the outset. Our interest will centre on designing a form of analysis in which we can maximise the capacity to detect the appearance and growth of structural errors in the candidate model as the data are processed in sequence. In other words, as the template of the candidate model is projected through this observed record we wish to know how it might snag on, and be distorted by, features of significance in the expanse between the tractable known (the template) and the unknowable truth. Some of these features may be similar to – yet nevertheless diverge significantly from – what was thought to be relatively well known. Others may turn out to be decidedly *not* the result of just chance events

---

[3] In the context of the innovations representation of equation 15.4, the operation of the prediction-correction format of the filter would have to be interpreted as follows: given the computed values for $\boldsymbol{K}$, the product $\boldsymbol{K}\boldsymbol{\varepsilon}(t\,|\,t_{k-1})$ in equation 15.4(a) would be defined only at the instants $t = t_k$, as *instantaneous* corrections to the values of the states $\boldsymbol{x}(t_k\,|\,t_k)$ *between* successive integrations over the intervals $t_{k-1} \rightarrow t_k$ and $t_k \rightarrow t_{k+1}$

emerging at random from the residue of what was previously thought to be not at all well known. In this latter, the elements of $\boldsymbol{K}$ fulfil exactly the same role in the description of the state dynamics of equation 15.4 as do the conventional parameters ($\boldsymbol{\alpha}$) appearing in $\boldsymbol{f}\{\cdot\}$ (and $\boldsymbol{h}\{\cdot\}$). They are markers, or logical tags, attaching to potential forces of more than pure chance acting in some as yet imprecisely described manner on the dynamic behaviour of the state variables. As with $\boldsymbol{\alpha}$, our task is to facilitate computation of these additional parameters as the structure of the model is placed under the stress of being exposed to the field observations. In principle, what we need is a recursive algorithm for tracking the trajectories of *both* $\boldsymbol{\alpha}$ and $\boldsymbol{K}$ with time, so that the structure of the model of the innovations representation of equation 15.4 can be adapted in the light of success and failure in successive attempts at predicting the next set of values of the observed outputs.

In order to illuminate better the various roles of $\boldsymbol{\varepsilon}$, $\boldsymbol{\alpha}$, and especially $\boldsymbol{K}$, as indices of structural error and change, let us imagine the following. A set of data is to be analysed, with the still to be designed recursive parameter [$\boldsymbol{\alpha}$, $\boldsymbol{K}$] estimation algorithm. *A priori* we presume that what is known ($\boldsymbol{\alpha}$) is known relatively well, so that $\boldsymbol{K}$ can be set systematically to zero, with the accompanying expectation that the innovations ($\boldsymbol{\varepsilon}$) will be small, if not negligible. After all, we should set out to expect success in applying our recursive scheme of prediction. As the data are processed, a pattern of relatively rapid, high-frequency fluctuations in $\boldsymbol{\varepsilon}$ will emerge, albeit perhaps exhibiting persistence and bias in some direction. Whatever the case, the high-frequency flutter of the rapid innovations (adaptations) will be continually fed back into the recursive predictions of the subsequently observed outputs. At a generally lower frequency, but nevertheless with a capacity for relatively swift change, the values assigned to $\boldsymbol{K}$ may undergo adaptation, reflecting possibly some integral of the fluctuations in $\boldsymbol{\varepsilon}$. For example, if the reconstructed estimates of some of the elements of this matrix were to veer away from essentially zero – as the historical record is processed sequentially (as in Stigter and Beck (1994) and Stigter (1997)) – this could imply that the one-step-ahead predictions of the model rest more on the unknown aspects of divergence between the immediate past predictions and observed behaviour, i.e., $\boldsymbol{\varepsilon}$, than on the supposedly known content of the model, symbolised by $\boldsymbol{\alpha}$ in [$\boldsymbol{f}$,$\boldsymbol{h}$]. In other words, $\boldsymbol{K}$ allows us now to illuminate sources of structural change and error residing in the largely {acknowledged unknown}, whereas similar nonstationary behaviour in the recursive estimates of $\boldsymbol{\alpha}$ permits tracing of potential change within what was previously {presumed known}, i.e., prior theory. The former, as might be expected, is less specific in its indications of structural error than the latter. For whereas the elements of $\boldsymbol{\alpha}$ are attached to each explicit, constituent hypothesis of the assembled knowns, the elements of $\boldsymbol{K}$ attach to the description of each state variable's dynamics, the presumed known aspects of which will in general comprise an agglomeration of several constituent hypotheses.

Figuratively, in the mesh of interactions in the template of the innovations representation of the model of the system's behaviour, the elements of $\boldsymbol{\alpha}$ mark out components of the explicit relationships among the state variables, while the elements of $\boldsymbol{K}$ mark out implicit relationships between each state included in the

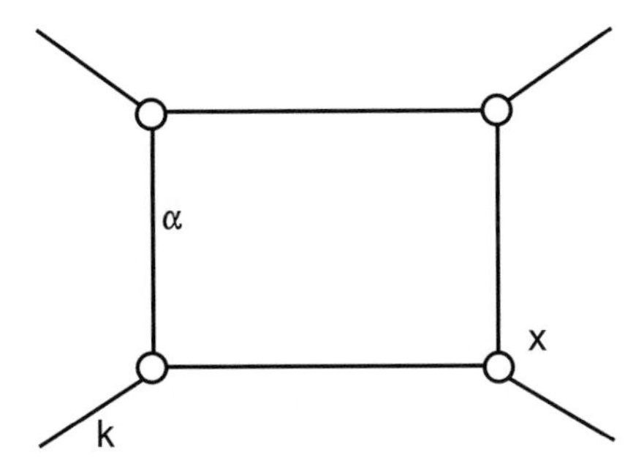

Fig. 15.2. Revised archetypal model structure (compare with Figure 5.5(b), for example), in which the elements k_{ij} of $\boldsymbol{K}$ have symbolically been attached to the branches emanating from the nodes of the structure.

model and all of that which has not been included. In Figure 15.2, the revised archetypal structure, the elements k_{ij} of $\boldsymbol{K}$ have symbolically been attached to the branches emanating from the nodes of the specified structure, extended thus into the space between the tractable known and the unknowable truth. As this structure is forced to negotiate a path through the observed behaviour our attention must be focused on the distortions thereby induced, that is, the apparent change of structure, in *both* $\boldsymbol{\alpha}$ and $\boldsymbol{K}$. The agent of this change – a bending moment applied, as it were, to each strut in the frame, including now the speculative, probing "antennae" reflected in the elements of $\boldsymbol{K}$ – should in part be driven by some integral of the fluctuations in the innovations (ε) over time.

In short, while our model structure should be well able to behave elastically when placed under load, what we are really interested in is deformation of a more plastic nature.

15.4 DESIGN FOR PLASTIC DEFORMATION

There should be little surprise about the choice of recursive estimation as the framework for solving the problem of apparent structural change as now constructed. For this very construction itself has been born of a long-standing fascination with filtering theory. Detecting apparent structural change, however, is not quite the same as the identification of model structure, as previously related, for example, in Beck (1994). What we need herein can be introduced by first recounting, in a historical context, what is *not* needed.

15.4.1 State Estimation and Model Structure Identification

To begin with, it will be fairly obvious that we do not need, primarily, an algorithm of *state* estimation. Apprehension of change in the structure of the system's behaviour will not be revealed in the "outer, surface" space of the model's state variables, but in the "inner, embedded" space of its parameters. Our goal, therefore, is not minimum-variance estimation of the values of the state variables and we would do well to recall that the gain matrix in a Kalman filter (and an extended Kalman filter), of which the matrix $\boldsymbol{K}$ is so strongly redolent, is computed precisely in pursuit of this goal (for example, Gelb (1974)). A matrix $\boldsymbol{K}$ defined algorithmically by such a goal function is not appropriate here. Above all, we do not need an algorithm with an intrinsic bias towards compensating for any significant mismatches between predicted and observed behaviour through superficial adaptation of the state variable predictions ($\boldsymbol{x}$) – as opposed to adaptation at a deeper level via the elements of $\boldsymbol{\alpha}$ and $\boldsymbol{K}$, i.e., the model's parameters. We do not want our aircraft to be jolted about by whatever disturbances of poor prediction may come its way, while its essential shape is retained entirely intact. Whatever goal function may be chosen to dictate the means by which $\boldsymbol{\alpha}$ and $\boldsymbol{K}$ are estimated, minimum-variance estimation of the states will *not* be it.

Second, we do not need the two-part procedure for attempting to solve the problem of *model structure identification* (Beck, 1986, 1994), although it is informative to reiterate here the conceptual framework in which this procedure is cast. For that (other) problem the goal is to construct first as rigidly specified a structure as possible – in Popperian terms, to make as bold an assembly of conjectures as possible (Figure 15.3(a)) – and then to test (or load) that structure to the point of failure as it is reconciled with the structure underlying the observed record. In particular, in seeking such a demonstrable refutation, as revealed by significantly non-stationary behaviour in the recursive estimates of the model's parameters, we wish to establish *not* whether the model as a whole is defective, but which of its constituent members might be inadequate (such as the oscillatory member indicated in Figure 15.3(b)).

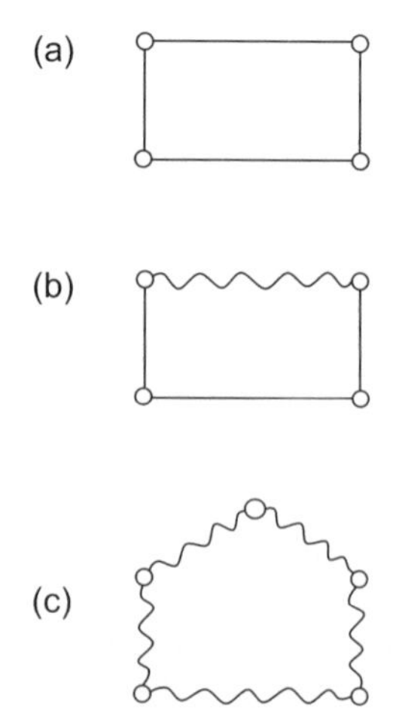

Fig. 15.3. Two-part procedure for attempting to solve the problem of model structure identification: (a) rigid specification of the prior structure; (b) exposure (plastic deformation) of defective member of the prior structure; and (c) speculative exploration with a candidate revised (posterior) structure.

Guided specifically by the outcome of this first step, the goal is then to revise the structure of the model in order to complete the second part of the procedure: of allowing the relatively flexible, more speculative, less boldly expressed model structure of Figure 15.3(c) to be moulded by whatever is the predominant shape of the structure apparently captured in the observed record.

The essential distinction between these two steps hinges upon assumptions about the material properties of the component members of the model. Roughly speaking, if one wishes to expose the abrupt failure of a member, the recursive estimation algorithm must not be primed with the means for any subsequent, smooth adaptation without clear, unequivocal failure. The parameters of the model should be fixed as truly invariant with time, albeit with constant values not known (*a priori*) with certainty. Their properties should be likened to those of a brittle, non-ductile material, capable of abrupt, demonstrable failure but with non-uniform strengths, i.e., able to withstand (without failing) different intensities of loading and stress. We are fortunate, therefore, in having a device for articulating such subtle notions within the recursive estimation scheme, namely, the variance–covariance matrix of prior model parameter (and state) estimation errors. Thus, a parameter attaching to a secure piece of prior theory, i.e., a parameter with a low prior uncertainty (or greater initial strength), should be able to withstand greater bending moments before failing, relative to a parameter attaching to a less secure item of theory.

If, on the other hand, one wishes to allow the model to be shaped by whatever is the predominant structure of the data (assumed to be constant with time yet different from that of the model), it must be endowed with a capacity for adaptation. Conventionally this can be accommodated in an extended Kalman filter by assuming the parameters to be stochastic processes (as opposed to random variables), and technically realised through the following elementary model, as an augmentation of the model of equation 15.1,

$$\dot{\alpha}(t) = \zeta(t) \tag{15.5}$$

Here, the variance of the stochastic disturbance ζ can be chosen to reflect the relative ductility (capacity to evolve without failure) of each constituent parameter in the model. The astute reader will now immediately protest that it may simply be impossible to arrive at any reasonable set of estimates for the elements of the variance–covariance matrix of ζ, and quite rightly so (Beck, 1994). It is precisely for this kind of reason that we have sought to develop an alternative – to what are strictly filtering algorithms – in order to better approach, however modestly, a solution to the problem of model structure identification (Stigter and Beck, 1994).

To summarise, an algorithm designed to achieve the goal of minimum-variance state estimation is not needed; nor is a procedure for exposing the brittle failure of structural elements of the model. Unlike the aircraft of our metaphor, we actually want to promote the potential for the structure of the model to undergo plastic deformation – without failure – which must then be detectable, and in as specific a manner as possible. We do not want a wing piece to be ruptured and torn way, leaving the aircraft incapable of any further flight. We might want instead to make use in some way of a structure composed of relatively ductile members, but without the accompanying mass of arbitrary assumptions thereby implied. In short, what we want is an algorithm designed for parameter estimation, directed by the goal of detecting smooth structural change in both what we presume to know well and what we acknowledge as not at all well known, subject to a set of constituent structural members with a spectrum of ductile material properties. Meeting this specification, let alone discerning whether it is an appropriate specification, is not something sought herein. We shall merely suggest the beginnings of a departure from what has hitherto proved to be the best of what can be achieved in resolving the somewhat different problem of model structure identification.

15.4.2 A Recursive Prediction Error Algorithm

The failure of filtering theory, in particular, the extended Kalman filter (EKF), to deliver satisfactory results in identifying the structure of a model is well documented (Beck, 1987, 1994). Such failure should hardly be surprising, for the algorithm was never specifically developed for the purposes of parameter estimation, in which terms the problem of model structure identification has been cast. In a significant, but curiously little utilised paper, Ljung (1979) isolated the intrinsic causes of this failure of the EKF to succeed as a recursive estimator of a model's parameters and proceeded to propose an algorithm that would eliminate them. The success of this new algorithm, albeit designed for parameter estimation (not for model structure identification or the detection of apparent structural change), was founded upon the innovations form of model set out in equation 15.4. Further, the algorithm is best appreciated in the present context as having been derived *not* from a background of filtering theory but as a recursive version of a more classical form of (non-recursive) optimisation scheme (Ljung and Söderström, 1983). One can therefore compose whatever is to be the goal directing the manner in which the structure of the model is to be reconciled with that underpinning the observed behaviour, and derive a recursive algorithm to suit this purpose, which is precisely what is needed.

When trained onto the problem of model structure identification the Recursive Prediction Error (RPE) algorithm performs really rather well (Stigter, 1997), especially in illuminating structural inadequacies similar to those set out in the motivating example given above, as we shall see. In its current form, and as used in one of the case studies presented below, this particular RPE algorithm is derived from the following objective function

$$J = \sum_{k=1}^{n} \varepsilon^T(t_k | t_{k-1})\varepsilon(t_k |_{k-1}) \tag{15.6}$$

whose minimisation is sought with respect to the parameters of the model, i.e., $\boldsymbol{\varphi}$, where $\boldsymbol{\varphi}$ is simply the vector containing the original model parameters, $\boldsymbol{\alpha}$, and all the elements of the matrix $\boldsymbol{K}$ that are to be estimated. Unlike a conventional filtering algorithm, such as the EKF, recursive correction of the estimates of these parameters is therefore *not* governed by the implied goal of obtaining estimates of the states ($\boldsymbol{x}$) and parameters with minimal errors of estimation.

In outline, given the innovations form of model representation (equation 15.4), derivation of a continuous-discrete version of the RPE algorithm is achieved as follows:

(i) The objective function of equation 15.6, at time t_k, is approximated by a first-order Taylor series about the current (nominal) estimate of the parameters, i.e., $\hat{\boldsymbol{\varphi}}(t_k|t_{k-1})$, and then differentiated with respect to $\boldsymbol{\varphi}$. This results in a recursive equation for $\hat{\boldsymbol{\varphi}}(t_k|t_{k-1})$ in terms, *inter alia*, of $\hat{\boldsymbol{\varphi}}(t_{k-1}|t_{k-2})$ and the Hessian matrix of second derivatives associated with the objective function.

(ii) A matrix of gradients of the one-step-ahead predictions of the outputs with respect to the parameters is introduced, ($\boldsymbol{\Psi}(t_k|t_{k-1})$), so that when the objective function is differentiated twice this leads to a second recursive equation, for the Hessian matrix, in which this matrix of gradients then appears.

(iii) The variance–covariance matrix of the parameter estimation errors ($\boldsymbol{P}_{\varphi\varphi}(t_k|t_{k-1})$) is introduced and defined as the inverse of the Hessian matrix such that, after some manipulation, this yields a recursive equation for $\boldsymbol{P}_{\varphi\varphi}(t_k|t_{k-1})$ in terms of $\boldsymbol{P}_{\varphi\varphi}(t_{k-1}|t_{k-2})$.

(iv) The model, i.e., the innovations representation of equation 15.4, is used to derive a computable expression for the matrix of gradients ($\boldsymbol{\Psi}(t_k|t_{k-1})$), which in turn requires a matrix of sensitivity coefficients, of the state predictions with respect to the parameters ($\boldsymbol{W}(t_k|t_{k-1})$), which can itself be generated from a recursive relationship involving $\boldsymbol{W}(t_{k-1}|t_{k-2})$. In fact, this last recursive equation is the discrete-time solution of the familiar adjoint equations of sensitivity analysis.

In detail, the derivation can be found in Stigter (1997).

Being recursive, this RPE algorithm is clearly not radically different from the equations of a Kalman filter, or extended Kalman filter. In summary, its principal advantages would appear to be: (i) performance that is considerably more robust than that of the EKF; (ii) fewer arbitrary, prerequisite (prior) assumptions, which have themselves bedevilled performance of the EKF; (iii) a better directed goal function; (iv) an associated algorithmic structure focusing on the properties of the parameters, not the states; and (v) the potential to facilitate discrimination of structural inadequacy/change in the {presumed known}, $\boldsymbol{\alpha}$, from that in the {acknowledged unknown}, $\boldsymbol{K}$. In short, we have an algorithm seemingly well designed for detecting plastic deformation in a model structure subjected to the stress of being reconciled with the field observations.

15.4.3 Some Material Properties of the Gain Matrix

At the heart of the RPE algorithm lies the following recursive adjustment of the estimates of the parameters

$$\hat{\varphi}(t_k) = \hat{\varphi}(t_{k-1}) + \boldsymbol{L}(t_k)\varepsilon(t_k \mid t_{k-1}) \tag{15.7}$$

Just as previously in filtering theory, great interest should now be centred upon the properties of the gain matrix $\boldsymbol{L}$. Put simply, it is a ratio of the {uncertainty attaching to the predicted outputs as derived from what is known well (the model)} and the {total uncertainty attaching to the predicted outputs}. The latter includes, in addition to the former, the uncertainty deriving from the innovations (ε), or that which is not known at all well – in some senses the computable portion of our {acknowledged unknown}. In contrast to the gain matrix of a conventional extended Kalman filter, there are two salient points to note. First, no uncertainty attaching to the one-step-ahead predictions of the state variables enters into the computation of $\boldsymbol{L}$. Second, whereas the variance-covariance matrix of ε can be estimated from the performance of the algorithm as it proceeds through the data, the filter requires a prior assumption to be made about the variance-covariance matrix of the output observation errors, i.e., $\boldsymbol{\eta}$ in equation 15.1. Together, these two features permit the evolution of $\boldsymbol{L}$ to be less dependent upon erroneous prior assumptions, which we know have been the downfall of earlier studies with the EKF (Beck, 1986, 1987, 1994). $\boldsymbol{L}$ will be more immediately and more exclusively dependent upon successful structural characterisation of the model, instead of being prone to the success or failure of its recursive *state* predictions. Its elements may even usefully reflect a relatively unprejudiced, posterior balance among the uncertainties attaching to our {presumed known} and our {acknowledged unknown}.

In the metaphor of the air-frame, the product $\boldsymbol{L}(t_k)\varepsilon(t_k \mid t_{k-1})$ in equation 15.7 can be likened to the strains in the parametric (φ) branches of the structure resulting from the stresses of the dynamic loading imposed through ε. Providing these analogs have been correctly translated from the metaphor into the abstract domain of structural change (in a model), they strongly imply that the elements of $\boldsymbol{L}$ must play a mediating role between an applied stress and the resulting strain. And this, by deduction, can only be a material property of the given parametric branch: something akin to the material properties of ductility or brittleness, or to the branch's capacity to resist plastic deformation.

15.5 CASE STUDIES

How exactly the foregoing interpretation of the material properties of $\boldsymbol{L}$ might assist us in our quest for an algorithm capable of addressing the issue of structural change, is not at all clear. Its further consideration is in any case a matter for the future. For what we must now demonstrate is the potential to obtain some kind of solution, in principle, to the two problems defined at the outset of the chapter. This will be achieved through two case studies – one hypothetical, the other real – using two variations on the basic theme of an RPE algorithm derived from the goal function of equation 15.6. Both case studies reflect facets of the motivating example introduced earlier.

15.5.1 Structural Error in the {Presumed Known}

In this first case study a hypothetical problem is constructed. In truth substrate (x_1) is metabolised by a biomass (x_2), yet the empirical observations of substrate behaviour are to be interpreted by a model with just a single state variable (x_1), as though there were chemistry (not biology) in this. We have, therefore, the following:

Interpretative Structure (Resolving Power 0):

$$\dot{x}_1^0(t|t_{k-1}) = (u_1^0(t)/\alpha_1^0)u_2^0(t) - (u_1^0(t)/\alpha_1^0)x_1^0(t|t_{k-1}) - \alpha_2^0 x_1^0(t|t_{k-1}) + k_1^0\varepsilon_1^0(t|t_{k-1}) \tag{15.8}$$

True Structure (Resolving Power 1):

$$\dot{x}_1^1(t) = (u_1^1(t)/\alpha_1^1)u_2^1(t) - (u_1^1(t)/\alpha_1^1)x_1^1(t) - \alpha_2^1 x_1^1(t)x_2^1(t)/[\alpha_3^1 + x_1^1(t)] \tag{15.9a}$$

$$\dot{x}_2^1(t) = -(u_1^1(t)/\alpha_1^1)x_2^1(t) + \alpha_4^1 x_1^1(t)x_2^1(t)/[\alpha_3^1 + x_1^1(t)] - \alpha_5^1 x_2^1(t) \tag{15.9b}$$

Error-corrupted observations of x_1 are generated simply as

$$y_1^1(t_k) = x_1^1(t_k) + \eta_1^1(t_k) \tag{15.10}$$

so that input–output data $[\boldsymbol{u}^0, y_1^0]$ can be provided for processing via the Interpretative Structure, as though the output were in fact obtained from a description consistent with the resolving power of that structure. Apart from the stochastic observation error sequence η_1^1 of equation 15.10, the True Structure of the system's behaviour is not subject to any other form of disturbance that would be classified as technically unobservable.

On this occasion, we may note that the structural error is of the following form: the {presumed known} of $-\alpha_2^0 x_1^0(t|t_{k-1})$ in equation 15.8 has been unfolded into the term $\{\alpha_2^1 x_1^1(t)x_2^1(t)/[\alpha_3^1 + x_1^1(t)]\}$ of equation 15.9(a) and the explicit dynamics of the second state, x_2^1, given by equation 15.9(b). There is no structural error subsumed under what we have labelled our {acknowledged unknown}, which would otherwise be associated with the amalgam of $k_1^0\varepsilon_1^0(t|t_{k-1})$ in the innovations representation of the Interpretative Structure.

Our problem is succinctly stated. Given the data $[\boldsymbol{u}^0, y_1^0]$ and given the Interpretative Structure of equation 15.8, is it possible to discern any indications of the

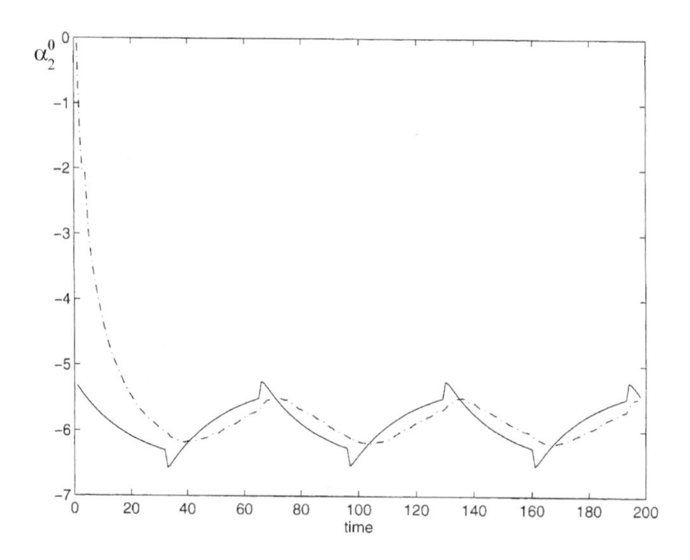

Fig. 15.4. Recursive estimates of α_2^0 in the Interpretative Structure of equation (15.8) shown as the dashed-dot line; true variations in this parameter, generated from the True Structure of equation (15.9) are shown by the continuous line.

underlying True Structure in the recursive estimates of the parameters φ obtained from an RPE algorithm?

Figure 15.4 provides the evidence for a response (Stigter, 1997). The recursive estimates of $\hat{\alpha}_2^0$ generated by an RPE algorithm, in which the Interpretative Structure has been embedded, are shown as the dash–dot line. After an initial transient, from its prior estimate of 0.0, α_2^0 clearly tracks what is known to be the true nature of its variation, i.e., $\alpha_2^1 x_2^1(t)/[\alpha_3^1 + x_1^1(t)]$ from the True Structure. The estimate (not shown) of the gain matrix element (k_1^0), after an initial transient fluctuation – from its prior value of 0.0 to just above 1.0 – settles at a value of 0.13 within just twenty or so sampling intervals, whereafter its relative sensitivity coefficient (within the matrix $\boldsymbol{W}$) is effectively negligible.

From this simple, hypothetical example – with admittedly little uncertainty entering into the propagation of the states and the observations thereof – we might tentatively conclude that the framework of the RPE algorithm does indeed have the capacity (a) to detect a change of structure and (b) to signal a correct classification of the type of structural error, as one due to our {presumed known} as opposed to our {acknowledged unknown}.

15.5.2 Structural Error in the {Acknowledged Unknown}

In reality, in our second case study (in the analysis of data from the River Cam), there is no access to the truth. We may conjecture, however, that it may have been similar to Structure II, i.e., equation 15.3, of our motivating example (Beck, 1975). Our

Interpretative Structure will be similar to that of the foregoing case study, likewise accounting for the behaviour of the substrate concentration:

Interpretative Structure (Resolving Power 0):

$$\dot{x}_1^0(t|t_{k-1}) = (u_1^0(t)/\alpha_1^0)u_2^0(t) - (u_1^0(t)/\alpha_1^0)x_1^0(t|t_{k-1}) - \alpha_2^0 x_1^0(t|t_{k-1}) + \alpha_3^0 + k_1^0 \varepsilon_1^0(t|t_{k-1}) \tag{15.11}$$

Input–output data $[\boldsymbol{u}^0, y_1^0]$ are taken from the original study (Beck, 1973).

In this instance, we cannot be at all sure of the nature of the structural error, but it seems most likely it was of the following form: the {acknowledged unknown}, recognised through the term $\{+k_1^0\varepsilon_1^0(t|t_{k-1})\}$ in equation 15.11, is to be partly filled by an expression – in the equation of state for x_1^0 – accounting for the way in which a supplementary state variable, denoted x_2^1, generates substrate in the river, according to an as yet to be specified second (additional) state equation. The speculation was that x_2^1, if it existed, would fulfil the role of detrital organic matter deriving from a phytoplankton population (as previously identified as the state x_5^1 in Structure II of the motivating example; Beck, 1975).

Given therefore the data $[\boldsymbol{u}^0, y_1^0]$ and given the **Interpretative Structure** of equation 15.11, is it possible to discern (a) anything (in the recursive estimates of the parameters $\boldsymbol{\varphi}$ obtained from an RPE algorithm) of a different structure underlying observed behaviour, and (b) any sense of whether this difference is logically associated with our {presumed known} or {acknowledged unknown}?

The outcome of testing the Interpretative Structure is shown in Figure 15.5 (Stigter and Beck, 1994). The recursive estimates of $\hat{\alpha}_2^0$ and $\hat{\alpha}_3^0$ are *relatively* invariant, although debatably so for the latter.[4] The trajectory for the estimate of k_1^0, on the other hand, shows a quite distinctive feature at about t_{40}, where it rapidly becomes significantly non-zero and positive. In this it is undoubtedly different from the corresponding result of the foregoing hypothetical case study. But given the results of Figure 15.5, *without* the benefit of a very great deal of hindsight (or foresight, for that matter), would one categorise the implied detection of a structural error as having to do more with the {presumed known} or {acknowledged unknown} of the Interpretative Structure? Perhaps the most one could conclude is that there is strong evidence of the latter, equivocal evidence of the former. This, however, is in the nature of real problems. At the least we appear to have an algorithmic framework with genuine potential for attacking the kinds of questions we have posed herein; it is therefore well worth further attention, perhaps along the following lines.

[4] The estimate shown in Figure 15.5(b) is certainly much more stationary than the equivalent estimate obtained from an EKF without the innovations term present in equation 15.11 (Stigter and Beck, 1994).

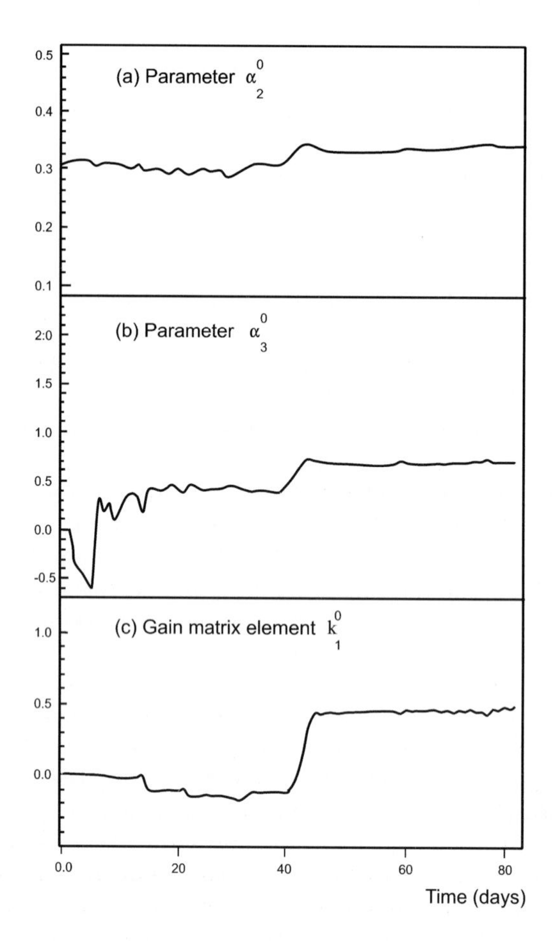

Fig. 15.5. Reconstructed estimates of the three parameters in the Interpretative Structure of equation (15.11): (a) parameter α_2^0 (day^{-1}); (b) parameter α_3^0 (gm^{-3} day^{-1}); and (c) gain matrix element k_1^0 (dimensionless).

15.6 REORIENTING THE GOAL FUNCTION

The goal function of equation 15.6 *may* be sufficient for our present purposes. Yet it makes no reference to controlling the speed of any resulting parametric change. It will not balance a good match of the observed behaviour against the capacity to illuminate plastic deformation of the model's structure. If we wanted to achieve this, we might speculate on a form of objective function as a means to do so, such as the following:

$$J'(t_k) = \sum_{\delta=-\tau}^{+\tau} \varepsilon^T(t_{k+\delta}|t_N)\mathbf{\Lambda}^{-1}\varepsilon(t_{k+\delta}|t_N) + \tilde{\boldsymbol{\varphi}}^T(t_{k+\delta}|t_N)\boldsymbol{P}_{\varphi\varphi}^{-1}(t_{k+\delta}|t_N)\tilde{\boldsymbol{\varphi}}(t_{k+\delta}|t_N) \quad (15.12)$$

in which τ specifies the dimensions of a moving interval of the observed record centred on the current sampling instant t_k, t_N denotes the end of this record, $\mathbf{\Lambda}$ is the variance–covariance matrix of the innovations sequence (ε), and $\tilde{\dot{\boldsymbol{\varphi}}}$ is an approximation of the rate of change of the parameters defined by

$$\tilde{\dot{\boldsymbol{\varphi}}}(t_{k+\delta}) = [\boldsymbol{\varphi}(t_{k+\delta+\Delta t}|t_N) - \boldsymbol{\varphi}(t_{k+\delta}|t_N)]/\Delta t$$

We hasten to add that this form of goal function is a conjecture. It is merely a succinct mathematical expression of how empirical experience would suggest we should remedy some algorithmic inadequacies. Some of its salient features, which will now be presented, may well prove inappropriate in practice.

First, the device of the moving interval defined by the choice of τ articulates the idea of a system undergoing evolution, or structural change, with a limited span of time over which relative invariance would be expected. In other words, we would not expect observed behaviour too distant from the present, either forward into the future or back into the past, to have much bearing on the way in which the model's parameters are adjusted according to this goal function (we have already encountered this issue in the preceding chapter; and it will surface again in the next). Second, conditioning computation of all the quantities on access to observations in the entire record, from t_0 through t_N, signals in principle the use of a smoothing algorithm, the power of which has already been well demonstrated in a similar context by Young and Beven (1994).[5] Third, $\mathbf{\Lambda}$ and $\boldsymbol{P}_{\varphi\varphi}$ act as weighting (or normalising) devices for the two opposed components of equation 15.12. Both should tend to decrease over the observation record. Excessive reduction in $\boldsymbol{P}_{\varphi\varphi}$ is probably responsible for the observed tendency of the structure of the model to crystallize around the first occurrence of a failure to match a change of structure underlying observed behaviour (Stigter and Beck, 1994). The effect of declining values of $\boldsymbol{P}_{\varphi\varphi}$ (with time) ought to counteract this tendency by weighting more strongly the contribution of any late-occurring parametric change to the revised objective function of equation 15.12, and desirably so.

At this point we should pause to note the following. Given equation 15.12, or some more tractable mathematical form of objective function reflecting the intent of this equation, an alternative RPE algorithm could be derived, in principle, along the lines of the procedure already set out above. Indeed, should this prove possible, the resulting algorithm would be one directed at the essence of the problem of detecting structural change. Another route to another form of desired algorithm is discernible, however, and it is not unrelated to the avenue just suggested. But in order to appreciate its appeal and potential we shall need first to rise well above the clutter of the detail.

---

[5] Since smoothing over the *entire* record of the data may be in conflict with the preceding requirement for a limited moving window in the goal function (of width just 2τ about the current time), one of these two desirable features of design might have to be put aside.

15.6.1 Bringing Out the Inner Parametric Space of the Model Structure

When it is natural for models to be cast in the state space, as in equation 15.1, it is just as natural for the algorithms of recursive estimation to be similarly cast. Our interest is in change in the parametric space, however. In the course of this chapter, we have progressively removed emphasis from the state space, transferring it to the parameter space. If we were to complete this conceptual process, it would end logically in the framework of an Unobserved Components (UC) model, with then access to some very attractive fixed interval smoothing (FIS) algorithms (for example, Young, 1999). There, the observed time-series (y) is assumed to be composed of several "unobserved components", such as trend, periodic, seasonal, irregular, or transfer-function derived components, each of which is treated in some way as a time-varying parameter ($\alpha(t)$).[6]

Here, if we were to follow the same path, the focus of our attention would shift from the equation of state, i.e., equation 15.1(a), to that of the observation equation, namely equation 15.1(b),

$$y(t) = \boldsymbol{h}\{\boldsymbol{x},\boldsymbol{\alpha};t\} + \boldsymbol{\eta}(t)$$

In fact, an observation equation *linear* in the parameters and formulated in discrete time (t_k) would be required, i.e., in the form of

$$y(t_k) = \boldsymbol{H}(\boldsymbol{x},\boldsymbol{u})\boldsymbol{\alpha}(t_k) + \boldsymbol{\eta}(t_k) \tag{15.13}$$

with the (time-varying) matrix $\boldsymbol{H}(\boldsymbol{x},\boldsymbol{u})$ obtained – in some manner (analytical or numerical) – from the structure of $\boldsymbol{h}\{\cdot\}$ and the equation of state for the original model, i.e.,

$$\mathrm{d}\boldsymbol{x}(t)/\mathrm{d}t = \boldsymbol{f}\{\boldsymbol{x},\boldsymbol{u},\boldsymbol{\alpha};t\} + \boldsymbol{\xi}(t)$$

The parameters $\boldsymbol{\alpha}$, together with the irregular elements, $\boldsymbol{\eta}$, would then be labelled the unobserved components of the model, which would now be given by equation 15.13. In essence, in the UC framework the original (state–space) model of the system is relegated merely to intervening in how the relationship between the parameters and the observations is articulated. The parameters of the naturally expressed model of the system's behaviour can thus be extracted from their secondary roles in the given functional relationships and placed in a pole position, around which all else, *including* the states, revolves. Given equation 15.13, only specification of a set of assumptions about the character of the parameter dynamics remains to complete the

[6] Of course, the distinction between what is a state and what a parameter is rather arbitrary. We are as familiar in everyday practice with a biomass concentration at equilibrium (a time-invariant state) as we are comfortable with permitting a temperature-dependent growth-rate constant to vary with the seasons (a time-varying parameter).

transformation. In the FIS algorithms of Young (1999) and Norton (1975) this is achieved through a Generalised Random Walk (GRW) model (Jakeman and Young, 1984),

$$\boldsymbol{\alpha}(t_k) = \boldsymbol{F}\boldsymbol{\alpha}(t_{k-1}) + \boldsymbol{G}\,(t_k) \tag{15.14}$$

which stems clearly from the same conceptual outlook as that of equation 15.5, and where $\boldsymbol{F}$ and $\boldsymbol{G}$ are appropriately dimensioned matrices.

Together, equations 15.14 and 15.13 constitute the model cast in the parameter space, both the dynamics of the parameters and their observations. The parameters have been brought out to the surface, while the states have been embedded within, as it were. Being linear, equations 15.14 and 15.13 are amenable to incorporation within the algorithms of linear, forward-pass filtering and backward-pass smoothing, of which the FIS is composed (Young, 1999). But this is not all. In Young and Pedregal (1999) it is shown how using any of the random walk models in the FIS algorithm is equivalent to weighting the data symmetrically about the current time t_k, where the effective "breadth" of the weighting window will be related to the propensity of a given parameter to vary with time. In fact it is further shown how the FIS algorithm is equivalent to a spline smoothing (deterministic regularisation) procedure, in which the variance of the residuals is minimised subject to the constraint of a given degree of smoothness in the estimated components of the signal, i.e., the estimated trajectories of $\boldsymbol{\alpha}$ (or, more generally, $\boldsymbol{\varphi}$) in the present context. Technically, this problem of constrained optimisation is transformed into the corresponding unconstrained problem by formulating a Lagrangian function. In effect, the FIS algorithm achieves its goal in a manner equivalent to minimising the value of an objective function comprising penalties attaching not only to deviations of the model's output from the data *but also* to excessive fluctuations in the parametric trajectories, which is essentially the intent of the suggested goal function of equation 15.12.

From this second path we have in principle a means of channelling, to the maximum extent possible, the information content of the observations $[\boldsymbol{u},\boldsymbol{y}]$ into the revealed parametric trajectories, $\hat{\boldsymbol{\alpha}}(t)$, subject to the constraint of the original model of the environmental system's behaviour. Plastic deformation of, or change in, the structure of the model, will thereby be illuminated. Indeed, it will be pinpointed to specific members (constituent hypotheses) of that structure.

15.7 CONCLUSIONS

We have been concerned with the problem of detecting non-stationarity in the structure of low-order, nonlinear models of the behaviour of environmental systems, where such structural change can be accommodated through the time-variable nature of the estimates of the model's parameters. This problem, central to the monograph and first defined in general terms in Chapter 4, has now been given specific substance and solved, at least in part, using a recently proposed recursive prediction error (RPE) algorithm (Ljung, 1979; Stigter and Beck, 1994; Stigter, 1997). Of particular significance is the way in which this algorithm – through the use of an innovations representation of the model – allows one to diagnose whether any apparent structural change is due to elements of what was *a priori* {presumed known} in the science base, or to elements of its complement, the {acknowledged unknown}. While the basic form of the RPE algorithm now requires both more extensive and more detailed testing, certain avenues for improvement are already discernible. Two such possible paths have been suggested. The first would involve re-orienting the goal function for reconciling the model with observed behaviour, in particular, by incorporating penalty costs on excessive parametric fluctuation. From this revised function a modified, and essentially better targeted, RPE algorithm might be derived. A second path might be cut from a rather different point of departure, by first transforming the natural expression of the system's dynamics in the state space into the less obvious parameter space of what is known as an unobserved components (UC) model (for signal extraction). Once cast in such a form, estimation of the all-important trajectories of the model's parameters may become amenable to the framework of fixed interval smoothing, the power of which is well illustrated in Young (1999).

If the developments of this chapter seem promising, so they are. Yet consider this. Structure II of the motivating example, which has the highest order of all the model structures introduced, contains just five state variables. This is hardly complex, although it is clearly an advance on the very low order models of the preceding chapter. Furthermore, tacit has been the presumption that all the states of the model would be accessible to observation, which we know is far from true for the higher-order models of, say, the global carbon cycle (Chapter 13) or tropospheric ozone formation (Chapter 17). Any model containing unobserved state variables will possess an intrinsic limitation if used to interpret a set of data for the detection of structural change. The sought-for adaptations in the model's parameter estimates may be preferentially channelled into the estimates of the unobserved states, in which domain a judgement on the presence or absence of structural change is less easily deliverable. After all, the core principle on which the entire discussion of this chapter has been based is this: estimates of quantities presumed not to change with time should not vary with time. If they do, this is revealing. Unobserved state variables are self-evidently intended to vary with time, so that application of this principle in their domain is rendered largely impotent.

Within the overall sweep of the monograph, the primary concern of this chapter has been that of *detecting* the seeds of change: figuratively, of apprehending that something within the space outside the branch-node template of the model in Figure 15.2 has impinged upon the template and begun to distort its shape. This will likewise occupy the discussion of the next chapter, albeit within a different algorithmic framework. But the next chapter is also about *pre-emptive forecasting* of the propagation of these seeds of change. If we can detect the onset of distortion, through the metaphor of Figure 15.2, that is, could we forecast its subsequent evolution? As we build back towards working with higher-order models – in fact, very high-order models (VHOMs) in Chapter 17 – we then ask the question, could we *design the structure of the model for discovery of our ignorance*? Could we contemplate specific designs of the template of Figure 15.2 that are better (faster) at detecting behaviour of significance, yet *not* in accordance with the best of our current science base (as encoded in the model structure)?

REFERENCES

Allen, P.M., 1990. Evolution, innovation, and economics. In: *Technical Change and Economic Theory* (G. Dosi, C. Freeman, R. Nelson, G. Silverberg and L. Soete, eds.). Pinter, London, pp. 95–119.

Ashby, M.F., 1987. Technology of the 1990s: Advanced materials and predictive design. *Philosophical Transactions of the Royal Society of London*, **A 322**, 393–407.

Beck, M.B., 1973. The application of control and systems theory to problems of river pollution. PhD Thesis, Department of Engineering, University of Cambridge.

Beck, M.B., 1975. The identification of algal population dynamics in a freshwater system. In: *Computer Simulation of Water Resources Systems* (G.C. Vansteenkiste, ed.). North-Holland, Amsterdam, pp. 483–494.

Beck, M.B., 1986. The selection of structure in models of environmental systems. *The Statistician*, **35**, 151–161.

Beck, M.B., 1987. Water quality modeling: a review of the analysis of uncertainty. *Water Resour. Res.*, **23**(8), 1393–1442.

Beck, M.B., 1994. Understanding uncertain environmental systems. In: *Predictability and Nonlinear Modelling in Natural Sciences and Economics* (J. Grasman and G. van Straten, eds.). Kluwer, Dordrecht, pp. 294–311.

Beck, M.B. and Halfon, E., 1991. Uncertainty, identifiability and the propagation of prediction errors: a case study of Lake Ontario. *J. Forecasting*, **10**(1&2), 135–161.

Beck, M.B. and Young, P.C., 1976. Systematic identification of DO-BOD model structure. *Proc. Am. Soc. Civil Eng., J. Environ. Eng. Div.*, **102**(EE5), 909–927.

Beck, M.B., Kleissen, F.M. and Wheater, H.S., 1990. Identifying flow paths in models of surface water acidification. *Rev. Geophys.*, **28**(2), 207–230.

Gelb, A. (ed.), 1974. *Applied Optimal Estimation*. MIT Press, Cambridge, Massachusetts.

Jakeman, A.J. and Young, P.C., 1984. Recursive filtering and the inversion of ill-posed causal problems. *Utilitas Mathematica*, **35**, 351–376.

Ljung, L., 1979. Asymptotic behaviour of the extended Kalman filter as a parameter estimator. *IEEE Trans. Autom. Contr.*, **24**, 36–50.

Ljung, L. and Söderström, T., 1983. *Theory and Practice of Recursive Identification*. MIT Press, Cambridge, Massachusetts.

Lloyd Smith, D. (ed.), 1990. *Mathematical Programming Methods in Structural Plasticity*. Courses and Lectures No **299**, International Centre for Mechanical Sciences, Springer, Berlin.

Malle, K.-G., 1994. Accidental spills – frequency, importance, control, and countermeasures. *Water Sci. Technol.*, **29**(3), 149–163.

Munro, J., and Lloyd Smith, D., 1972. Linear programming duality in plastic analysis and synthesis. In: *Computer Aided Structural Design. Proceedings of a Symposium*. Peter Peregrinus, Stevenage, Vol. 1, pp. A1.22–A1.54.

Norton, J.P., 1975. Optimal smoothing in the identification of linear time-varying systems. *Proc. Inst. Elect. Eng.*, **122**, 663–668.

Stigter, J.D., 1997. The development of a continuous-discrete recursive prediction error algorithm in environmental systems analysis. PhD dissertation, University of Georgia, Athens, Georgia.

Stigter, J.D. and Beck, M.B., 1994. A new approach to the identification of model structure. *Environmetrics*, **5**(3), 315–333.

Young, P.C., 1999. Nonstationary time series analysis and forecasting. *Progr. Environ. Sci.*, **1**, 3–48.

Young, P.C. and Beven, K.J., 1994. Data-based mechanistic modelling and the rainfall-flow non-linearity. *Environmetrics*, **5**(3), 335–363.

Young, P.C. and Pedregal, D., 1999. Recursive and *en-bloc* approaches to signal extraction. *J. Appl. Statistics*, **26**, 103–128.

Environmental Foresight and Models: A Manifesto
M.B. Beck (editor)

CHAPTER 16

Detecting and Forecasting Growth in the Seeds of Change

J. Chen and M.B. Beck

16.1 INTRODUCTION

Our studies of the global carbon cycle (Chapter 13), the longer-term response of catchment hydrology to land-cover change (Chapter 14), and the assimilation of easily degradable organic matter in a river system (Chapter 15), have prepared the ground for us now to ask what must be the most audacious of all the questions we could have posed in this monograph. Is there any possibility whatsoever of detecting the onset of apparent structural change – in the domain of a model's parametric space – before it manifests itself more palpably at the macroscopic level of the conventionally observed state variables? And if such change were detectable, could we entertain any hope of exploring the forecast consequences of propagating these seeds of change and dislocation into the future? One's immediate response to such questions is likely to be resoundingly in the negative. But the disbelief behind this reaction requires suspension. The questions as posed may not eventually be answered, even if they were answerable. But granting them the time to be given attention – in formulating the means for their analysis, at least in principle – can lead to fruitful insights, albeit in directions somewhat different from the intent of the questions as originally posed.

In the present chapter we shall continue with our case study of the River Cam, employing it this time to assist us in reflecting on these profoundly difficult questions. To some considerable extent we think we know the answers to them, in this specific instance. The total record of the observations spans 80 days (during the summer of

1972). Before the results of the analysis now reported herein behaviour of the system at the start of the record was as expected; at about day 40, however, a rather abrupt change of structure was thought to occur; and this change was believed to have been wrought primarily by the action of solar radiation on a (conjectured) population of algae passing slowly through the system. What we shall uncover in this chapter has a significantly different interpretation. It is born of the challenge of seeking to identify the seeds of the oncoming change well before its manifestation, as a dislocation in the behaviour of the model's state variables around day 40.

On this occasion, we shall not use recursive estimation as the algorithmic means of revealing parametric variation. Instead, a form of controlled random search (CRS) (Price, 1979) is applied successively to moving windows of the observed record. We shall argue that the purpose of system identification, or more commonly model calibration, is to detect and to diagnose the possible paths of structural evolution in the observed behaviour of the system, and that this might be perceived as deviations from the core, average, structurally invariant characterisation of behaviour, the customary product of calibration. Behaviour "at the fringe", between core, average behaviour and the highly implausible and bizarre, may thus assume very considerable significance, as we now relate.

16.2 AN IMPORTANT VARIATION ON THE BASIC THEME OF STRUCTURAL CHANGE

We assume first our familiar form of describing the dynamics of the environmental system, i.e.,

$$\mathrm{d}\boldsymbol{x}(t)/\mathrm{d}t = \boldsymbol{f}\{\boldsymbol{x},\boldsymbol{u},\boldsymbol{\alpha};t\} + \boldsymbol{\xi}(t) \tag{16.1a}$$

with outputs observed at discrete instants in time (t_k),

$$\boldsymbol{y}(t_k) = \boldsymbol{h}\{\boldsymbol{x},\boldsymbol{\alpha};t_k\} + \boldsymbol{\eta}(t_k) \tag{16.1b}$$

in which $\boldsymbol{f}$ and $\boldsymbol{h}$ are vectors of nonlinear functions, $\boldsymbol{u}$, $\boldsymbol{x}$, and $\boldsymbol{y}$ are the input, state, and output vectors, respectively, $\boldsymbol{\alpha}$ is a vector of model parameters, $\boldsymbol{\xi}$ and $\boldsymbol{\eta}$ are notional representations respectively of those attributes of behaviour and output observation that are not to be included in the model in specific form, and t is continuous time.

As has now become the norm, we assert that the arrangements of the logical interactions among all the variables in this macroscopic map of the relevant science base are in truth not constant. If we impose the pragmatic constraint of $\boldsymbol{f}$ and $\boldsymbol{h}$ being invariant with time, in principle the parameters $\boldsymbol{\alpha}$ in equation 16.1 must accordingly be viewed as stochastic processes. The introductory discussion and definition of apparent structural change in Chapter 4 requires therefore the assumption that $\boldsymbol{\alpha}$ should satisfy the following condition in the *absence* of apparent structural change:

$$\mathrm{d}\boldsymbol{\alpha}(t)/\mathrm{d}t = \mathbf{0} \tag{16.2}$$

or $\boldsymbol{\alpha}(t) = \boldsymbol{c}$, a single, constant value. This is the complement of equation 4.7. Taken literally – which was not the intent in Chapter 4 – the condition of equation 16.2 is unworkably strict. Indeed, interpreting it at its limit, any arbitrarily small, microscopic perturbation of $\boldsymbol{\alpha}$ away from $\boldsymbol{c}$ has the capacity, in principle, to induce a macroscopic manifestation of structural change in the state space ($\boldsymbol{x}$). In practice, moreover, uncertainty in the empirical observations and the constituent hypotheses of the model will preclude the identification of such arbitrarily small deviations from invariance in the values of the model's parameters. They might be "real", but we would be unable to discover and demonstrate this. We shall have to explore the possibility of developing a more relaxed, more pragmatic, condition for defining change.

Given that estimates of the parameters will be surrounded by uncertainty, we nevertheless assert that this parametric uncertainty will be bounded. Suppose, then, that equation 16.2 is relaxed to the following pair of constraints:

$$\mathrm{d}\bar{\boldsymbol{\alpha}}(t)/\mathrm{d}t = \mathbf{0} \tag{16.3a}$$

and

$$\boldsymbol{\alpha}(t) \in [\boldsymbol{c} - 0.5\boldsymbol{\Delta}, \boldsymbol{c} + 0.5\boldsymbol{\Delta}] \tag{16.3b}$$

where $\bar{\boldsymbol{\alpha}}$ is the time average of the parameters (over a period ΔT, significantly less than the interval of the entire empirical record T), while $\boldsymbol{\Delta}$ is the span of the maximum allowable range of plausible values for these parameters. The first of these acknowledges the inevitable uncertainty of any practical test of the presence/absence of structural change. But on its own it is flawed. For it permits acceptance of unlimited values (or unlimited excursions in the values) of $\boldsymbol{\alpha}$, as long as its mean over ΔT is invariant. It would allow us to accept a model as not undergoing apparent structural change when it contains parameters approximating white-noise processes with infinite variance, for example. Equation 16.3(a) alone appears to be overly relaxed, hence the incorporation of equation 16.3(b) into the conditions required for accepting that no change of structure has occurred.

For a model designed to replicate aggregate, average behaviour of the system it is therefore required merely that the averages of the parameters over time should be constant, while their true values, wandering through time as stochastic processes (such as a random walk), should remain within certain pre-specified bounds, defined by the choice of $\boldsymbol{\Delta}$. If the conditions of equation 16.3 can be found to hold – in practice, for the estimates $\hat{\boldsymbol{\alpha}}$ of the parameters and for bounds $\boldsymbol{\Delta}$ judged to be reasonable and appropriate – we might safely conclude, as a point of reference, that no apparent change of structure resides in the observed record.

16.2.1 Types of Structural Change

In effect, the birth and death of state variables over time is at the core of the definition of structural change in Chapter 4 (and as elaborated further in the preceding chapter). Within a model of fixed structure, wherein the definition of the state vector $\boldsymbol{x}$ cannot vary with time, such apparent structural change – brought about because the model is bound to be an approximation of the truth – should be manifest in reconstructed estimates $\hat{\boldsymbol{\alpha}}$ of the model's parameters that must necessarily vary with time. This is proof positive, and the essential definition, of apparent structural change. We shall label this a structural change of *Type I*.

From the conditions of equation 16.3 a second manifestation of structural change is possible. Below the resolving power of the model, or within the space of features not included in the model's structure, changes of potential significance are taking place all the time. Within the model, at some relatively macroscopic level of description, the parameters $\boldsymbol{\alpha}$ would in truth experience a degree of temporal variation, or flutter and drift; if these are of low amplitude relative to the bounds $\boldsymbol{\Delta}$, they might be considered essentially inconsequential. What we now conceive of is this. Such parametric flutter and drift may already, in fact, be of growing significance and might well become of consequence in the future. They are of consequence not necessarily because of their rising amplitude, but because the behaviour of the system is evolving in a manner increasingly prone to – or increasingly sensitive to – some of the parameters experiencing the low-amplitude flutter. Whereas such flutter was "always" present, macroscopic behaviour might not previously have been influenced by it to any significant extent; the point is, it may well become so. This, then, is structural change of a rather different character, which we shall label as *Type II*.

We can have proof positive of a *Type I* apparent change of structure. Chapter 15 has been devoted to the means of eliciting such evidence from the observed record, in the form there of a continuous trace of estimates $\hat{\boldsymbol{\alpha}}(t)$ over time t. The setting of that approach within the context of filtering theory provides a degree of averaging over time (now made formally a part of equation 16.3(a)); and mere visual inspection of the trajectories of the recursive parameter estimates discriminates effectively against any significance of the low-amplitude fluctuations in these reconstructed estimates (as reflected in the terms of equation 16.3(b)). So too was Chapter 12 similarly devoted, in that instance, through the juxtaposition of distributions for acceptable estimates $\hat{\boldsymbol{\alpha}}(\Delta T_j)$ at several different spans of time ΔT_j – before and after a fire or drought – and even in the face of significant (if not gross) uncertainty.[1]

Proof positive of a *Type II* apparent change of structure may be more elusive. For a start, lest we had forgotten this, the conditions of equations 16.3 are cast in the relative remoteness of the "inner" parameter space ($\boldsymbol{\alpha}$), not the more immediately accessible "outer" state space ($\boldsymbol{x}$) of the system's behaviour. And how exactly should we choose the bounds $\boldsymbol{\Delta}$, embedded in this remote, inner space? For the time being we shall have to adopt a less direct approach to constructing a test for a *Type II* structural change. As we do so it will be helpful to keep the following in mind. Detecting a *Type I* change should, in principle, become apparent because of something having been left out of the model – in the space between the tractable known (the model) and the unknowable truth, as we have said in the preceding chapter. For

[1] In fact, in a manner computationally similar to what will follow in this chapter.

a change of *Type II*, whatever is important as the agent of such change, it resides already within the inner structure of the model, but its influence over the behaviour of the more immediately accessible state (and observed) behaviour is changing with time.

16.2.2 Detecting the Seeds of Change

Conventionally, calibrating a model is the pursuit of an average parameterisation that is unchanging with time, consistent with the goal of having a calibrated model capable of replicating aggregate, average behaviour of the system (presumed to be structurally invariant). In principle, this goal can be expressed formally as a matter of minimising the following criterion,

$$e = \|\hat{y} - \bar{y}\| \tag{16.4}$$

in which $\hat{y}$ is the simulated response and $\bar{y}$ is the average of the true behaviour of the system. $\bar{y}$ is, of course, not identical with behaviour as observed (y), but has the following form of relationship with y

$$y = \bar{y} + \delta y_{\zeta} + \eta \tag{16.5}$$

where δy_{ζ} is the micro-scale variability of the system's behaviour about the aggregate average of $\bar{y}$. δy_{ζ} is due, we argue, to low-amplitude parametric flutter and drift, themselves arising from, for example, taxonomic and spatially differentiated characteristics of the system's properties at scales lower than the resolving power of the model. Substituting for $\bar{y}$ from equation 16.5 in equation 16.4 we may have therefore

$$e = \|\hat{y} - (y - \eta)\| + \|\delta y_{\zeta}\| \tag{16.6}$$

By convention it is the first of the two terms on the RHS of this equation that is to be minimised in reconciling the model with the observed behaviour, i.e., simulated responses should fall within the range of values where (subject to uncertainty) the measurements are thought to lie. If they do not, the inference to be drawn from the second term in equation 16.6 is that the consequences of the fine-grained micro-scale variability of the system's behaviour are more substantial than presumed *a priori*. And while one might view this as counter-productive from the customary perspective of successfully calibrating the model, such an outcome is important for understanding the potential for variability and structural change in behaviour in the future.

To summarise, no matter how good the conceptualisation of the model, when juxtaposed with the real thing we must inevitably find that this reality is much more variegated than the macroscopic homogeneity of the model. The goal and procedure of model calibration are conventionally focused on deriving an average parameterisation. The microscopic variability underlying behaviour at the fringes, as it were, of what was observed in the past, is pushed thus to one side – and thereby excluded from further, subsequent consideration, albeit unintentionally. But as Allen (1990) has argued so forcefully, such variability should be kept within the frame of consideration (for forecasting), since it is precisely therefrom that the potential for structural change in the future may spring.

How then should the consequences of this variability be detected, and the variability itself suitably characterised for the purposes of forecasting? Our immediate response to this challenge is to suggest that, with this outlook, what becomes of great interest during identification is *not* behaviour at the core of the "average", nor is it behaviour well beyond the "bounds of plausibility" (amidst the outliers) but that which lies in between these two domains – at the "fringe" of the barely tenable. Cast in terms of the residual errors ε of mismatch between observed (y) and simulated behaviour ($\hat{y}$), curiosity should be focused upon which crucial elements of the model's parametric behaviour ($\boldsymbol{\alpha}$) – its low-amplitude flutter and drift, or more substantial change – is associated with residuals possessing a norm $\|\varepsilon\|$ satisfying the following conditions

$$\|\boldsymbol{\eta}\| < \|\boldsymbol{\varepsilon}\| < \|\delta\mathbf{y}_{\zeta} + \boldsymbol{\eta}\| \tag{16.7}$$

as illustrated in Figure 16.1. What in the map of the science base encoded in the model, we could ask, is key to discriminating whether simulated behaviour is generated within these fringes or within the core, average behaviour (in the envelope of tolerable observation errors)? In other words, in the spirit of a Regionalised Sensitivity Analysis (RSA; Hornberger and Spear, 1980; Spear and Hornberger, 1980; and Chapter 11), we would wish to know which are the parameters $\{\boldsymbol{\alpha}^K\}$ resulting from such a test, for it these that are potentially the markers of a *Type II* change of structure.[2] In particular, we would have an interest in how the content of $\{\boldsymbol{\alpha}^K\}$ varies as a function of assumptions about the magnitude of $\delta\mathbf{y}_{\zeta}$ and, in the light of equation 16.3(b), the magnitude of the parametric bounds $\boldsymbol{\Delta}$. More ambitiously, in the spirit of the recursive estimation studies of the preceding chapter, we could search for evidence of parametric change of a more specific character, of a *Type I* structural change, that is, associated with the residuals satisfying equation 16.7.[3]

To detect these changes of structure is one thing, to relate them to a cause is quite another, a challenge already summarised in the preceding chapter and elsewhere (Beck, 1978, 1986, 1994). In the case study that now follows, we shall on this occasion relate them to the system's input disturbances, $\boldsymbol{u}$. For while structural change may be revealed through parametric variation, its causes lie elsewhere, within $\boldsymbol{u}$ (and arguably exclusively so). The challenges, then, are first to identify those perturbations in $\boldsymbol{u}$ potentially capable of making the model (structurally) inconsistent with the observed

[2] Detection of interesting features within the framework of an RSA, that is, through juxtaposing what is required to generate "fringe" as opposed to "average" behaviour.

[3] Detection of interesting features, through emphasising observed "fringe" behaviour at the expense of "average" behaviour (within a goal function such as that of equation 15.12 of the preceding chapter).

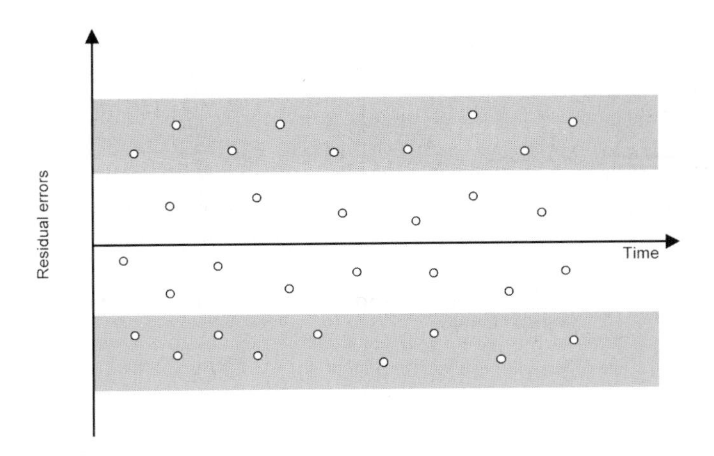

Fig. 16.1. Defining behaviour at the fringe: what is important is the identification of those elements of the model key to discriminating between fringe behaviour (the shaded grey domains) and core, average behaviour — between the inner bounds of the grey domains — or bizarre, outlier behaviour, beyond the uppermost and lowermost bounds of the grey domains.

record and, second, to identify those states especially sensitive to such perturbations. Ergo, success in detecting the seeds of structural change is a function of the richness of abnormal input perturbations. This seems obvious, of course: extreme events are really revealing. Here, however, if there were such extreme events, we would spot them directly, without having to go through – what some may consider – the rigmarole of this discussion of structural change. The point is rather that "abnormal" means "abnormal, subtle and/or weak", *not* "abnormal and extreme". Dramatic inconsistency between the model and observed record is obvious; but it is not the sum total of what we are labouring to identify in this chapter.

16.3 CASE STUDY

Figure 16.2 shows the data for four of the five observed input disturbances ($\boldsymbol{u}$) of the system, i.e., flow, water temperature, hours of incident sunlight, and the concentration of biochemical oxygen demand (BOD) at the upstream boundary of a 4.5 km stretch of the River Cam. A fifth observed input is that of the concentration of dissolved oxygen (DO) at this upstream boundary, while just two of the system's state variables are available as observed outputs (y), these being the concentrations of BOD and DO at the downstream boundary of the stretch of river. Put simply, our challenge is to detect the onset of the supposed structural change – previously thought not to have been apparent before t_{40} – prior to t_{36} and to explore the possibility of characterising the propagation of this dislocation as a function of $\boldsymbol{u}$. We shall begin our response within the framework of just a single-state model of the BOD dynamics, but subsequently extend our analysis to the two-state problem.

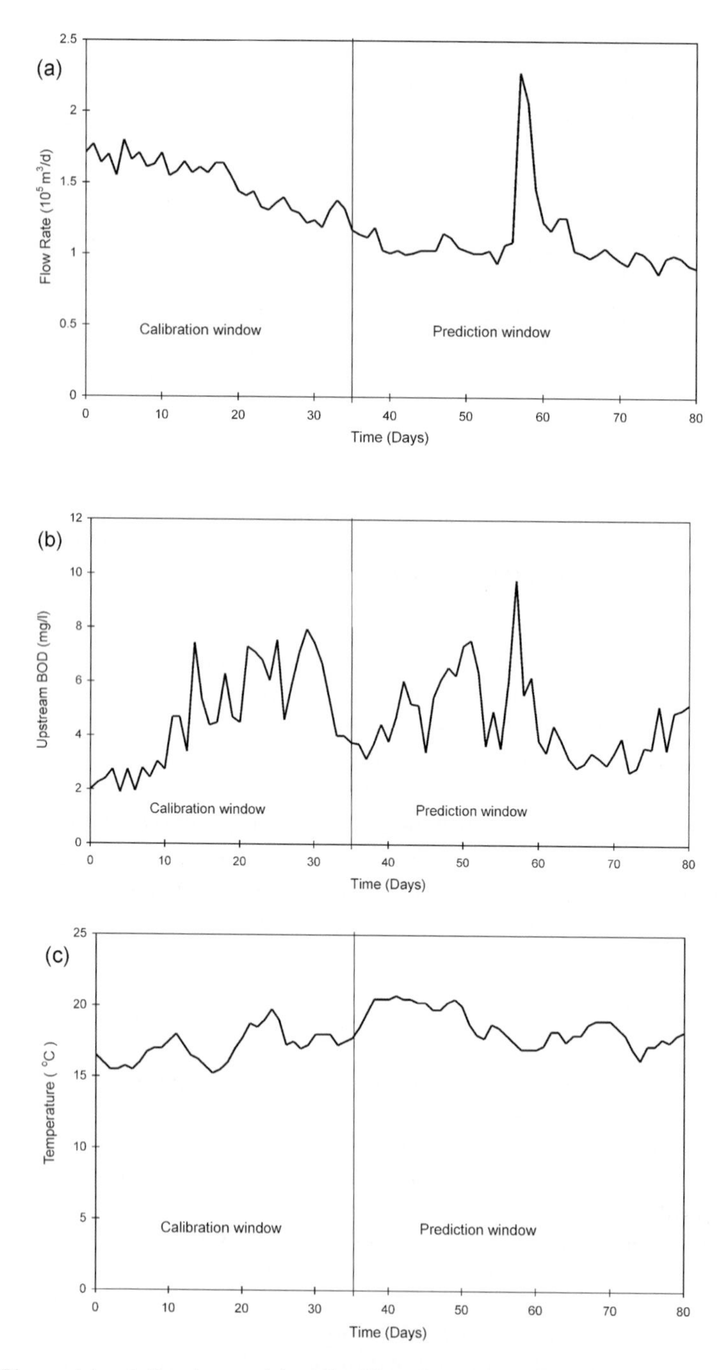

Fig. 16.2. Observed input disturbances (***u***) of the River Cam during the summer of 1972: (a) stream discharge (m^3 day^{-1}); (b) concentration of BOD at the upstream boundary of the reach (gm^{-3}); (c) temperature (°C).

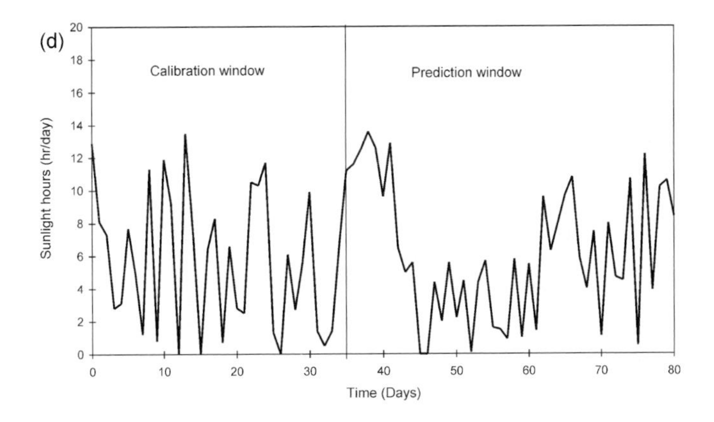

Fig. 16.2 (continued). (d) hours of sunlight per day.

16.3.1 Single-state Model of BOD Dynamics

Assuming the lumped-parameter approximation of a continuously stirred tank reactor (CSTR) for representing solute transport through a stretch of river, one of the simplest structures for a model of the dynamics of BOD (x_1) is given as follows

$$\dot{x}_1(t) = -\alpha_1 x_1(t) + s_1\{\boldsymbol{u}(t)\} \tag{16.8}$$

in which α_1 is the rate constant for biochemical degradation of the BOD and s_1 summarises the effects of solute transport into and out of the reach of river, conceptualised for the time being as comprising just a single CSTR element.

Our first task is to identify some approximation of core, average behaviour, simply achieved by choosing α_1 so as to minimise the sum of squared residual errors of mismatch (ε_1) between observed (y_1) and estimated ($\hat{y}_1$) downstream BOD concentration. In this particular instance minimisation is sought using a Controlled Random Search (CRS) procedure. From the sample set of good parameterisations found from this procedure, a single "optimal" estimate of the parameter, $\hat{\alpha}_1$, can be selected, as that candidate associated with the minimum value of the squared error criterion. Given $\hat{\alpha}_1$, an optimal trajectory for the (observed) downstream BOD concentration may be generated and the sample mean and variance (σ^2) of the attaching residuals computed. Armed with these various components a range of specifications of the system's core (average), invariant behaviour can be constructed according to the following:

$$\hat{y}_1(t_k) - w_l(t_k)\beta\sigma \leq y_1^i(t_k) \leq \hat{y}_1(t_k) + w_u(t_k)\beta\sigma \tag{16.9}$$

in which $y_1^i(t_k)$ is a member at time t_k of an acceptable candidate trajectory (i) of replicated behaviour, w_l and w_u are weights attaching to the lower and upper bounds

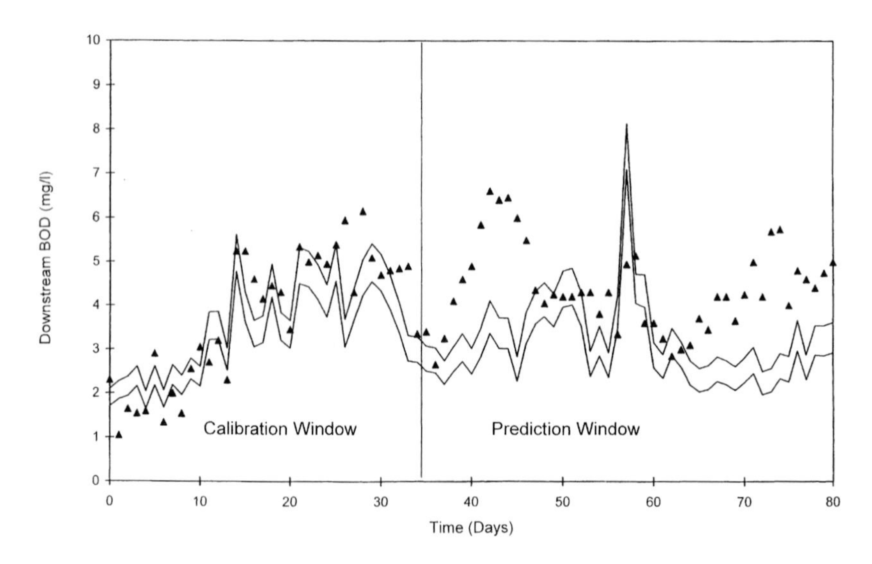

Fig. 16.3. Maxima and minima of range of trajectories for simulated downstream BOD concentration (gm^{-3}) over periods of both calibration and prediction; observations shown as solid triangles.

of behaviour respectively (which, if not assigned identical values, may be used to take account of any bias in the residual errors), and β reflects a measure of the uncertainty associated with the average trajectory, as identified from the error-corrupted observations. In practice, β will be taken to be 1.64 in what follows, indicating that 90% of all the observations should fall within the span of behaviour defined by equation 16.9, and to be covered by candidate parameterisations of the model (under a normal distribution), while w_l and w_u are both set to 1.0.

Figure 16.3 shows the resulting range (maxima and minima) of acceptable model trajectories, $\{y_1^i(t_k)\}$, over the first 35 days of the record (the calibration phase), where parameterisation was implemented through a Latin hypercube strategy, sampling within the range 0.0 to 1.5 (day^{-1}) for the parameter α_1. These results show no obvious structural flaw. Yet, given the collection of acceptable candidate parameterisations of the model deriving from this phase, and assuming knowledge of the requisite input perturbations ($\boldsymbol{u}$) over the remainder of the record, the model clearly fails to generate an ensemble of acceptable projections into the future (the prediction phase, from t_{36} onwards in Figure 16.3). Whether we wish to interpret this as an illustration of the folly of model-building is not material to the present argument. Our point is rather to suggest that the purpose of calibration should not be merely that of matching past behaviour but also that of exploring the potential of the system to experience a change in the structure of its behaviour.

Two pieces of evidence hinting at such change are already available. First the optimal estimate of the BOD decay-rate constant, $\hat{\alpha}_1$, is uncharacteristically close to zero (outside the range of values generally considered normal; Bowie et al., 1985). Perhaps there was another source of BOD in the system – for example, from the

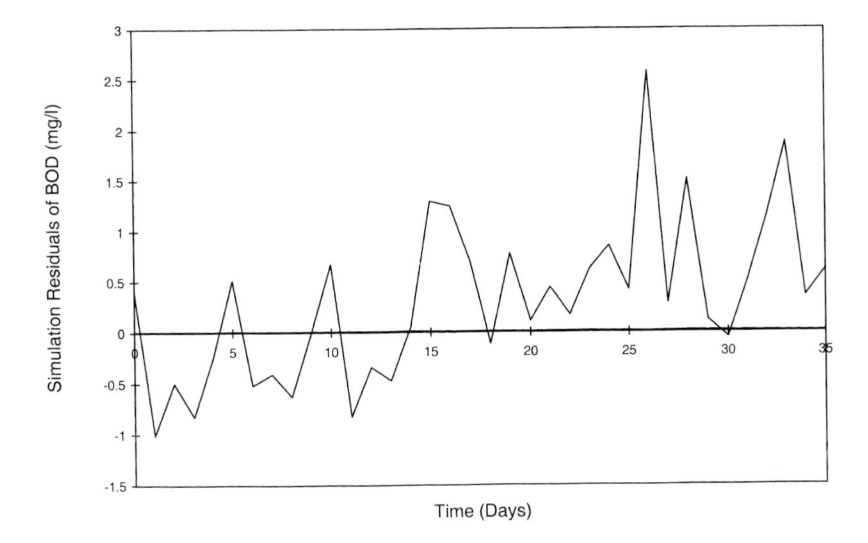

Fig. 16.4. Residual errors of mismatch for the optimal (time-invariant) candidate parameterisation of the single-state model for downstream variations of BOD concentration.

scouring of sediments or the detritus of biological growth in the river – yet one unaccounted for in the model. Second, the residual errors associated with the optimal parameterisation of the model seem to exhibit a trend (Figure 16.4), possibly something of growing magnitude and significance for the future (beyond t_{36}, that is). We are therefore encouraged to enquire further into these potential anomalies.

Parametric change

We are, of course, profoundly interested in changes with time in the estimates of the model parameters. On this occasion, however, we shall put aside the obvious choice of using the schemes of recursive estimation underpinning the discussions of Chapter 15, in favour of a more robust form of moving-window procedure. In this the window is of fixed length (N sampling intervals, say), an invariant (mean) estimate of the parameter is sought for each application of the window, and the position of the window in the observed time series is moved one sampling interval at a time, i.e., the first window might include the observations over the interval $[t_0,t_N]$, the second $[t_1,t_{N+1}]$, and so on. The usual considerations apply in choosing N (as they did in Chapter 14). It should be neither too short, in order to have reasonable confidence in the resulting parameter estimate, nor too long, such that any tendency of the parameters to change with time is thereby obscured. When N is chosen as 10 sampling intervals, and the CRS procedure outlined above is applied to minimising the conventional sum-of-squared-errors function, the trajectory of reconstructed estimates of the BOD decay-rate constant, $\hat{\alpha}_1(t_k)$, are as given in Figure 16.5(a). Strictly speaking, k denotes here the time of the sampling instant at the beginning of

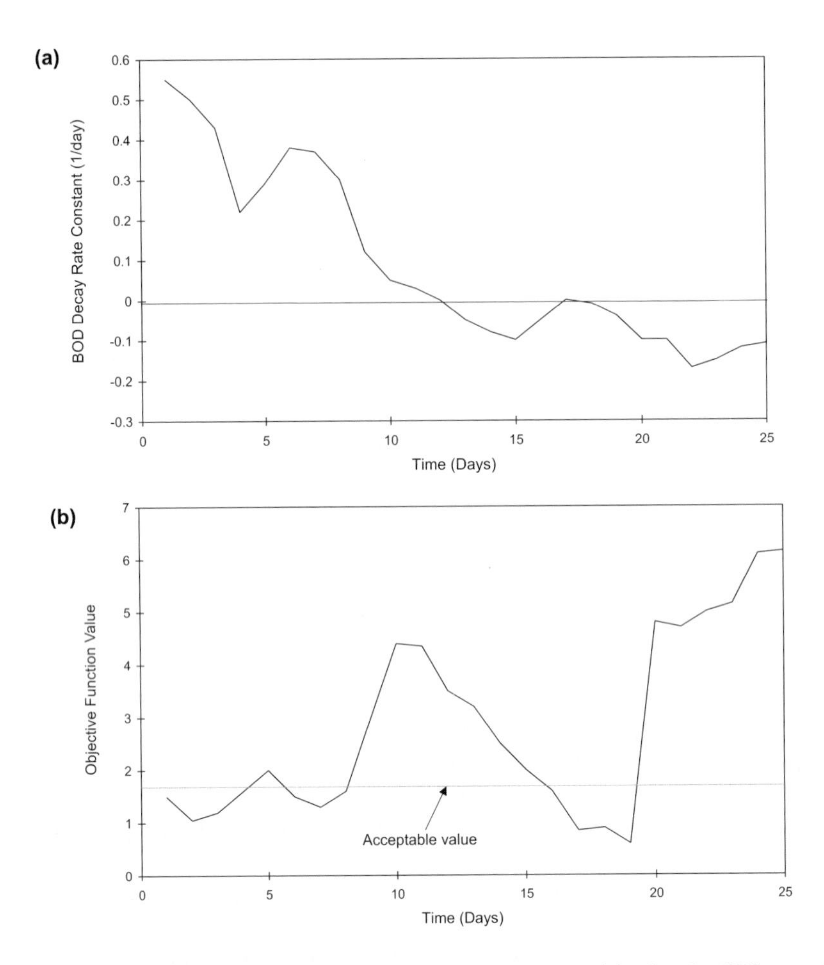

Fig. 16.5. Identification of parametric change in the single-state model using the CRS procedure: (a) temporal variation in estimated BOD decay-rate constant (day^{-1}); and (b) accompanying temporal variation in the minimum value of the squared-error objective function.

the window. Over this first part of the record the reconstructed estimate of the BOD decay rate constant has declined to less than zero, corroboration indeed of a source of BOD not accounted for in the model. In concert with detection of this structural flaw in the model, Figure 16.5(b) suggests an accompanying growth in the sum-of-squared-errors function used for estimating the model's single parameter.

16.3.2 Two-state Model of BOD-DO Dynamics

Pre-emptive detection of a change of structure in the dynamics of the BOD concentration can be investigated by placing the above single-state model in the wider web of observed interactions, notably through its interaction with the state of DO in the river. The goal remains the same, merely the framing of the question and the evidence brought to bear in answering it have been modified and amplified.

Equation 16.8 can be augmented with a second state equation, for the downstream DO concentration, $x_2(t)$, to give

$$\dot{x}_1(t) = -\alpha_1 x_1(t) + s_1\{\boldsymbol{u}(t)\} \tag{16.10a}$$

$$\dot{x}_2(t) = -\alpha_1 x_1(t) + \alpha_2(C_s\{\boldsymbol{u}(t)\} - x_2(t)) + s_2\{\boldsymbol{u}(t)\} \tag{16.10b}$$

in which α_2 is the reaeration rate constant, C_s is the saturation concentration of DO (a function of temperature and flow, i.e., variables labelled as inputs $\boldsymbol{u}$ in this study), and s_2 accounts for the effects of solute transport. If we remove the restriction of approximating solute transport phenomena by a single CSTR, by allowing our model structure to comprise m such elements, these equations become

$$\dot{x}_{j1}(t) = -\alpha_1 x_{j1}(t) + s_{j1}\{\boldsymbol{u}(t), \boldsymbol{x}(t)\} \tag{16.11a}$$

$$\dot{x}_{j2}(t) = -\alpha_1 x_{j1}(t) + \alpha_2(C_s\{\boldsymbol{u}(t)\} - x_{j2}(t)) + s_{j2}\{\boldsymbol{u}(t), \boldsymbol{x}(t)\} \tag{16.11b}$$

for $j = 1, ..., m$

with m being treated as a parameter to be identified and where the solute transport terms will be functions of $\boldsymbol{u}$ when $j = 1$, otherwise $\boldsymbol{x}$.

In this broader setting of prior theory and empirical evidence, calibration of the model yields subsequent projections of the variations of BOD beyond t_{36} "into the future" (Figure 16.6(a)), little different from those generated in an identical manner with the single-state model of equation 16.8 (Figure 16.3).Yet performance in respect of matching simulated with past observed BOD behaviour, i.e., over $t_0 \rightarrow t_{35}$, has been enhanced relative to that of the single-state model, albeit in a less than obvious sense. For while the quality of the match is comparable with the previous model's performance, in terms of its outer state space, the span of acceptable values attaching to the CRS-generated sample of estimates of the BOD decay rate constant – the model's inner parametric space – turns out to be lower for the two-state model. Performance of the two structures, in terms of matching history and making projections into the future, is roughly identical; but the more complex (second) model structure, containing a larger number of parameters, has less uncertainty attaching to the estimated (acceptable) values of those parameters. In this sense the second structure might arguably be deemed the preferred candidate structure. But since the sample of subsequently generated predictions is no better than that of the single-state model, one might again be tempted to cite such results as evidence of the folly of model calibration. Our point is rather that the purpose of calibration should be directed towards interpreting past behaviour as, in principle, capable of supporting one or more plausible paths of evolution in the structure of the system's behaviour, as revealed in changes with time of the estimates of the model's parameters.

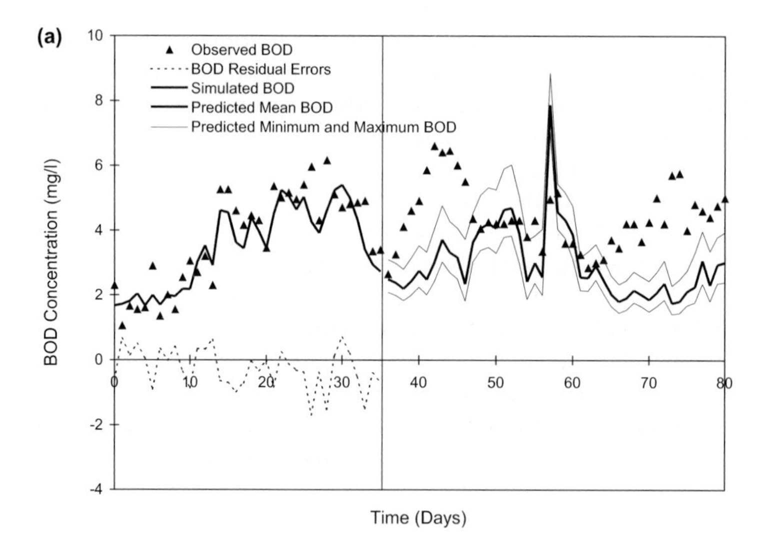

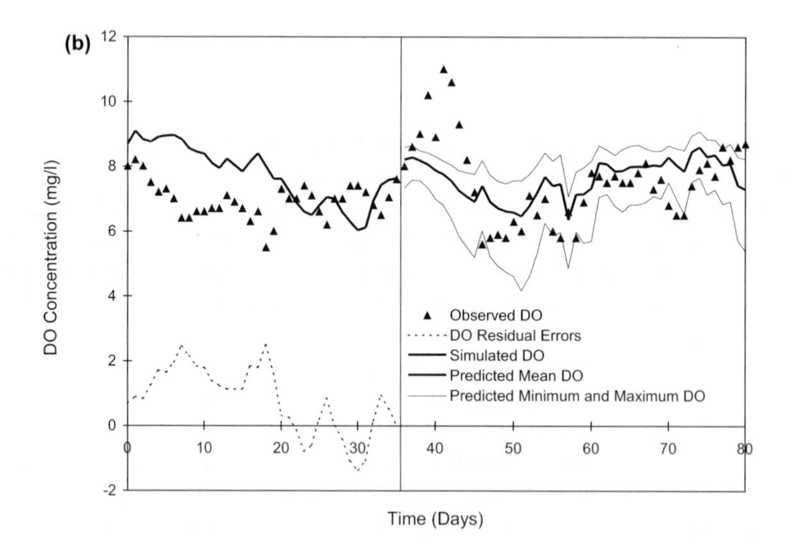

Fig. 16.6. Performance of the two-state model of BOD-DO interaction: (a) downstream BOD concentration (gm^{-3}); and (b) downstream DO concentration (gm^{-3}).

But before proceeding to an examination of possible parametric change, we should reflect on the collateral performance of the two-state model in its interpretation of the DO dynamics, as shown in Figure 16.6(b). There is mixed success: a more persistent, larger residual error over the calibration phase for DO (relative to matches of the BOD trajectories of Figure 16.6(a)), yet greater consistency of match over the phase of prediction, in particular, from t_{65} onwards. Why, then, should there have been such an over-estimation of the observed downstream DO concentration over the first twenty days of the record? There are only two factors to point to: an erroneously over-estimated rate of reaeration of the river, or an equally erroneously under-estimated rate of consumption of DO. For the first of these, the reaeration rate constant (α_2) is primarily a function of stream discharge and temperature, while the saturation concentration of DO (C_s) is here largely determined by fluctuations in stream temperature. Variations in both stream discharge and temperature over this period, however, are quite *in*consistent with overly high values of α_2 and C_s: stream discharge (Figure 16.2(a)) is in decline, tending thus to lower the capacity for oxygen transfer into the water; while temperature shows a modest transient peak around t_{10} (Figure 16.2(c)), which would ordinarily be associated with a lower value of C_s and hence a lower value of stream DO at that time. The results of the model signally fail to track this expectation (in Figure 16.6(b)). On the balance of this evidence, over-estimation of the rate of reaeration of the river can be dismissed as the cause of the model's over-estimated trajectory of DO for the first twenty days of the record.

Parametric change

If there are plausible paths of evolution latent in the past observed behaviour – paths, that is, deviating significantly from core, average, invariant behaviour for that period – their presence might be revealed, if anywhere, in the "fringe" domain of the barely tenable, as expressed pictorially in Figure 16.1 and algebraically according to equation 16.7. In the present case study it has not been possible to realise the conceptual intent of equation 16.7 in a suitable computational form. Instead, we have exploited the continuum of behaviour definitions embraced, in principle, in equation 16.9, whose bounds may be continuously varied as a function of $(w_l(t_k)\beta\sigma)$ and $(w_u(t_k)\beta\sigma)$. That is to say, samples of good parameterisations can be repeatedly generated from the CRS procedure as these terms are varied across a given range for β, from 0.5–3.0 in the following, with w_l being treated as identical to w_u (and again set to 1.0). Our search would then be directed towards any significant differences in the model's parameterisation as the continuum of behaviour definitions so defined passes out of the core domain, through the span of the fringe, and on into inclusive coverage of the highly improbable. Such a surrogate test, as realised in this way through equation 16.9, must clearly be inferior in its power to detect the seeds of structural change, for those parameterisations good for the fringe must also be good for the core, since they have been generated under a definition of past behaviour covering both (not merely the fringe alone, or the core alone, as intended in Figure 16.1).

Taking again a calibration window of days t_0 through t_{35}, Figure 16.7(a) shows how the distribution of estimates of the (invariant) BOD decay-rate constant ($\hat{\alpha}_1$) changes as a function of the expanding bounds of the behaviour definition for that period. To provide perspective, when these bounds cover essentially only core behaviour, i.e., $w(t_k)\beta = 0.5$, i.e., the bounds cover $\pm 50\%$ of one standard deviation of the primary, residual fitting errors, the values of $\hat{\alpha}_1$ are tightly focused around 0.30.

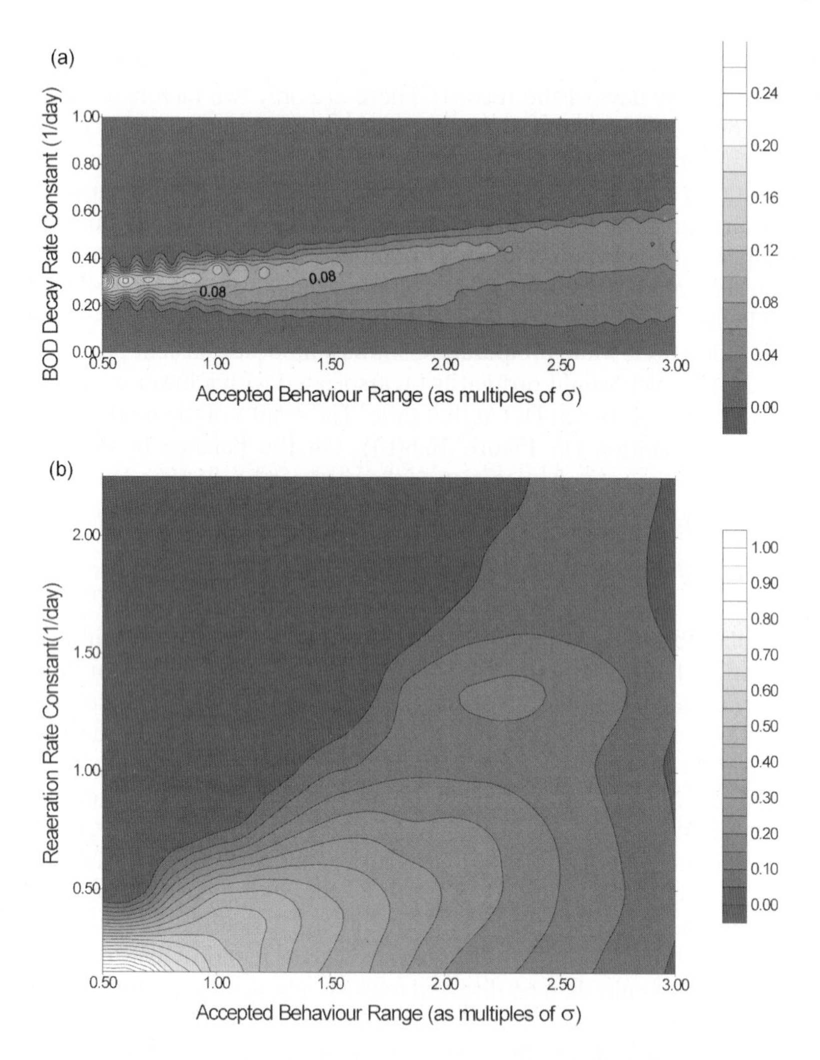

Fig. 16.7. Probability distributions for parameters identified using the CRS procedure for progressively expanded behaviour definitions ($w\beta\sigma$): (a) BOD decay-rate constant (day^{-1}); and (b) reaeration-rate constant (day^{-1}).

When behaviour covers more or less everything, i.e., $w(t_k)\beta = 3.0$, they are more widely distributed, between roughly 0.20 and 0.60. At first sight these numerical results are themselves somewhat surprising. In contrast to the comparable results for the single-state model, there is no hint here of the BOD decay-rate constant being required to be equal to zero under any span of the behaviour definition. In other words, when the singular hypothetical mechanism of BOD decay is embedded within the wider setting of the coupled BOD-DO dynamics – wherein the observed state of DO is at a relatively high level and, therefore, relatively insensitive (in principle) to the workings of this mechanism – its status appears to have changed from being inconsequential (if not antipathetic to convention) to that of significant and in line with convention. But on further reflection, it seems clear and unsurprising that these distributions of the BOD decay-rate constant must be an artefact of the goal of matching the observed DO, enabled (up to a point) by choosing higher values of $\hat{\alpha}_1$ and lower values for the estimates of the reaeration-rate constant ($\hat{\alpha}_2$), which indeed proves to have been the case (in Figure 16.7(b)). Under almost all spans of the behaviour definition there is a reasonable probability of the reaeration-rate constant being chosen as negligibly small, yet the mean stream discharge over these first weeks of the observed record was at the highest of all its levels (implying thus the highest of rates of reaeration for the entire observed record). The third parameter in the model, m, the number of CSTR elements chosen to characterise transport phenomena, is found to be essentially inconsequential. It will simply be treated as having the value of one in the subsequent exercises and will not be the subject of any further examination.

Neither result in Figure 16.7 would persuade us to conclude that there is any particular span of data – observed at the fringe – within which some incipient departure from the core, average, structure of the system's behaviour might be revealed. On this occasion, the test is inconclusive and again its very implementation is potentially confused with other spurious factors.

Within the confines, then, of what has been included in the model as known about, these results are discomforting. In whichever way the structure of the model is distorted in order to match the structure underpinning behaviour as observed (over the initial period of the record), a satisfactory interpretation of the consequences is not achieved. Something – palpable in this period, *before* its obvious manifestation after t_{40} – has been left out of the structure of the model. This "something" must look like an extra source of BOD, and it must also produce oxygen. Anything more incisive about the nature and causative stimulae of the "something" has not yet been revealed, neither in the easily accessible domain of the observed state variables nor in the less immediate space of reconstructed estimates of the model's parameters.

Our last test of the two-state model is directed at revealing changes with time in these parameter estimates, in the same manner as described above for the single-state model. Figure 16.8 shows the resulting parametric variations for a moving window of calibration (N) of 12 sampling intervals. Just as in the single-state model, the estimates of the BOD decay-rate constant decline over time, although here the distribution becomes noticeably better defined towards the end of the calibration period (Figure 16.8(a)).[4] The estimates of the reaeration-rate constant exhibit much less regularity in the pattern of their trajectory through time, but are biased towards zero, as expected (Figure 16.8(b)). Between windows 10 and 18, that is, from t_{12} to t_{29}, knowledge of suitable values for the reaeration rate constant is

---

[4] This is not a function of having processed a larger volume of the data by that point; the window of calibration is the same for each block in time for which an estimate of the parameter has been generated.

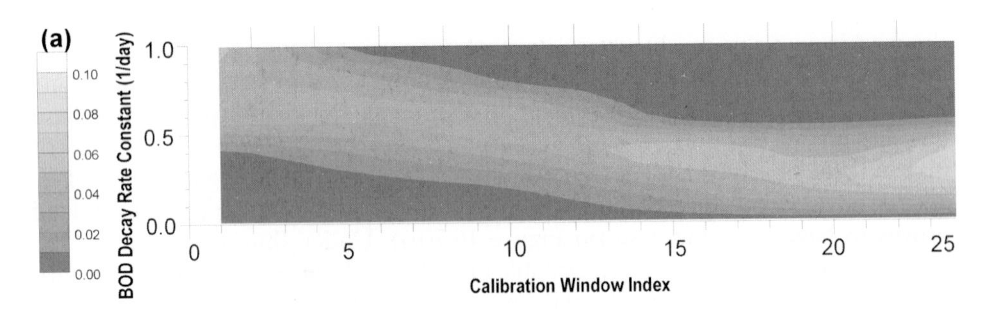

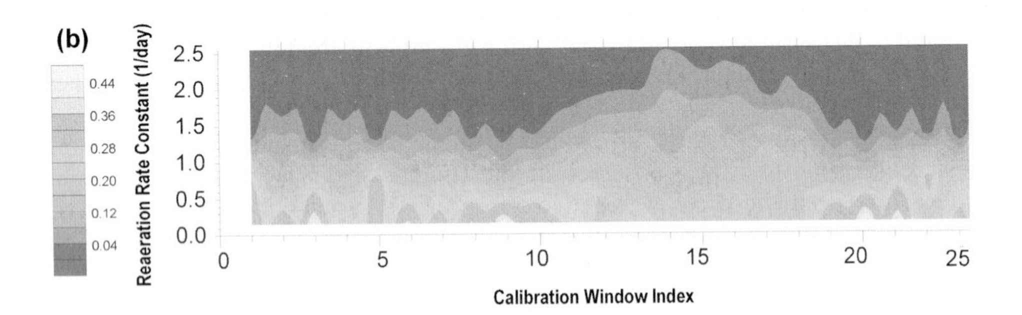

Fig. 16.8. Identification of parametric change in the two-state model using the CRS procedure: (a) temporal variation in probability distribution of estimated BOD decay-rate constant (day^{-1}); (b) temporal variation in probability distribution of estimated reaeration-rate constant (day^{-1}).

especially uncertain (and indeed the residual errors of fit are substantially larger than at other times). Since this is also the period in which the BOD decay-rate constant is most markedly lowered in value, both here (Figure 16.8(a)) and for the single-state model, the seeds of any structural change might be latent in the empirical evidence recorded from t_{12} to t_{29}, in particular. Looking back at Figure 16.2, we observe that, first, the stream flow declines at its fastest over this portion of the calibration period, while, second, the temperature undergoes its most significant excursions, experiencing both its minimum and maximum values for the period of calibration.

If then the structure of the system's behaviour is changing with time, one might argue that the best possible chance of extrapolating into the future from the period of calibration will be had from exploiting the most recent characterisation (parameterisation) of the model's structure, i.e., that available on the threshold of this future, from the results of the last window of calibration. This fails to be the case. The projections over the prediction horizon beyond t_{36} for the downstream concentration of BOD (Figure 16.9) are barely an improvement upon those obtained from the very first test with the single-state model (Figure 16.3). In this instance, our judgement would be that the current location on the parametric trajectory matters *not at all*. What matters is some understanding of, insight into, or approximate mathematical

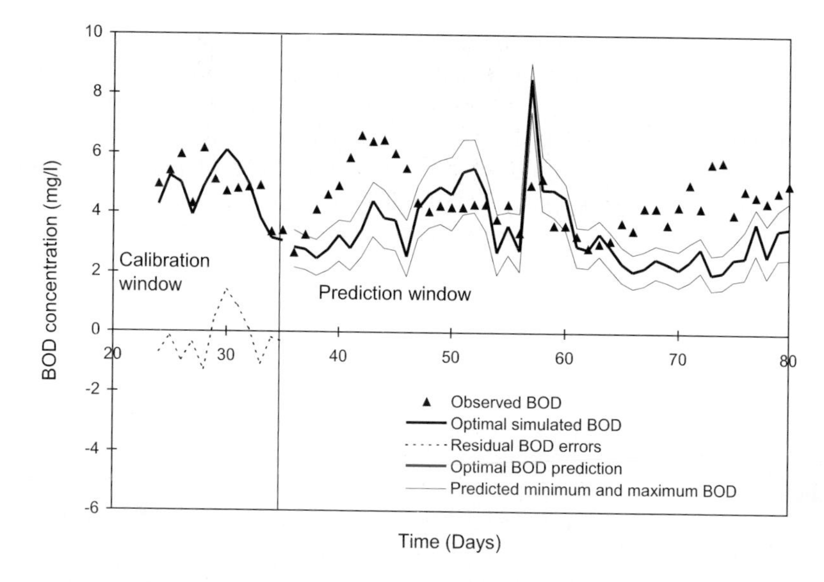

Fig. 16.9. Projection into the prediction period using the model and candidate parameterisations for the last window of calibration, i.e., that encompassing the current time (t_{35}).

characterisation of, the path by which the structure of the system – as interpreted through the vehicle of the model – came to be at that location in the evolution of its behaviour.

16.3.3 Propagation of the Seeds of Structural Change

We have found that the behaviour of water quality in the River Cam is subject to the workings of an unknown process, of potentially growing significance; that this something must generate both oxygen and BOD; and that its putative workings are in some way correlated with a declining streamflow and significant excursions of water temperature about some nominal level (as possible input stimulae). In accordance with all the previous studies of the Cam data, these features are strongly redolent of the behaviour of a population of algae moving through the reach of river. But what we have not detected in the outcomes of the foregoing explorations is a need to look for a correlation between the something and the observed sequence of hours of incident sunlight during $t_0 \rightarrow t_{35}$. This is curiously different, not least because all the previous hypotheses for capturing the subsequently manifest consequences of the latent process, which are strongly implied over $t_{38} \rightarrow t_{54}$ and $t_{65} \rightarrow t_{80}$, have been based on this input as the primary causative agent (as recorded originally in Beck (1978), for example).

Had we been standing on the threshold of the future as it was then, at t_{35}, looking back on this revelation of the first signs of an oncoming change of structure, buried in the now partially interpreted evidence going back to t_0, what would we have done in an attempt to propagate the growth in these seeds of change? For this is *the* question.

In general terms, any response must search for an adequate characterisation of the path along which the behaviour of the system might have been evolving, from as far back into the past as possible up to the threshold of prediction. One such candidate characterisation is the additional term incorporated into the following state equation for the downstream BOD concentration (whether this be assumed to be part of a single- or a two-state model structure),

$$\dot{x}_1(t) = -\alpha_1 x_1(t) + \alpha_3 (\bar{u}_1/u_1(t))\alpha_4^{u_2(t)-\bar{u}_2} + s_1(t) \tag{16.12}$$

Here u_1 and u_2 are respectively the observed stream discharge and temperature, whose mean values over the period of *calibration* are $\bar{u}_1$ and $\bar{u}_2$ respectively, and α_3 and α_4 are additional parameters, to be estimated jointly with the other parameters of the model, prior to the exercise of prediction. In this instance, we chose not to modify the DO state equation, which remains as before, as expressed in equation 16.10(b). The precise form of the additional term in equation 16.12 was chosen on the following basis (assuming we are poised at time t_{35}): hitherto, the evidence has been accumulating in favour of the behaviour of the system being observed to be drifting (over the calibration window) towards a pattern inconsistent with the previous candidate model structures, whether these be for BOD alone or for the interaction between BOD and DO; this drift appears to be correlated with the fact that the system is experiencing significant changes in the inputs ($\boldsymbol{u}$) bearing upon it; amongst the most apparent of such changes are those in the stream discharge and temperature; summer is approaching, with the (reasonable) prospect of further changes of significance in these two quantities; and last, in view of the Arrhenius relationship, it is customary for the influence of temperature variation to be expressed in an exponential form.

Whether one chooses to treat the BOD dynamics as a single-state process or embed equation 16.12 in the two-state framework proves to be immaterial. Results for the single-state model alone are therefore shown in Figure 16.10. These may be compared directly with the reference set of results from the original single-state model (equation 16.8) given in Figure 16.3. Calibration of the model, and subsequent computation of the sample of trajectories projected into the future (beyond t_{35}) have been achieved in the same manner as previously, i.e., with again the sequences of all input perturbations ($\boldsymbol{u}$) being assumed identical with those actually observed over the period of prediction. Close inspection of these results reveals the following. First, while the performance of the revised model over the period of calibration may seem essentially identical with that of the original model in Figure 16.3, the range of the sample of acceptable trajectories for the revised structure is both narrower and in general either above or towards the upper bound of the trajectories of the original model. Although not shown, this difference is echoed in a residual fitting-error sequence, which is largely without a trend over the calibration period, and estimates of the BOD decay-rate parameter that are significantly higher (for the revised model

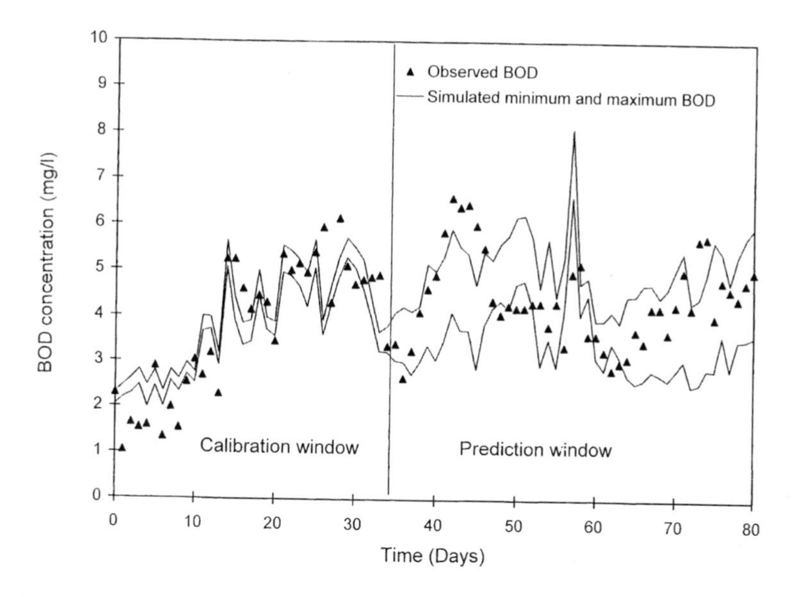

Fig. 16.10. Performance of the single-state model of BOD-DO interaction as given by equation 16.12.

relative to the original). Second, since the modified single-state structure has three parameters to be estimated (as opposed to the single, original parameter), their identifiability has been impaired, such that the greater uncertainties attaching to the samples of posterior estimates lead to clearly wider ranges of predicted BOD concentration over $t_{36} \rightarrow t_{80}$ (Figure 16.10). What we have here, in the light of the narrower range of calibrated trajectories of BOD, must therefore be a case of non-uniquely identified individual parameters that are yet strongly cross correlated among themselves. In effect, the amalgams of pairs, or the triplet, of parameters are well identified, but not their constituent elements. Third, these wider bounds notwithstanding, the projected propagation of the seeds of structural change capture better what was observed to happen, in particular over $t_{65} \rightarrow t_{80}$.

To summarise, we assert that the results of Figure 16.10 should be treated broadly as a success; perhaps all the more notably so, because the origin of this success – the additional term of equation 16.12 above – resides in a composite hypothesis rather different from that underpinning the substantial weight of the prior expectations of all the earlier interpretations of the Cam data. Yet let us quibble with this assertion. For example, had we continued with an adaptive approach to the problem up to t_{50}, say, and then looked back again at the possible paths of evolution in the structure of the system's behaviour towards that (later) threshold of prediction, the probability of being pointed in the direction of the hours of sunlight as a possible causative agent ought to have been greatly increased. For the encouraging projected upward trend over $t_{36} \rightarrow t_{50}$ in Figure 16.10 is in fact significantly slower and too lengthy in its duration relative to what was observed.

16.4 CONCLUSIONS

Is there, then, any possibility whatsoever of detecting the onset of apparent structural change before it manifests itself more palpably at the macroscopic level of the conventionally observed state variables? And if such change were detectable, could we entertain any hope of exploring the forecast consequences of propagating these seeds of change and dislocation into the future? From the case study of this chapter, we conclude that there *is* hope of responding to something akin to these questions, and with the prospect of deriving useful insights.

In order to ponder such questions we have had to think more carefully about the almost too dogmatic, unsubtle definition of what constitutes apparent structural change (in Chapters 4 and 15). Structural change can be conceived of in two rather different ways: (i) as being revealed through changes over time in the estimates of the model's parameters, as the model is forced to track the pattern of observed behaviour, or as (figuratively) the template of the model (the {presumed known}) snags on features of significance in the {acknowledged unknown} in the space between the model and the truth – a *Type I* change of structure (the subject of Chapter 15); and (ii) as being revealed through a changing sensitivity of the model's behaviour to the ubiquitous flutter and drift in parts of its parametric space, parts already included in the {presumed known}, in principle – a *Type II* change of structure. At times this low-amplitude flutter and drift is significant, in terms of its manifestation in the state space (or its influence over the system's observed responses); at times it is insignificant. It is there all the time, but is not always tangible. The behaviour of the system moves into and out of domains where it becomes more or less sensitive to such drift and flutter.

Having conceived of this distinction in the types of apparent change in the structure of behaviour, the obligation is then to propose the means of detecting either type, in particular, a *Type II* change. Our suggestion has been to construct a reference trajectory, i.e., to begin by adopting the conventional strategy of calibration – to identify core, average behaviour – so that behaviour in the fringe (between the core and the outlying bizarre) can be illuminated by reflection against this reference. And where there is behaviour at the fringe, therein should be sought the seeds of structural change, as they may reside within the model's assembly of parameters. Calibration ceases thus to be a matter merely of matching past behaviour; it should also be about exploring the potential for the system to undergo structural change. Our case study of the Cam sought to test out this procedure (as well as implementing, in a different algorithmic manner, the strategy of testing for a *Type I* change illustrated in the preceding chapter).

The outcome has been as follows. Realising the full intent of the test for identifying which parameters may be key (as opposed to redundant) in discriminating between the giving of core and the giving of fringe behaviour has proved elusive. It is obvious from the manner in which we have specified this test that it is born of the Regionalised Sensitivity Analysis seen so often elsewhere in the monograph. The problem resides in being unable to specify core and fringe definitions of behaviour in a manner clearly cleaving the two apart. The computational results on this issue, in particular, are therefore somewhat inconclusive. But through this probing of the data, from yet another methodological perspective, conditions have been created under which a novel interpretation of the impending change of structure can arise – and be successfully evaluated in terms of its capacity to forecast propagation of the seeds of this change into the future. Significantly, what counts in this case study – in the sense of forecasting ahead into the future – is not the *current position* of the image of the system's structurally evolving behaviour on the threshold of prediction, but a formal characterisation of the *path from the past* by which it came to be in that position.

REFERENCES

Allen, P.M., 1990. Evolution, innovation, and economics. In: *Technical Change and Economic Theory* (G. Dosi, C. Freeman, R. Nelson, G. Silverberg and L. Soete, eds.). Pinter, London, pp. 95–119.

Beck, M.B., 1978. Random signal analysis in an environmental sciences problem. *Appl. Math. Modelling*, **2**, 23–29.

Beck, M.B., 1986. The selection of structure in models of environmental systems. *The Statistician*, **35**, 151–161.

Beck, M.B., 1994. Understanding uncertain environmental systems. In: *Predictability and Nonlinear Modelling in Natural Sciences and Economics* (J. Grasman and G. van Straten, eds.). Kluwer, Dordrecht, 294–311.

Bowie, G.L., Mills, W.B., Porcella, D.B., Campbell, C.L., Pagenkopf, J.R., Rupp, G.L., Johnson, K.M., Chan, P.W.H. and Gherini, S.A., 1985. Rates, constants, and kinetics formulations in surface water quality modeling. Report EPA/600/3-85/040, US Environmental Protection Agency, Athens, Georgia.

Hornberger, G.M. and Spear, R.C., 1980. Eutrophication in Peel Inlet, I, Problem-defining behaviour and a mathematical model for the phosphorus scenario. *Water Res.*, **14**, 29–42.

Price, W.L., 1979. A controlled random search procedure for global optimisation. *The Computer Journal*, **20**(4), 367–370.

Spear, R.C. and Hornberger, G.M., 1980. Eutrophication in Peel Inlet, II, Identification of critical uncertainties via generalised sensitivity analysis. *Water Res.*, **14**, 43–49.

Environmental Foresight and Models: A Manifesto
M.B. Beck (editor)

CHAPTER 17

Probing the Shores of Ignorance

R.L. Dennis, J.R. Arnold and G.S. Tonnesen

17.1 INTRODUCTION

The challenges we face when confronting problems in complex environmental systems are ones of developing new approaches to help foresee the unwanted futures, which may result from our current actions and of anticipating any possibly imminent, unexpected discontinuities in the behaviour of these systems. Further, we desire a better – more complete, more accurate, more precise – understanding of the potential sources of plausible future changes in environmental conditions (the goal of Chapter 11, in particular). Finally, and distinctively so, we must have clear guidance concerning the empirical measurements we might make at present to clarify the potential causes of change and help determine the actions we might take now to avoid unwanted futures.

To serve these purposes we seek to make the process of gathering data and building an understanding of the environmental system less *ad hoc*. In the past, complex high-order models were developed simply to reflect the current perception of the system's behaviour in as much faithful detail as possible. Essentially we have waited for discrepancies to appear, which have then brought to light some facet of our ignorance. What we need is something more pre-emptive and more incisive, that is to say, more sensitive to the internal functions of the system and their responses to man's impacts. What we need too is something more diagnostic, in which we can selectively and carefully probe those processes and mechanisms of the system having the most promise of capturing, in principle, useful and quantifiable measures of structural change.

Higher-order models (HOMs; the subjects variously of Chapters 7, 9, 11, 12, and 13) have significant potential for helping with these challenges to understanding and forecasting environmental change. Their strength must self-evidently lie in the (relative) completeness of their assembly of the micro-scale theory of the environmental system, which can thus be an engine of explanation and a potential source of understanding any system change, even when that change is not immediately apparent. But making the most of this potential will call for active and directed experimentation within the virtual, laboratory world of the HOM. In this chapter, we argue therefore that our ability to foresee the future and forecast the effects of current and future changes depends on having the right micro-scale theoretical constituents and having them correctly incorporated in the model. This is essentially the issue of the model's degree of well-formedness (Kuhn, 1996) – not a trivial matter to assess, since the environmental system is necessarily an open system and our ignorance, including ignorance of nascent constituent theories, is pervasive. The well-formedness issue for a HOM is one of whether the model's structure is complete enough – the {presumed known} is sufficient – to encompass the richness of future potential change, and of whether the physical system being modelled is represented with an accuracy sufficient for the intended use of the model. Ignorance about the system – the {acknowledged unknown} – could mean that the true sources driving structural change in the system may not be represented in the model at all, or that self-cancelling errors in the mechanisms which *are* included are masking the true system behaviour. Nevertheless, we wish to exploit the power of the virtual reality produced by the micro-scale theory embodied in the HOMs, even while recognizing our ignorance of both the physical and the theoretical-numerical systems from which it is derived.

The questions we would like to answer, therefore, are these: what, uniquely, might the *complexity* of a HOM contribute to fruitful exploration of a system's potential to experience the dislocations of structural change in the future; and how can we proceed beyond mere coping with this complexity to actually exploiting it to our advantage. We must cope with the fact that the massive, dense assembly of micro-scale theory must be reconciled with sparsely available observations, yet with precision sufficient to home in on the shortcomings of that theory and indeed then to rectify them. We argue we shall be able to exploit the complexity of the HOM to our advantage, providing we can:

- acquire a high-level understanding of the potential sources of future change;
- analyze the extent of our ignorance; and
- specify better, more insightful empirical investigations of the physical environmental system based on what can be simulated within the HOM.

17.1.1 HOMs, Explanation, and Prediction

The task of environmental modelling of interest here is forecasting future conditions and events that may result not only from current conditions but also from projected future changes. The way we perceive currently the interconnections of the system's state variables is not immutable, however; acknowledging and studying the potential for such apparent structural change are the essence of this monograph. Observing a complex environmental system will not provide an explanation of how the system comes to be as observed. Explanation must be derived from the model and will be constructed on the basis of all the micro-scale theory embodied in the HOM, even though the ability to comprehend and express this explanation may have to be achieved through a relatively macroscopic high-level conceptual description. We need this explanation because we must reason, before the event (and in terms of the model's same micro-scale theory), about the loci of potentially surprising dislocations in future behaviour.

If we can somehow demonstrate that the HOM is not logically inconsistent with macro-scale observation, we have a basis (a degree of support) for endorsing the absence of inconsistency in the "unpacked" micro-scale components of the model. In general, reconciling the model with field data is cast in terms of just a subset of the model's state variables; the behaviour of the unpacked components cannot be separately observed in the real world. For our purposes, however, our need is to be reassured that these many constituent process mechanisms – whose net, aggregate effects determine the course of the state variables – are sufficiently sound, as the basis both of explanation and prediction. Ideally, the explanations must be sufficiently precise – scientifically precise – to support a prediction from the model that can be checked. In Popperian terms, the prediction must be falsifiable (Popper, 1992). Our understanding of the system will thereby be extended, when a measurement, made subsequent to a prediction by the model, can be explained using the theory represented in the model. Precision – or the boldness of explanation and prediction – does not necessarily imply correctness. Most explanations are merely tentative hypotheses about the behaviour of the system and about possible sources of observed or predicted change. As such, they are suitable for guiding further investigation and may be useful in spite of not necessarily being subsequently confirmed by observations. That is to say, it is possible in some circumstances to derive explanatory use from a model that does not always yield confirmed predictions (Oreskes, 1998). Likewise, it remains altogether possible for a model to predict well but explain poorly or not at all – the result, for example, of significant errors in the various parts of the model, whose effects nevertheless tend to cancel each other out. Nonetheless, we wish to minimise the number of errors in the model, compensating or otherwise, so that its explanations will be robust and meaningful. In this way, useful model explanations from HOMs reflect Kuhn's idea of the predictive capacity, scope, and fertility of useful scientific theories (Kuhn, 1996).

Ultimately, we desire what Toulmin has called the "forecasting technique which not only works, but works for explicable reasons" (Toulmin, 1961). And while we cannot insist that a numerical model of a geophysical problem, such as chemistry in a fluid flow, be proved or shown to be real or true, i.e., "validated" (see Oreskes et al., 1994), we can insist that it at least be reliable – as a guide for taking action – and its output explicable. To judge the degree of this reliability, we must have a model which is much more transparent to users and evaluators than has previously been the case, so that explanations for predictions do not seem to come from a mysterious and opaque, black box, but can be seen to have been made for explicable reasons.

17.2 PROBING OUR IGNORANCE OF THE UNPACKED PARTS OF THE HOM

The issues, of course, are those we expressed in Chapter 5, in particular, those depicted in Figure 5.8. They are the problems of how to project illumination into the complex inner workings of what may be an extremely large, black box and, then, of how to comprehend the findings, at some higher level of easily communicable, conceptual insight. They are also problems of generating indices, or some measures of quantities within the model, that can better indicate the potential of the system to undergo change. What the HOM affords us, both uniquely and unlike the much more restricted and more costly exercises of developing novel technologies for observing the real world, are wide-ranging opportunities to "instrument" the model – as we shall call it – and thus learn about the inner workings and interactions of its unpacked constituents.

Our approach to probing the shores of our ignorance is based on our experience with nonlinear air quality models (AQMs). It has three supporting and interlocking principles:

- the guiding insight of a high-level conceptual description – such as the causal descriptions of the now well-known "ocean conveyor-belt", or even the yet better known "greenhouse effect" – of the fundamental processes and elemental pathways of the system;
- instrumentation of the model to reveal its finely detailed internal mechanisms and process structure;
- development of new "probes", i.e., measures or indices that can be routinely constructed from combinations of readily observed environmental state variables, whose purpose will be to reveal, more incisively, the real system's potential to experience structural change.

17.2.1 *A High-level Conceptual Description*

The high-level conceptual model defines the strategy of our experimentation in the laboratory world of the HOM by codifying our understanding of the system and its process interactions in basic statements at a very high and generalised descriptive, or conceptual, level of explanation. Specific details of behaviour are organised and generalised into broad features. In practice, this description should provide a clear, concise, useful mental model of what we understand about the system, which we then write out or express diagrammatically. The conceptual model provides the big picture: the essence without the confusing details. The advantage of looking at the system from this macroscopic perspective is that we can more easily identify fundamental or controlling process interactions or pathways. The elements and interactions revealed in this way can then be organised into specific terms in a process taxonomy. The process taxonomy in turn forms one basis for development of tests of the individual, unpacked parts of the model – the third supporting element of our approach – by structuring hypothesis generation, the model-derived explanations, and the associated testing of the model designed to probe our ignorance about the system and its future. Without the coherent overview afforded by this macroscopic conceptual model it is entirely possible to miss testing crucial parts of the system, by focusing too narrowly on one element, or by using a simplified but distorted representation of several processes.

17.2.2 Instrumenting the HOM

The purpose of model instrumentation is to make explicit the details of how the component processes represented in the higher-order model lead to the predicted outcome (state) variables. They are elements of the model at a *much* finer scale than that of the high-level conceptual description introduced above. Typically, for AQMs the calculated concentrations of chemical species in the atmosphere, which are retained at the end of each computational step, represent only the net effect of the coupled processes; no attempt is made to separate and determine the contributions to the final result of individual component processes. Instrumenting the HOM amounts to modifying its source code in order to receive as output both the contribution from each process and the species concentrations, thus revealing the component rates acting on each state variable. That is to say, we thereby make explicit how the component processes regulate the species transformations leading to the predicted concentrations of the outcome variables. In fact, what constitutes a component (or unpacked part) may be understood at more than one level of disaggregation.

Making explicit the contributions of the individual, inner mechanisms of the overall black box of the HOM substantially increases our ability to understand the processes of pollutant transport and transformation. Note that in AQMs any predicted concentration is really a non-unique solution: it is entirely possible for more than one combination of constituent process magnitudes to give the same outcome variable concentration. Revealing the behaviour of these processes by instrumenting the model helps identify whether predictions may have been produced through compensating errors or other unacceptable anomalies in the model. Moreover, sensitivity and uncertainty studies, typically performed simply by perturbing model inputs and observing outputs, treat the model as a black box. Instrumenting the model allows us instead to clarify the sources of change lying behind the sensitivity result. In this way, the model's explanatory power is greatly increased and its predictions more conclusively confirmed.

17.2.3 Developing New Diagnostic Measures Using the Conceptual Mental Model and the Instrumented HOM

In some way we must harness the vastly different scales of enquiry, guided by the macroscopic conceptual description (on the one hand) and implemented at the level of the microscopic parts of the HOM (on the other). The goal is to construct measures (the diagnostic probes) that track or indicate how the system is changing in its response space. The probes will usually be particular combinations of environmental state variables; and to track the system across its response space the complete model – the entire HOM, that is – is indispensable. The probes are intended, in effect, to test the key constituent hypotheses undergirding the construction of the HOM. In our example, we shall be testing the AQM's simulation of atmospheric chemistry in a fluid flow. Here, in contrast to the case of pure chemistry, where theoretical understanding and laboratory confirmation have (for all practical purposes) established foundational scientific facts, there can be no unambiguous confirmation of theory with data. Unlike classical laboratory science, in which reaction A can be isolated and studied under tightly controlled conditions in a beaker, in the real world a multiplicity of reactions are proceeding in parallel in an essentially unconfined moving fluid. Our challenge is to address the conjoined issues of reconciling the micro-scale theoretical constituents in the model system with but sparsely available observations, which by themselves do not necessarily probe or measure process rates.

17.3 CASE STUDY OF TROPOSPHERIC PHOTOCHEMISTRY

HOMs are currently in use for examining several issues in which chemistry in a fluid flow is of principal concern, for example: global climate change; stratospheric ozone (O_3) depletion and the stratospheric impact of high-altitude commercial transport; and the photochemical air quality system of the urban and regional troposphere. In all of these ignorance has played a role in shaping the way in which the problem has been addressed with HOMs, as we have been at pains to point out in our case history of the ozone problem in Chapter 9. For the case of tropospheric photochemistry, ignorance derives from our lack of full understanding of the internal workings of the chemical system and our inability to probe thoroughly the system's behaviour. Important details of chemical products and cycling are missing; and this is especially a problem, since intermediate reactants become products further down the chain of photochemical reactions. There are even cases where primary products have been omitted from the conceptual model: new data from computational chemistry suggested product formation not previously considered, but subsequently discovered in smog chamber experiments, wherein predictions had indicated they should indeed be present (Bartolotti and Edney, 1995; Jeffries, 1996, Yu and Jeffries, 1997). It is not possible to estimate the time it might take to describe the tropospheric photochemical system fully starting from the first principles of basic chemistry (Madronich and Calvert, 1989, 1990; Saunders et al., 1997; Derwent et al., 1998); it remains an open question whether this approach will ever advance sufficiently far to reduce our ignorance of the system to reasonably negligible proportions. Our argument is clearly that reducing ignorance can also be pursued from the opposite end of the spectrum of representation, from the high-level conceptual insights gained from experimentation with HOMs.

The case of the tropospheric photochemical environment is typical of the several other complex systems discussed in this monograph. Much of the development of the basic ideas underpinning the associated HOMs has, of necessity, taken place for conditions of high concentrations of species, indicative of the circumstances over large urban centres. Extrapolation from these conditions to moderately sized urban and rural conditions is undoubtedly akin to extrapolating beyond our current understanding into largely novel situations. Importantly, the case of tropospheric ozone represents a system not considered idle scientific curiosity by society. On the contrary, society has asked science to comprehend its behaviour in order to make changes to effect an outcome. In addition to trying to avoid an unwanted future, then, society has been trying to mitigate and move from an unwanted present. The purpose of photochemical modelling is directed at returning the environment to a previous state through institution of a strategy of controls on the two precursor emissions, of nitrogen oxides (NO_X) and volatile organic compounds (VOCs or HCs for hydrocarbons), to effect reductions in ambient concentrations of either, or both.

17.3.1 Understanding Tropospheric Photochemistry

At a Relatively Macroscopic Level

Experimental work in environmental smog chambers and with early numerical models more than 20 years ago showed that the chemistry of O_3 formation and accumulation is highly nonlinear (Dodge, 1977). That is to say, although changing either NO_X or VOC emissions can alter the production of O_3 ($P(O_3)$), changes in $P(O_3)$ are not monotonic with changes in the emissions precursors, especially for changes in NO_X. In some circumstances NO_X reductions lead counter-intuitively to *increases* in $P(O_3)$ and ultimately to higher O_3 concentrations ($[O_3]$). To use, then, an AQM for successfully forecasting a photochemical system's O_3 response to emissions changes, we must know both the position of that system on its O_3 response surface and the shape of that response surface. Figure 17.1 shows the surface of *peak* $[O_3]$ response (during the daily cycle) to various levels of NO_X and VOC emissions from a simulation for Atlanta, Georgia, made using a trajectory model (description of the model and its setup can be found in Tonnesen and Dennis, 2000a). The isopleth lines of the response surface are derived by fitting contours to the peak predicted $[O_3]$ in multiple model simulations using different NO_X and VOC emissions. Note that many different combinations of NO_X and VOC can produce the same $[O_3]$. For example, 20 parts per billion (ppb) NO_X and 400 parts per billion carbon (ppbC) VOC produce 100 ppb O_3 in this simulation, as do 80 ppb NO_X and 450 ppbC VOC. Solutions to the

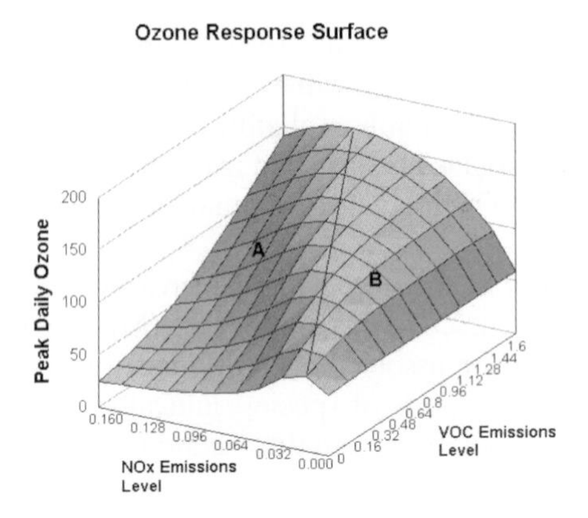

Fig. 17.1. Depiction of the nonlinear ozone response surface through plotting daily peak ozone concentrations as predicted by a photochemical box model for a large range of combinations of VOC and NO_X emissions levels.

mass continuity equation governing $P(O_3)$ are therefore non-unique and can be reached through many different combinations of the equation's positive and negative *constituent* terms, hence the need for instrumenting the model.

The bold line drawn across the surface of Figure 17.1 is the ridgeline of maximum $[O_3]$ in the simulation and divides the response surface into two domains. In these two domains, $P(O_3)$ is limited in different ways: to the right of the ridgeline (domain B) $P(O_3)$ is limited by NO_X availability and reductions in NO_X decrease $P(O_3)$, while VOC reductions have little influence; to the left of the ridgeline (domain A) radical availability from VOCs limits $P(O_3)$ and reductions in NO_X increase $P(O_3)$, while VOC reductions reduce $P(O_3)$. The shape of the O_3 response surface, such as that in Figure 17.1, is in fact determined by the nonlinearities in $P(O_3)$, which change over time as the NO_X and VOC levels change throughout the day and from one location to another. The O_3 system response will also be modified by perturbations in the model other than in NO_X and VOC emissions levels, such as by changes in physical parameters of the driving meteorology or in specific details of the chemical mechanism. Uncertainties such as these may in fact alter the spacing and shape of the response surface contour lines, affecting both the change in $[O_3]$ and the forecasted change in O_3 for future conditions. Consequently, the sensitivity of O_3 to emissions control of these two precursors remains a key variable to the vexing problem of secondary oxidant formation. Forecasting that sensitivity is the central problem for regulators seeking successful control strategies for O_3; and it is the motivating factor in probing the model at various scales of representation. Figure 17.1 plays a central role in summarising the status of the system to be managed.

At a Relatively Microscopic Level

The structure of the O_3 response surface in Figure 17.1 depicts explicitly how $P(O_3)$ in the troposphere varies with VOC and NO_X concentrations. But in addition to the

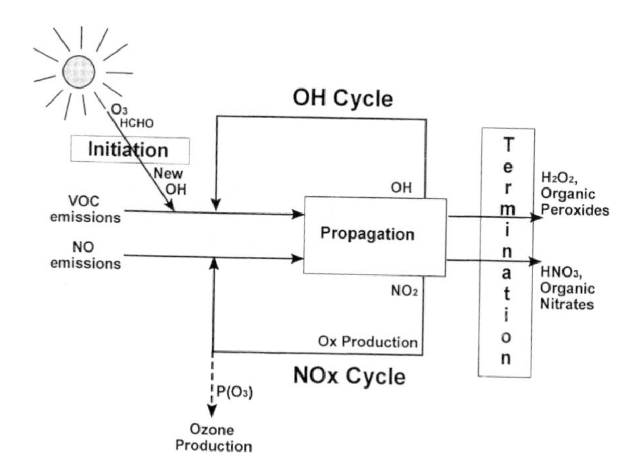

Fig. 17.2. High-level conceptual description of tropospheric photochemistry showing the major elements of radical initiation, propagation, and termination and the resultant OH and NO_X cycles.

concentrations of these two precursors, $P(O_3)$ is governed by the linked initiation, propagation, and termination of a few important radical species that preferentially attack either VOC or NO_X. Concentrations of all species in the pollutant mix change throughout the day and from one site to another. The schematised diagram in Figure 17.2, our high-level conceptual description, shows the relationships of the most important radical species and the emitted precursors. This conceptual model is further described here; it is our strategic insight into understanding the system's behaviour.

$P(O_3)$ is initiated by creation of the hydroxyl radical (OH) through photolysis of O_3 or nitrous acid (HONO) as follows:

$$O_3 + hv \rightarrow O(^1D) + O_2 \tag{17.1}$$

$$O(^1D) + H_2O \rightarrow 2OH \tag{17.2}$$

$$HONO + hv \rightarrow OH + NO \tag{17.3}$$

Other, multi-step pathways, such as the photolysis of formaldehyde (HCHO) and its subsequent reactions, also exist for creating OH through intermediary compounds from species present in urban settings.

Once created, the OH radical can react with carbon monoxide (CO) or VOC to produce the peroxy radicals HO_2 and RO_2 as in equation 17.4 below. At high [NO], peroxy radicals react with NO to split an oxygen–oxygen bond, thereby creating an odd oxygen (O_X) in the form of NO_2, as in equation 17.5:

$$OH + CO + O_2 \rightarrow CO_2 + HO_2 \tag{17.4}$$

$$HO_2 + NO \rightarrow OH + NO_2 \tag{17.5}$$

Note that equation 17.5 also recreates the original OH, which is then available for propagation through additional cycles. We define O_X as the sum of all species that can act as reservoirs for atomic oxygen: primarily NO_2, O_3, and peroxyacetylnitrate (PAN), with some contribution from various short-lived radical species and organic nitrates and nitric acid (HNO_3). A very fast, dynamic equilibrium (termed the pseudo-photostationary-state equilibrium) exists between NO_X and O_3 such that at high [NO], the O_X preferentially inter-converts to NO_2 (titration) and at low [NO] the O_X may be converted to O_3 by:

$$NO_2 + h\nu \rightarrow NO + (O^3P) \tag{17.6}$$

$$(O^3P) + O_2 \rightarrow O_3 \tag{17.7}$$

Note here that equation 17.6 recreates the original NO, making it too available for propagation through additional cycles.

$P(O_3)$ may proceed through 17.1–17.7 in an autocatalytic cycle without terminating either an OH or a NO_X molecule. At conditions with high $[NO_2]$ this photochemical cycle can be stopped, since OH radicals preferentially attack NO_2 to produce relatively inert HNO_3, i.e.,

$$OH + NO_2 \rightarrow HNO_3 \tag{17.8}$$

thereby effectively removing OH and NO_2 from the reaction system in one step. The O_3–NO_X equilibrium causes [NO] to decrease as $[O_3]$ increases, or as NO_2 is terminated by equation 17.8 (and other reactions that convert NO_X to other nitrogen-containing compounds unreactive on time scales relevant to the urban setting). At low levels of NO this photochemical reaction series will also be stopped when peroxy radicals preferentially self-terminate to produce peroxides, such as for HO_2,

$$HO_2 + HO_2 \rightarrow H_2O_2 \tag{17.9}$$

instead of recreating the OH by equation 17.5. These termination reactions end the propagation of O_X through the radical-NO_X cycles and shut down local chemical $P(O_3)$.

From the perspective of Figure 17.2, the photochemical process starts with the initiation of new radicals primarily from O_3, HCHO and HONO. Subsequent propagation of radicals through the system as HO_2 radicals, formed following OH attack on VOCs and the recreation of the OH radical, sets up an OH cycle defined by the number of times the OH is regenerated before there is a termination of HO_2 or OH. There is a companion cycle set up with the conversion of NO to NO_2 in which NO is recreated or cycled until it is terminated. The number of molecules processed by this cycling, expressed in concentration terms, is huge compared to the individual species concentrations at any given time.

Systems with a low ratio of VOC/NO_X emissions are therefore found in the domain to the left of the ridgeline of Figure 17.1 (domain A) where $P(O_3)$ is limited by the availability of radicals. Under conditions of high $[NO_X]$ in this radical-limited domain, NO_2 reacts with OH and terminates to HNO_3, as in equation 17.8 above, removing both OH and NO_2, which limits $P(O_3)$ by reducing the production of OH. Furthermore, excess NO in this high NO_X region titrates O_3 back to NO_2 thereby reducing O_3 photolysis as a source of new OH for additional cycling. For photochemical systems in these radical-limited conditions, the efficiency of O_3 production per NO_X terminated is low, and $P(O_3)$ is more responsive to reductions in VOC than in NO_X.

Conditions to the right of the ridgeline (domain B in Figure 17.1), with high VOC/NO_X emissions ratios, present a very different case, however. There, [NO] is relatively low, allowing peroxy radicals to self-terminate as in equation 17.9 above. This reduces the number of times OH can be propagated and so lowers the efficiency of $P(O_X)$ per radical. Termination of NO_2 with OH according to equation 17.8 is also reduced in this domain because the higher VOC/NO_2 ratio causes NO_2 to compete less effectively for the available OH radicals. Hence in these cases, although the efficiency of $P(O_3)$ per NO_X terminated is high, less NO_X is available for reaction, resulting in lower $P(O_3)$ and a lower final $[O_3]$. In contrast to the radical-limited domain to the left of the ridgeline, in this NO_X-limited domain $P(O_3)$ is more responsive to reductions in emissions of NO_X than of VOC.

Given this basic, process-level understanding, the O_3 ridgeline on the response surface in Figure 17.1 can now be explained – in terms of the VOC–NO_X–OH cycles of the process-level conceptual model – as the region of maximum OH production and propagation. The process-level viewpoint also makes it clear that the system is very dynamic. As NO_X and VOC levels change throughout the day, the photochemical system will change its state and move over the response surface from a more radical-limited to a more NO_X-limited state, frequently crossing the ridgeline into the other domain.

In short, the reader unfamiliar with tropospheric photochemistry will readily appreciate the effort involved in distilling down all the detail of the underlying HOM in order to arrive at the foregoing succinct, conceptual insights.

17.3.2 Model Representations

AQMs must include and process information about pollutant emissions and meteorological transport functions, but computational limitations dictate that this be balanced with the need for full and accurate description of the chemistry in order to represent conditions and events in the troposphere correctly. This means that some of the more than 200 chemical species regularly identified in urban air (Jeffries et al., 1989), together with their individual reaction processes and products, must be excluded so that the whole reaction set can be reduced to a size solvable on even the largest and fastest commercially available supercomputers. In the current generation of AQMs solving the chemistry requires approximately 80% of the computing time; simplifying the chemical representation and solution can therefore result in a significant economy.

The chemistry of AQMs generally employs from 30 to 60 species. Inorganic species are functionally unique and no generalisation of their chemistries is possible; consequently, current-generation AQMs must use at least these 14 model species explicitly: NO_X ($NO + NO_2$); NO_3; N_2O_5; HNO_3; HNO_4; HONO; $O(^3P)$ and $O(^1D)$; O_3; OH; HO_2; H_2O_2; and CO. Organic species, from 2-carbon ethane up to 10-carbon aromatics, can be generalised and grouped so that their representation is much simplified, using thus from 15 to 40 species in explicit but condensed form, or they may be simplified by generalised representations in any of three common schemes (Jeffries, 1995):

- Carbon Bond (Whitten et al., 1980): using a complete representation of all carbon in a reacting system using nonphysical, created mechanism species to represent the carbon-carbon bond types rather than the actual species molecule;
- Surrogate Species (Lurmann et al., 1987): using chemistry of one explicit species to stand as a surrogate for the chemistry of all other species in its chemical class;
- Lumped Species (Stockwell et al., 1986): similar to the Surrogate Species approach, but where the model species characteristics are determined before each simulation by a mole fraction weighting of the characteristics of the individual explicit reactions for the range of species lumped together in the surrogate.

17.3.3 Taxonomy of Diagnostic Measures

As we have said, the high-level conceptual mental model must be used in conjunction with the instrumented numerical model (the HOM) in order to develop our proposed diagnostic measures for evaluating the model's predictive performance. Such *in situ* testing of a model's processes, as it were, can be performed either internally with one model, across several models, or in comparisons using specially collected aerometric data which emphasise atmospheric processes (see for example Parrish et al., 1993; Trainer et al., 1993). A taxonomy of diagnostic measures can usefully be constructed using the terms and relations presented in the schematic diagram of O_3 formation in Figure 17.2.

The *first level* of the taxonomy (of Table 17.1) holds the initial and final components of the photochemical relations (shown in Figure 17.2) associated with the two encompassing elements of initiation and termination. A breakdown of termination into its two constituent pathways and into the speed of termination (airmass ageing) provides additional, high-level insight regarding system processing. The competition between the two termination pathways of radical cycling between the self-combining of peroxy radicals and of the conversion of $OH + NO_2$ to the relatively inert nitrogen-containing products (NO_Z) provides the diagnostic power for probing the general state of the system propagating the radicals.

The *second level* of the taxonomy holds the photochemical process groupings, including the related concepts of OH propagation efficiency (*Pr*(OH)) and chain length (Cl(OH)). Pr(OH) is the fraction of OH recreated for each OH radical entering the photochemical cycle. Pr(OH) is always less than 1, because termination reactions producing peroxides or converting NO_X to NO_Z remove some fraction of the radicals as well. Recall from equations 17.1 through 17.7 that the cycle of OH radicals, from the attack on VOCs and production of HO_2, is closely bound to the NO $\rightarrow NO_2$ conversion, such that excess availability of one species (NO) ensures OH propagation when the excess availability of the other, NO_2, leads to OH termination. Pr(OH) is approximated by the product of the two halves of these connected cycles, which together propagate the OH. This relation can be represented as:

$$f(OH+HC) \times f(HO_2+NO).$$

Due to the very fast, dynamic equilibrium of the pseudo-photostationary steady-state that obtains among OH–VOC–NO_X, the two propagation terms have an opposite dependence on NO_X. As NO_X increases, the fraction of HO_2 being recreated as OH increases, so that the fraction of OH attacking HCs to create HO_2 must decrease. The net result of the interrelation of these cycles is that Pr(OH) is maximised for some intermediate level of NO_X maximising the product of these two terms. In so doing, it maximises $P(O_X)$ and creates the O_3 ridgeline. Cl(OH) is the average number of times a new radical cycles through the reaction series before being removed in a termination reaction. OH initiation, Cl(OH), and Pr(OH) are related such that Cl can be calculated as 1/(1–Pr(OH)) and total OH production (P(OH)) is the product

$$Cl(OH) \times \text{OH initiation}.$$

The *third level* of the process taxonomy (of Table 17.1) holds the integrated response of the system, such as that represented on the O_3 response surface shown in Figure 17.1.

Table 17.1

Taxonomy of photochemical diagnostic elements

Individual Component Aspects	Radical Initiation Radical Termination Competition Between Termination Pathways Airmass Ageing
Process Aspects	OH Production Radical Propagation Radical Propagation Efficiency OH Chain Length NO_X Chain Length $P(O_3)$ Efficiency per NO_X Termination
Response Surface Aspects	System State Relative to Ridge Line Location of Ridge Line in Response Space Slope of Radical-Limited Response Surface Slope of NO_X-Limited Response Surface

17.4 INSTRUMENTING THE MODEL

17.4.1 Background

The objective of model instrumentation is to make explicit the inner workings of the model, the constituent behaviours of the unpacked parts of the model resulting in outcome (state) variable predictions at the end of each major time step. In order to reach this objective we must have a technique for quantifying the contributions of individual processes to model predictions, to make explicit hidden balances, and to open up hidden processes, cycles, and process pathways and process interactions. Our approach to instrumenting photochemical models is based on the pioneering work of Jeffries and his colleagues at the University of North Carolina (Jeffries, 1995; Tonnesen and Jeffries, 1994; Jang et al., 1995a,b). We describe the basic approach and then show examples of different types of investigations of model predictions with respect to chemical processes.

Complex photochemical models are almost exclusively Eulerian, i.e., fixed-grid models. The governing equation for these models is typically the species continuity equation of the form:

$$\frac{\partial C_i}{\partial t}+\frac{\partial(uC_i)}{\partial x}+\frac{\partial(vC_i)}{\partial y}+\frac{\partial(wC_1)}{\partial z}+\frac{\partial}{\partial z}\left(-K_z\frac{\partial C_i}{\partial z}\right)=S_i-R_i+P_i-L_i+Q_i$$

where C_i is the concentration of species i; u, v, and w are the wind velocity components in the x, y, and z directions, respectively; K_z is the turbulent diffusion coefficient; S_i is the source due to emissions for species i; R_i is the removal term for species i *via* various processes ,e.g., dry and wet deposition; P_i is the chemical production term for species i; L_i is the loss rate of species i *via* gas-phase chemical reactions; and Q_i is the production and/or loss of species i by cloud or aqueous-phase chemical processes.

For chemically reactive species, this results in a system of partial differential equations (PDEs) predicting the temporal change in the species concentrations due to the various chemical and physical processes. Photochemical models are typically configured to produce as output only the concentration fields reflecting the cumulative effect of all processes. Instrumentation of the model is a purposeful and careful intervention in which additional code, new solvers, or whole modules, are added to the model to produce the supplemental outputs needed to track the contributions of individual processes to model predictions. The approach to instrumentation is slightly different when dealing with the effects of the physical processes, e.g., transport by winds or vertical mixing by eddy diffusion; it is also different when dealing with the net effect of the chemistry, as opposed to making explicit the details of the chemical transformations calculated from the nonlinear chemical mechanisms of the model. We term the former integrated process rates (IPR) and the latter integrated reaction rates (IRR).

Instrumenting the model for the effects of the physical processes (IPR) takes advantage of the fact that the photochemical Eulerian models use operator splitting to solve the species continuity equation. In this, the system of PDEs is solved by separating the continuity equation for each species into several simpler PDEs or ordinary differential equations (ODEs) that calculate the effect of only one or two processes on the species concentration. The change in species concentration computed by the operator from the initial to the final concentration is equivalent to the integrated process rate for that time step. This does not require additional computation, since the integration must be achieved in any case. Hence, for the operators involved in operator splitting, instrumenting the model means saving the change in species concentration due to each operator for each basic model time step. We use a sub-unit of the advection time step as the basic time step. If the effect of two processes is being calculated by an operator, then it may be necessary to integrate the process rates separately, change to another algorithm, or use mass balance information to separate the required process information.

Instrumenting the model to make explicit the chemical transformations involved in producing the final concentration (IRR) requires working closely with the chemical mechanism and solver in the model. While the techniques are slightly different, the issue here is the same as with instrumenting the physical processes in the model. The numerical solvers of the set of nonlinear, coupled ODEs used to describe the chemical interactions compute only the net effect on the species concentrations as a function of time. To instrument the model, the rates of the individual chemical reactions must be integrated, using the reaction rate r_i for each reaction i. The change in mass caused by reaction, M_i,is expressed as follows:

$$M_i(t_2) = M_i(t_1) + \int_{t_1}^{t_2} r_i \mathrm{d}t$$

with $M_i(t_0) = 0$ at the start. The units of r_i are concentration-min^{-1}, the units of M_i are concentration, and the values represent the mass throughput of reaction i. The value of M_i represents the total throughput of the reaction and can be used with the appropriate stoichiometry to determine the amount of an individual species that is produced or consumed by the reaction. As the chemistry solver marches through time, M_i accumulates the integrated mass. Since processes vary with time, the value of M_i is periodically output and reset to zero to track the chemical processing. For the examples presented here, we use a one-hour processing step for the integrated reaction-rate output.

Several photochemical models have been instrumented based on this approach of Tonnesen and Jeffries (1994). For the work underpinning this chapter we have used output from one model we have instrumented, the Regional Acid Deposition Model (RADM). A highly modular "select-and-play" implementation has been structured into a more modern, quasi-object oriented AQM, i.e., the EPA's Models-3/CMAQ, the Community Multiscale Air Quality model. Description of that implementation can be found in Gipson (1999).

17.4.2 Illustrations from RADM

The following examples illustrate some of the capabilities of an instrumented model using output and analysis from RADM computations. Four examples illustrate the ability to (1) quantify processes, not merely outcome variable concentrations, (2) illuminate the difference between a species production rate and its final concentration, (3) reveal hidden interactions, and (4) provide explanations for non-intuitive predictions.

Quantification of Photochemical Processes

Several key attributes of photochemical processes stemming from the conceptual model and highlighted in the taxonomy are neither state variables nor output variables. These attributes correspond to rates or cycling amplifications, or other features gauging the system's dynamic functioning. The model instrumentation quantifies these process measures (or process "resultants"), thereby providing valuable diagnostic information, one example being the OH chain length in a RADM simulation shown in Figure 17.3 for one hour on 1 August 1988. Reconstructing these measures from typical outcome variables is arduous at best, and nearly impossible or impossible at worst. The OH chain length "response" surface is the inverse of the ozone response surface (as a function of NO_X and VOC levels). Like ozone, the chain length is non-unique, i.e., many different combinations of NO_X and VOC can produce the same value for *Cl*(OH). Figure 17.3 shows a fair degree of spatial structure, some of which is not intuitively obvious. But having this picture of the chain length available provides valuable insight into the local state of the system, which may be confirmed with additional instrumented output. Fortunately, we believe we

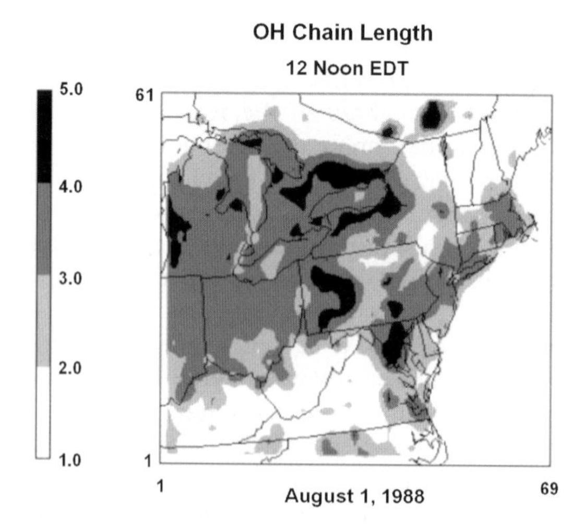

Fig. 17.3. Isopleths of the OH chain length within the eastern United States domain at 12 noon on 1 August 1988, as predicted and output by the instrumented 20-km grid Regional Acid Deposition Model (RADM).

have found a way to approximate the OH chain length using outcome variables, as we shall see shortly.

The fraction of HO_2 reacting with NO, $f(HO_2+NO)$, is another example of the type of diagnostic information which would be impossible to reconstruct from output species concentrations. $f(HO_2+NO)$ defines the system's location relative to the ridge line: it will be on the radical-limited side for fractions greater than 0.94 and on the NO_X-limited side for fractions smaller than 0.90. Figures 17.4(a), (b), and (c)

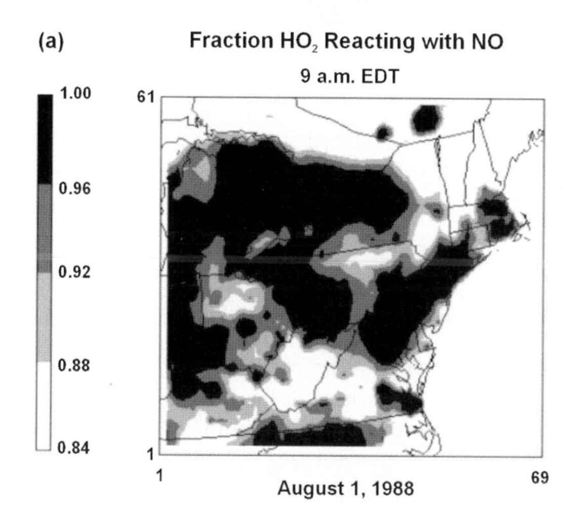

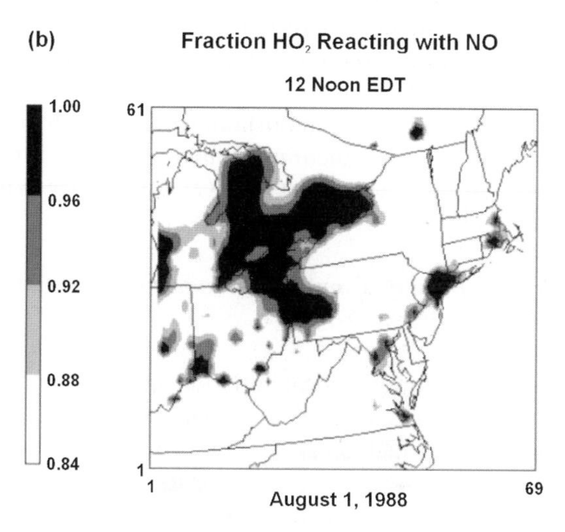

Fig. 17.4. Isopleths of the fraction of HO_2 reacting with NO as output by the instrumented 20-km RADM for three different daylight hours on 1 August 1988, illustrating the progression of the photochemical system, especially outside the large urban centres, from radical- to NO_X-limited conditions: (a) at 9 a.m., (b) at 12 noon, (c) see next page.

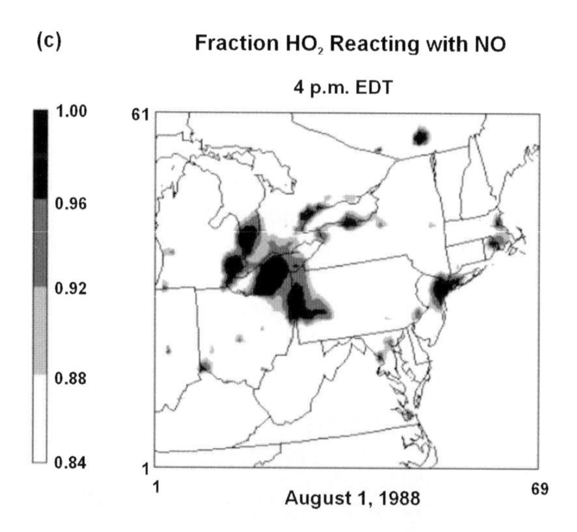

Fig. 17.4 continued. (c) At 4 p.m. Values of the fraction greater than 0.94 indicate the system is radical-limited; values below 0.90 indicate the system is NO_X-limited.

show this measure and the state of the photochemical regime across space at three times of the day, respectively, early morning, mid-day, and afternoon. The system generally starts off radical-limited almost everywhere (Figure 17.4(a)). By mid-day broad regions outside major urban areas are NO_X -limited (Figure 17.4(b)), and by afternoon (Figure 17.4(c)) only the core urban areas with very large NO_X emissions may continue to be in the radical-limited regime. Importantly, we can capture this measure during the base simulation rather than resorting to a very large and cumbersome series of sensitivity runs. Equally importantly, different chemical mechanisms in the model could give different predictions of the future state for the same future scenario. Instrumented output, such as $f(HO_2+NO)$ is necessary to help explain why the two predicted outcomes are different.

Illuminating the Difference Between Constituent Production Rates and Aggregate Environmental Variable Outcomes

In our high-level conceptual model, radical initiation and propagation are important features governing the photochemical system. One significant source of radical initiation is photolysis of O_3 which subsequently produces OH. The rates of OH production and propagation are very high relative to the OH concentration at any given time because of the short lifetime of the OH radical, as illustrated in Figures 17.5, 17.6, and 17.7. The OH production rate (P(OH)) is shown in Figure 17.5. Given its very short lifetime in the atmosphere, production and loss rates of the OH radical are approximately equal and its concentration can be calculated using the steady-state budget equation:

$$[OH] = P(OH)/L_{OH}$$

where L_{OH} is the reaction frequency of OH, which includes all propagation and termination reactions. Figure 17.6 shows the reaction frequency, L_{OH}, and Figure 17.7 shows the [OH] for the same time period as P(OH) in Figure 17.5. The resulting [OH] is orders of magnitude smaller than the production rate; the latter, P(OH), is

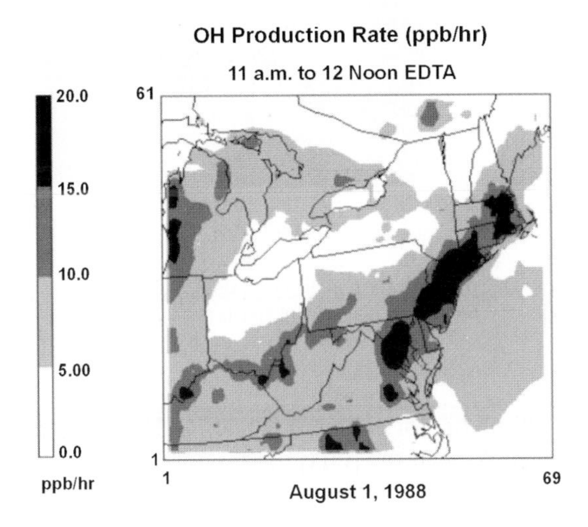

Fig. 17.5. Isopleths of the rate of OH production integrated between 11 a.m. and 12 noon on 1 August 1988, as output by the instrumented 20-km RADM. Units are ppb (parts per billion) per hour.

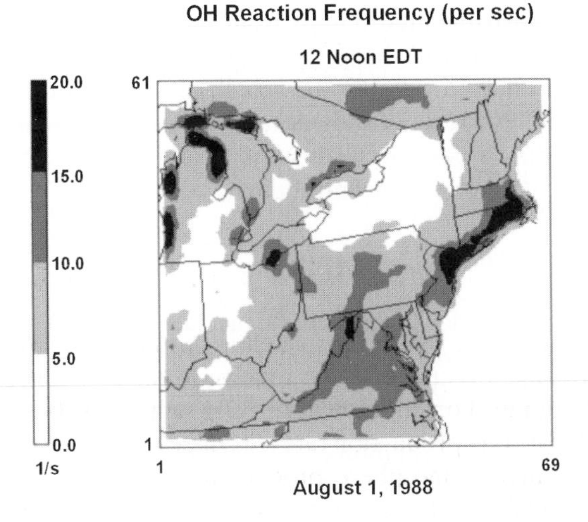

Fig. 17.6. Isopleths of the OH reaction frequency at 12 noon on 1 August 1988, as output by the instrumented 20-km RADM.

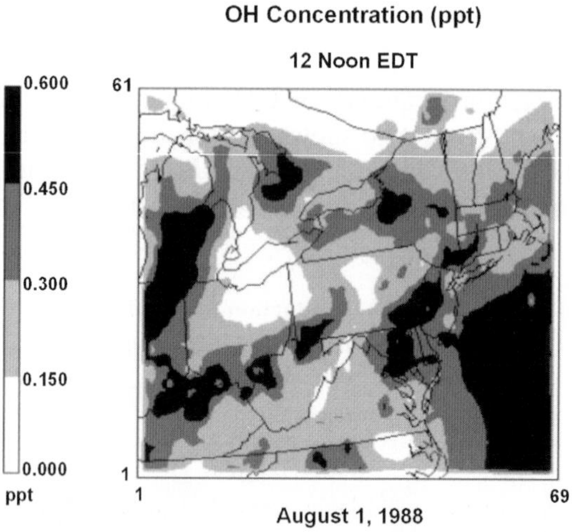

Fig. 17.7. Isopleths of the OH concentration at 12 noon on 1 August 1988, as predicted by the 20-km RADM. Units are ppt (parts per trillion).

measured in units of ppb/hr, for example, while [OH] is measured in units of ppt. Furthermore, simply looking at the [OH] field is not informative with respect to the reactivity of the photochemical system. For one thing production rates can be associated with high reaction frequencies due to a positive feedback between reaction frequency and production rate – a given [OH] value can be associated with either low or high OH production rates to various degrees. The [OH] field is not correlated with reactivity, i.e., with $P(O_X)$, which is the driving force in the system we wish to understand. Instrumented output from the model gives us a significantly more insightful picture of what happens where and when in the photochemical system. This type of output opens up the spatial and temporal functioning of the system through the ability to examine unpacked, constituent rate terms, and allows us to go well beyond what is possible from simply examining the resulting aggregate outcome variables, e.g., species concentrations.

Exposing Hidden Interactions

Figure 17.8 shows a time series of total HNO_3 production, HNO_3 production via the $OH+NO_2$ termination pathway, and the HNO_3 ambient concentration for a model grid cell downwind of New York City. Given that the total production is larger than the ambient concentration in this cell, this is a source area that is exporting HNO_3 to the surrounding region. Interpretation of the level of HNO_3 predicted or measured at a site, usually involves only the OH termination pathway, because that is the dominant pathway during the day-time and OH dominates current thinking about photochemistry. However, at this site during the day the OH termination pathway

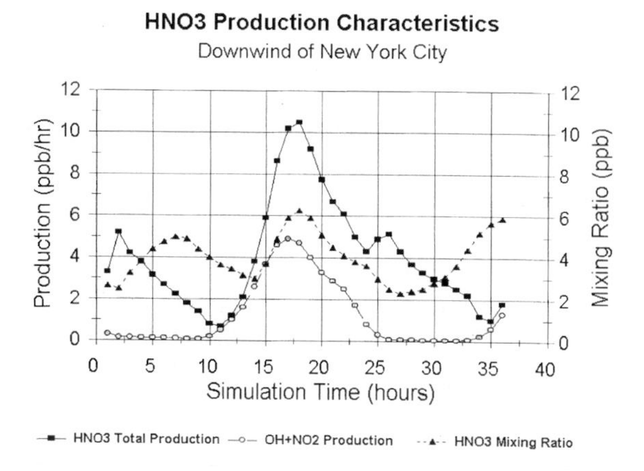

Fig. 17.8. 36-hour time series centred on 30 July 1988, of total HNO_3 production (solid line, squares), production of HNO_3 from the $OH+NO_2$ reaction (dashed line, circles), and HNO_3 ambient concentration as mixing ratio (dotted line, triangles) for a 20-km cell downwind of New York City, as predicted by the instrumented RADM.

accounts for only half of the HNO_3 production; the other half is almost entirely the result of an HNO_3 production pathway involving the nitrate radical NO_3. This additional pathway is absent from most scientists' mental models. For them, the nitrate radical path remains a hidden interaction and their interpretations of the sources of HNO_3 are likely distorted because of this omission. At night, when there is no $OH+NO_2$ termination, there is still significant HNO_3 production involving N_2O_5 and particles, as a loss mechanism for NO_2 and a production mechanism for HNO_3. However, this production pathway is not well understood (this model prediction is an instantiation of an hypothesis) and its importance appears to be very much underrated, much as the influence of NO_3 in general has been. The night-time "hump" in the HNO_3 concentration trajectory in Figure 17.8 shows that the importance of HNO_3 production during the night can be inferred from an analysis of outcome variables. There are no distinguishing features for bringing out the influence of the nitrate radical, however, so that mental models – or high-level conceptual descriptions – will remain focused on the hydroxyl radical. Without output from the instrumented HOM the importance of the nitrate radical would probably have remained obscured for some time to come.

Providing Explanations of Counter-intuitive Predictions

The instrumented model is a valuable tool to investigate the processes behind the outcome variables predicted by the model. The assumption is that the theoretical concepts programmed into the models have sufficient validity, relative to real-world processes, for the insights gained to inform diagnostic interpretations of model behaviour, even where model predictions are wrong. This is, then, an important logical distinction between the trustworthiness of the model as gauged by quantities that can have been observed in the past, i.e., history matching, and its trustworthiness according to the elemental parts of which it has been constructed. The instrumented model output has been shown to be important to the understanding and explanation of model sensitivity analysis results, even when only one parameter is changed.

Often the mental or less-sophisticated model used as the intuitive basis for creating the predicted response of the system contains a limited set of assumptions and considerations. The change in one pathway is assessed, but not changes in the myriad set of pathways in a highly interconnected system. A change in one pathway in a connected system creates changes in other pathways in response, potentially changing the balance of processing from one pathway to another, or altering the rate of processing across the set of pathways. The intuitive expectation is that of the initial change but not any resultant changes. Thus, the prediction from the HOM can appear to be counter-intuitive. Our experience with photochemical systems is that more often than not our intuition is not in line with model results. Our inability to "second-guess" a complex system is a continually humbling experience. The only way to understand the model prediction and reconcile it with intuition is to examine the internal processing of the model by instrumenting it. To make this concrete we provide an example in Section 17.6.3. What we shall find there is that some aspects of the system response, upon reflection, should have been anticipated, but others were not at all obvious and we could not have understood them without the instrumented model. The instrumented model is a key to schooling our intuition.

17.5 CONNECTING MODEL PROBES TO AMBIENT FIELD OBSERVATIONS

From the conceptual model and taxonomy of tropospheric photochemistry several avenues can be suggested for probing AQMs using individual chemical components, photochemical process groupings, and response-surface measures. The important point is that our understanding of photochemical theory embodied in the macroscopic, conceptual model guides the development of diagnostic tests of the model using these more refined, microscopic elements and suggests how the tests might be interpreted. These first tests and interpretations are then cross-checked using process-analysis results from the instrumented model. Here, then, we shall construct a compendium of model diagnostic measurements associated with specific elements in the taxonomy and then assess the viability of performing tests of the model based on these measures using the sparse, variable, and uncertain ambient data available at present.

17.5.1 Associating Field Observations with Diagnostic Model Measures

The *first level* of the taxonomy (Table 17.1) defines several individual steps in the overall photochemical process which can be examined. Key among them are the steps of radical initiation and radical termination. Field observations can be identified which, in combination, allow examination of these individual process steps, for example:

- radical initiation pathways using comparisons for O_3, HCHO, HONO, peroxides, $J(NO_2)$, $J(O_3)$, where J represents the spectral photolysis rate of these compounds, and other spectral irradiance observations (radical initiation is the sum of individual contributions from each of the initiating species);
- radical termination pathways using comparisons of HNO_3, PAN, and organic nitrates and peroxides (radical termination is the sum of the species terminated that involve either NO_X, i.e., NO_3^-, HNO_3, PAN, and organic nitrates, or combinations of radicals, such as organic peroxides and H_2O_2);
- the balance between radical initiation and termination;
- competition between radical termination pathways, by comparing the production of HNO_3 and other nitrates to that of H_2O_2 and other peroxides;
- speciation of NO_Z (the sum of species terminated involving NO_X) to compare competition between NO_Z termination pathways;
- airmass ageing pathways using relative fractions of NO_X *versus* NO_Y, and PAN and total nitrates *versus* NO_Y.

The *second level* of the taxonomy in Table 17.1 examines process grouping, production and cycling. Key concepts are the OH propagation efficiency (Pr(OH)), the OH chain length (Cl(OH)), and the NO_X chain length. Combinations of observations can be defined, allowing examination of these process groups, although the definition requires work with the instrumented model, as summarised here. Recall, therefore, our earlier discussion of section 17.3.3, in which we described the relationship between OH and NO cycling, and demonstrated how Pr(OH) could be approximated as the product of the fraction of OH reacting with VOCs, f(OH+HC), and the fraction of HO_2 reacting with NO, $f(HO_2+NO)$. The two fractions, f(OH+HC) and $f(HO_2+NO)$, are measures of processing but cannot be measured directly in the field. Using the photochemical theory embodied in describing mathematically the set of reactions behind the fractions, we sought to devise diagnostic measures for these two elements of the conceptual model that could be constructed from observations made in the field. The resulting indicators so constructed, $I(NO,t\text{-}RO_2)$ for $f(HO_2+NO)$ and $I(HC,NO_2)$ for f(OH+HC), become the operational diagnostic measures. The two indicators were constructed by listing all chemical reactions involving the relevant species and dropping any minor terms, from which the following expressions for the operational diagnostic measures, the indicators, emerge:

$$I(\mathrm{NO}, t\text{-}\mathrm{RO_2}) = \frac{k_{\mathrm{HO_2+NO}} * [\mathrm{NO}]}{k_{\mathrm{HO_2+NO}} * [\mathrm{NO}] + k_{\mathrm{RO_2+HO_2}} + k_{\mathrm{HO_2+O_3}} * [\mathrm{O_3}]}$$

$$I(\mathrm{HC}, \mathrm{NO_2}) = \frac{\sum k_{\mathrm{OH+HC}_i} * [\mathrm{HC}_i]}{\sum k_{\mathrm{OH+HC}_i} * [\mathrm{HC}_i] + k_{\mathrm{OH+NO_2}} * [\mathrm{NO_2}]}$$

With these two expressions we can approximate *Pr*(OH) and also *Cl*(OH). The concept of the NO_X chain length can be approximated by examining the day-by-day association between O_3 and the NO_X termination products (NO_3^-, HNO_3, PAN, and organic nitrates) for mid-day temporal stratifications. This approximation, however, is always attenuated by deposition loss processes, an artifact which needs therefore to be removed before a true measure can be obtained.

With regard to the *third level* of the taxonomy (Table 17.1) we can take the measures developed for the propagation efficiencies and adapt them to examine the response surface. In fact, historically the measures were developed in reverse order because of the photochemical community's interest in indicators of the ozone ridgeline. Beginning with the concept of radical propagation, *Pr*(OH), and its two terms, *f*(OH+HC) and *f*(HO_2+NO), from our high-level conceptual model – using an instrumented 1-D (box) model – we then determined that isopleths of constant *Pr*(OH) fractions, calculated for the two terms each representing half of the OH cycle, are very nearly parallel across most of the O_3 response space. Thus, the O_3 ridgeline cutting across the response surface corresponds to a relatively constant value of the two propagation terms; however, the value is not completely constant because of the small influences of several minor termination reactions. As a result, a *Pr*(OH) value can be identified for the two terms which will identify the O_3 ridgeline. That is to say, these two halves of the OH cycle can be treated as two process-oriented measures used to indicate the distance of a photochemical system from the ridgeline on the response surface. This hypothesis proved successful when tested with 1-D and 3-D instrumented models using sensitivity runs with a variety of changes in input emissions designed expressly to produce different shifts of the photochemical system on the response surface.

Hence, with the indicators *I*(NO, *t*-RO_2), and *I*(HC, NO_2) we now have two explicit gauges of the system's position on the O_3 response surface, and, consequently, its relative distance from the [O_3] ridgeline. Moreover, our conceptual model suggests another possible gauge of the system's position – either to the left of the ridgeline and so in the radical-limited domain, or to its right in the NO_X-limited one – that make use of the system's very fast, dynamic pseudo-photostationary steady state for O_X. The new gauge (or indicator) of system position is O_3/NO_X.

These three measures, i.e., *I*(NO, *t*-RO_2), *I*(HC, NO_2), and O_3 /NO_X, probe an important attribute of AQMs, namely, the sensitivity of their O_3 predictions to controls on VOCs and the oxides of nitrogen (NO_X) and, most significantly, on either of the two necessary emissions precursors (the focus of attention of *all* management applications). Together these measures help to indicate the photochemical system's place on the O_3 response surface relative to the ridgeline; they are therefore most useful for policy-related control strategy predictions from the models. These derived indicators were then tested in the instrumented 1-D and 3-D models to demonstrate their fidelity with respect to the more accurate model process measures for determining placement of the system on the response surface. They were also tested in a series of model simulations to characterise their degree of precision across space and time (see Tonnesen and Dennis, 2000a; 2000b). In testing these three (and other) indicators recently proposed by other researchers (Sillman, 1995; Milford et al., 1994; Kleinman, 1994; Chang et al., 1997) we concluded that none of the measures indicates the relative sensitivity of the photochemical system with respect to precursor-emission changes with great precision. However, we believe that the promise of this technique and power of these specific probes has been shown, and that much can be learned about the model and the physical system if we continue to study them.

17.5.2 What Observations Can We Make at Present?

A hallmark of the diagnostic measures is that they involve combinations of several quantities, either outcome (state) variable concentrations or the unpacked process terms in the state equations. Although potential indicators and gauges in new and different combinations are still being developed, the absolute number of chemical species involved in all possible measures has stabilised. It is likely, then, that we now know of which quantities the diagnostic measures will be composed, even if we have yet to compose them. There are, nevertheless, some quantities to which we still lack observable access, especially process rates (as opposed to species concentrations) in the atmosphere. Yet this notwithstanding, the list of companion field measurements required for diagnostic model probing is a daunting one. Which diagnostic measures, then, are actually feasible at present? There are two issues here:

- the availability of the suite of needed field observations; and
- the coherence and representativeness of these measurements.

Current Observation Availability

Not surprisingly, our experience suggests that there can be no single definitive test constructed from any elements on any level of the process taxonomy. Rather, to be useful for revealing true functioning of the system, diagnostic tests will need to probe as many different aspects of both the real and virtual (modelled) systems as achievable with available analytical (observing) technology. Wherever possible, more than one way of probing a particular element or process of either system will be preferred. All possible measures – hence all species contained in them – will be needed to develop a picture of the true functioning of the model. Although we can develop priorities about which species need to be included, in the end we should be striving to have all of the necessary diagnostic species measured at the same time.

In recent years new, state-of-the-science research instrumentation has been developed and put into production. It is capable of observing many of the key species and variables needed to set up the diagnostic probes described above. This represents a significant advance on the part of the analytical community, since these observations are made over very short time scales and often at extremely low concentrations. We consider this a list of what would be needed for a full diagnostic probing of the (real) photochemical system of the polluted urban and regional troposphere, ordered here by how readily available such observations in the ambient air are at present:

- Species or variables easily observed accurately: O_3, NO, CO, HCHO, PAN(s);
- Species or variables observed accurately, but with difficulty: NO_2, NO_3, HNO_3, NO_Y, H_2O_2, HONO, organic peroxides, some alkyl nitrates, $J(O_3)$, $J(NO_2)$;
- Species for which analytical techniques are being developed and tested, and are moving slowly into more general use, if not routine production: OH, HO_2, total RO_2, speciated RO_2;
- Missing classes: alkyl nitrates for biogenic hydrocarbons, oxygenated hydrocarbons, and (as we have noted above) process rates.

While only observations in the first rank are routinely made in current networks at ambient concentrations, a significant number of key species are within reach of becoming more routine, both at the surface and aloft on aircraft platforms.

Coherence and Representativeness of Observations

In addition to the availability of appropriate observations it is critical to assess the viability of developing measures based on the behaviour of combinations of outcome variables or species to probe process formulations in the model. Not only is there the difficulty in making the observations, leading to sparse data in space and time, there is also tremendous spatial and temporal variability in the data because of the multiple interactions of pathways along time scales from seconds to minutes, both meteorologically and chemically. Thus, there are important issues of representativeness when considering these ambient measures. The meteorological contribution to this variability has proven to be quite difficult to filter out, making it very difficult to extract a chemical signal, and the discernment of long-term chemical trends in atmospheric data has been fraught with difficulty. Furthermore, because smog chambers are artificial environments generally at concentrations far above regional backgrounds, their results are not straightforward to interpret even though they are controlled experiments. It is not surprising, then, that there are real concerns about whether one could ever discern systematic chemical behaviour from *in situ* data; data that are inherently sparse and represent only a very small window of insight into all that is occurring in the atmosphere. Is the chemical theory only in evidence at the micro-scale of seconds and centimetres, or can it be evident at a more macro-scale of hours and days? Is chemical coherence, the "signal", allowing the theory to be extracted, or destroyed by meteorological or other sources of variability?

An investigation of relatively recent, intensive, field-study data using one of the new diagnostic measures suggests the chemical functioning of the system has a coherence in its behaviour that allows a signal to be discerned, i.e., behaviour that is not completely chaotic. This is the measure of the NO_X cycling, defined as the number of O_3 molecules created per molecules of NO_X termination products, or NO_Z, produced. The chemical system character of the measure is made evident by first plotting the daytime observed $[O_3]$ against the daytime observed $[NO_Z]$ at the same site across a multiple-day period encompassing large variations in daily $[O_3]$. This produces a scatterplot of daytime $[O_3]$ *versus* $[NO_Z]$. Two such scatterplots are shown in Figure 17.9 for a surface site with observations from a sequential period of 24 days in the summer of 1988. In Figure 17.9(a) all hourly observations for the day-time photochemically active period (10:00 am to 4:00 pm) are shown. In Figure 17.9(b) the 10:00 am to 4:00 pm observations have been averaged to show the photochemical behaviour across days. For any one day, the ratio of the $[O_3]$ to $[NO_Z]$ is a measure of the NO_X cycling on that day. It is possible to estimate a well-defined best-fit line through the set of points, illustrated by the running median line in Figure 17.9. This line depicts the coupled relationship between O_3 and NO_Z for a variety of conditions. The slope of the tangent of the best-fit line at a given level of $[O_3]$ defines the incremental change in O_3 associated with an incremental change in NO_Z as conditions change from one day to the next. The slope is the operational best-estimate measure of the NO_X cycling at various levels of $[O_3]$. If the chemistry were not coherent from one daily realisation to the next, a well-defined best-fit line could not be constructed. The clear, empirical realisation of the coupled relationship between O_3 and NO_Z, shown in Figure 17.9, indicates there is a coherent signal from the chemical system that is systematic across days. In fact, because this measure involves air-mass history, the signal here is derived from the behaviour across days, not within days. This coherence of the signal has been observed by others as well (see for example, Parrish et al., 1993; Trainer et al., 1993; Bertman et al., 1995; O'Brien et al., 1995; Kleinman et al., 1996a,b; Daum et al., 1996; Ridley et al., 1998; and Frost et al., 1998). We note that this test is strictly valid only for NO_X-limited conditions and that the results presented here would be contaminated by air masses that are radical-limited, i.e., air masses in which NO_X titration is occurring. However, we are able to screen out those data from radical-limited regimes that would contaminate the interpretation using the O_3/NO_X measure described above, thereby improving our interpretive ability.

Sampling designs often concentrate on just one type of episode, that most associated with damaging conditions. What we have learned from a diagnostic perspective is that the chemical signal is best clarified from the perspective of the constituent process rates when the full range of photochemical forcing is used, extending from the system being (nearly) turned off photochemically to being at full production. Modulation of the chemical system in terms of O_3 production is sufficiently large for us to observe the results. The plots show that meteorology, while it may create an alteration or interference, does not destroy the signal produced by the chemical forcing function. We can use meteorology to advantage to systematically modulate the photochemical forcing and the resultant chemical signal. We also know (e.g., Cohn and Dennis, 1994; Frost et al., 1998; Ridley et al., 1998; and Kleinman et al., 1996a,b) that the signal is evident at both the surface and aloft. This fact too can be used to advantage, because there will be less air-surface exchange interference aloft.

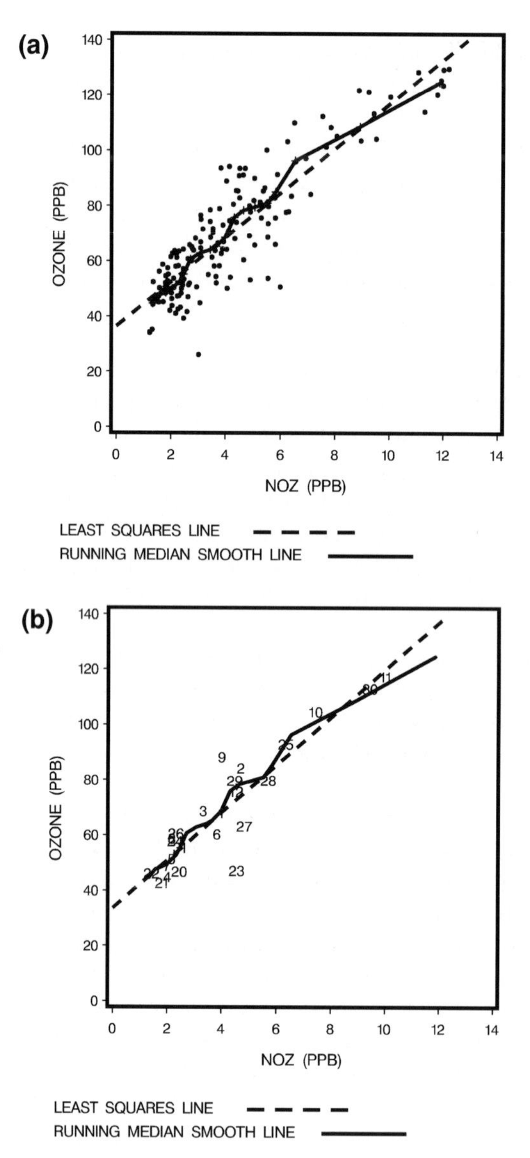

Fig. 17.9. A plot of measured O_3 *versus* the termination products of NO_X oxidation, NO_Z ($= NO_Y - NO_X$), for daylight hours between 10 a.m. and 4 p.m., inclusive, for the 24-day period of 20 July to 12 August 1988: (a) all hours, (b) the 10 a.m.–4 p.m. average for each day.

The general conclusion is that for chemistry in a fluid flow, such as the atmosphere, there is a sufficient coherence to the chemical signal for our method to work. Hence, sparse (in space and time) micro-scale testing has value for probing our understanding of components of the system and of the overall system, and probing that understanding as embodied in the HOMs.

17.6 DESIGN FOR DISCOVERY: PROBING A REGIONAL PHOTOCHEMICAL MODEL

To summarise, in the foregoing we have achieved the following:

- set out a new approach for testing HOMs, describing how the high-level conceptual model of the physical system and its component processes is central to building the required understanding of that system and its potential for future change, which we then illustrated with the example of AQMs and tropospheric O_3 forecasting;
- described and illustrated the process of instrumenting an AQM with additional code, to reveal previously hidden component process interactions (within the HOM) and make them available for analysis;
- characterised the means by which probes of these newly revealed processes can be used as *in situ* diagnostic tests of model behaviour, which we depicted with examples using combination measures that are – or may soon be – routinely observable in the (real world) of the polluted urban and regional troposphere.

Upon these foundations we wish now to illustrate the utility of our approach for probing our ignorance using three examples from an instrumented AQM. In other words, we seek to provoke apprehension (or discovery) of what – of significance – may have been omitted from the structure of the HOM, in spite of its seeming "comprehensiveness".

The first of our examples examines the model's ability to forecast future conditions of regional O_3 using relationships between model-predicted outcome variables in various combinations designed (hopefully) to assist in learning more about the system's behaviour. The second example demonstrates our approach to assessing the extent of our ignorance regarding the radical propagation pool, from a macroscopic perspective, as opposed to the perspective of working from detailed, elemental theoretical constructs, which is under investigation elsewhere (Madronich and Calvert, 1989; Saunders et al., 1997). The third example reinforces the importance of working with an instrumented model, where process interactions are explicitly shown, using results from a sensitivity study on a key rate constant in the chemical mechanism.

17.6.1 Indicators of Structural Change: O_3 **Versus** *NO_Z under NO_X-limited Conditions*

Using the high-level conceptual model (Figure 17.2) and the taxonomy of diagnostic measures (Table 17.1), we have explained and demonstrated how NO_X cycling controls $P(O_3)$ in the regional chemical environment. We expect from photochemical theory and our understanding of the O_3 response surface (Figure 17.1) that the efficiency of NO_X cycling with regard to $P(O_3)$ will change with changing $[NO_X]$. If the AQM does not capture this change in cycling efficiency with changing concentration, then its [O_3] predictions as a function of future changes in NO_X levels will be wrong and the AQM will not be a reliable policy tool for forecasting O_3 response to the economic growth and regulatory controls assumed for future conditions.

The plotted pairs of [O_3] and [NO_Z] corresponding to a range of observed O_3 across a number of days trace out a locus of points that can be fitted with a curve ("the curve") as crudely represented by the median line in Figure 17.9(b). The curve is a smooth representation of the systematic relation of [O_3] to [NO_Z] as an expression, or a best-estimate measure, of the NO_X cycling efficiency for the various levels of [O_3]. The day-to-day changes in NO_X emissions are relatively small. Hence, for current emissions conditions the changes in [O_3] and [NO_Z] levels are a consequence of changes in [NO_X] levels (its availability) and processing caused by day-to-day differences in physical processes, such as meteorological transport and mixing, or ultraviolet radiation. As [NO_X] availability increases from low to moderately high values the NO_X cycling efficiency slowly decreases, that is, the incremental increase of [O_3] slowly decreases per incremental increase in [NO_Z] as [O_3] levels increase and the curve traced out is concave.

In the NO_X-limited domain of the O_3 response surface, both O_3 and NO_Z respond in parallel, and essentially exclusively, to changes in NO_X emissions: they rise and fall together. [NO_X] availability can be changed by changes in NO_X emissions, as well as by physical processes. It turns out that because the chemical system is coherent, the change in [O_3] and [NO_Z] stemming from a change in NO_X availability will be essentially the same whether the change is due to a change in NO_X emissions or a change caused by the physical processes. Hence, the relationship between [O_3] and [NO_Z] illuminated by a variety of NO_X availability conditions, a smooth curve through the locus of points in Figure 17.9, also provides a "demonstration" of how this photochemical system would respond to anthropogenic changes in NO_X levels – either increases due to growth, or decreases due to controls. Under NO_X-limited conditions and for the same meteorological condition, NO_X-emissions changes of the size that can be expected in one or two decades will produce changes in resultant [O_3] that will move in a direction on the [O_3] *versus* [NO_Z] plot predefined by the curve. This is illustrated in Figure 17.10(a). The heads of the "pins" in this figure are at the current-condition base-case daily (10 am to 4 pm) average values of [O_3] and [NO_Z]; the pinpoint is at the location to which this pair has migrated for a 50% uniform increase in NO_X emissions. Moreover, the response of O_3 production and NO_X-cycling efficiency, and resultant [O_3], will migrate up the predefined curve for NO_X-emissions increases and down for NO_X-emissions decreases. This migration of the system's behaviour along the curve is shown in Figure 17.10(b), where the system responses to a 50% decrease, a 50% increase, and a 100% increase, respectively, in region-wide NO_X emissions are shown. Thus, the slope and shape of the coupled relationship between [O_3] and [NO_Z] provides an indication of how the system and resultant [O_3] is expected to change for a change in NO_X emissions. A comparison between the predicted and observed coupled relationship provides a means of probing the predictive ability of the model.

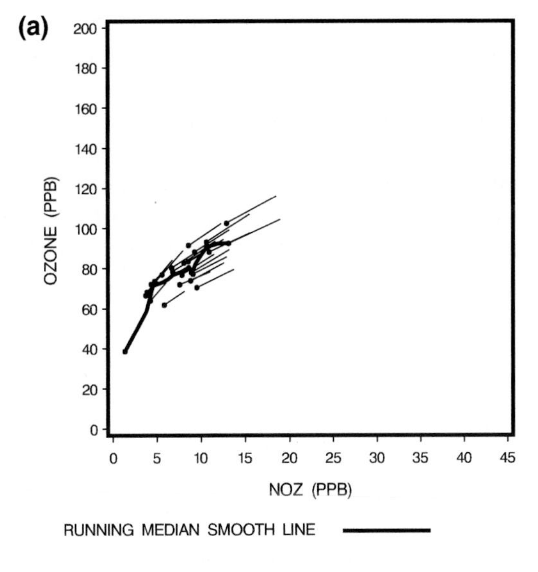

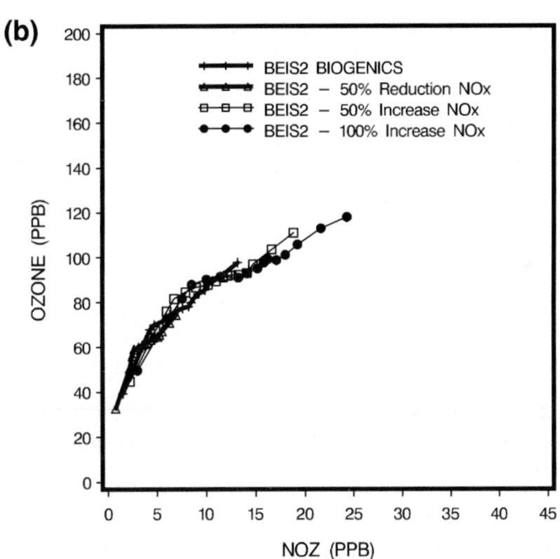

Fig. 17.10. A plot of predicted O_3 *versus* the termination products of NO_X oxidation, NO_Z, showing that the local slope indicates the trajectory of the system response to a change in NO_X emissions when conditions are NO_X-limited: (a) dot at base case and end of arrow at position of system after a 50% increase in NO_X emissions, (b) the progression of best-fit lines as the system moves from a state with NO_X emissions 50% below the base case through to a state with NO_X emissions 100% above the base case.

The comparison between the modelled and measured relationships, however, must be undertaken with care. The curves in Figures 17.9 and 17.10 depicting the coupled relationship between O_3 and NO_Z do not provide a true estimate of the NO_X cycling efficiency, because this relation is based on *in situ* ambient measurements.

This means that the seemingly direct relation between NO_X and NO_Z and O_3 is actually influenced by other sources and sinks, such as the deposition of HNO_3 (a sink) or night-time heterogeneous production of HNO_3 (a source). The relation can be seriously contaminated with episodes of O_3 titration by NO when a radical-limited photochemical system encounters a large regional plume of NO_X from a power plant. We noted above that the ridgeline indicator (O_3/NO_X) successfully distinguishes such conditions. Independent tests of the other process influences are necessary to incorporate uncertainty appropriately into comparisons of the model-predicted and measured two-way relationship.

17.6.2 The Radical Maintenance Pool: Measures of OH Propagation

Probes of the functioning of the radical propagation pool have already been discussed. In particular, the production of OH, *P*(OH), is approximated by the product $I(\mathrm{HC}, \mathrm{NO_2}) \times I(\mathrm{NO}, t\text{-}\mathrm{RO_2})$, while the chain length of OH cycling, *Cl*(OH), is related to *Pr*(OH) as 1/(1–*Pr*(OH)). Since reliable observations of NO_2 and RO_2 are not yet routinely available, except at a few specialised sites, it may be possible to compare the exact calculation of *Pr*(OH) and *Cl*(OH) in the model with observation-derived approximations at these few locations. From such comparisons we might begin to learn how well we have estimated the effects on all VOCs in the system without explicitly representing all of them in the chemical mechanism. In this way, we could gauge how well we have circumvented the limitations of our ignorance of the particular details of radical cycling (using the lumped and surrogate mechanisms already defined above). Making these comparisons is quite important, since it has recently been suggested – arguing from small data sets collected at remote sites – that aerosols may be removing a major fraction of the OH radical pool from participating in any gas-phase chemistry (Cantrell et al., 1996). If this proves to be the case, then this would be an example of where the models – through our ignorance, internal compensating errors, and excessive calibration – have been producing absurd predictions without the correct supporting explanation. Consequently, without correct explanation, our confidence in the model's reliability for forecasting would be low.

17.6.3 Process-oriented Testing of Component Mechanisms: A Newly Proposed OH+NO_2 Reaction-Rate Constant

The reaction $OH + NO_2 \rightarrow HNO_3$ plays a significant role in photochemistry, because it is thought to be the primary termination reaction for both NO_X and OH radicals. Uncertainty analyses in box models have shown it to be one of the most important reactions affecting peak [O_3] (Gao et al., 1995, 1996). New observations (Donahue et al., 1997), which suggest the value for the rate currently in wide use is 10–30% too high, have therefore given cause for some concern. Given our high-level conceptual description of tropospheric chemistry, and its accompanying representation in the constituent mechanisms of the HOM, it was expected that decreasing the value of the rate constant to this extent would produce significant and notable increases in the model's predicted O_3 fields. It was further thought that OH and NO_2 concentrations would increase, that the production efficiency of O_3 per molecule of NO_X terminated to NO_Z would also increase, and that together these would lead to the expected increases in predicted O_3. It would also be possible in some cases for a self-reinforcing, positive feedback on OH to ensue, up to a point, since O_3 photolysis is a significant source of OH in the rural environment.

We explored these hypotheses (suggested by our conceptual model) by changing the chemical mechanism of the instrumented models through a 20% reduction in the rate constant for this reaction and carried out a sensitivity analysis with the model (details are reported in Tonnesen, 1999). The effects of the new rate constant were smaller than anticipated and much less than proportional to the reduction in the rate constant. Ozone increased only by between 2% and 6% for typical rural conditions, and by between 6% and 16% for typical urban conditions. There were at least three major explanations for the less-than-proportional response of the system, and all of these were revealed solely as a result of the model instrumentation and subsequent analysis of internal model processes. First, concentrations of both OH and NO_2 did in fact increase and this caused the decrease in the actual rate of termination (i.e., HNO_3 formation) to be smaller than the reduction in the rate constant. Second, highly reactive VOCs are always consumed more rapidly than are the slowly reacting ones, such as CO, CH_4 and the alkanes. In typical urban settings, however, the increased OH attack on VOCs, enabled by the lower rate constant, was dominated by attack on the slower-reacting VOCs, since they are present in substantially larger concentrations. This results in a lower $P(O_3)$ since the production efficiency of the slowly reacting VOCs is lower than that of the fast-reacting ones. Consequently, the increase in O_3 was less than proportional to the increase in OH that did occur, because the photochemical system was shifted down to a lower overall $P(O_3)$ efficiency. And third, in non-urban, rural settings a negative (or compensating) feedback resulted whereby the system shifted up toward increased termination of NO_X by pathways other than OH + NO_2. These other pathways included increased termination by NO_3 and to organic nitrates during the day, and increased heterogeneous termination to HNO_3 at night. In short, reductions in termination rates were partly offset by increases in concentrations; increases in productivity were mitigated by the eventual involvement of more slowly reacting VOCs; and other non-OH pathways to termination increased their rates, holding down increases in concentrations. Thus, the system did not remain static. Different processing elements, not directly connected to the reaction whose rate had been changed, adjusted to the change in conditions (a re-balancing), such that the net effect on $[O_3]$ was quite muted and less than hypothesised. This demonstrates our early contention that it remains immensely difficult to implement "self-contained", isolated changes to the structure of the model. Furthermore, it would not have been possible to see the changes induced in the model by the new reaction rate constant from the output concentrations alone, using a non-instrumented model. The instrumented model is the only means of investigating absolute and relative reaction rates in the chemistry.

17.7 CONCLUSIONS

Complex systems, with their accompanying complex models (the HOMs, as we have labelled them here), are simply not easy to comprehend. In order to explain the approach introduced in this chapter we have been forced to enter into the detail of tropospheric photochemistry, primarily to reassure the specialist reader of the rigour of our proposals, leaving thus the general reader doubtless perplexed in places. And this potential for residual confusion lingers in spite of the substantial effort invested in distilling all the complexity down to the two essential schemas of Figures 17.1 and 17.2. It is extraordinarily difficult to convey the potential to exploit complexity to our advantage, without entering into the labyrinth of that (discipline-specific) complexity.

Standing back from the detail – as much as is reasonable – we have seen in this chapter an illumination of the inner workings of a HOM; we have sought to shed light on quantities, mechanisms, the cogs, and the wheels behind the outward face of the "machine". We have called this *instrumenting* the model. And if we can use the metaphor of a torch, the manner in which the beam of the torchlight has been projected into the inner space of the model has been guided by a *high-level conceptual description* of the system's behaviour (Figure 17.2). The workings of the elemental, unpacked parts of the machine can thus be observed. This alone is not sufficient, however. We have not the prospect of being able to observe the workings of all these parts in the real world; yet – we have speculated – there might be the prospect of observing the behaviour of combinations, or small agglomerations, of some of the parts (in the real world), guided by the key features in what is, in effect, a transformed inner space of the HOM. These key features we have called *probes*. In other words, illumination of the inner workings of the HOM, guided by the high-level conceptual description of Figure 17.2, suggests more incisive ways of gauging (observing) the propensity of the real world, i.e., the photochemistry of the troposphere over a polluted urban region, to respond (one way or the other) to economic growth and regulatory emissions reductions.

This is not the normal reductionism it might at first sight seem. For we do not believe the elemental parts can be observed and studied in complete isolation from the whole. They are to be studied within the context of the whole, whether this "whole" be the virtual or the real world; all the other cogs and wheels are in motion as the behaviour of just the single component is scrutinised. There can be discovery about the virtual world of the HOM, precisely because it has been rendered opaque and "mysterious" by its very complexity. And once something has been discovered therein, the occurrence of this something can be sought in the real world – providing we have the requisite technology of observation and (ideally) with a view to uncovering yet something other, should the something originally sought prove a mere chimera from the virtual world. But this is discovery – uncovering of our ignorance – through a process of reverberation, back and forth between the real and the virtual worlds. The outcome is ultimately a different form of experimental design for enquiry into the real world. This is not the same as the outcome from the analysis of reachable futures (in Chapter 11), which pointed to the (few) key unknowns worthy of further investigation, as the pivots on which the coming to pass of some target behaviour might critically turn. Our approach in this chapter points rather towards novel ways of observing the real world, almost setting an agenda for innovation in instrumentation technology.

Our approach, in the larger sense, is dramatically different from that of Chapter 14, for example. There, the domain of past observed behaviour was the point of departure into the discovery of slightly more refined – somewhat more general – relationships for the impacts of changes of land cover and land use on a catchment's hydrological response: an incremental step up from the lower-order state-parameter structure of $[\boldsymbol{x}^{-n},\boldsymbol{\alpha}^{-n}]$ to that of $[\boldsymbol{x}^{-n+1},\boldsymbol{\alpha}^{-n+1}]$. Our point of departure has been from the massive, dense assembly of micro-scale theory in the HOMs of tropospheric air quality, beginning with the myriad, unpacked, elemental parts, $[\boldsymbol{x}^{+n},\boldsymbol{\alpha}^{+n}]$, and migrating incrementally towards the more macroscopic probes cast in the transformed inner space of, say, $[\boldsymbol{x}^{+n-1},\boldsymbol{\alpha}^{+n-1}]$. The implication of convergence is obvious: towards a compatible order of representation from the two ends of the spectrum (of theory and observation); set somewhere around $[\boldsymbol{x}^{0},\boldsymbol{\alpha}^{0}]$; and designed to emulate the kind of thought-provoking discourse found in the discussion of competing candidate conceptualisations of the macroscopic dynamics of the global carbon cycle in Chapter 13. The monograph celebrates a variety of perspective and approach. Here, at the end of book, it must be abundantly clear that addressing the questions of detecting the seeds of structural change – at the earliest possible moment – and of arresting their further propagation, will not yield to attack from just a single perspective.

REFERENCES

Bartolotti, L.J. and Edney, E.O., 1995. Density functional theory derived intermediates from the OH initiated atmospheric oxidation of toluene. *Chem. Phys. Lett.*, **254**, 119–122.

Bertman, S.B., Roberts, J.M., Parrish, D.D., Buhr, M.P., Goldan, P.D., Kuster, W.C. and Fehsenfeld, F.C., 1995. Evolution of alkyl nitrates with air mass age. *J. Geophys. Res.*, **100**, 22,805–22,813.

Cantrell, C.A., Shetter, R.E., Gilpin, T.M., Calvert, J.G., Eisele, F.L. and Tanner, D.J., 1996. Peroxy radical concentrations measured and calculated from trace gas measurements in the Mauna Loa Observatory Photochemistry Experiment 2. *J. Geophys. Res.*, **101**, 14,653–14,664.

Chang, T.Y., Chock, D.P., Nance, B.I. and Winkler, S.L., 1997. A photochemical extent parameter to aid ozone air quality management. *Atmos. Environ.*, **31**, 2787–2794.

Daum, P.H., Kleinman, L.I., Newman, L., Luke, W.T., Weinstein-Lloyd, J., Berkowita, C.M. and Busness, K.M., 1996. Chemical and physical properties of plumes of anthropogenic pollutants transported over the North Atlantic during the North Atlantic Regional Experiment. *J. Geophys. Res.*, **101**, 29,029–29,042.

Dennis, R.L., Arnold, J.R., Tonnesen, G.S. and Li, Y., 1999. A new response surface approach for interpreting Eulerian air quality model sensitivities. *Comp. Phys. Commun.*, **117**, 99–112.

Derwent, R.G., Jenkin, M.E., Saunders, S.M. and Pilling, M.H., 1998. Photochemical ozone creation potentials for organic compounds in Northwest Europe calculated with a Master Chemical Mechanism. *Atmos. Environ.*, **32**, 2429–2441.

Dodge, M.C., 1977. Combined use of modeling techniques and smog chamber data to derive ozone precursor relationships. In: *Proceedings of the International Conference on Photochemical Oxidant Pollution and Its Control* (B. Dimitriades, ed.). U.S.EPA. Report No. EPA-600/3-77-001b, Office of Research and Development, Research Triangle Park, NC, USA, pp. 881–889.

Donahue, N.M., Dubey, M.K., Mohrschladt, R., Demerjian, K.L. and Anderson, J.G., 1997. High-pressure flow study of the reactions OH + NO_X → $HONO_X$: errors in the falloff region. *J. Geophys. Res.*, **102**, 6159–6168.

Frost, G.J., Trainer, M., Allwine, G., Buhr, M.P., Calvert, J.G., Cantrell, C.A., Fehsenfeld, F.C., Goldan, P.D., Herwehe, J., Hübler, G., Kuster, W.C., Martin, R., McMillen, R.T., Montzka, S.A., Norton, R.B., Parrish, D.D., Ridley, B.A., Shetter, R.E., Walega, J.G., Watkins, B.A., Westberg, H.H. and Williams, E.J., 1998. Photochemical ozone production in the rural southeastern United States during the 1990 Rural Oxidants in the Southern Environment (ROSE) program. *J. Geophys. Res.*, **103**, 22,491–22,508.

Gao, D., Stockwell, W.R. and Milford, J.B., 1995. First-order sensitivity analysis for a regional-scale gas-phase chemical mechanism. *J. Geophys. Res.*, **100**, 23,153–23,166.

Gao, D., Stockwell, W.R. and Milford, J.B., 1996. Global uncertainty analysis of a regional-scale gas-phase chemical mechanism. *J. Geophys. Res.*, **101**, 9107–9119.

Gipson, G., 1999. Science Algorithms of the EPA Models-3 Community Multiscale Air Quality (CMAQ) Modeling System: Chapter 16, Process Analysis, edited by Byun, D.W. and J.K.S. Ching, Report No. EPA/600/R-99/30, U.S. Environmental Protection Agency, Office of Research and Development, Washington, D.C.

Jang J.C., Jeffries, H.E., Byun, D. and Pleim, J.E., 1995a. Sensitivity of ozone to model grid resolution—I. Application of high-resolution regional acid deposition model. *Atmos. Environ.*, **29**, 3085–3100.

Jang J.C., Jeffries, H.E. and Tonnesen, S. 1995b. Sensitivity of ozone to model grid resolution – II. Detailed process analysis for ozone chemistry. *Atmos. Environ.*, **29**, 3085–3100.

Jeffries, H.E., Sexton, K.G. and Arnold, J.R., 1989. Validation Testing of New Mechanisms with Outdoor Chamber Data, Vol. 2: Analysis of VOC Data for the CB4 and CAL Mechanisms. U.S.EPA Report No. EPA-600/3-89-010b, Office of Research and Development, Research Triangle Park, NC USA.

Jeffries, H.E., 1995. Photochemical air pollution. In: *Composition, Chemistry, and Climate of the Atmosphere* (H.B. Singh, ed.). Van Nostrand-Reinhold, New York, pp. 308–348.

Jeffries, H.E., 1996. Personal communication.

Kleinman, L.I., 1994. Low- and high-NO_x tropospheric photochemistry. *J. Geophys. Res.*, **99**, 16,831–16,838.

Kleinman, L.I., Daum, P.H., Lee, Y-N., Springston, S.R., Newman, L., Leaitch, W.R., Banic, C.M., Isaac, G.A. and MacPherson, J.I., 1996. Measurement of O_3 and related compounds over southern Nova Scotia 1. Vertical distributions. *J. Geophys. Res.*, **101**, 29,043–29,060.

Kleinman, L.I., Daum, P.H., Springston, S.R., Leaitch, W.R., Banic, C.M., Isaac, G.A., Jobson, B.T. and Niki, H. 1996. Measurement of O_3 and related compounds over southern Nova Scotia 2. Photochemical age and vertical transport. *J. Geophys. Res.*, **101**, 29,061–29,074.

Kuhn, T.S., 1996. *The Structure of Scientific Revolutions*, 3rd Edn. University of Chicago Press. Chicago, IL, USA.

Lurmann, F.W., Carter, W.P.L. and Conyer, L.A., 1987. A Surrogate Species Chemical Reaction Mechanism for Urban Scale Air Quality Simulation Models, Vol. 1: Adaptation of the Mechanism, U.S.EPA Report No. EPA-600/3-87-014a, Office of Research and Development, Research Triangle Park, NC USA.

Madronich, S. and Calvert, J.G., 1989. The NCAR master mechanism of gas phase chemistry – Version 2.0, NCAR Technical Note NCAR/TN-333+STR, National Center for Atmospheric Research, Boulder, CO.

Madronich, S. and Calvert, J.G., 1990. Permutation reactions of organic peroxy radicals in the troposphere. *J. Geophys. Res.*, **95**, 5697–5715.

Milford, J., Gao, D., Sillman, S., Blossey, P. and Russell, A.G., 1994. Total reactive nitrogen (NO_y) as an indicator for the sensitivity of ozone to NO_x and hydrocarbons. *J. Geophys. Res.*, **99**, 3533–3542.

O'Brien, J.M., Shepson, P.B., Muthuramu, K., Hao, C., Niki, H. and Hastie, D.R., 1995. Measurements of alkyl and multifunctional organic nitrates at a rural site in Ontario. *J. Geophys. Res.*, **100**, 22,795–22,804.

Oreskes, N., 1998. Evaluation (not validation) of quantitative models. *Environ. Health Perspect.*, **106**(Suppl. 6), 1453–1460.

Oreskes, N., Shrader-Frechette, K. and Belitz, K., 1994. Verification, validation, and confirmation of numerical models in the earth sciences. *Science*, **263**, 641–646.

Parrish, D.D., Buhr, M.P., Trainer, M., Norton, R.B., Shimshock, J.P., Fehsenfeld, F.C., Anlauf, G.K., Bottenheim, J.W., Tang, Y.Z., Wiebe, H.A., Roberts, J.M., Tanner, R.L., Newman, L., Bowersox, V.C., Olszyna, K.J., Bailey, E.M., Rodgers, M.O., Wang, T., Berresheim, H., Roychowdhury, U.K. and Demerjian, K.L., 1993. The total reactive oxidised nitrogen levels and the partitioning between the individual species at six rural sites in eastern North America. *J. Geophys. Res.*, **98**, 2927–2939.

Popper, K.R., 1992. *Conjectures and Refutations: The Growth of Scientific Knowledge*, 5th Edn. Routledge. Cambridge, UK.

Ridley, B.A., Walega, J.G., Lamarque, J-F., Grahek, F.E., Trainer, M., Hübler, G., Lin, X. and Fehsenfeld, F.C., 1998. Measurements of reactive nitrogen and ozone to 5-km altitude in June 1990 over the southeastern United States. *J. Geophys. Res.*, **103**, 8369–8388.

Saunders, S.M., Jenkin, M.E., Derwent, R.G. and Pilling, M.H., 1997. Report summary: World Wide Web site of a Master Chemical Mechanism (MCM) for use in tropospheric chemistry models. *Atmos. Environ.*, **31**, 1249.

Sillman, S., 1995. The use of NO_Y, H_2O_2, and HNO_3 as indicators for ozone–NO_X–hydrocarbon sensitivity in urban locations. *J. Geophys. Res.*, **100**, 14,175–14,188.

Stockwell, W.R., 1986. A homogeneous gas-phase mechanism for use in a regional acid deposition model. *Atmos. Environ.*, **20**, 1615–1632.

Tonnesen S. and Jeffries, H.E., 1994. Inhibition of odd oxygen production in the carbon bond four and generic reaction set mechanisms. *Atmos. Environ.*, **28**, 1339–1349.

Tonnesen, G.S., 1999. Effects of uncertainty in the reaction of the hydroxyl radical with nitrogen dioxide on model-simulated ozone control strategies. *Atmos. Environ.*, **33**, 1587–1598.

Tonnesen, G.S. and Dennis, R.L., 2000a. Analysis of radical propagation efficiency to assess ozone sensitivity to hydrocarbons and NO_x. Part 1: Local indicators of instantaneous odd oxygen production sensitivity. *J. Geophys. Res.*, **105**, 9213–9225.

Tonnesen, G.S. and Dennis, R.L., 2000b. Analysis of radical propagation efficiency to assess ozone sensitivity to hydrocarbons and NO_x. Part 2: Long-lived species as indicators of ozone concentration sensitivity. *J. Geophys. Res.*, **105**, 9227–9241.

Toulmin, S.E., 1961. *Foresight and Understanding, An Enquiry into the Aims of Science*. Harper Books, New York, NY, USA.

Trainer, M., Parrish, D.D., Buhr, M.P., Norton, R.B., Fehsenfeld, F.C., Anlauf, G.K., Bottenheim, J.W., Tang, Y.Z., Wiebe, H.A., Roberts, J.M., Tanner, R.L.,, Newman, L., Bowersox, V.C., Meagher, J.F., Olszyna, K.J., Rodgers, M.O., Wang, T., Berresheim, H., Demerjian, K.L. and Roychowdhury, U.K., 1993. Correlation of ozone with NO_Y in photochemically aged air. *J. Geophys. Res.*, **98**, 2917–2925.

Whitten, G.Z., Hogo, H. and Killus, J.P., 1980. The carbon bond mechanism: A condensed kinetics mechanism for photochemical smog. *Environ. Sci. Technol.*, **14**, 690–700.

Yu, J. and Jeffries, H.E., 1997. Atmospheric photooxidation of alkylbenzenes-I: carbonyl product analyses. *Atmos. Environ.*, **31**, 2261–2280.

PART IV
EPILOGUE

Environmental Foresight and Models: A Manifesto
M.B. Beck (editor)

CHAPTER 18

Parametric Change as the Agent of Control

K.J. Keesman

18.1 INTRODUCTION

A generic description of a system's behaviour has inputs $\boldsymbol{u}$, states and parameters $[\boldsymbol{x},\boldsymbol{\alpha}]$, and output responses, $\boldsymbol{y}$. For simplicity, let us assume there is essentially no difference between $\boldsymbol{x}$ and $\boldsymbol{y}$; target future behaviour can be specified in terms of either, i.e., as $\bar{\boldsymbol{x}}_f(t^+)$. In general, some of the inputs may be considered disturbances simply impinging upon the system and beyond our manipulation, while others are quantities whose values we can choose at will, which we call controls. Control theory concerns itself with answering the question: what pattern of controls ($\boldsymbol{u}$) over some span of time will transfer the state from its present given value to another desired value, subject to appropriate assumptions about the nature of the model's parameters ($\boldsymbol{\alpha}$). Method upon method is available for solving this problem; and solutions may look like the trajectory in Figure 2.3 of Chapter 2, for the (input) fossil fuel flux required to stabilise atmospheric carbon dioxide concentration at some target level $\bar{\boldsymbol{x}}_f(t^+)$ by the year 2300.

In the discussion gathered around equation 6.1 in Chapter 6 we have transposed this question to ask: first, what pattern of variations of $\boldsymbol{\alpha}$ over time will bring about some specified change in the state of the system, subject to appropriate assumptions about the nature of the inputs ($\boldsymbol{u}$); and, then, what methods from control theory might be available to deliver an answer. This latter is the subject of the present chapter. But we shall have to approach it by first introducing some pre-requisite, elementary concepts from control theory and then, in this instance, treating $\boldsymbol{\alpha}$ as formally similar to $\boldsymbol{u}$ (although this is not the only way of interpreting "parametric change as the agent of control"). Aligning thus the parameters of the model with the inputs is a conceptual "shift" akin to that first introduced by Lee (1964) when he treated the concept of time-varying parameters as being unobserved states in the state-space framework.

18.2 FEEDBACK

Few large-scale environmental systems have been the subject of studies in controlling their dynamic behaviour. Geophysics and control theory have not traditionally been close disciplinary neighbours and their goals are conventionally very different: the one (geophysics) is concerned primarily with understanding and describing the nature of a system; the other is driven by the need to maintain a system in some desired state, or to steer it to some future desired state, irrespective of how well its behaviour is understood.

Feedback is familiar to all of us, however, both in our everyday lives and as the essence of keeping the implementation of a plan on course. In spite of this, the management of environmental systems is most often thought of from the perspective of a system considered to be operating in an open-loop format, without feedback, that is (Figure 18.1). Here Σ, the block representing the system, could consist of a number of subsystems, while $\boldsymbol{u}$ is the collection of "causes" and z the collection of "effects". Without resorting to any use of feedback, "best management practices" (derived from our knowledge of Σ) would be implemented through changes of the causes so that, given complete knowledge of the system's behaviour (Σ), the desired effects ought to be brought about. At any point in time, no information is used from checks made on the status of the effects – in particular, whether they match what is wanted – to alter the manipulations being implemented through the causes. We do not, of course, have perfect understanding of the nature of the system; our models M, in which appear $[\boldsymbol{x},\boldsymbol{\alpha}]$, are not identical with Σ. And the causes we can manipulate are not, as we have said above, the only disturbances impinging upon the system. The open-loop approach is unlikely therefore to be entirely successful.

To cope with such model and input uncertainty, as well as other more general uncertainties, the closed-loop system description of Figure 18.2 has been introduced, where the signals $\boldsymbol{u}$, $\boldsymbol{w}$, $\boldsymbol{y}$ and z are, in general, vector-valued functions of time. The components of $\boldsymbol{w}$ are now defined as all the extraneous inputs – disturbances (uncontrollable inputs, such as solar radiation or wind), references (desired behaviour trajectories), sensor noise (including sampling errors), and so on – leaving $\boldsymbol{u}$ as the vector of manipulable, *control inputs*. The (weighted) signals we want to

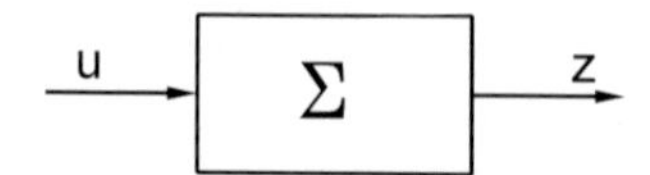

Fig. 18.1. Block diagram of an open-loop system.

control, for instance, differences between references and system outputs, or controller outputs (costs of management policies), are all collected in z. The vector y, therefore, contains all the *measured outputs* available to the controller. This closed-loop system representation lies at the heart of many impressively successful procedures for designing controllers for technical, engineered, systems (see, for instance, Kwakernaak, 1993). But for environmental systems, given their very large time constants and the large disturbances affecting them, the concept seems less promising. Automated feedback, and hence a large portion of control theory, grew out of the need to implement control actions at speeds significantly faster than the normal speeds of human reactions to observed events. We clearly have more than a few seconds or minutes to respond to our apprehension of the seeds of structural change in the global climate system, for example. We must ask therefore: how can feedback really be relevant to problems with very large time constants, such as in the greenhouse gas system, when we are dealing with sampling and control strategies at a much higher frequency? Our discussion will return to this question, but after the concept of feedback has been further elaborated for a class of simple systems.

To describe the basic relationships in a feedback control system in greater detail, the scheme for a single-input, single-output (SISO) system is shown in Figure 18.3, where the principal components of a control system are explicitly represented. From this figure, significantly, we can derive relationships between the three external signals – the reference input (r), input disturbance (d), including a *feedforward* control signal, and sensor noise (n) – and three signals that are of interest in feedback control design, namely, the tracking error (e), controlling input (u) and controlled system output (y). In linear feedback control design the sensitivity function ($S(s)$), that is, the transfer function from r to e, and the complementary sensitivity function

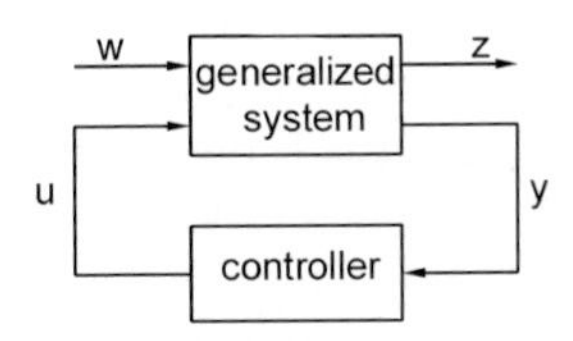

Fig. 18.2. System with closed-loop control.

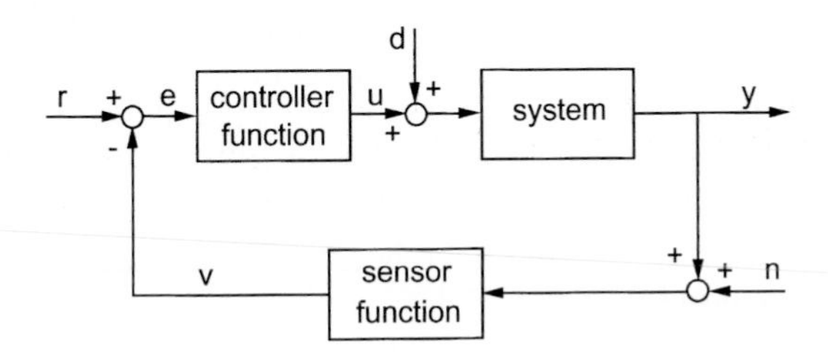

Fig. 18.3. Basic form of a feedback loop.

($T(s)$), from r to y or from d to u, where $T(s) = 1 - S(s)$, play important roles, since they always appear in some weighted form in the relationships between the external signals and e, u or y. Here s is the Laplace transform variable.The basic objectives for design of the feedback control system, which can be expressed in terms of the sensitivity and complementary sensitivity functions, are:

- to guarantee internal stability, which implies that bounded external signals guarantee bounded internal signals u, y and v;
- to obtain good tracking features, which implies small tracking error e and thus small $S(s)$; and
- to obtain good disturbance rejection, which requires small $T(s)$.

Consequently, there is a trade-off between tracking and disturbance rejection, which is basically a trade-off between making both the sensitivity and the complementary sensitivity functions small. In fact, this trade-off is the main issue in feedback control system design (as apparent in the texts of Morari and Zafiriou (1989) and Doyle et al. (1992)).

So, what does this all have to do with our problem of controlling *structurally changing* systems, as defined in Chapter 4? Let us first observe that so far little attention has been given to controller design in the face of uncertain system descriptions, i.e., uncertain models M. During the development of the model we know that structural errors will appear due to simplifications, such as averaging and the aggregation of variables, or through neglecting some elements of the system's dynamics, and so on (just as we have discussed at length in Chapter 4). For the design of control systems under model uncertainty, H_∞ control theory has provided some attractive procedures, which are certainly not restricted to the linear (e.g., van der Schaft, 1991) or SISO cases (e.g., Morari and Zafiriou, 1989). H_∞ control for the design of robust control systems deals with the minimisation of the peak amplitude of certain weighted *closed-loop* frequency response functions, such as $S(s)$ and $T(s)$. Hence, roughly speaking, the emphasis is on minimizing the maximum gain of the system instead of minimizing the magnitudes of certain signals, as is the case in classical Linear Quadratic (LQ) control theory. If we are therefore able to specify the model's uncertainty, which unlike the conventional time-invariant case is now the result of *expected* changes in the system dynamics, in either structured *via* parameter bounds or unstructured form, then H_∞ theory can provide a controller which is robust against these changing dynamics. In this context unstructured means the uncertainty is represented by an additional bounded modelling error term. To turn to our advantage the tale related in Chapter 3 – of the channel in Holland, whose weir became clogged with grass clippings blown into it from nearby mowing activities – the design of an H_∞ controller for flow through the channel could accommodate such *expected* structural change. It could have been taken into account, for example, in defining appropriate bounds on time delays and time constants. This would allow us, then, to control the system in an appropriate way, in spite of changing dynamics, but only inasmuch as these structural changes can be foreseen, or, as will be the focus below, *desired*.

If, on the other hand, the changes are unexpected, then we can for instance choose the bounds to be much larger, in the hope that this will suffice. In the end, however, the best we can do is to learn from the system itself, i.e., we should try to identify these changes from observed data, and adapt the control strategy accordingly. Combining the dual functions of learning and (feedback) control is technically what distinguishes feedback from adaptive control (as discussed in Chapters 3, 5 and 15). For environmental systems, most changes are unforeseen, so that *change detection* is a real issue; and that – in a nutshell – is what much of the remainder of this monograph has been about.

18.2.1 Large Time Constants and Low Sampling/control Frequencies

The original statement of the problem to be addressed herein (in Chapter 6), caricatured $\boldsymbol{u}$ as representing the high-frequency flutter of economic and regulatory cycles of control, relative to the slow, low-frequency evolution of the environmental system. But our problem does not have to be constrained in this manner. In fact, choosing the parameter $\boldsymbol{\alpha}(t)$ as the "control" vector to be computed – the central aspect of our transposed problem – can be viewed instead as merely transferring attention from high- to low-frequency inputs. Furthermore, since $\boldsymbol{\alpha}(t)$ is a time-varying parameter vector, it may depend on features of the system's behaviour below (or outside) the resolving power of the model. If then $\boldsymbol{\alpha}^{o}(t)$ has been computed, as some "optimal" trajectory for the parametric variations, we could, in principle, even try to find a high-frequency (true) control input ($\boldsymbol{u}$), associated with a fast sub-process, which allows the desired parameter trajectory to be realised.[1] This approach is known as cascade control (depicted in Figure 18.4), in which decomposition of time scales is essential. It merits further discussion.

In Figure 18.4, two systems (or sub-processes), Σ_1 and Σ_2, are indicated. Σ_1 has slow dynamics and is influenced by the "control" input α, or we could say by $\alpha_r = \alpha^{o}(t)$, the reference optimal trajectory of α, if the second process (Σ_2) is controlled such that its output (α) meets the reference value. Thus, the secondary loop with controller C_2 and process Σ_2 has to realise the desired values of the parameter/control trajectory $\alpha(t)$. The dynamics of the secondary loop are fast, or at least faster than the dynamics of the primary loop. The major benefit of cascade control is now that disturbances arising within the secondary loop can be attenuated by the controller C_2, before they can affect the output of the primary loop (y), a notion that will be important to solving our problem of parametric change as the agent of control.

At this point, therefore, it can be concluded that for a restricted class of environmental problems with structurally changing dynamics, appropriate controllers can be designed. The class is restricted, since it only covers systems with expected or desired changes and, in practice, more complex design requirements usually have to be met.

---

[1] Although it is not obvious – for the moment – why one might want to do this, it will become so later, at the close of the chapter.

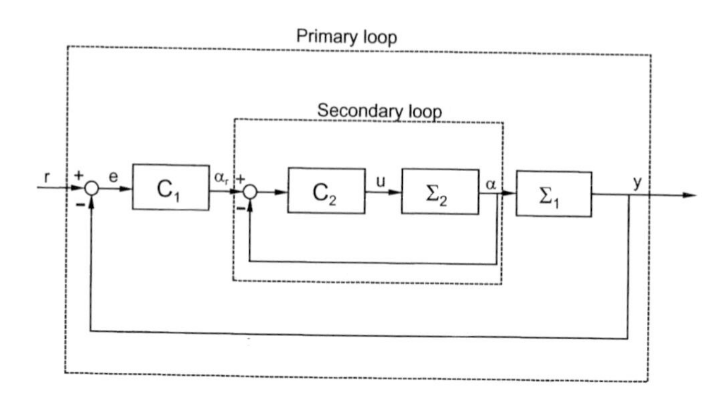

Fig. 18.4. Cascade control.

In addition, the main limitation of H_∞ theory for controller design is that it can not deal with hard system constraints. In the context of climate change, for example, these hard constraints are a result of limitations in the political scope for taking control actions; in general, a country will only reduce its input fluxes to a specific limit. Furthermore, hard constraints will also appear when norms on specific gas concentrations must be fulfilled, with the result that, its appeal notwithstanding, H_∞ controller design is not fully applicable for environmental systems with changing dynamics. We shall have to examine an optimisation-based feedback controller design that can accommodate both our more complex requirements (of structural change) and hard (political) constraints. The key aspect of feedback to be remembered, however, is that it has only been introduced because of the uncertainty in the model and inputs.

18.3 MODEL-BASED PREDICTIVE CONTROL

In order to deal with the more complex specifications for control system synthesis in constrained dynamical systems Mayne and Polak (1993) have formulated a canonical, non-smooth optimisation problem for the design of finite-dimensional closed-loop controllers. We can label this a "constraint optimiser" for convenience and note that it can be incorporated into the feedback loop of Figure 18.3 to form the optimisation-based feedback control system of Figure 18.5. This, in turn, allows us now to proceed to the idea of model-based predictive control (MBPC), as a subset of optimisation-based feedback control, as in Figure 18.6. Such an arrangement is referred to as an internal model scheme, where the error signal, v, containing system disturbances, model uncertainties, as well as sensor noise, is fed back. Feedback of the error signal instead of the measured output y, as shown in the previous schemes, is a distinctive characteristic of MBPC methods. In its general form, when applied to the problem of structurally changing systems in which α is considered as the control input vector, MBPC starts from the supposition that the system being controlled is described by our customary state-space model (equation 6.1, for example),

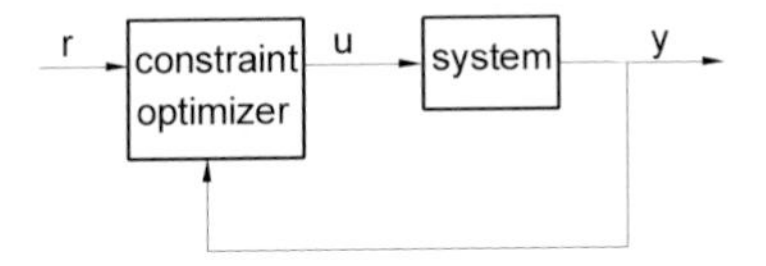

Fig. 18.5. Optimisation-based control system.

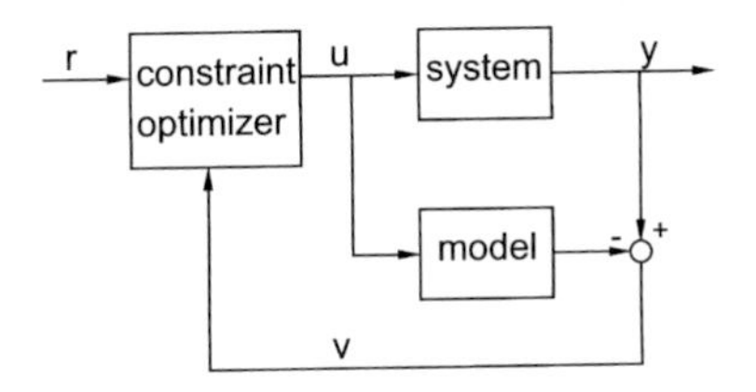

Fig. 18.6. Model-based predictive control system.

$$\mathrm{d}\boldsymbol{x}(t)/\mathrm{d}t = f\{\boldsymbol{x}(t), \boldsymbol{u}(t), \boldsymbol{\alpha}(t)\} \tag{18.1}$$

subject to state and control constraints

$$\boldsymbol{x}(t) \in \boldsymbol{X};\ \boldsymbol{u}(t) \in \boldsymbol{U};\ \boldsymbol{\alpha}(t) \in \boldsymbol{A} \tag{18.2}$$

where domain $\boldsymbol{X}$ will define the region of "reachable futures" and $\boldsymbol{U}$ the set of presumed (known), prespecified controls. For our transposed control problem, therefore, at time t the MBPC will choose that set of parameters $\boldsymbol{\alpha}$ minimising

$$\int_t^{t+T} F\{\boldsymbol{x}(\tau), \boldsymbol{u}(\tau), \boldsymbol{\alpha}(\tau)\}\mathrm{d}\tau \tag{18.3}$$

over the finite prediction horizon T under the constraints given by 18.2 on the interval $[t, t+T]$. $F\{\cdot\}$ represents, for instance, some function of the misfit between realised and target future behaviour over the time interval $[t, t+T]$. Notice from this formulation that there is no (practical) restriction on T, that is $T \in (0,\infty)$. The optimising control $\alpha^{o}(\cdot)$ is then applied to the system over the first discrete interval of time $[t, t+\delta]$, where $\delta < T$ is the sampling interval. At the time instant $t+\delta$ this procedure is repeated, again optimising over the horizon T (to $t+T+\delta$). It should be emphasised that this concept of "receding horizon control" introduces a non-linear feedback loop, since in general $\alpha^{o}(\cdot)$ depends on $\boldsymbol{x}(t)$ in a non-linear way.

In addition to this receding horizon strategy, MBPC is also characterised by the application of a control horizon $T_\alpha \leq T$, such that over the interval $[T_\alpha, T]$ the control is assumed to be constant, thereby reducing the order of the optimisation problem and the order of the computational burden. In discrete-time the MBPC concept can be illustrated as in Figure 18.7. The notion of the system's target future behaviour has thus been introduced in a quite natural fashion. Hence, *if* we are able to predict the behaviour of structurally changing environmental systems, a control strategy focused

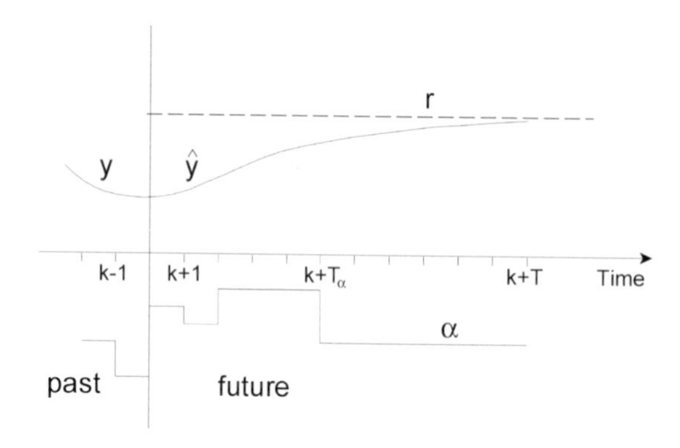

Fig. 18.7. Putting the principle of model-based predictive control to work (conceptually) in solving the problem of computing those changes of the parameter values (α) required to transfer the state of the system (denoted as $\hat{y}$) from its present position to some desired future position (r).

on mitigating, or adapting to, these changes can be found by solving equation 18.3 subject to 18.2 for time instants on the interval $[t, t+T]$. The trajectory sketched for $\alpha(t)$ in Figure 18.7 captures the essence of our search. It is this "change in the science base", or apparent structural change, that causes the future response $\hat{y}(t^+)$ to converge towards the target behaviour, here denoted as r.

In practice, however, the structurally changing behaviour of the system is not well known and, thus, it is crucial to acknowledge the uncertainties arising from this ignorance. One approach is to predefine a number of possible future (state) trajectories and to identify the associated parameter/control trajectories $\boldsymbol{\alpha}(\cdot)$ recursively (as in Keesman and Jakeman, 1997). Alternatively, uncertainty in the regulatory controls can be incorporated into the problem formulation and a min–max control scheme specified, in which $\boldsymbol{\alpha}$ is chosen so as to *minimise* the *maximum* value of the function F of equation 18.3 attained over all choices for $\boldsymbol{u}$ (the controls) belonging to some allowed domain Ξ. This min–max problem formulation can be interpreted in terms of a *worst-case* design. The presumed input controls are chosen so as to maximise the function F, i.e., driving the system thereby to the most undesirable future states, for example. The choice of $\boldsymbol{\alpha}$ then minimises the worst-case situation occurring, subject to the freedom assigned to $\boldsymbol{u}$; it makes the most of a bad situation, as it were.

18.4 CONCLUSIONS: BUT THE VERY BEGINNINGS OF AN APPROACH

In closing, let us provide a brief example, designed to give the abstraction of the foregoing somewhat greater immediacy. Keesman and van Straten (1989) and van Straten and Keesman (1991) have discussed a problem of eutrophication in a lake in the Netherlands (echoing our case history of Lake Erie in Chapter 7 and, indeed, the analysis of reachable futures in Chapter 11). The ultimate goal of the problem was to limit the amount of phosphorus in the lake, to some target level, say $\bar{x}_f(t^+)$. From the approach formulated in equations 18.2 and 18.3, it might have emerged that the reachability of this target implied a reduction in the background concentration of phosphorus in the sediment, treated in the model (quite typically) as a parameter (α), under the *assumed* control of a "flushing" strategy (u) for this (Dutch) lake. In other words, presuming the choice of policy action, it might have been possible, in principle, to have peered some distance from the present time t into the future ($t+T$, in Figure 18.7) and discerned – from the shape of the resulting required trajectory for α, in particular, in the vicinity of the present – the more precise character of the oncoming apparent change of structure in the system's behaviour. Furthermore, if we allow ourselves the licence of just a little more speculation, given the long-term strategy of the policy action (of flushing, as opposed to sediment dredging or biomanipulation of the lake, for instance), we might then have facilitated the means of developing the tactical flushing actions designed to make the trajectory of α come to pass in the immediate future, perhaps along the lines of the cascade control scheme of Figure 18.4.

In this brief chapter, then, we have taken a variation on the conventional theme of control system design, transposing the roles fulfilled by the model's parameters ($\boldsymbol{\alpha}$) and the system's controls ($\boldsymbol{u}$) in that problem formulation. Given the preamble of Part I of the monograph, and of Chapter 6, in particular, our question has been a simple one: could we, in principle, find a technique for solving our transposed problem. The answer, we believe, is "yes". The use and significance of any result that could be so computed for a given environmental system remain to be discerned; our intention has been merely to open the door into this domain of enquiry, so to speak. Yet there are broader issues of value here. The forms of logic and ways of conceptual thinking mobilised in control theory have rarely been turned to advantage on the subjects of environmental foresight and management, such as those of how communities adapt their understanding of their cherished pieces of the environment, together with their evolving hopes and fears for these environmental futures. It is not that we believe the behaviour of society, in relation to the environment, can be replicated in some mechanistic, engineering manner. It is rather that some of the conceptual tools of thinking about the control, stability, robustness, and coherence of (technical) dynamic systems can assist in unscrambling the nature of the man-environment relationship.

REFERENCES

Doyle, J.C., Francis B.A. and Tannenbaum, A.R., 1992. *Feedback Control Theory*. MacMillan, New York.

Keesman, K.J. and Jakeman, A.J., 1997. Identification for long-term prediction of rainfall–streamflow systems. In: *Proc. 11th IFAC Symp. on System Identification*, Fukuoka, Japan, 8–11 July, Vol. 3, pp. 1519–1523.

Keesman, K.J. and van Straten, G., 1990. Set-membership approach to identification and prediction of lake eutrophication. *Water Resour. Res.*, **26**(11), 2643–2652.

Kwakernaak, H., 1993. Robust control and H_∞-optimization – tutorial paper. *Automatica*, 29(2), 255–273.

Lee, R.C.K., 1964. *Optimal Identification, Estimation and Control*. M.I.T. Press, Cambridge, Mass.

Mayne, D.Q. and Polak, E., 1993. Optimization based design and control, In: *Proc. 10th IFAC World Congress*, Sydney, Vol. 3, pp. 129–138

Morari, M. and Zafiriou, E., 1989. *Robust Process Control*. Prentice-Hall, London.

van der Schaft, A., 1992. L_2-gain analysis of nonlinear systems and nonlinear state feedback H_∞ control. *IEEE Trans. Autom. Control*, **37**, 770–784.

van Straten, G. and Keesman, K.J., 1991. Uncertainty propagation and speculation in projective forecasts of environmental change: a lake eutrophication example. *J. Forecasting*, **10**, 163–190.

Environmental Foresight and Models: A Manifesto
M.B. Beck (editor)

CHAPTER 19

Identifying the Inclination of a System Towards a Terminal State from Current Observations

A.V. Kryazhimskii and M.B. Beck

19.1 INTRODUCTION

In this closing chapter, perhaps paradoxically, we shall open up something of another new potential avenue for addressing the issues of exploring environmental change of central concern to this monograph. Given has been a model (a computational encoding of the current science base), the observed or experienced record of past behaviour, and the imagined – feared or desired – target future behaviour. With these we have been able to construct (and answer) a number of questions. In Chapter 11, for example, we asked on which elements of the map of the current science base, with all its flaws and uncertainties, might the reachability of the given target futures quintessentially turn? For it is these key elements, not the many other scientific unknowns, to which the scarce resources of further empirical study should be allocated as a matter of priority. In the preceding chapter the question was instead: what changes in the science base – as opposed to what changes of policy – would be required to transfer the behaviour of the system from its present state to some future target state? Throughout this book, in fact, our concern has been to enquire into the nature of past behaviour, in particular, to identify any seeds of imminent dislocations in the structure of the system's dynamics, as a function of having first contemplated what it might take to bring about some pattern of behaviour imagined for the future.

In this chapter we pose yet another question, again within the domain of our general concern, and seek to answer it with a form of a model rather different in character from what has gone before. We take as our point of departure a binary classification of the behaviour of an environmental system. That is to say, there is a target sub-space of behaviour and its complement, not-the-target behaviour, typified by the separation of behaviour observed in the past from all else, or by the imagination of behaviour in the future that is either "essentially similar" to that of the past or "radically different" therefrom. Such a binary classification has appeared frequently in the foregoing (most notably in Chapter 11). It was indeed essential to the way in which the Regionalized Sensitivity Analysis of Hornberger, Spear, and Young was originally conceived and developed (Young et al., 1978; Hornberger and Spear, 1980; Spear and Hornberger, 1980). Of course, it is not hard to think of problem situations requiring more than merely a binary classification, for instance, of future behaviour that is "desired", "feared", and their complement "neither feared nor desired". For the time being, however, we shall restrict attention to binary classifications, in particular, binary outcomes in the future. This allows us to adopt what we shall call a binary model of behaviour, which in turn permits us to draw upon the ideas and mathematics of path-dependent stochastic process theory (as reflected in the work, for example, of Arthur et al., 1987). The model in this case is not therefore any reasonably direct encoding of the map of scientific knowns and unknowns, as elsewhere in the book. It is a model of the probabilities of transition from one category of behaviour at one point in time to either of the two categories of behaviour at the next point in time. We acknowledge that in due course this may not prove to be the best line of attack. Herein our goal is merely to present the essence of the approach, to set up the machinery with which to solve the problem as currently constructed. We shall deliberately avoid theoretical considerations – apart from a brief description of the basic model – focusing instead on a simple, but non-trivial, example, its stepwise analysis, and finally an extension of this analysis to a preliminary study of the observed dynamics of rodent populations in the vicinity of Chernobyl in the Ukraine.

In short, we wish to answer the following question. Does access to a brief window of observed past behaviour of the environmental system give any indication of whether that system has embarked upon a path towards some domain of behaviour that may be feared, "surprising", or desired in the future? If the probability of reaching the target future is greater *given* access to this empirical record, *relative* to the prior, maximally uninformed position of no access to these data (in spite of all the uncertainties in both the model and the data), there is the distinct possibility the data conceal within themselves the seeds of an imminent structural change. We believe this specific construction of the problem to be another potentially fruitful re-arrangement of the elements surrounding our capacity to detect, in general, and at the earliest possible moment, the seeds of structural change – as they begin to propagate through the observed record of the past behaviour of an environmental system. That the window of past observations should be "brief" is inevitable when we are confronting problem contexts, such as the impacts of climate change, whose characteristic time-constants are inherently large.

19.2 BINARY MODELS AND PROBLEM DEFINITION

The binary classification of the system's futures into "collapse" and "survival" implies that tendencies towards these futures should be observable in short-term displacements of the system. The alternative paths of the state of the system into the future can be thought of, therefore, as sequences of incremental changes also classified in a binary fashion, say, "–/+" (as suggested for encoding the dynamics of natural processes in Kryazhimskii et al., 1996). This "–/+" encoding scheme can clearly be linked to path-dependent stochastic processes and put to use in two ways: (a) without reference to any historical observations (model-based); and (b) with reference to a set of historical observations (model and data-based). Due to the crude, binary, structure of the model, neither (a) nor (b) can pretend to predict a particular future, or to provide a numerical value of a probability of the system's terminal state. Yet one can usefully compare (or rank) some of the numbers emerging from such a scheme: the ("prior") probability of collapse/survival generated from the model alone – in the absence of any conditioning upon historical observations – and the ("posterior") probability of collapse/survival generated from the model conditioned upon the historical observations. If the "posterior" probability of, say, collapse is greater than the "prior", the conclusion is that the historical observations may contain within them the seeds of change towards collapse (at least relative to the maximally uninformed condition of not having had access to such information). The model may be viewed then as collapse-oriented. This rough characterisation certainly holds for a set of models. In other words, it is model robust, an observation that is crucial. For uncertain environmental systems, robustness, i.e., insensitivity to assumptions (in particular about the models' parameters), is perhaps a single measure of reliability of any judgement on the system's performance.

Retrospective analysis of the collections of collapse- and survival-oriented models can reveal relationships among the parameters of the models, from which it may then be possible to make a categorical judgement on the inclination of the system's behaviour (towards collapse or survival). Otherwise, it may still be possible to identify the degree to which one tendency will dominate the other.

19.2.1 Hypotheses

We make two hypotheses about the behaviour of the environmental system:

Hypothesis 1: The fate of the system in the far distant future is governed by just two radically different terminal states, *collapse* or *survival*.

Hypothesis 2: Local changes in time towards collapse or survival are positively correlated with those in the past. More specifically, if the local changes towards, say, collapse dominate in the past, then most likely this tendency will be maintained in the immediate future; conversely, if there is no clear domination in these past local changes, any subsequent local change will be highly random.

We presume thus that the local changes are classified in a binary way: every local change is oriented either towards collapse or towards survival. The local changes may be small but, as time progresses, one of the tendencies dominates and the system will move towards collapse or survival.

Remark: A fundamental question is this: given a local change, how can one identify whether this change is oriented towards collapse or survival? Sometimes the orientation is clearly identifiable (as in the illustration presented below). On other occasions this will not be so, presenting us then with a non-trivial challenge. At this early stage in developing our approach, we shall simply assume that we have a method to identify the orientation of these local changes towards collapse/survival.

19.2.2 Designing a Binary Model

In line with the theory of path-dependent stochastic processes, we specify Hypotheses 1 and 2 by assuming a *binary* dynamical model of the system. The model operates over an infinite number of time periods. These are identified with periods in real time, over which the local changes in the system are normally apparent (say, months, years, or decades). We shall denote these periods by integers, k. The model's transitions between the periods are of two categories, "–" and "+". A "–" transition is associated with a local change oriented towards collapse, and a "+" transition with a local change towards survival. In each period, k, a finite string s of length m (where m is fixed) is used to characterize those features of the system's past having an impact on the system's local change between the periods k and $k+1$. The string is composed of the "–/+" transitions realized sequentially between the periods $k-m$ and $k-m+1$, $k-m+1$ and $k-m+2$,..., and, finally, $k-1$ and k. It gives rise to the model's transition between the periods k and $k+1$. We shall call s the *state* of the model in period k. The model represents the system's evolution as an infinitely growing sequence of minuses and pluses (symbolizing the system's local changes), while the model's state is the m-long moving window always adjoint to the latest minus/plus in the sequence.

Treating Hypothesis 2 in a somewhat extreme manner, we assume that a strong dominance of "–" transitions in the state s automatically causes a "–" at the next transition. We register the strong dominance of "–" transitions in s if n^-, the number of minuses in s, is no smaller than some critical number m^-. Symmetrically, if n^+, the number of pluses in s, is no smaller than some critical number m^+, the strong domination of "+" transitions is registered in s and the next transition is determined to be "+". In the intermediate situation, in which $n^- < m^-$ and $n^+ < m^+$, the next transition is "–" with a probability r^- and "+" with a probability r^+. Since both probabilities are assumed to depend on the state s, it will be appropriate to refer to r_s^- instead of r^- and r_s^+ instead of r^+. We set $r_s^- = 1$ when $n^- \geq m^-$ and $r_s^+ = 1$ when $n^+ \geq m^+$. To avoid a situation of self-contradiction in which $n^- \geq m^-$ while $n^+ \geq m^+$, we assume that the sum of m^- and m^+ does not exceed the state length, i.e., $m^- + m^+ \leq m$. Since the probabilities of the next transition being "–" and "+" are complementary, i.e., $r_s^+ = 1 - r_s^-$, we shall operate with r_s^- only.

Thus, the dynamics of a binary model are formally characterized by the following parameters:

(i) the *state length*, m,

(ii) the *collapse critical level*, m^-, and the *survival critical level*, m^+, satisfying $m^- + m^+ \leq m$, and

(iii) the family of probabilities of the "–" transitions, r_s^-, where s, the family index, runs through the set of all model states, i.e., all m-long permutations of pluses and minuses. The latter set will further be denoted as S.

It is required that $r_s^- = 1$ if n^-, the number of minuses in s, is no smaller than m^-, and $r_s^- = 0$ if $n^+ = m - m^-$, the number of pluses in s, is no smaller than m^+.

In the following analysis, such binary models will be used for the assessment of possible system evolutions starting from a finite period, say, $k = 0$, that precedes a record of observed local changes. As long as no data on the actual evolution before period 0 are available, the analysis will invoke a (hypothetical) probability distribution of the model states in period 0. Therefore we extend the definition of a binary model by supplementing (i), (ii), (iii) with the following attribute:

(iv) the *initial distribution*, q, defined as a probability measure on S. In what follows, we shall write q_s for the initial probability of a state s (the value of q at s).

19.2.3 An Elementary Illustration

Let us imagine we are exploring the dynamics of a species, A, in some spatial domain. Species A interacts with the environment, E. The latter comprises a number of factors (climate, other species, etc.) having different impacts on A. Common dynamical patterns for the number, N, of the species A are oscillatory. To be specific, we assume that N has a cycle with a period, normally of three years. Two years of growth are followed by a year of decline, after which N is brought back to approximately the same level as at the beginning of the growth period. If, for some reason, the period of decline covers two years instead of one, the population of A gradually becomes extinct and the A-in-E system can be said to experience a structural change. We identify it as *collapse*. If, conversely, the growth in N is observed during three years instead of two, the species A gradually becomes dominant (thus causing, for example, a competitor species to become extinct). This we identify as *survival*. Collapse- and survival-oriented local changes are obviously identifiable. Growth in N is identified as a local change oriented towards survival, while decline in N is oriented towards collapse.

One can easily design a binary model of the system. We identify the model's time periods, k, with years. A "+" transition and a "–" transition between periods k and

$k + 1$ are associated with, respectively, growth and decline in N between the years numbered k and $k + 1$. A three-year stable cycle in the system's behaviour implies that the last three years in the system's history have a principal impact on the system's behaviour next year. In the model, we assume that each transition is correlated with the preceding three most recent. The model's state length, m, is therefore 3. The collapse critical level, m^-, is obviously 2 (two "–" transitions in strict sequence imply irreversible decline into collapse). The survival critical level, m^+, is 3 (three "+" transitions imply survival).

Thus, to summarize:

$$\begin{aligned} &\text{state length:} && m = 3,\\ &\text{collapse critical level:} && m^- = 2, \qquad (19.1)\\ &\text{survival critical level:} && m^+ = 3. \end{aligned}$$

The model's states, s, are $(+++)$, $(-++)$, $(+-+)$, $(++-)$, $(--+)$, $(-+-)$, $(+--)$, $(---)$. The assumption $m^- = 2$ implies that in every state with no less than two minuses, the probability r_s^- is 1, so that

$$r^-_{--+} = 1, \quad r^-_{-+-} = 1, \quad r^-_{+--} = 1, \quad r^-_{---} = 1. \qquad (19.2)$$

According to the assumption $m^+ = 3$ we have $r_s^- = 0$ whenever the state s comprises three pluses:

$$r^-_{+++} = 0. \qquad (19.3)$$

For all states s with one minus, the probability r_s^- lies strictly between 0 and 1.

We need, therefore, to consider how the probabilities might be assigned, as follows. Under a stable regime the binary behaviour of this hypothetical system will evolve as follows, starting with the state $(++-)$, reflecting a full three-year cycle being followed by a "+" reflecting the first year of growth in the next cycle. This transition has a high probability, R. We set $r^+_{++-} = R$; equivalently, $r^-_{++-} = 1 - R$. The state $(+-+)$ models the last two years in the previous cycle followed by the first year in the next cycle. In the steady regime this state is followed by a "+" with a high probability. Again, we set $r^+_{+-+} = R$, or $R^-_{+-+} = 1 - R$. Finally, $(-++)$ corresponds to the last year in the previous cycle followed by the first two years in the next cycle. This state gives rise to a "–" with a high probability. We assume $r^-_{-++} = R$. Thus,

$$r^-_{-++} = R, \quad r^-_{+-+} = 1 - R, \quad r^-_{++-} = 1 - R. \qquad (19.4)$$

The value R, located strictly between 0 and 1, is a probability of a stable environment, in which the number of the species A performs normal cycling. This probability is high, i.e., R is close to 1. The value $1 - R$ is a probability of a perturbation in the environment, one capable of breaking the cycle and pushing the species A towards collapse or survival (as defined above).

Remark: Environmental perturbations which give rise to a "–" where a "+" would be expected in normal cycling are different from those which give rise to a "+" instead of a "–". Such different perturbations may have different probabilities although this has been ignored here. If we took it into account, we would end up with a more complicated family of transition probabilities, r_s^-.

The probability distribution, q, for the initial state of the system must reflect the investigator's guess as to the evolution of the system in the last three years (in which no observations were made). The most plausible assumption about the unobserved past is that the system has been in a stable cycle. The model's three-period initial state must then be among (–++), (+–+), (++–). Let us give a nonzero probability to each of these states, so that q satisfies

$$q_{-++} > 0, \quad q_{+-+} > 0, \quad q_{++-} > 0, \quad q_{-++} + q_{+-+} + q_{++-} = 1, \tag{19.5}$$

$$q_{---} = 0, \quad q_{--+} = 0, \quad q_{-+-} = 0, \quad q_{+--} = 0, \quad q_{+++} = 0. \tag{19.6}$$

The binary model is now specified and we are in a position to consider an illustrative simulation of the system's evolution. For example, let (...++–++–) be the record of the model's transitions up to period k. In period k the model is in the state (++–). According to equation (19.4), the model's transition to period $k + 1$ is "–" with the (low) probability $1 - R$ and "+" with the (high) probability R. Let the actual transition be "+". Now the transition record is (... ++–++–+), and the model is in the state (+–+). Again "+" comes with the (high) probability R and "–" with the (low) probability $1 - R$. This time, however, let the actual transition be "–". The record changes to (... ++–++–+–) and the model's state to (–+–). This state has two minuses, i.e., the collapse critical level $m^- = 2$ is reached. The next transition is automatically "–". The new record is (... ++–++–+– –). There are two minuses in (+– –). The system makes another "–" making (... ++–++–+– – –) and passing to the state (– – –). All the subsequent transitions are "–" and the system irreversibly moves towards collapse.

19.2.4 Problem Definition

Let one or several sequential local changes in the system's state be observed. The observer's task (in what follows, we identify ourselves with the observer) is to detect whether a tendency ultimately towards collapse or towards survival is currently taking place. We shall invoke our binary models in order to approach an answer to this problem, building upon the foregoing illustrative example.

19.3 IDENTIFICATION OF INCLINATION FROM CURRENT OBSERVATIONS

Armed thus with an understanding of how the binary model functions, we proceed in this section to set out the principles of our approach to answering the question of central concern herein: to what extent does access to a small window of current and recent past observations of the system's behaviour inform our appreciation of whether the system is inclined ultimately towards collapse or survival?

19.3.1 Step 0: Construction of the Binary Model

We assume that a real system under investigation is described by a binary model with state length 3, collapse critical level 2, and survival critical level 3; thus, the relations (19.1)–(19.3) are assumed. Moreover, in all situations in which neither of the critical levels is crossed, the probability of the "–" transition, r_s^-, is assumed to take a constant value, r. Thus,

$$r^-_{-++} = r, \quad r^-_{+-+} = r, \quad r^-_{++-} = r \quad (0 < r < 1). \tag{19.7}$$

Remark: Recall that in the model discussed in subsection 19.2.3, the transition probabilities are defined differently (19.4). In terms of the underlying A-in-E system (subsection 19.2.3), the relations (19.7) reflect the situation in which the environment E has a constant probability (modelled as r) of causing a decline in the number of the species A in each year after a normal three-year cycle. This situation is less realistic than that discussed in subsection 19.2.3 but facilitates our analysis. Our goal is to illustrate our method on an extremely simple, though nontrivial, example.

The initial probability distribution, q, in the model is defined by equations (19.5) and (19.6) as before. Thus, the initial transition – from period 0 to period 1 – can be either a "–" or a "+", each with a nonzero probability.

In what follows, we shall refer to the described model as M.

Remark: Note that the transition probability r in (19.7) and the values of the initial probability q in (19.5) are not specified. We shall retain this parametric uncertainty and, moreover, utilize it in our subsequent analysis.

19.3.2 Step 1: Model-based Analysis (No Observations)

The first step in our analysis makes no reference to the observation data. We shall compute the probability of collapse, p^-, in the model M, where M attains a collapse whenever in some period its state contains no less than two minuses. Starting from this period, M thereafter makes only "–" transitions and goes to collapse deterministically.

We shall call any state with no less than two minuses a *pre-collapse* condition (for *M*).

Let us estimate $p^{-}(k)$, the probability of *M*'s pre-collapse in period k. We denote by $p_s(k)$ the probability of the state s occurring in period k (more accurately, the probability that the model M has the state s in period k). Let us now introduce the probability column vector

$$p(k) = \begin{matrix} p_{+++}(k) \\ p_{-++}(k) \\ p_{+-+}(k) \\ p_{++-}(k) \\ p_{--+}(k) \\ p_{-+-}(k) \\ p_{+--}(k) \\ p_{---}(k) \end{matrix}$$

Obviously,

$$\begin{aligned} p(1) &= Zq, \\ p(2) &= Zp(1) = Z^2 q, \\ &\dots \\ p(k) &= Zp(k-1) = Z^k q, \end{aligned} \qquad (19.8)$$

where Z is the matrix of the transition probabilities,

$$Z = \begin{matrix} +++ & -++ & +-+ & ++- & --+ & -+- & +-- & --- & \\ 1 & 1-r & 0 & 0 & 0 & 0 & 0 & 0 & +++ \\ 0 & 0 & 1-r & 0 & 0 & 0 & 0 & 0 & -++ \\ 0 & 0 & 0 & 1-r & 0 & 0 & 0 & 0 & +-+ \\ 0 & r & 0 & 0 & 0 & 0 & 0 & 0 & ++- \\ 0 & 0 & 0 & 0 & 0 & 0 & 0 & 0 & --+ \\ 0 & 0 & r & 0 & 1 & 0 & 0 & 0 & -+- \\ 0 & 0 & 0 & r & 0 & 1 & 0 & 0 & +-- \\ 0 & 0 & 0 & 0 & 0 & 0 & 1 & 1 & --- \\ +++ & -++ & +-+ & ++- & --+ & -+- & +-- & --- & \end{matrix}$$

Here the element at the intersection of column s with row s' stands for the probability of the s-to-s' state transition of M from one period to another. By (19.8) we have in particular

$$p_{---}(k) = p_{+--}(k-1) + p_{---}(k-1),$$

$$p_{+--}(k) = p_{-+-}(k-1) + rp_{++-}(k-1),$$

$$p_{-+-}(k) = p_{--+}(k-1) + rp_{+-+}(k-1),$$

and $p_{--+}(k) = 0$ $(k \geq 1)$. The probability of pre-collapse in period k is

$$p^-(k) = p_{---}(k) + p_{+--}(k) + p_{-+-}(k) + p_{-+-}(k) + p_{--+}(k). \tag{19.9}$$

Hence,

$$\begin{aligned} p^-(k) = {} & p_{+--}(k-1) + p_{---}(k-1) + \\ & p_{-+-}(k-1) + rp_{++-}(k-1) + \\ & p_{--+}(k-1) + rp_{+-+}(k-1) \\ = {} & p^-(k-1) + r\,((p_{+-+}(k-1) + p_{++-}(k-1)). \end{aligned}$$

We see that $p^-(k)$ grows as k grows. Consequently, there is a limit,

$$p^- = \lim_{k\to\infty} p^-(k).$$

This limit represents the probability for M to enter a pre-collapse state in some period k, which is equivalent ultimately to its reaching a terminal state of collapse. This probability of collapse, p^-, can be expressed in terms of the limit matrix

$$Z^\infty = \lim_{k\to\infty} Z^k$$

provided the latter exists. Indeed, equations (19.8) and (19.9) yield

$$p^- = p_{--+} + p_{-+-} + p_{+--} + p_{---}. \tag{19.10}$$

In other words, p^- is the sum of four elements of the column vector $p = [p_{+++}, p_{-++}, p_{+-+}, p_{++-}, p_{--+}, p_{-+-}, p_{+--}, p_{---}]^T$, where p is the limit probability vector,

$$p = Z^\infty q. \tag{19.11}$$

Similarly, the probability of survival, p^+, in the model M is represented as

$$p^+ = p_{+++}. \tag{19.12}$$

In our case the limit matrix Z^∞ exists and can be computed explicitly to be as follows (omitting the details of the derivation):

$$Z^{\infty} = \begin{array}{cccccccc|c} +++ & -++ & +-+ & ++- & --+ & -+- & +-- & --- & \\ 1 & \tau_{-++}^{+++} & \tau_{+-+}^{+++} & \tau_{++-}^{+++} & 0 & 0 & 0 & 0 & +++ \\ 0 & 0 & 0 & 0 & 0 & 0 & 0 & 0 & -++ \\ 0 & 0 & 0 & 0 & 0 & 0 & 0 & 0 & +-+ \\ 0 & 0 & 0 & 0 & 0 & 0 & 0 & 0 & ++- \\ 0 & 0 & 0 & 0 & 0 & 0 & 0 & 0 & --+ \\ 0 & 0 & 0 & 0 & 0 & 0 & 0 & 0 & -+- \\ 0 & 0 & 0 & 0 & 0 & 0 & 0 & 0 & +-- \\ 0 & \tau_{-++}^{---} & \tau_{+-+}^{---} & \tau_{++-}^{---} & 1 & 1 & 1 & 1 & --- \\ +++ & -++ & +-+ & ++- & --+ & -+- & +-- & --- & \end{array}$$

where

$$\tau_{-++}^{---} = \frac{r^2(2-r)}{1-r(1-r)^2},$$

$$\tau_{+-+}^{---} = \frac{r^2(1-r)+r}{1-r(1-r)^2}, \tag{19.13}$$

$$\tau_{++-}^{---} = \frac{r(2-r)}{1-r(1-r)^2},$$

$$\tau_{-++}^{+++} = 1-\tau_{-++}^{---}, \quad \tau_{+-+}^{+++} = 1-\tau_{+-+}^{---}, \quad \tau_{++-}^{+++} = 1-\tau_{++-}^{---}.$$

Therefore, for the limit probability vector p (19.11) we have

$$p_{---} = q_{-++}\tau_{-++}^{---} + q_{+-+}\tau_{+-+}^{---} + q_{++-}\tau_{++-}^{---},$$

$$p_{+++} = q_{-++}\tau_{-++}^{+++} + q_{+-+}\tau_{+-+}^{+++} + q_{++-}\tau_{++-}^{+++},$$

$$p_s = 0, \quad s \neq (+++), (---).$$

Then by (19.10) $p^- = p_{---}$, and we arrive at the next formula for the probability of collapse in the model M:

$$p^- = q_{-++}\tau_{-++}^{---} + q_{+-+}\tau_{+-+}^{---} + q_{++-}\tau_{++-}^{---}, \tag{19.14}$$

Remark: Note that $p_{---} + p_{+++} = 1$; equivalently (see (19.12)), $p^- + p^+ = 1$. In other words, the model M meets either collapse or survival with probability 1, which is in agreement with Hypothesis 1.

19.3.3 Step 2: Model and Data-based Analysis

We have now the probabilities of the system terminating in either collapse or survival, in the absence of any conditioning of the model and analysis of the observed behaviour of the system. Our interest, of course, is centred on how these probabilities change when the foregoing analysis has access to an observed sequence of local transitions, g. In the following we consider the particular observation record: $g = (-+)$, and our goal is to compute $p^{-}[-+]$, the conditional probability of collapse under the observation g. Here the new elements of g indicate a "–" transition in period 1 and a "+" transition in period 2. First, for every state s, we shall compute $q_s[-+]$, the conditional probability s in period 2 subject to the observation $(-+)$. Thus, we shall find the vector $q[-+]$, the conditional probability distribution of the model's states in period 2. Then we shall calculate $p^{-}[-+]$ using the same argument as in calculating p^{-}, in which the initial distribution, q, will be replaced by the conditional one, $q[-+]$.

Let us define a random event $E_{-+}(2)$ by the condition that the two-step model's transition from period 0 to period 2 is $(-+)$. Obviously, this happens if and only if in period 2 the model's three-period state is ended by $(-+)$. The probability of $E_{-+}(2)$ is expressed through $p(2)$, the probability vector for period 2, as follows:

$$p_{-+}(2) = p_{+-+}(2) + p_{--+}(2).$$

Now we address the relation $p(2) = Z^2 q$ (see (19.8)). Note that the row $(--+)$ in the transition matrix Z is zero, hence, this row is zero in the matrix Z^2, too. Consequently, $p_{--+}(2) = 0$. Therefore the probability of $E_{-+}(2)$ is $p_{+-+}(2)$. Referring to Bayes' rule, we find that $q_s[-+]$, the conditional probability of the state s in period 2 subject to the observation $(-+)$ (or, equivalently, subject to the random event $E_{-+}(2)$) has the form:

$$q_{+-+}[-+] = 1, \quad \text{otherwise} \quad q_s[-+] = 0 \quad (s \neq (+-+)). \tag{19.15}$$

By analogy with the preceding model-based analysis, with no access to observations, simply replacing q by the conditional distribution $q[-+]$ (composed of the conditional probabilities $q_s[-+]$) leads to a formula for $p[-+]$, the conditional limit probability vector, identical to (19.11):

$$\mathrm{p}[-+] = Z^{\infty} q[-+].$$

Due to the form of the matrix Z^{∞}, and (19.15), the nonzero components of $p[-+]$ are

$$p_{+++}[-+] = \tau_{+-+}^{+++}$$

$$p_{---}[-+] = \tau_{+-+}^{---}.$$

The latter probability describes the sought conditional probability for M to enter collapse:

$$p^{-}[-+] = \tau_{+-+}^{---}. \tag{19.16}$$

19.3.4 Step 3: Comparing the Results From Steps 1 and 2

The next step is the key to our analysis. We compare the unconditional and conditional probabilities of collapse, p^- and $p^-[-+]$, in the model M. Namely, we check if the inequality

$$p^-[-+] > p^- \tag{19.17}$$

holds. If it does, we can claim that the observed $(-+)$ sequence of local changes in the system (identified with the $(-+)$ double transition in the model) increases the probability of collapse in the model. Our straightforward conclusion is then: the model M *registers an inclination towards collapse*. If, on the contrary,

$$p^-[-+] < p^-, \tag{19.18}$$

we claim that the observed $(-+)$ record decreases the probability of collapse in the model (respectively, increases the probability of survival) and, therefore, the model M *registers an inclination towards survival*. Finally, if $p^-[-+] = p^-$, the model M *registers no inclination*, although this is obviously exceptional.

19.3.5 Step 4: Separating the Space of Models

Such a categorical (yes/no) algorithm, as just described, would look quite satisfactory if we were to overlook an important but confounding detail: the output of the algorithm is strongly model-dependent. Implicitly, the algorithm functions as follows: given a model, M, characterized by a particular q, the probability distribution of the initial state, and r, the probability of the environment causing decline in population A, we check which of the inequalities (19.17) or (19.18) is satisfied; accordingly, we register an inclination towards either collapse or survival. Let us, however, ask ourselves: are we sure that, say, the transition probability, r, in the model is 0.69 as opposed to 0.75 or 0.81, or that the initial probability q_{+-+} is 0.34, not 0.41 or 0.28? The answer is clearly "no, we are not sure". A model of a highly uncertain and poorly understandable environmental system should be uncertain. More specifically, at the start of the analysis, the model's parameters should accommodate a wide range of possibilities. In fact, in this sense we shall treat all combinations of model parameters (equivalently, all models from the class selected at Step 0) as equally acceptable – at the start.

The data-based analysis carried out for *all* candidate (concurrent) models gives a structure to their set. The simplest such structure induced by implementing our model and data-based algorithm for *all* models is the partition of the space of models into two domains, in which tendencies towards collapse and survival are registered. Let us now describe this partition.

We focus on the inequality (19.17), under which a tendency to collapse is registered. The formulae (19.14) and (19.16) for p^- and $p^-[-+]$ transform (19.17) into

$$(1-q_{+-+})\,\tau^{---}_{+-+} > q_{-++}\,\tau^{---}_{-++} + q_{++-}\,\tau^{---}_{++-}.$$

Substituting (19.13), we express this inequality in terms of the original parameters, r (the transition probability) and q (the initial distribution). After simple transformations we obtain

$$(q_{-++} + q_{+-+} - 1)\,r^2 + (1 - 2\,q_{-++} - q_{+-+} + q_{++-})\,r + 1 - q_{+-+} - 2q_{++-} > 0.$$

Since $q_{++-} = 1 - q_{-++} - q_{+-+}$ (see (19.5)), we end up with

$$q_{++-}r^2 + (q_{-++} - 2q_{++-})\,r - q_{-++} + q_{++-} < 0. \tag{19.19}$$

The inequality (19.19) is equivalent to (19.17).

Thus, to summarize, the model M registers an inclination towards collapse if the inequality (19.19) holds, and an inclination towards survival if the strict inequality opposite to (19.19) is satisfied.

19.3.6 Step 5: Posterior Assumptions

A new binary structure in the model space reveals relations between the parameters, which are key for making a categorical judgement on a dominating tendency. If we are able to decide which of these relations is preferable, i.e., back fitting with the real situation, we can set preferences for the tendencies.

Let us find the key relations associated with the partition of our model space (Step 4). We focus on the inequality (19.19) characterizing the collapse-oriented models. First, we note that at $r = 1$ the left-hand side in (19.19) vanishes, and its derivative with respect to r is q_{-++}, i.e., positive. Hence, the inequality (19.19) holds for all r sufficiently close to 1. At $r = 0$, the left-hand side in (19.19) takes the value $-q_{-++} + q_{++-}$. If $q_{-++} > q_{++-}$, (19.19) holds for all r between 0 and 1. This simple analysis leads us to the following summary:

Assertion 1: *If we assume that $q_{-++} > q_{++-}$, then no matter what the transition probability, r, is, the observation of $(-+)$ increases the probability of the model to collapse, and, therefore, is identified as the system's inclination towards collapse.*

If, on the contrary, $q_{-++} < q_{++-}$, then (19.19) is violated at $r = 0$, and there is an r_0 located strictly between 0 and 1 such that for $r > r_0$ (19.19) holds, and for $r < r_0$ the strict inequality opposite to (19.19) is satisfied. The latter inequality is equivalent to the inequality (19.18) justifying the registration of a tendency to survival.

Assertion 2: *If we assume that $q_{-++} < q_{++-}$, then there is a critical r_0 between 0 and 1 such that:*

(i) *for all transition probabilities $r > r_0$, the observation of $(-+)$ increases the probability of the model towards collapse and, therefore, is identified as the system's inclination towards collapse; and*

(ii) *for all transition probabilities $r < r_0$, the observation of $(-+)$ decreases the probability of the model towards collapse and, therefore, is identified as the system's inclination towards survival.*

We see that a relation between the parameters q_{-++} and q_{++-} is crucial. Assertion 1 says that if, in our posterior analysis of the model space, we find that in fact $q_{-++} > q_{++-}$, we have good evidence for saying that a tendency to collapse is taking place. If, conversely, a result of our posterior analysis is $q_{-++} < q_{++-}$, Assertion 2 tells us to look at the transition probability r and decide whether it is large or small. If we regard r as large, we end up with a conclusion that a tendency to collapse is taking place; if we assume r to be small, we register a tendency to survival.

Finally, if we find ourselves unable to judge whether $q_{-++} > q_{++-}$ or the opposite holds, and whether in the latter situation r is large or small, then we must explore the partition of the model space in more detail. This brings us to the next step in our analysis.

19.3.7 *Step 6: Dominance Analysis*

In Assertion 2, r_0 is a root of a quadratic equation obtained from (19.19) by replacing "$<$" with "$=$":

$$q_{++-}r^2 + (q_{-++} - 2q_{++-})\,r - q_{-++} + q_{++-} = 0.$$

Resolving this quadratic equation with respect to r, we find:

$$r = 1 - \frac{q_{-++}}{q_{++-}}. \tag{19.20}$$

Now we can represent the partition of the model (parameter) space graphically. The space of the parameters q_{-++}, q_{++-}, r is a prism, Π (Figure 19.1). The base of Π is a triangle, Δ, on the (q_{-++}, q_{++-}) plane, which is determined by the inequalities $q_{-++} > 0$, $q_{++-} > 0$, $q_{-++} + q_{++-} < 1$. The height of Π is 1. The triangle Δ is split into two (equal) smaller triangles, Δ^- characterized by $q_{-++} > q_{++-}$, and Δ^+ characterized by $q_{-++} < q_{++-}$. That part of the prism Π located above Δ^+ contains a surface, Π^0, described by equation (19.20). The domain Π^- in the prism Π, which lies above the triangle Δ^- and above the surface Π^0 comprises those candidate conditions of the model's parameters, for which $(-+)$ is identified as a tendency to collapse. The complementary domain, Π^+, in Π, which is located between the triangle Δ^+ and surface Π^0, comprises those parameters, for which $(-+)$ is identified as a tendency to survival.

Figure 19.1 shows that the set of collapse-oriented models (parameters), Π^-, is broader than that of the survival-oriented models, Π^+. A majority of models "vote" for collapse.

Could we now claim categorically that there is a tendency to collapse without having to make any posterior assumptions (as at Step 5)? Being strictly cautious and

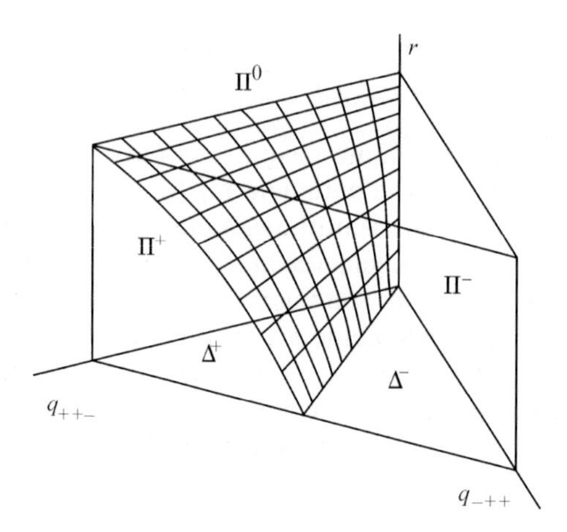

Fig. 19.1. The surface Π^0 separating the model space under the (−+) observation record. Collapse- and survival-oriented models are located, respectively, in the volumes Π^- and Π^+.

treating the partition as an imprecise reflection of a real tendency in the space of the models, a considered response would be:

Assertion 3: *If (−+) is observed, a tendency to collapse dominates.*

Remark: In our example, we deliberately study the two-step observation record (−+), for which it is not so easy to say immediately by intuition what tendency dominates. For a one-step observation record (−) intuition immediately identifies "a tendency to collapse". Indeed, our analysis given this observation results in: *if (−) is observed, all models register a tendency to collapse.* A similar statement holds for the observation record (+), to the effect that all models register an inclination towards survival.

Assertion 3 cannot but inspire a further question: "*How strong* is the dominance?"

To answer this question, we must provide some quantities that may serve as *collapse dominance indices*, i.e., that would be able to characterize the degree to which a tendency to collapse dominates. Let us characterize the degrees of dominance of a tendency to collapse as *absolute*, *strong*, *moderate*, *discernible*, *weak*, and *negative* (the latter meaning that a tendency to survival dominates). Each degree will correspond to some "conventional" subinterval in the interval of all admissible values of a collapse dominance index. The subintervals do not intersect, and a higher degree of dominance is attributed to a subinterval with larger values. For the simplest of the collapse-dominance indices, we take

$$c^- = \frac{V(\Pi^-)}{V(\Pi^+)} \tag{19.21}$$

where $V(\Pi^-)$ and $V(\Pi^+)$ are the volumes of the collapse- and survival-oriented domains Π^- and Π^+, respectively, in Figure 19.1. The greater is c^-, the stronger is the dominance of the set of the collapse-oriented models. When all models are survival-oriented, c^- takes its minimum value, 0; its maximum value, ∞, is reached when all models are collapse-oriented. The neutral value, 1, corresponds to the situation in which c^- does not identify any tendency (the sets of collapse- and survival-oriented models have equal volumes). Thus, the interval of admissible values of c^- is the half-line of all non-negative reals. The index c^- registers some degree of collapse dominance whenever $c^- > 1$. If we assume the qualification of the degrees of dominance of a tendency to collapse (with respect to the index c^-) given in Table 19.1, we may compute the actual value of c^-. Referring to equation (19.20) for the separation surface Π^0, we represent the volume $V(\Pi^+)$ as an integral,

$$V(\Pi^+) = \int_0^{1/2} \mathrm{d}y \int_y^{1-y} \left(1 - \frac{y}{x}\right) \mathrm{d}x;$$

here, and in what follows, x and y are the brief notations for q_{++-} and q_{-++}, respectively. The integral is computed explicitly, giving for our example

$$V(\Pi^+) = 0.24 - 0.5\,(\log 2 - 0.5) = 0.144\ldots .$$

The volume of the whole parameter space is

$$V(\Pi) = 0.5.$$

Therefore,

$$V(\Pi^-) = V(\Pi) - V(\Pi^+) = 0.5 - 0.144\ldots = 0.355\ldots ,$$

and

$$c^- = 0.355/0.144 = 2.46\ldots .$$

Table 19.1 indicates a moderate dominance of the tendency to collapse.

Table 19.1

Values of c^-	Dominance
$c^- > 10$	absolute
$3 < c^- \leq 10$	strong
$1.5 < c^- \leq 3$	moderate
$1.1 < c^- \leq 1.5$	discernible
$1 < c^- \leq 1.1$	weak

Now we can supplement Assertion 3 by a quantitative observation:

Assertion 4: *The set of the collapse-oriented models is about* 2.5 *times larger in volume than that of the survival-oriented models. Equivalently, given a uniform probability distribution on the model space* Π, *the probability of a randomly chosen model to register collapse is about* 2.5 *times higher than registering survival. The index* c^- (19.21) *shows a moderate dominance of the tendency to collapse.*

Let us consider another index of collapse dominance. By introducing a *collapse indicator* of a model M,

$$w(M) = \frac{p^-[-+](M)}{p^-(M)} \tag{19.22}$$

we can explicitly indicate the dependence of the conditional and unconditional probabilities, $p^-[-+]$ (19.16) and p^- (19.14), on the model M. Assuming a uniform probability distribution in the model space, we treat $w(M)$ as a random variable, and define a collapse-dominance index, c^-, as the expectation of $w(M)$. The interval of admissible values of c^- is the half-line of all non-negative reals. The greater is c^-, the higher is, on average, the dominance of the conditional probability of collapse over the unconditional. We again qualify the degrees of dominance of a tendency to collapse (with respect to the index c^-) by Table 19.1. An integral representation of c^- is

$$c^- = \frac{1}{V(\Pi)} \int_0^1 \mathrm{d}r \int_0^1 \mathrm{d}y \int_0^y \frac{-r^2 + r + 1}{(x-1)r^2 + (1-2x+y)r + 1 + x - y} \mathrm{d}x. \tag{19.23}$$

Computations result in

$$c^- = 1.298...$$

and Table 19.1 registers a discernible dominance of the tendency to collapse.

We end up with

Assertion 5: *The expected value of the collapse indicator* $w(M)$ (19.22) *is about* 1.3. *In other words, the observation of* $(-+)$ *raises the probability of collapse about* 30% *on average. The index* c^- (19.23) *shows a discernible dominance of the tendency to collapse.*

Our further example of a collapse indicator of a model M is

$$w(M) = p^-[-+](M) - p^-(M). \tag{19.24}$$

Again, define a collapse dominance index, c^-, as the expectation of $w(M)$ under the uniform probability distribution in the model space, the admissible values of c^- vary

Table 19.2

Values of c^-	Dominance
$c^- > 0.8$	absolute
$0.6 < c^- \leq 0.8$	strong
$0.3 < c^- \leq 0.6$	moderate
$0.1 < c^- \leq 0.3$	discernible
$0 < c^- \leq 0.1$	weak

between –1 and 1. Dominance of the tendency to collapse is registered whenever $c^- > 0$. Assuming thus the qualification of the degrees of dominance of the tendency to collapse (with respect to c^-) given in Table 19.2, the relations (19.16), (19.14) and (19.13) of our specific example allow us to specify the collapse indicator (19.24) as

$$w(M) = w(x, y, r) = (x + y)\frac{r^2(1-r)+r}{1-r(1-r)^2} - y\frac{r^2(2-r)}{1-r(1-r)^2} - x\frac{r(2-r)}{1-r(1-r)^2}.$$

Then

$$c^- = \frac{1}{V(\Pi)}\int_0^{1/2} \mathrm{d}y \int_y^{1-y} \mathrm{d}x \int_y^{1-y/z} w(x, y, r)\,\mathrm{d}r, \qquad (19.25)$$

from which we obtain in this instance

$$c^- = 0.308...$$

allowing us to conclude with:

Assertion 6: *The expected value, c^-, of the difference between the probability of collapse conditioned by the observation of* $(-+)$ *and the initial (unconditional) probability of collapse is about* 0.31. *The index c^-* (19.25), *shows a moderate dominance of the tendency to collapse.*

Thus, two of the three collapse-dominance indices register a moderate dominance of the tendency to collapse, while one registers a discernible dominance.

Remark: Unlike the collapse-dominance indices (19.25) and (19.23), the index c^- (19.21) was introduced without the usage of collapse indicators. In fact, c^- utilizes a collapse indicator $w(M)$ which equals 0 if M is survival-oriented, and 1 otherwise. Indeed, $c^- = a_1/a_2$ where $a_1 = V(\Pi^-)/V(\Pi)$, $a_2 = V(\Pi^+)/V(\Pi)$. Obviously, a_1 is the average of $w(M)$ and $-a_2$ is the average of $-1 + w(M)$.

Through this analysis we have assembled some preliminary machinery with which to make our problem – of discerning an ultimate inclination from a short window of recent observations – numerically tractable. Vital to this analysis are two points:

(i) acknowledgement of the great uncertainty in the parameterisation of the model, M; and

(ii) the "relativistic" nature of the analysis.

That is to say, our interest focuses on whether access to observations makes outcomes relatively more or less likely than deductions made in the absence of the observations. We are now in a position to embark upon a genuine case study.

19.4 CASE STUDY: RODENT POPULATION IN THE VICINITY OF CHERNOBYL, UKRAINE

Mezhzherin (1996) has recently reported on what seems to have been a significant change of structure in the fluctuating populations of three species of small rodents in a hornbeam forest near Chernobyl in the Ukraine. The catastrophe at the Chernobyl Nuclear Power Station (in 1986) was followed by a dramatic increase in the amplitude of the oscillations of population density (see Figure 19.2). While Mezhzherin concerned himself with identifying the change that brought about such a massive environmental disturbance, our enquiry focuses on the inclination of the population prior to the Chernobyl disaster.

Some irregular variations in the amplitude (though not so strong) are seen in the 1978–1985 (pre-collapse) period. Theoretically, these could be indicative of a tendency to collapse; yet the hypothesis that the population was drifting towards collapse before the catastrophe seems less plausible than the hypothesis that collapse was determined by the catastrophe. What then is our more systematic analysis likely to reveal?

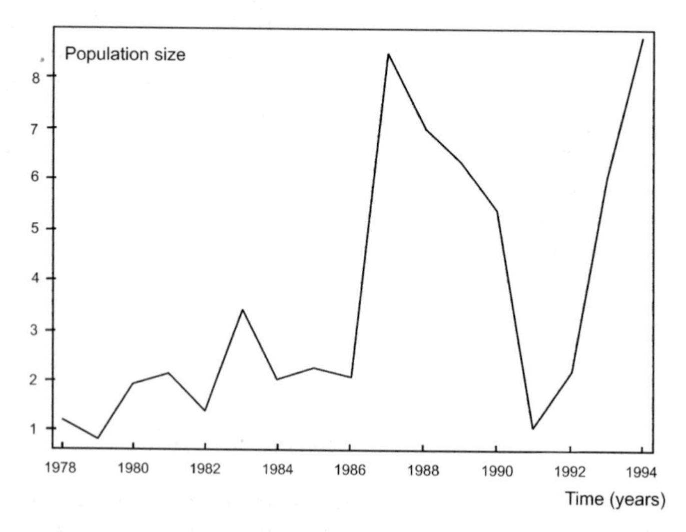

Fig. 19.2. Dynamics of standardized values of the population densities.

The binary model in this case operates with a yearly interval starting in 1978. We take four years, which is the approximate lifetime of the rodents, for the state length, m. The "+" and "–" transitions between years k and $k + 1$ are determined via the population densities in years $k + 1$, k, $k - 1$ and $k - 2$ as follows. Let us denote these densities as x_{k+1}, x_k, x_{k-1} and x_{k-2} and then denote by a_{k+1} the maximum of the deviations of the density in year $k + 1$ from the densities in any of the preceding years k, $k - 1$ and $k - 2$. More accurately, we define a_{k+1} to be the maximum of $|x_{k+1} - x_k|$, $|x_{k+1} - x_{k-1}|$ and $|x_{k+1} - x_{k-2}|$. We identify a transition from k to $k+1$ as "–" if a_{k+1} exceeds a *collapse threshold* α; otherwise, this transition is identified as "+". We assume that if 4 "–" ("+") transitions follow one by one during a 4-year lifetime of one generation, the population reaches collapse (survival) deterministically. Thus, we take 4 for the collapse critical level, m^-, and the survival critical level, m^+. We also assume that r_s^-, the probability of a "–" transition from state s, does not depend on s, so that $r_s^- = r$ (given s is not composed only of "–" or only of "+" transitions). Finally, we assume that the initial state is favourable for survival, in other words, the "+" transitions dominate in the initial state. More accurately, we define the initial distribution, q, by associating a probability of 1/4 with each of the four states containing one "–" transition, and a probability 0 to all states containing more "–" transitions. The model parameters are α, the collapse threshold, and r, the transition probability. We let α and r range over appropriate intervals, thus giving us a family of models suitably acknowledging the uncertainties of the situation. We regard all models in the family as equally admissible descriptions of the process analyzed.

We consider the interval between 0.1 and 0.9 as the admissible interval for r. We associate the admissible interval for α with the variety of observed values of the maximum density variation a_{k+1} in the year-to-year transitions over the period 1981–1986. The lowest value of a_{k+1} (1981–1982) is close to zero and the highest (1982–1983) is about 2. The average is around 1. We assume that "weak" collapse happens for a_{k+1} greater than 1. The situation where a_{k+1} is greater than 2 (the highest observed value of a_{k+1} in 1981–1986) is viewed as "sufficiently strong" collapse. If a_{k+1} assumes a value above 3, "strong collapse" is reached (in reality a_{k+1} averaged around 6). We let the collapse threshold α lie between 1 and 3. Thus, all grades of collapse from "weak" to "strong" are reflected in the models. In this manner we here completed Step 0 of our identification analysis (as in subsection 19.3.1).

Four records of observed "+/–" transitions are analyzed:

- Record 1: 1981–1982, 1982–1983
- Record 2: 1981–1982, 1982–1983, 1983–1984
- Record 3: 1981–1982, 1982–1983, 1983–1984, 1984–1985

For each model M in the family, characterized by parameters α and r, we perform Steps 1 (model-based analysis) and 2 (model and data-based analysis) and compute a collapse indicator

$$w(\alpha, r) = \frac{p_l^-(\alpha, r)}{p^-(\alpha, r)}$$

where (in line with (19.22)) $p^-(\alpha,r)$ is the unconditional probability of collapse for the model M and $p_l^-(\alpha,r)$ is the probability of collapse for M conditioned on the observation record l. The model is collapse-oriented if $w(\alpha,r) > 1$ and survival-oriented if $w(\alpha,r) < 1$. The shape of the surface $w = w(\alpha,r)$ above the $[\alpha,r]$ parameter rectangle $[1,3] \times [0.1,0.9]$ gives us a picture of how the family of models is split into the collapse- and survival-oriented domains. That part of the rectangle, above which the surface exceeds 1, represents the domain of collapse-oriented models; that part of the square, above which the surface is lower than 1, represents the domain of survival-oriented models. This split completes Steps 3 (comparison) and 4 (splitting the space of models). Figures 19.3, 19.4 and 19.5 show the collapse indicator surfaces $w = w(\alpha,r)$ found numerically for the observation records 1, 2 and 3, respectively.

Step 6 of the analysis (dominance analysis) identifies the degree of collapse dominance as the extent to which the area of the collapse-oriented domain is larger than that of the survival-oriented domain (see (19.21)). For record 1 (with 2 transitions observed) the collapse- and survival-oriented domains are approximately equal in area (Figure 19.3). *No inclination is registered*. For record 2 (in which 3 transitions are observed) the collapse-oriented domain clearly dominates in area (Figure 19.4). *A tendency to collapse is seen*. For record 3 (4 transitions observed) the collapse-oriented domain dominates totally (Figure 19.5). *There is a clear tendency to collapse*. Note that in Figure 19.5 the collapse indicator surface goes below 1 only within a narrow strip adjoining the right border of the rectangle corresponding to α greater than 2. Since 2 is the highest observed value of the maximum density variation a_{k+1}, if $\alpha > 2$, all transitions are identified as "+"; hence, the associated models are necessarily survival-oriented. In other words, for record 3 all models, which are not *a priori* survival-oriented, are collapse-oriented.

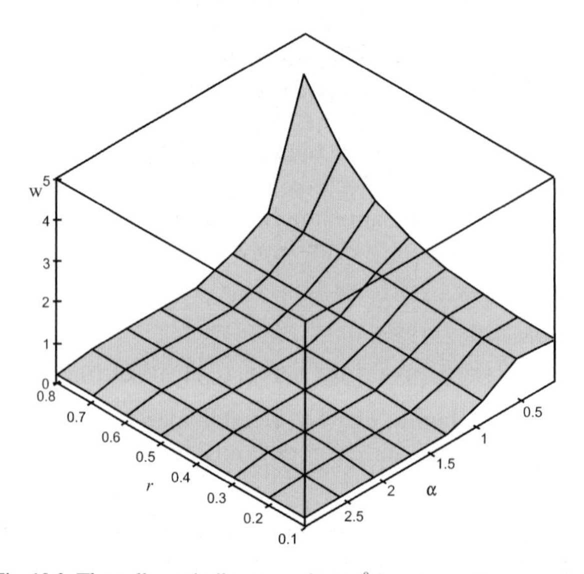

Fig. 19.3. The collapse indicator surface Π^0 for observation record 1.

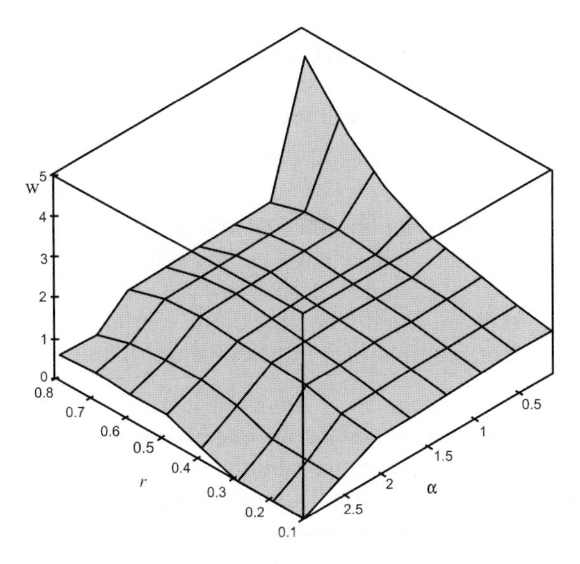

Fig. 19.4. The collapse indicator surface Π^0 for observation record 2.

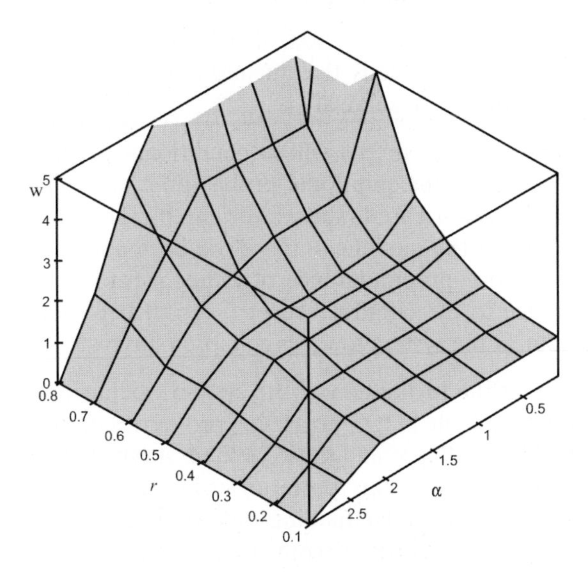

Fig. 19.5. The collapse indicator surface Π^0 for observation record 3.

It is remarkable that the inclination towards collapse becomes more apparent as the length of the observation record grows from 2 (record 1) to 4 (record 3). Moreover, our analysis supports the quite counterintuitive hypothesis that the population was drifting towards collapse *even before* the Chernobyl disaster.

19.5 THE METHOD: IN RETROSPECT AND PROSPECT

To summarise, the essence of our proposed method is as follows:

1. For the given environmental system a set of candidate binary models is chosen. The models are comprised of a Markov chain describing changes in the state of the system from one point in time to the next. A procedure (logical or computational) for transcribing changes in the system's behaviour over a single period of time into the "+" and "–" transitions required for the operations of the binary model must be specified. The state of the system will normally be defined as a finite history of multiple past transitions, up to and including that at the current time.

2. The model is assumed to be subject to uncertainty, notably in respect of the values assigned to its transition probabilities and the probabilities of its initial state. There can thus be a multiplicity of candidate parameterisations of the model. For each such candidate parameterisation an unconditional probability of collapse (survival) and a conditional probability of collapse (survival) are computed. Conditioning is here assumed to be according to a set of observed "+/–" transitions in the behaviour of the system.

3. For each candidate parameterisation of the model, whether the conditional probability of collapse exceeds the unconditional probability is noted (and likewise for the outcome of survival).

4. Those realisations of the model whose conditional probabilities of collapse (survival) exceed their unconditional probabilities of collapse (survival) are judged to be collapse-oriented (survival-oriented). The total set of candidate parameterisations of the model can therefore be separated into two sub-sets, such that the relative preponderance of either will provide an indication of whether reference to the empirical record of past behaviour predisposes the future behaviour of the system towards collapse (survival).

5. Information about the inclination of the system's behaviour towards collapse (survival) is now available in the (posterior) properties of the candidate parameterisations yielding collapse and those yielding survival. There is therefore an opportunity for examining this posterior evidence – of reconciling it with whatever other evidence, understanding, or knowledge is available for the given environmental system – in order to come to a closing judgement on whether the system is predisposed towards collapse or survival.

6. In the event that there is no such supplementary evidence available for corroboration or refutation of the outcome of the analysis, we may proceed to establish dominance – of the tendency towards collapse over that towards survival – through the use of some proposed numerical indices. These simply manipulate the relative dimensions of the two sub-sets of collapse- and survival-oriented candidate parameterisations of the model. We have suggested that various

intervals of the resulting indices might be associated with linguistic descriptors, for example, that a tendency to collapse is "strongly" or "weakly" apparent.

The supreme advantage of a binary classification of transitions and outcomes is the substantial compression it affords in terms of grappling with an otherwise immensely rich variety of possible trajectories of the system's behaviour through time (from the past into the future). In the analysis of environmental systems use of a binary classification of outcomes, in the sense of describing some pattern of behaviour deemed acceptable/unacceptable or feared/desired, has arisen quite naturally. We have herein merely introduced some mathematics for manipulating this binary feature. The price to be paid for this, however, is the unfamiliar abstraction of the resulting binary model, with respect to what is normally encountered in the study of environmental systems, and the consequent challenge of providing a mapping between the abstract "+/–" transitions of the model and the more immediate appreciation of behaviour in terms of changes with time in rodent population numbers, aquatic nutrient concentrations, and so forth. In this chapter we have accordingly set up the prototypes of our analysis using hypothetical examples and a case study in which the local transitions in the present towards collapse or survival in the future are about as self-evident as it is possible to make them (without the subsequent analysis then being trivially obvious). These mappings, from the changes of species levels in an ecological system to the abstraction of a "+/–" transition in a binary model, have not been implemented in an especially systematic manner herein. But it is clear they are crucial, since they provide the link between the power of the analysis to generate insights and the scope for the analyst to apprehend these insights in the physical, non-mathematical world of practical interest for our future well-being. What is more, we ourselves acknowledge the suggestion that our interventions in determining what constitutes a current transition towards collapse (survival) might cause the subsequent mathematical manipulations merely to arrive at the outcomes of these (our) prior judgements. It will therefore be vital to explore ways in which assigning the "+/–" transitions in the binary model can be associated as closely as possible with the workings of a more conventional model of the given environmental system, i.e., a more familiar computational encoding of the current science base. One such possibility might be to use a conventional differential-equation model of the system's dynamics, as predominantly presumed throughout this book, in order to establish whether the minimum time (τ_k) for the system to reach the target behaviour (from the present) diminishes as we pass from the current time period (t_k) to the next (t_{k+1}), i.e., whether $\tau_{k+1} < \tau_k$. If this is so, we could classify such a transition as "–", otherwise as "+".

19.6 CONCLUSIONS

As we have come to address environmental problems of ever greater scope and spatial extent, so the windows of past empirical observation have become ever briefer *relative* to the dominant, larger time constants in the behaviour of the systems of interest. The issue of extrapolating a future from the record of the past, of expressing something about the entire mosaic from observation of just a single tile, as it were, has become increasingly urgent. In such a context, the question of specific interest to this chapter has been cast as follows: given a model, and given a set of historical observations, is the behaviour of the system more inclined towards entry into some target domain of the state space in the future when the model is reconciled with the data than when the data, in effect, are ignored? Furthermore, how may the answers generated be made robust in the face of the uncertainties attaching inevitably to both the model and the record of past behaviour?

Our first conclusion is that this particular phrasing of the question *is* indeed a fruitful way of attacking, yet again, the problems of general concern in contemplating structural change in the future behaviour of an environmental system. Especially noteworthy is the manner in which the question seeks to establish *not* an absolute probability for the occurrence of some future event but a ranking of two probabilities relative to each other.

Our second conclusion is that the mathematical machinery for solving this problem works. It is to be noted in this respect that its application in a preliminary study of the cyclical dynamics of three rodent populations in the vicinity of Chernobyl in the Ukraine suggests that these populations had already entered a potentially "unstable" state before the massive (environmental) disturbance of the failed nuclear reactor. This inference is not necessarily obvious from other conventional examinations of the empirical record.

Last, the rather speculative nature of both the problem formulation and its method of solution is freely acknowledged. Current results are encouraging, but there is every reason for us to maintain a healthy scepticism regarding the likelihood of success in addressing other, more fully defined problems of structural change in the behaviour of environmental systems. Mapping the manifestations of such changes in the readily observable real world – of population numbers and pollutant concentrations – into the abstraction of the "+/–" transitions of the binary model is, *inter alia*, an issue of critical importance. This, together with application of the current solution procedure to problems with more extensive data than those of the Chernobyl study, will be some of the priorities for furthering this line of enquiry.

ACKNOWLEDGEMENT

One of us (AVK) wishes to acknowledge that this research was supported in part by the Russian Foundation for Basic Research, Project # 00-01-00682.

REFERENCES

Arthur, W.B., Ermoliev, Y.M. and Kaniovski, Y.M., 1987. Adaptive growth processes modeled by urn schemes. *Kibernetika*, **6**, 49–57 (in Russian); *Cybernetics*, **23**, 779–789 (English translation).

Hornberger, G.M. and Spear, R.C., 1980. Eutrophication in Peel Inlet, I, Problem-defining behaviour and a mathematical model for the phosphorus scenario. *Water Res.*, **14**, 29–42.

Kryazhimskii, A.V., Maksimov, V.I., Solovyov, A.A. and Chentsov, A.G., 1996. On a probabilistic approach to a quantitative description of dynamics of natural processes. *Problems of Control and Informatics*, **1**(2), 23–41 (in Russian).

Mezhzherin, V.A., 1996. The specific nature of ecological monitoring. *Russian J. Ecology*, **27**, 79–84.

Spear, R.C. and Hornberger, G.M., 1980. Eutrophication in Peel Inlet, II, Identification of critical uncertainties via generalised sensitivity analysis. *Water Res.*, **14**, 43–49.

Young, P.C, Hornberger, G.M. and Spear, R.C., 1978. Modelling badly defined systems – some furthur thoughts. In: *Proceedings SIMSIG Conference, Australian National University, Canberra*, pp. 24–32.

Contributing Authors – Biosketches

J.R. Arnold

J.R. Arnold is Associate Scientist with UCAR, the University Corporation for Atmospheric Research, assigned to the Atmospheric Sciences Modeling Division, a model development and applications research unit jointly operated by the U.S. National Oceanic and Atmospheric Administration and the U.S. Environmental Protection Agency. In 1996 Dr. Arnold received his Ph.D. from the Department of Environmental Sciences and Engineering at the University of North Carolina at Chapel Hill. His research interests include: experimental design, implementation, and outcome analysis for evaluation of multi-scale observational and numerical air quality models; statistical and mechanistic analysis of models and their chemistry, with development of new diagnostic metrics using ambient aerometric and specialized experimental data (field intensives, environmental chambers, and kinetics studies); and the development and implementation of rational methods for incorporating model results into environmental decision-making.

M.B. Beck

Bruce Beck is Professor and Eminent Scholar in the Warnell School of Forest Resources at the University of Georgia, where he holds the Wheatley–Georgia Research Alliance Chair of Water Quality and Environmental Systems. He is also currently Visiting Professor in the Department of Civil and Environmental Engineering at the Imperial College of Science, Technology and Medicine in London. Professor Beck holds a first degree in Chemical Engineering from the University of Exeter (1970) and a Ph.D. in Control Engineering from King's College, Cambridge (1973). His current research interests include environmental systems analysis, structural change and the identification of model structure, applications of process control in biological systems of wastewater treatment, control in microbial ecosystems, and infrastructure requirements for sustainable cities. He is founder and presently Chair of the Specialist Group on Systems Analysis and Integrated Assessment of the International Water Association.

K.J. Beven

Keith Beven has been Professor and Head of the Hydrology and Fluid Dynamics Group at Lancaster University since 1991. Previously he held positions at the

University of Leeds, Institute of Hydrology, and University of Virginia and has held visiting positions at Colorado State University, Princeton University, University of California at Santa Barbara, and EPFL Lausanne, Switzerland. In 1991 he received the Horton Award of the American Geophysical Union and in 1995 was made an AGU Fellow. In 1999/2000 he held the International Francqui Chair at the Katholiek Universiteit, Leuven, Belgium. He has served on the NERC Freshwater Sciences Committee, French National Hydrology Committee, and on subject review committees in the US, France and Sweden. He is well known for his work in distributed rainfall-runoff modelling, including the development of TOPMODEL, and has recently published a text on Rainfall-Runoff Modelling (Wiley, 2001). He was also instigator of the Generalised Likelihood Uncertainty Estimation (GLUE) methodology. He has published 200 papers and edited 5 other books. In 2001 he received the James Dalton Medal of the European Geophysical Society.

J. Chen

Jining Chen is Professor and Head of the Department of Environmental Science and Engineering at Tsinghua University in Beijing, P.R. China. Professor Chen holds an honours first degree in Environmental Engineering from Tsinghua University (1986) and a Ph.D. in Environmental Systems Analysis from Imperial College, London (1993). His current research interests include environmental systems analysis, identification of environmental models, water resources and environmental policy, integrated river basin planning and management, non-point source pollution control and sustainable cities. He is presently a member of the governing boards of several Chinese technical associations, as well as being Deputy Chair of the Chinese Environmental Engineering Society.

R.L. Dennis

Robin Dennis is a Senior Program Manager in the Atmospheric Sciences Modeling Division (ASMD) of the Air Resources Laboratory of the National Oceanic and Atmospheric Administration. Members of ASMD are on permanent assignment to the National Exposure Research Laboratory of the U.S. Environmental Protection Agency. Dr. Dennis has a Ph.D. in Physics from the University of Wisconsin at Madison (1972). His current research interests include applications of advanced models to assessment of environmental conditions and the forecasting of environmental change, with recent attention being devoted to linking air and water models for coastal, estuarine assessments. Dr Dennis's research also includes diagnostic evaluation and probing of photochemical and acidic deposition regional/urban air quality models and characterisation of their strengths and weaknesses from the perspective of both the application of these models and the design of diagnostic field studies for their empirical evaluation. He is a science co-chair of the Atmospheric Chemistry and Modeling Team of NARSTO, a tri-national, North American public/private partnership whose purpose is to coordinate and enhance policy-relevant scientific research and assessment of tropospheric pollution behaviour.

G.M. Hornberger

George Hornberger is a Professor in the College of Arts and Sciences at the University of Virginia, where he holds the Ernest H. Ern Chair in Environmental Sciences. Professor Hornberger holds a first degree in Civil Engineering from Drexel University (1965) and a Ph.D. in Hydrology from Stanford University (1970). Major areas of interest are catchment hydrology and hydrochemistry and transport of colloids and bacteria in porous media. His current research is aimed at understanding how hydrological processes affect the transport of dissolved and suspended constituents through catchments, soils, and aquifers. Professor Hornberger is a Fellow of the American Geophysical Union and of the Association for Women in Science. He is a recipient of the Robert E. Horton Award of the Hydrology Section of AGU, the Biennial Medal for Natural Systems of the Modelling and Simulation Society of Australia, the John Wesley Powell Award for Citizen's Achievement from the USGS, and the Excellence in Geophysical Education Award from AGU. He is a member of the U.S. National Academy of Engineering.

A.J. Jakeman

Tony Jakeman is Professor, Centre for Resource and Environmental Studies, and Director of the Integrated Catchment Assessment and Management Centre, The Australian National University. He has been an Environmental Modeller with the Centre for Resource and Environmental Studies for 25 years and has over 250 publications in the open literature, half of these in refereed international journals. His current research interests include integrated assessment methods for water and associated land resource problems, as well as modelling of water supply and quality problems including in ungauged catchments. Other scientific and organisational activities include: Editor-in-Chief, Environmental Modelling and Software (Elsevier); President, International Environmental Modelling and Software Society; President, Modelling and Simulation Society of Australia and New Zealand, Inc.; Director, International Association for Mathematics and Computers in Simulation; and regularly a member of scientific advisory committees of international conferences.

K.J. Keesman

Karel Keesman (born 1956) has been Associate Professor in the Systems and Control Group at Wageningen University since 1991. He received his Ph.D. degree in 1989 from the University of Twente, the Netherlands, for his work on set-membership identification and prediction of ill-defined systems, with applications to a water quality system. From May 1989 to December 1990 he worked as a research fellow on batch process control in the Process Control Group of the Department of Chemical Engineering at the University of Twente. His research interests include (robust) identification and control of uncertain dynamic systems, in particular environmental and biotechnical systems. Typical examples of these systems are: water quality and hydrological systems, and waste-water treatment and composting plants.

T. Kokkonen

Teemu Kokkonen is a Ph.D. student in the Laboratory of Water Resources at the Helsinki University of Technology, Finland. In the course of his studies he has spent one and a half years in the Centre for Resource and Environmental Studies at the Australian National University, Canberra. His Ph.D. research seeks to identify controls exerted by land use and landscape attributes on runoff processes, and developing simulation models for incorporating such controls into runoff predictions. Teemu Kokkonen holds the degrees of Master of Science in Technology and Licentiate of Technology from the Helsinki University of Technology.

A.V. Kryazhimskii

Arkadii Kryazhimskii is Principal Research Scholar with the Steklov Institute of Mathematics of the Russian Academy of Sciences in Moscow. He is also Professor at Moscow State University. Professor Kryazhimskii received his first degree in Mathematics from the Faculty of Mathematics and Mechanics of the Urals State University in Ekaterinburg (1971), from where he subsequently obtained his Candidate of Physics and Mathematics (1974) and his Doctorate of Physics and Mathematics (1981). From 1993–1996 he was Senior Scholar with the International Institute for Applied Systems Analysis, where he led the Dynamic Systems project. Professor Kryazhimskii has research interests in the fields of control theory, dynamic games, inverse problems in dynamics and ill-posed problems, with applications to economic and ecological systems. He is co-author of the monograph on *Inverse Problems for Ordinary Differential Equations* (published in 1995).

D.C.L. Lam

David Lam is Senior Research Scientist and Project Chief, Integrated Watershed Modelling and Management, at the National Water Research Institute, Environment Canada, Canada Centre for Inland Waters, Burlington, Ontario, Canada. He currently holds an Adjunct Professorship at the University of Guelph (Computing and Information Sciences) and was also an Adjunct Professor at McMaster University (Civil Engineering) and at the University of Waterloo (Computer Science) in Canada. Dr. Lam received a B.Sc. degree (1968) from the University of Hong Kong and an M. Math. (1970) and a Ph.D. (1974) in Computer Science from the University of Waterloo. His research interests include environmental prediction and decision support systems, integrated watershed and lake water quality modelling for environmental applications in climate change, watershed acidification, lake eutrophication, and contaminant transport problems. He served as Chair (1996–1999) of the Task Committee on Climate Change Effects on Lake Hydrodynamics and Water Quality, American Society of Civil Engineering.

D. Lloyd Smith

Dr David Lloyd Smith is currently College Tutor at the Imperial College of

Science, Technology and Medicine, University of London, with responsibilities relating to the welfare and discipline of some 10,000 students. He is Reader in Structural Mechanics of the University of London and Head of the Systems and Mechanics Research Section, Department of Civil and Environmental Engineering, Imperial College. His particular research interests are in the plastic behaviour of structures, both for extreme static loading and for dynamic or impact effects. This encompasses the related computational mechanics, and in particular mathematical programming methods, for describing and predicting structural response to such loading environments. Through collaboration within a Research Section of gifted colleagues and students, his research output has been presented in some seventy publications.

O.O. Osidele

Olufemi Osidele gained his Ph.D. in Environmental Systems Analysis at the Warnell School of Forest Resources, University of Georgia. He holds a BSc in Civil Engineering from the University of Ife, Nigeria (1987), and a Masters in Hydrology from Imperial College, University of London (1992). His current research work involves the application of sampling-based uncertainty analysis methodology to an analysis of reachable environmental futures. Within the methodology the natural sciences of limnology and hydrology are integrated with the social sciences, via mathematical models, and with citizens' values and elicited stakeholder perceptions. He is a member of the Specialist Group on Systems Analysis and Integrated Assessment of the International Water Association.

S. Parkinson

Stuart Parkinson is Chair of Scientists for Global Responsibility, a UK science and ethics organisation. Until recently he was a Research Fellow at the Centre for Environmental Strategy at the University of Surrey, UK, working in the area of environmental systems analysis and its applications in environmental policy. He worked on issues such as climate change, materials use and recycling, and the environmental impacts of agriculture. He was an expert reviewer for the Third Assessment Report of the Intergovernmental Panel on Climate Change, published in 2001, and co-author of the book *Flexibility in Climate Policy: Making the Kyoto Mechanisms Work*. He obtained a Ph.D. in climate systems modelling in 1995, and a first class honours degree in physics and electronic engineering in 1990, both from Lancaster University, UK.

W.M. Schertzer

William Schertzer is a Research Scientist in the Aquatic Ecosystems Impacts Research Branch of the National Water Research Institute at the Canada Centre for Inland Waters. He currently holds an Adjunct Professorship in the College of Physical and Engineering Science, (Computing and Information Science) at the University of Guelph. He holds an Honours B.A from the University of Windsor (1969) and an M.Sc. specializing in Physical Climatology from McMaster

University (1975). He is certified as a Professional Hydrologist (PH) from the American Institute of Hydrology (1988). His current research interests include modelling of lake heat and mass exchanges, thermal responses of lakes to meteorological forcing and climate change impacts on water quality and aquatic ecosystems. He has served as a member and co-editor on the ASCE Task Committee for Climatic Effects on Lake Hydrodynamics (1996–1999). Current professional responsibilities include serving on the Executive of the Canadian Meteorological and Oceanographic Society (CMOS) and on the Council of the Canadian Foundation for Climate and Atmospheric Sciences (CFCAS).

J.D. Stigter

Hans Stigter is an Assistant Professor in the Systems and Control Group at Wageningen University and Research Centre. He holds a Masters degree in Applied Mathematics from the University of Twente, The Netherlands, and a Ph.D. degree in Environmental Systems Analysis from the University of Georgia, in Athens, Georgia, USA. He has also worked as a Post-doctoral Associate at the University of Leuven, Belgium. His current interests include on-line greenhouse control, system identification and optimal input design, recursive state and parameter reconstruction algorithms, and the application of control theory in a study on the effect of grazing in African ecosystems.

G.S. Tonnesen

Gail Tonnesen is an Assistant Research Engineer at the University of California, Riverside, and manager of environmental modeling at the College of Engineering's Center for Environmental Research and Technology. Dr. Tonnesen received a B.S. in Chemical Engineering from Michigan State University (1983) and M.S. and Ph.D. degrees in Environmental Engineering from the University of North Carolina at Chapel Hill in 1995. Dr. Tonnesen's research focuses on numerical modeling of the chemistry and transport of trace species in the troposphere, including both fundamental research and model applications. Research activities include: improving the understanding of chemical processes in the troposphere, particularly the budgets of the free radical and odd nitrogen species that control the formation of urban- to regional-scale air pollutants; development of process analysis methods to evaluate results of model simulations; investigation of the usefulness of indicator ratios for assessing ozone sensitivity to precursor emissions; and methods to evaluate source impacts and area of influence of precursor emissions.

O. Varis

Olli Varis is a Docent and Academy Fellow at Helsinki University of Technology, Laboratory of Water Resources, Finland. He has a Ph.D. in Water Resources Management from Helsinki University of Technology and an M.Sc. in Limnology from the University of Helsinki. He has been working with the Asian Institute of Technology, Bangkok, the International Institute for Applied Systems Analysis,

Laxenburg, Austria, and the United Nations University/World Institute for Development Economics Studies, Helsinki. He has a broad experience of environmental modelling, and is the author of more than 170 scientific articles. His current research is concentrated on the interconnections of water, food, poverty, environment, and urbanisation in Africa and Asia.

P.C. Young

Peter Young is Professor of Environmental Systems in the Environmental Sciences Department, and Director, Centre for Research on Environmental Systems and Statistics, at Lancaster University, UK. He is also an Adjunct Professor of Environmental Systems in the Centre for Resource and Environmental Studies at the Australian National University (ANU), Canberra. Prior to this he was Head of the Environmental Science Department at Lancaster (1981–1988); Professorial Fellow at the Australian National University and Head of the Systems Group in the Centre for Resource and Environmental Studies, Institute of Advanced Studies (1975–1981); University Lecturer in the Department of Engineering, Cambridge, and Fellow of Clare Hall, Cambridge (1970–1975). He holds a first degree in Aeronautical Engineering from Loughborough University and M.A., Ph.D degrees from Cambridge University. He completed an apprenticeship with British Aerospace (1958–1963) and worked as a civilian research and development engineer for the United States Navy at China Lake, California (1968–1969). His many research interests include: nonstationary/nonlinear time series analysis and forecasting; data based mechanistic modelling of stochastic systems; and the design of multivariable control systems. This research covers diverse areas of application from ecology and environmental science, through biological sciences and engineering, to business and socio-economics.

Subject Index